Lecture Notes in Computer Science 14425

Founding Editors

Gerhard Goos
Juris Hartmanis

Editorial Board Members

Elisa Bertino, *Purdue University, West Lafayette, IN, USA*
Wen Gao, *Peking University, Beijing, China*
Bernhard Steffen, *TU Dortmund University, Dortmund, Germany*
Moti Yung, *Columbia University, New York, NY, USA*

The series Lecture Notes in Computer Science (LNCS), including its subseries Lecture Notes in Artificial Intelligence (LNAI) and Lecture Notes in Bioinformatics (LNBI), has established itself as a medium for the publication of new developments in computer science and information technology research, teaching, and education.

LNCS enjoys close cooperation with the computer science R & D community, the series counts many renowned academics among its volume editors and paper authors, and collaborates with prestigious societies. Its mission is to serve this international community by providing an invaluable service, mainly focused on the publication of conference and workshop proceedings and postproceedings. LNCS commenced publication in 1973.

Qingshan Liu · Hanzi Wang · Zhanyu Ma ·
Weishi Zheng · Hongbin Zha · Xilin Chen ·
Liang Wang · Rongrong Ji
Editors

Pattern Recognition and Computer Vision

6th Chinese Conference, PRCV 2023
Xiamen, China, October 13–15, 2023
Proceedings, Part I

Editors
Qingshan Liu
Nanjing University of Information Science and Technology
Nanjing, China

Zhanyu Ma
Beijing University of Posts and Telecommunications
Beijing, China

Hongbin Zha
Peking University
Beijing, China

Liang Wang
Chinese Academy of Sciences
Beijing, China

Hanzi Wang
Xiamen University
Xiamen, China

Weishi Zheng
Sun Yat-sen University
Guangzhou, China

Xilin Chen
Chinese Academy of Sciences
Beijing, China

Rongrong Ji
Xiamen University
Xiamen, China

ISSN 0302-9743 ISSN 1611-3349 (electronic)
Lecture Notes in Computer Science
ISBN 978-981-99-8428-2 ISBN 978-981-99-8429-9 (eBook)
https://doi.org/10.1007/978-981-99-8429-9

© The Editor(s) (if applicable) and The Author(s), under exclusive license
to Springer Nature Singapore Pte Ltd. 2024

This work is subject to copyright. All rights are reserved by the Publisher, whether the whole or part of the material is concerned, specifically the rights of translation, reprinting, reuse of illustrations, recitation, broadcasting, reproduction on microfilms or in any other physical way, and transmission or information storage and retrieval, electronic adaptation, computer software, or by similar or dissimilar methodology now known or hereafter developed.
The use of general descriptive names, registered names, trademarks, service marks, etc. in this publication does not imply, even in the absence of a specific statement, that such names are exempt from the relevant protective laws and regulations and therefore free for general use.
The publisher, the authors, and the editors are safe to assume that the advice and information in this book are believed to be true and accurate at the date of publication. Neither the publisher nor the authors or the editors give a warranty, expressed or implied, with respect to the material contained herein or for any errors or omissions that may have been made. The publisher remains neutral with regard to jurisdictional claims in published maps and institutional affiliations.

This Springer imprint is published by the registered company Springer Nature Singapore Pte Ltd.
The registered company address is: 152 Beach Road, #21-01/04 Gateway East, Singapore 189721, Singapore

Paper in this product is recyclable.

Preface

Welcome to the proceedings of the Sixth Chinese Conference on Pattern Recognition and Computer Vision (PRCV 2023), held in Xiamen, China.

PRCV is formed from the combination of two distinguished conferences: CCPR (Chinese Conference on Pattern Recognition) and CCCV (Chinese Conference on Computer Vision). Both have consistently been the top-tier conference in the fields of pattern recognition and computer vision within China's academic field. Recognizing the intertwined nature of these disciplines and their overlapping communities, the union into PRCV aims to reinforce the prominence of the Chinese academic sector in these foundational areas of artificial intelligence and enhance academic exchanges. Accordingly, PRCV is jointly sponsored by China's leading academic institutions: the Chinese Association for Artificial Intelligence (CAAI), the China Computer Federation (CCF), the Chinese Association of Automation (CAA), and the China Society of Image and Graphics (CSIG).

PRCV's mission is to serve as a comprehensive platform for dialogues among researchers from both academia and industry. While its primary focus is to encourage academic exchange, it also places emphasis on fostering ties between academia and industry. With the objective of keeping abreast of leading academic innovations and showcasing the most recent research breakthroughs, pioneering thoughts, and advanced techniques in pattern recognition and computer vision, esteemed international and domestic experts have been invited to present keynote speeches, introducing the most recent developments in these fields.

PRCV 2023 was hosted by Xiamen University. From our call for papers, we received 1420 full submissions. Each paper underwent rigorous reviews by at least three experts, either from our dedicated Program Committee or from other qualified researchers in the field. After thorough evaluations, 522 papers were selected for the conference, comprising 32 oral presentations and 490 posters, giving an acceptance rate of 37.46%. The proceedings of PRCV 2023 are proudly published by Springer.

Our heartfelt gratitude goes out to our keynote speakers: Zongben Xu from Xi'an Jiaotong University, Yanning Zhang of Northwestern Polytechnical University, Shutao Li of Hunan University, Shi-Min Hu of Tsinghua University, and Tiejun Huang from Peking University.

We give sincere appreciation to all the authors of submitted papers, the members of the Program Committee, the reviewers, and the Organizing Committee. Their combined efforts have been instrumental in the success of this conference. A special acknowledgment goes to our sponsors and the organizers of various special forums; their support made the conference a success. We also express our thanks to Springer for taking on the publication and to the staff of Springer Asia for their meticulous coordination efforts.

We hope these proceedings will be both enlightening and enjoyable for all readers.

October 2023

Qingshan Liu
Hanzi Wang
Zhanyu Ma
Weishi Zheng
Hongbin Zha
Xilin Chen
Liang Wang
Rongrong Ji

Organization

General Chairs

Hongbin Zha	Peking University, China
Xilin Chen	Institute of Computing Technology, Chinese Academy of Sciences, China
Liang Wang	Institute of Automation, Chinese Academy of Sciences, China
Rongrong Ji	Xiamen University, China

Program Chairs

Qingshan Liu	Nanjing University of Information Science and Technology, China
Hanzi Wang	Xiamen University, China
Zhanyu Ma	Beijing University of Posts and Telecommunications, China
Weishi Zheng	Sun Yat-sen University, China

Organizing Committee Chairs

Mingming Cheng	Nankai University, China
Cheng Wang	Xiamen University, China
Yue Gao	Tsinghua University, China
Mingliang Xu	Zhengzhou University, China
Liujuan Cao	Xiamen University, China

Publicity Chairs

Yanyun Qu	Xiamen University, China
Wei Jia	Hefei University of Technology, China

Local Arrangement Chairs

Xiaoshuai Sun	Xiamen University, China
Yan Yan	Xiamen University, China
Longbiao Chen	Xiamen University, China

International Liaison Chairs

Jingyi Yu	ShanghaiTech University, China
Jiwen Lu	Tsinghua University, China

Tutorial Chairs

Xi Li	Zhejiang University, China
Wangmeng Zuo	Harbin Institute of Technology, China
Jie Chen	Peking University, China

Thematic Forum Chairs

Xiaopeng Hong	Harbin Institute of Technology, China
Zhaoxiang Zhang	Institute of Automation, Chinese Academy of Sciences, China
Xinghao Ding	Xiamen University, China

Doctoral Forum Chairs

Shengping Zhang	Harbin Institute of Technology, China
Zhou Zhao	Zhejiang University, China

Publication Chair

Chenglu Wen	Xiamen University, China

Sponsorship Chair

Yiyi Zhou	Xiamen University, China

Exhibition Chairs

Bineng Zhong	Guangxi Normal University, China
Rushi Lan	Guilin University of Electronic Technology, China
Zhiming Luo	Xiamen University, China

Program Committee

Baiying Lei	Shenzhen University, China
Changxin Gao	Huazhong University of Science and Technology, China
Chen Gong	Nanjing University of Science and Technology, China
Chuanxian Ren	Sun Yat-Sen University, China
Dong Liu	University of Science and Technology of China, China
Dong Wang	Dalian University of Technology, China
Haimiao Hu	Beihang University, China
Hang Su	Tsinghua University, China
Hui Yuan	School of Control Science and Engineering, Shandong University, China
Jie Qin	Nanjing University of Aeronautics and Astronautics, China
Jufeng Yang	Nankai University, China
Lifang Wu	Beijing University of Technology, China
Linlin Shen	Shenzhen University, China
Nannan Wang	Xidian University, China
Qianqian Xu	Key Laboratory of Intelligent Information Processing, Institute of Computing Technology, Chinese Academy of Sciences, China
Quan Zhou	Nanjing University of Posts and Telecommunications, China
Si Liu	Beihang University, China
Xi Li	Zhejiang University, China
Xiaojun Wu	Jiangnan University, China
Zhenyu He	Harbin Institute of Technology (Shenzhen), China
Zhonghong Ou	Beijing University of Posts and Telecommunications, China

Contents – Part I

Action Recognition

Action Recognition

Learning Bottleneck Transformer for Event Image-Voxel Feature Fusion Based Classification

Chengguo Yuan, Yu Jin , Zongzhen Wu, Fanting Wei, Yangzirui Wang, Lan Chen(✉), and Xiao Wang

Anhui University, Hefei 230601, Anhui, China
{e21301283,E02114335,E02114336}@stu.ahu.edu.cn,
{chenlan,xiaowang}@ahu.edu.cn

Abstract. Recognizing target objects using an event-based camera draws more and more attention in recent years. Existing works usually represent the event streams into point-cloud, voxel, image, etc., and learn the feature representations using various deep neural networks. Their final results may be limited by the following factors: monotonous modal expressions and the design of the network structure. To address the aforementioned challenges, this paper proposes a novel dual-stream framework for event representation, extraction, and fusion. This framework simultaneously models two common representations: event images and event voxels. By utilizing Transformer and Structured Graph Neural Network (GNN) architectures, spatial information and three-dimensional stereo information can be learned separately. Additionally, a bottleneck Transformer is introduced to facilitate the fusion of the dual-stream information. Extensive experiments demonstrate that our proposed framework achieves state-of-the-art performance on two widely used event-based classification datasets. The source code of this work is available at: https://github.com/Event-AHU/EFV_event_classification.

Keywords: Event Camera · Graph Neural Networks · Transformer Network · Bottleneck Fusion

1 Introduction

Recognizing the category of a given object is a fundamental problem in computer vision. Most of the previous classification models are developed for frame-based cameras, in other words, these recognition models focus on encoding and learning the representation of RGB frames. With the rapid development of deep learning, frame-based classification achieves significant improvement in recent years. Representative deep models (e.g., the AlexNet [1], ResNet [2], and Transformer [3]) and datasets (e.g., ImageNet [4]) are proposed one after another. However, the recognition performance in challenging scenarios is still far from unsatisfactory, including heavy occlusion, fast motion, and low illumination.

© The Author(s), under exclusive license to Springer Nature Singapore Pte Ltd. 2024
Q. Liu et al. (Eds.): PRCV 2023, LNCS 14425, pp. 3–15, 2024.
https://doi.org/10.1007/978-981-99-8429-9_1

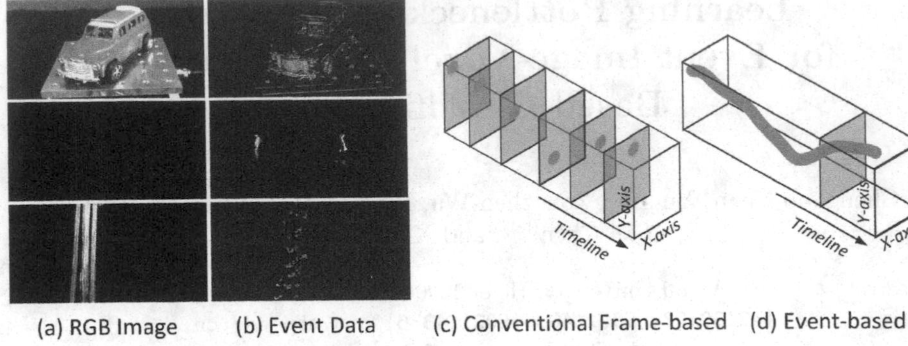

Fig. 1. Comparison of the frame- and event-based cameras https://youtu.be/6xOmo7Ikwzk. (a, b) shows representative samples in regular scenarios, low-illumination, and fast motion. (c, d) illustrates the different types of raw data representation of frame- and event-based cameras.

To improve object recognition in challenging scenarios, some researchers have started leveraging other sensors to obtain more effective signal inputs, thus enhancing recognition performance [5]. Among them, one of the most representative sensors is the event camera, also known as DVS (Dynamic Vision Sensor), which has been widely exploited in computer vision [6–8]. This paper focuses on using event cameras for object recognition. As shown in Fig. 1, different from the frame-based camera which records the light intensity for each pixel simultaneously, the event camera captures pulse signals asynchronously based on changes in light intensity, recording binary digital values of either zero or one. Typically, an increase in brightness is denoted as an ON event, while a decrease corresponds to an OFF event. An event pulse signal can be represented as a quadruple (x, y, t, p), where x, y represents the spatial position information, t represents the timestamp, and p represents the polarity, i.e., ON/OFF event. Many works demonstrate that the event camera performs better in High Dynamic Range (HDR), high temporal resolution, low latency response, and strong robustness. Therefore, utilizing event cameras for object recognition is a research direction that holds great research value and practical potential.

Recently, researchers have already conducted studies on object recognition using event cameras and have proposed various approaches to address this task, including CNN (Convolutional Neural Network) [9], GNN (Graph Neural Network) [10], Transformer [3], etc. Although these methods have achieved good accuracy by representing and learning events from different perspectives, they are still limited by the following aspects: **Firstly**, they rely on a single event representation form, such as images, point clouds, or voxels, which may limit the expressiveness and versatility of the learned features. Different event representation forms may capture different aspects of the data, and using only one representation may lead to the loss of valuable information. **Secondly**, the current methods are constrained to using only one of the deep learning architectures, such as CNNs, GNNs, or Transformers, for feature learning. Each architecture has its strengths and limitations in capturing different types of patterns and dependencies in data. By restricting the choice to a single architecture, the methods may not fully exploit the potential benefits and complementary strengths of different architectures. To address these limitations, future research should explore approaches that can integrate multiple event representation forms and leverage the combined power of different deep learning architectures. This could involve developing novel fusion techniques or hybrid architectures that can effectively capture and leverage diverse features and dependencies present in event data. By doing so, we can potentially enhance the performance and flexibility of event-based object recognition methods.

To address the aforementioned issues, in this work, we propose an effective dual-stream event information processing framework, referred to as EFV, as shown in Fig. 2. Specifically, we first transform the dense event point cloud signals into event images and event voxel representations. For the input of image frames, we utilize advanced spatiotemporal Transformer networks to learn spatiotemporal features. For voxel input, considering the sparsity of events, we employ a top-k selection method to sample meaningful signals for constructing a structured graph, and then use GNN (Graph Neural Network) to learn these volumetric structured features. Importantly, we introduce the Bottleneck Transformer to integrate these two types of feature representations, which are ultimately input to the dense layer for classification. It is easy to find that our proposed EFV possesses the characteristics of efficient event information processing, integration of multiple feature representations, spatiotemporal modeling capability, consideration of event sparsity, and accurate classification capability.

To sum up, the main contributions of this work can be concluded as the following two aspects:

- We propose an effective framework for recognition in event-based cameras, utilizing Event Image-Voxel feature representation and fusion.
- The introduction of the Bottleneck Transformer enables the interaction and fusion of dual-stream information, leading to improved recognition results.

2 Related Work

In this section, we give an introduction to Event-based Recognition[1], Graph Neural Networks, and Bottleneck Transformer.

Event-Based Recognition. Current research on event-based recognition can be divided into three distinct streams: CNN-based [9], SNN (Spiking Neural Networks)-based [11,12], and GNN-based models [13–15]. For the CNN-based models, Wang et al. [9] proposed an event-based gait recognition (EV-gait) method, which effectively removes noise via motion consistency. SNN is also utilized for encoding the event stream in order to achieve energy-efficient recognition. A kind of highly efficient conversion of ANN to SNN method is put

[1] https://github.com/Event-AHU/Event_Camera_in_Top_Conference.

forward by Peter and others [16], the method involves the balance of the weights and thresholds, while achieving lower latency and requiring fewer operations. In [17], a sparse backpropagation method for SNN was introduced by redefining the surrogate gradient function form. Fang et al. [18] propose spike element-wise (SEW) ResNet to implement residual learning for deep SNNS, while proving that SEW ResNet can easily implement identity mapping and overcome the vanishing/exploding gradient problem of Spiking ResNet. Wang et al. propose a hybrid SNN-ANN framework for RGB-Event based recognition by fusing the memory support Transformer and spiking neural networks, termed SSTFormer [19]. Jiang et al. propose to aggregate the event point and voxel using absorbing graph neural networks for event-based recognition [20].

For point cloud based representation, Wang et al. [21] treat the event stream as a set of 3D points in space-time, i.e., space-time event clouds, and adopt the PointNet [22] architecture, which directly takes the point cloud as input and outputs the class label for the entire input or each point segment/part label for each input point. Xie et al. [23] propose VMV-GCN, a voxel-wise graph learning model designed to integrate multi-view volumetric data. Li et al. [24] introduce a Transformer network to directly process event sequences in its native vector tensor format to effectively represent the temporal and spatial correlations of input raw events, thereby generating effective spatio-temporal features for the task. Different from previous works, this paper designs an event recognition method based on Transformer and graph convolutional neural network, which transmits bimodal information through a specific method and learns a unified feature representation, so as to represent event data more effectively.

Graph Neural Networks. One notable application of GNNs in event data recognition is gait recognition. Wang et al. [15] propose a 3D graph neural network specifically designed for gait recognition. The model leverages the graph structure to capture the spatial and temporal dependencies in gait patterns. Bi et al. introduce the concepts of residual Graph Convolutional Neural Networks (RG-CNN) and Graph2Grid blocks [13,14], which exploit graph structure to extract and exploit spatial and temporal information from event data. The Asynchronous, Event-based Graph Neural Networks (AEGNN) proposed by [25] addresses the processing of events as "evolving" spatio-temporal graphs. In the field of object recognition, Li et al. [26] introduce SlideGCN, a GNN-based model that focuses on fast graph construction using a radius search algorithm. Different from previous works, we adopt a graph Convolutional Neural Network (GCN) to process the graph data and connect the outputs of GCN and ST-Transformer module for accurate event-based pattern recognition.

Bottleneck Transformer. The traditional Transformer model has the problem of large computation and memory overhead when processing large-size images. Srinivas et al. [27] propose a novel network architecture called Bottleneck Transformer, which achieves dimensionality reduction of spatial attention by introducing a "bottleneck layer" between high-resolution and low-resolution representations of input features, thereby reducing computational consumption and increasing model scalability. Li et al. [28] introduce a local multi-head self-attention mechanism and a novel position encoding method to solve the scalability bottleneck of Transformer under GPU memory constraints. Nagrani et al. [29] propose a multimodal bottleneck converter (MBT) and guided the bottlenecks in it to connect across modes. Song et al. [30] propose a new model BS2T that captures long-range dependencies between pixels in HS images by leveraging the self-attention mechanism in Transformers. In addition, we innovatively introduce bottleneck Transformer to promote the fusion of dual-stream information and improve the performance of module fusion.

3 Our Proposed Approach

3.1 Overview

Given an input event stream consisting of hundreds of thousands of events, our approach involves several steps to enhance the representation. Initially, we employ event frame stacking and voxel construction techniques to generate event frame and voxel representations, respectively. Subsequently, we utilize two intermediate representations, namely event frame and voxel graph, to capture the spatio-temporal relationships within the event stream. To further improve the feature descriptors for event frame and graph-based event representation, we propose a novel dual branch learning network. Finally, we combine these representations to create a comprehensive representation for event data, enabling effective recognition. The overall framework is depicted in Fig. 2. In the following sections, we provide a detailed explanation of each module.

3.2 Network Architecture

Input Representation. Considering the large amount of data and computational complexity, it is necessary to employ some down-sampling techniques to reduce the number of events. In this paper, we adopt two kinds of sampling techniques to obtain the compressed event representations. We first transform the asynchronous event flow into the synchronous event images by stacking the events in a time interval based on the exposure time. We also employ voxelization to obtain voxel representations. Specifically, given the original event stream $\mathcal{E}$ with range H, W, T, we divide the spatio-temporal 3D space into voxels with the size of each voxel being h', w', t'. Hence, each voxel generally contains several events and the resulting event voxels in spatio-temporal space are of size $H/h', W/w', T/t'$. In practice, the above voxelization usually still produces tens of thousands of voxels. In order to further reduce the number of voxels and alleviate the effect of noisy voxels, we also adopt a voxel selection process to select top K voxels based on the number of events contained in each voxel. Let $\mathcal{O} = \{o_1, o_2 \cdots o_K\}$ denote the collection of the final selected voxels. Each event voxel o_i is associated with a feature descriptor $\mathbf{a}_i \in \mathbb{R}^C$ which integrates the attributes (polarity) of its involved events. Hence, each $o_i \in \mathcal{O}$ is represented as: $o_i = (x_i, y_i, t_i, \mathbf{a}_i)$, where x_i, y_i, t_i denotes the 3D coordinate of each voxel.

Graph Neural Networks for Event Voxel Encoding. We similarly construct a geometric neighboring graph $G^o(V^o, E^o)$ for voxel event data $\mathcal{O}$. To be specific, each node $v_i \in V^o$ represents a voxel $o_i = (x_i, y_i, t_i, \mathbf{a}_i) \in \mathcal{O}$ which is described as a feature vector $\mathbf{a}_i \in \mathbb{R}^C$. The edge $e_{ij} \in E^o$ exists between node v_i and v_j, if the Euclidean distance between their 3D coordinates is less than a threshold R. We adapt Gaussian Mixture Model (GMM), convolution to learn the effective representations for voxel graph. To be specific, in each GCN layer, each event node v_i aggregates the features from its adjacency nodes as

$$f'_d(v_i) \leftarrow \sigma\Big(\sum_{v \in V} \omega_d(v_i, v) f(v)\Big), d = 1, 2 \cdots D \tag{1}$$

where $\sigma(\cdot)$ denotes the activation function, such as ReLU. V denotes adjacency nodes of v_i. $\omega_d(v_i, v)$ denote the learnable convolution kernel weights. Finally, we adapt average graph pooling to get the global representation of voxel graph.

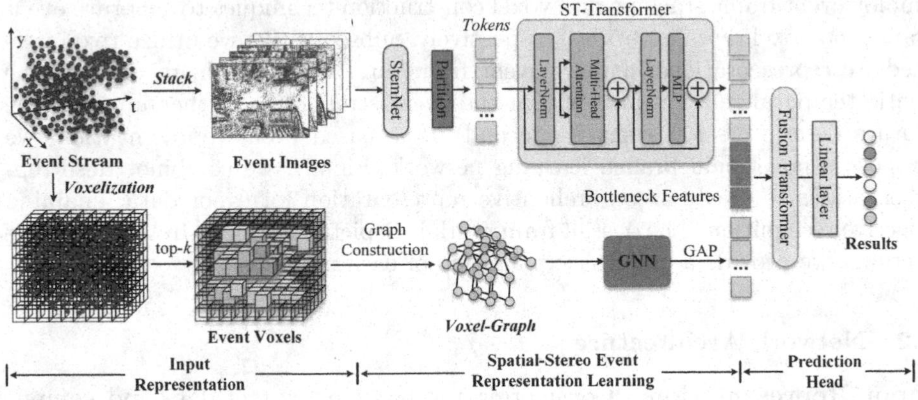

Fig. 2. An overview of our proposed Image-Voxel Feature Learning framework for event-based recognition.

Spatial-Temporal Transformer for Event Frame Encoding. After a series of data augmentation, each video sample obtained T event frames with a size of $H \times W$. We extract initial CNN features and embed event frames through StemNet (ResNet18 [2] is used in our experiments). After obtaining the initial features, we designed an ST-Transformer module to further achieve a better representation of spatio-temporal information. The proposed module consists of multi-head self-attention (MSA), MLP, and Layernorm (LN). As shown in Fig. 2, T event frames are divided into N patches in spatial dimension, therefore, the $T \times N$ tokens can be obtained. We add learnable location encoding to these tokens and feed them into the ST-Transformer module to fully extract the enhanced spatio-temporal features, as shown in Eq. 2 and Eq. 3:

$$Y = X^{in} + MSA(LN(X^{in})) \tag{2}$$

$$X^{out} = Y + MLP(LN(Y)) \tag{3}$$

Bottleneck Transformer. In order to achieve the interaction between Event Images and Event Voxels information representations and learn a unified spatio-temporal context data representation. We also designed the Fusion Transformer module and introduced the Bottleneck mechanism. Specifically, let $X^{image} \in \mathbb{R}^{T \times N \times d}$ and $X^{voxel} \in \mathbb{R}^{1 \times d}$ represent the outputs of the previous ST Transformer and GNN modules, respectively. We first collect the $T \times N$ image and $T \times N$ randomly initialized Bottleneck tokens together and feed them to Fusion Transformer which includes multi-head self-attention (MSA) and MLP submodule, i.e.,

$$F^1 = [X^{image}, X^{bottleneck}] \in \mathbb{R}^{2 \times T \times N \times d} \tag{4}$$

$$\widetilde{F}^1 = FusionTransformer(F^1) \tag{5}$$

We then split $\widetilde{F}^1$ into two parts, i.e., the images feature representation $\widetilde{F}^{image}$ and the bottleneck feature representation $\widetilde{F}^{bottleneck}$. The latter one will be concatenated with X^{voxel} and fed into the Fusion-Transformer module for interactive learning of the two representations. Similarly,

$$F^2 = [X^{bottleneck}, X^{voxel}] \in \mathbb{R}^{(T \times N+1) \times d} \tag{6}$$

$$\widetilde{F}^2 = FusionTransformer(F^2) \tag{7}$$

Finally, we concatenate both $\widetilde{F}^2$ and $\widetilde{F}^{image}$ together and flattened them into a feature representation. After that, we utilize a two-layer MLP to output the final class label prediction, as shown in Fig. 2. We adopt the Negative Log Likelihood Loss function [31] to train the whole network.

4 Experiment

4.1 Dataset and Evaluation Metric

In this work, we utilized two datasets, namely DVS128-Gait-Day [10], N-MNIST [32], and ASL-DVS [14], to evaluate our proposed model. Here is a brief introduction to these datasets:

- **ASL-DVS** [14]: This dataset consists of 100,800 samples, with 4,200 samples available for each letter. The focus was on the 24 letters representing the handshapes of American Sign Language. Each video in this dataset has a duration of approximately 100 ms. The author captured these samples using an iniLabs DAVIS240c camera under realistic conditions.
- **DVS128-Gait-Day** [10] dataset is proposed for event-based gait recognition. It contains 4,000 videos corresponding to 20 classes. 20 volunteers are recruited for data collection using a DVS128 Dynamic Vision Sensor (the pixel resolution is 128 × 128).
- **N-MNIST** [32] dataset is obtained by recording the display equipment when visualizing the original MNIST (28 × 28 pixels). The ATIS event camera is used for the data collection and each event sample lasts about 10 ms. There are 70,000 event files for this dataset, the training and testing subset contains 60,000 and 10,000 videos, respectively. The resolution of this dataset is 28×28.

Note that the top-1 and top-5 accuracy are employed as the evaluation metrics throughout our study.

4.2 Implementation Details

Our proposed dual-stream event-based recognition framework can be trained in an end-to-end manner. The initial learning rate is set as 0.001 and multiplied by 0.1 for every 60 epochs. We select eight frames for each video sample and divide each frame into eight tokens. For the constructed voxel graph, the threshold R is set to 2. The scale of the voxel grid is (4. 4. 4) for the ASL-DVS dataset. We select 512 voxels as the graph node for the structured graph representation learning. Our code is implemented using Python 3.8 and trained on a server with RTX3090 GPUs.

Table 1. Results on the ASL-DVS [14] dataset.

EST [33]	**AMAE** [34]	**M-LSTM** [35]	**MVF-Net** [36]	**EventNet** [37]
0.979	0.984	0.980	0.971	0.833
RG-CNNs [14]	**EV-VGCNN** [38]	**VMV-GCN** [23]	**EV-Gait-3DGraph** [9]	**Ours**
0.901	0.983	0.989	0.738	0.996

4.3 Comparison with Other SOTA Algorithms

As shown in Table 1, previous works already achieve high performance on the ASL-DVS [14] dataset. For example, the EST [33] (0.979), AMAE [34] (0.984), M-LSTM [35] (0.980), and MVF-Net [36] (0.971). Note that the GCN-based model, VMV-GCN [23], achieves better results, i.e., 0.989 on the top-1 accuracy. Thanks to the spatial-temporal feature learning and fusion network proposed in this work, we set new state-of-the-art performance on this dataset, i.e., 0.996. On the N-MNIST [32] dataset, as shown in Table 2, we also achieve SOTA performance compared with recent strong models. These comparisons fully validated the effectiveness of our proposed framework for event-based recognition. We provide two figures to better illustrate our results, as shown in the left subfigure of Fig. 3.

Table 2. Results on the N-MNIST [32] dataset.

EST [33]	**M-LSTM** [35]	**MVF-Net** [36]	**Gabor-SNN** [39]	**EvS-S** [26]
99.0	98.6	98.1	83.7	99.1
HATS [39]	**EventNet** [37]	**RG-CNNs** [14]	**EV-VGCNN** [38]	**Ours**
99.1	75.2	99.0	99.4	98.9

4.4 Ablation Study

To help researchers better understand the method we proposed, in this section, we conduct comprehensive experiments of component analysis on the DVS128-Gait-Day dataset and ASL-DVS dataset to check their influence on the overall model.

Component Analysis. Table 3 shows the effect of using different components on experimental results. In this part, we didn't use the Bottleneck, and the dataset we use is DVS128-Gait-Day. **Event image only** indicates that we only transform the event flow into the synchronous event images, which gets the result of 95.2. **Event voxel only** indicates that we only employ voxelization to obtain the compressed event representation and it achieves 98.0. We also use Event image and Event voxel together when obtaining event representation, denoted by **Event Image+Voxel**. It gets the result of 98.7. We can easily draw the conclusion by comparing the above three cases that using Event Image and Event voxel together can achieve higher performance, which reflects the effectiveness of our method.

Effect of Bottleneck. In this paper, we use Bottleneck Transformer to enhance the performance when fusing the modules. As shown in Table 4, **w/o Bottleneck Feature** means we do not feed the learning token into Bottleneck. It gets the result of 98.5. **w/o FusionFormer** means we do not use Fusion Transformer before the linear layer, in other words, we use the all components proposed in this paper except for the Bottleneck, and the result is 98.3. We found that compared with these two experiments, the result has increased after introducing the Bottleneck which indicates that the Bottleneck is a better choice for our framework. At the same time, a comparison with the experimental results in Table 1 shows that the experimental result drops when there is no Bottleneck, which also demonstrates Bottleneck has a positive effect on our proposed model (Table 3).

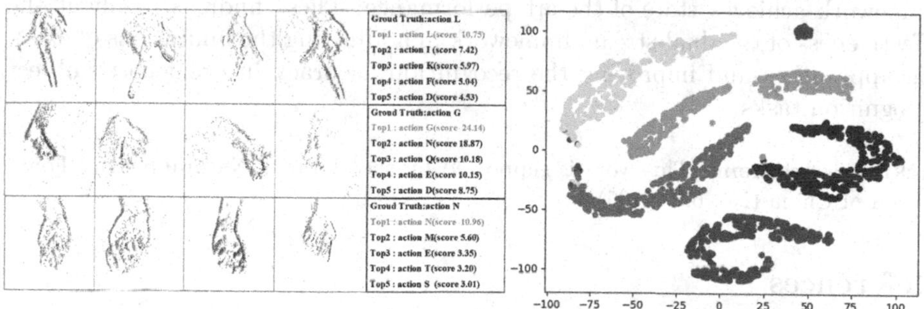

Fig. 3. Visualization of top-5 recognition results and feature distribution on the ASL-DVS dataset.

Table 3. Ablation study on DVS128-Gait-Day dataset [10].

Index	Component	Results
1	Event image only	95.2
2	Event voxel only	98.0
3	Event Image + Voxel	98.7

Table 4. Ablation study on ASL-DVS [14].

Index	Component	Results
1	w/o Bottleneck Feature	98.5
2	w/o FusionFormer	98.3

4.5 Parameter Analysis

The storage space of our proposed method is 220.3 MB. Our model spends 16.7 ms for each video in ASL-DVS dataset.

5 Conclusion

Previous event-based recognition approaches typically represented event streams as point clouds, voxels, or images, and employed various deep neural networks to learn feature representations. However, these approaches are usually challenged by monotonous modal expressions and the design of the network structure. To overcome these challenges, this paper introduces a novel dual-stream framework for event representation, extraction, and fusion. The proposed framework simultaneously models two common representations: event images and event voxels. By leveraging Transformer and Structured Graph Neural Network (GNN) architectures, spatial information and three dimensional stereo information can be learned separately. Moreover, the introduction of a bottleneck Transformer facilitates the fusion of the dual-stream information. Extensive experiments were conducted to evaluate the performance of our framework, using two widely used event-based classification datasets. The results demonstrate that our proposed framework achieves state-of-the-art performance. These findings highlight the effectiveness of the dual-stream framework in addressing the limitations of existing approaches and improving the recognition accuracy in event-based object recognition tasks.

Acknowledgement. This work is supported by the National Natural Science Foundation of China (No. 62102205).

References

1. ul Hassan, M.: AlexNet ImageNet classification with deep convolutional neural networks (2018)
2. He, K., Zhang, X., Ren, S., Sun, J.: Deep residual learning for image recognition. In: Proceedings of the IEEE Conference on Computer Vision and Pattern Recognition, pp. 770–778 (2016)
3. Vaswani, A., et al.: Attention is all you need. Adv. Neural Inf. Process. Syst. **30** (2017)

4. Krizhevsky, A., Sutskever, I., Hinton, G.E.: ImageNet classification with deep convolutional neural networks. Commun. ACM **60**(6), 84–90 (2017)
5. Sun, Z., Ke, Q., Rahmani, H., Bennamoun, M., Wang, G., Liu, J.: Human action recognition from various data modalities: a review. IEEE Trans. Pattern Anal. Mach. Intell. **45**(3), 3200–3225 (2023)
6. Wang, X., et al.: VisEvent: reliable object tracking via collaboration of frame and event flows arXiv preprint arXiv:2108.05015 (2021)
7. Tang, C., et al.: Revisiting color-event based tracking: a unified network, dataset, and metric, arXiv preprint arXiv:2211.11010 (2022)
8. Zhu, L., Wang, X., Chang, Y., Li, J., Huang, T., Tian, Y.: Event-based video reconstruction via potential-assisted spiking neural network. In: Proceedings of the IEEE/CVF Conference on Computer Vision and Pattern Recognition, pp. 3594–3604 (2022)
9. Wang, Y., et al.: EV-Gait: event-based robust gait recognition using dynamic vision sensors. In: Proceedings of the IEEE/CVF Conference on Computer Vision and Pattern Recognition, pp. 6358–6367 (2019)
10. Wang, Y., et al.: Event-stream representation for human gaits identification using deep neural networks. IEEE Trans. Pattern Anal. Mach. Intell. **44**, 3436–3449 (2021)
11. Fang, H., Shrestha, A., Zhao, Z., Qiu, Q.: Exploiting neuron and synapse filter dynamics in spatial temporal learning of deep spiking neural network, arXiv preprint arXiv:2003.02944 (2020)
12. Fang, W., Yu, Z., Chen, Y., Masquelier, T., Huang, T., Tian, Y.: Incorporating learnable membrane time constant to enhance learning of spiking neural networks. In: Proceedings of the IEEE/CVF International Conference on Computer Vision, pp. 2661–2671 (2021)
13. Bi, Y., Chadha, A., Abbas, A., Bourtsoulatze, E., Andreopoulos, Y.: Graph-based object classification for neuromorphic vision sensing. In: Proceedings of the IEEE/CVF International Conference on Computer Vision, pp. 491–501 (2019)
14. Bi, Y., Chadha, A., Abbas, A., Bourtsoulatze, E., Andreopoulos, Y.: Graph-based spatio-temporal feature learning for neuromorphic vision sensing. IEEE Trans. Image Process. **29**, 9084–9098 (2020)
15. Wang, Y., et al.: Event-stream representation for human gaits identification using deep neural networks. IEEE Trans. Pattern Anal. Mach. Intell. **44**(7), 3436–3449 (2021)
16. Diehl, P.U., Neil, D., Binas, J., Cook, M., Liu, S.-C., Pfeiffer, M.: Fast-classifying, high-accuracy spiking deep networks through weight and threshold balancing. In: 2015 International Joint Conference on Neural Networks (IJCNN), pp. 1–8. IEEE (2015)
17. Perez-Nieves, N., Goodman, D.: Sparse spiking gradient descent. Adv. Neural Inf. Process. Syst. **34**, 11 795–11 808 (2021)
18. Fang, W., Yu, Z., Chen, Y., Huang, T., Masquelier, T., Tian, Y.: Deep residual learning in spiking neural networks. Adv. Neural Inf. Process. Syst. **34**, 21 056–21 069 (2021)
19. Wang, X., et al.: SSTFormer: bridging spiking neural network and memory support transformer for frame-event based recognition arXiv preprint arXiv:2308.04369 (2023)
20. Jiang, B., Yuan, C., Wang, X., Bao, Z., Zhu, L., Luo, B.: Point-voxel absorbing graph representation learning for event stream based recognition arXiv preprint arXiv:2306.05239 (2023)

21. Wang, Q., Zhang, Y., Yuan, J., Lu, Y.: Space-time event clouds for gesture recognition: from RGB cameras to event cameras. In: 2019 IEEE Winter Conference on Applications of Computer Vision (WACV), pp. 1826–1835. IEEE (2019)
22. Qi, C.R., Su, H., Mo, K., Guibas, L.J.: PointNet: deep learning on point sets for 3d classification and segmentation. In: Proceedings of the IEEE Conference on Computer Vision and Pattern Recognition, pp. 652–660 (2017)
23. Xie, B., Deng, Y., Shao, Z., Liu, H., Li, Y.: VMV-GCN: volumetric multi-view based graph CNN for event stream classification. IEEE Robot. Autom. Lett. **7**(2), 1976–1983 (2022)
24. Li, Z., Asif, M.S., Ma, Z.: Event transformer, arXiv preprint arXiv:2204.05172 (2022)
25. Schaefer, S., Gehrig, D., Scaramuzza, D.: AEGNN: asynchronous event-based graph neural networks. In: Proceedings of the IEEE/CVF Conference on Computer Vision and Pattern Recognition, pp. 12 371–12 381 (2022)
26. Li, Y., et al.: Graph-based asynchronous event processing for rapid object recognition. In: Proceedings of the IEEE/CVF International Conference on Computer Vision, pp. 934–943 (2021)
27. Srinivas, A., Lin, T.-Y., Parmar, N., Shlens, J., Abbeel, P., Vaswani, A.: Bottleneck transformers for visual recognition. In: Proceedings of the IEEE/CVF Conference on Computer Vision And Pattern Recognition, pp. 16 519–16 529 (2021)
28. Li, S., et al.: Enhancing the locality and breaking the memory bottleneck of transformer on time series forecasting. Adv. Neural Inf. Process. Syst., **32** (2019)
29. Nagrani, A., Yang, S., Arnab, A., Jansen, A., Schmid, C., Sun, C.: Attention bottlenecks for multimodal fusion. Adv. Neural Inf. Process. Syst. **34**, 14 200–14 213 (2021)
30. Song, R., Feng, Y., Cheng, W., Mu, Z., Wang, X.: BS2T: bottleneck spatial-spectral transformer for hyperspectral image classification. IEEE Trans. Geosci. Remote Sens. **60**, 1–17 (2022)
31. Miranda, L.J.: Understanding softmax and the negative log-likelihood. ljvmiranda921. github. io (2017)
32. Orchard, G., Jayawant, A., Cohen, G.K., Thakor, N.: Converting static image datasets to spiking neuromorphic datasets using saccades. Front. Neurosci. **9**, 437 (2015)
33. Gehrig, D., Loquercio, A., Derpanis, K.G., Scaramuzza, D.: End-to-end learning of representations for asynchronous event-based data. In: Proceedings of the IEEE/CVF International Conference on Computer Vision, pp. 5633–5643 (2019)
34. Deng, Y., Li, Y., Chen, H.: AMAE: adaptive motion-agnostic encoder for event-based object classification. IEEE Robot. Autom. Lett. **5**(3), 4596–4603 (2020)
35. Cannici, M., Ciccone, M., Romanoni, A., Matteucci, M.: A differentiable recurrent surface for asynchronous event-based data. In: Vedaldi, A., Bischof, H., Brox, T., Frahm, J.-M. (eds.) ECCV 2020. LNCS, vol. 12365, pp. 136–152. Springer, Cham (2020). https://doi.org/10.1007/978-3-030-58565-5_9
36. Deng, Y., Chen, H., Li, Y.: MVF-Net: a multi-view fusion network for event-based object classification. IEEE Trans. Circuits Syst. Video Technol. **32**(12), 8275–8284 (2021)
37. Sekikawa, Y., Hara, K., Saito, H.: EventNet: asynchronous recursive event processing. In: Proceedings of the IEEE/CVF Conference on Computer Vision and Pattern Recognition, pp. 3887–3896 (2019)

38. Deng, Y., Chen, H., Chen, H., Li, Y.: EVVGCNN: a voxel graph CNN for event-based object classification, arXiv preprint arXiv:2106.00216, vol. 1, no. 2, p. 6 (2021)
39. Sironi, A., Brambilla, M., Bourdis, N., Lagorce, X., Benosman, R.: HATS: histograms of averaged time surfaces for robust event-based object classification. In: Proceedings of the IEEE Conference on Computer Vision and Pattern Recognition, pp. 1731–1740 (2018)

Multi-scale Dilated Attention Graph Convolutional Network for Skeleton-Based Action Recognition

Yang Shu, Wanggen Li(✉), Doudou Li, Kun Gao, and Biao Jie

Anhui Normal University, Wuhu, China
{yangshu,xchen}@ahnu.edu.cn

Abstract. Due to the small size, anti-interference and strong robustness of skeletal data, research on human skeleton-based action recognition has become a mainstream. However, due to the incomplete utilization of semantic information and insufficient time modeling, most methods may not be able to fully explore the connections between non-adjacent joints in the spatial or temporal dimensions. Therefore, we propose a Multi-scale Dilated Attention Graph Convolutional Network for Skeleton-Based Action Recognition (MDKA-GCN) to solve the above problems. In the spatial configuration, we explicitly introduce the channel graph composed of high-level semantics (joint type and frame index) of joints into the network to enhance the representation ability of spatiotemporal features. MDKA-GCN uses joint-level, velocity-level and bone-level graphs to more deeply mine the hidden features of human skeletons. In the time configuration, two lightweight multi-scale strategies are proposed, which can be more robust to time changes. Extensive experiments on NTU-RGB+D 60 datasets and NTU-RGB+D 120 datasets show that MDKA-GCN has reached an advanced level, and surpasses the performance of most lightweight SOTA methods.

Keywords: Action Recognition · Multi-scale · Semantic Information · Dilated Attention · Lightweight

1 Introduction

In recent years, the task of action recognition has become one of the most attractive topics in the field of artificial intelligence, especially human action recognition (HAR) is widely used in various fields such as human or object interaction, video surveillance systems and healthcare systems [1], providing accurate judgment analysis and understanding of human actions for machinery and equipment in these fields, playing a crucial role in the development and progress of artificial intelligence.

Early on, research on human skeleton action recognition is mainly through deep neural network models to learn the correlation of human actions in time and space. In these models, the performance of human skeleton action recognition

© The Author(s), under exclusive license to Springer Nature Singapore Pte Ltd. 2024
Q. Liu et al. (Eds.): PRCV 2023, LNCS 14425, pp. 16–28, 2024.
https://doi.org/10.1007/978-981-99-8429-9_2

based on graph convolutional networks (GCN) [2] is better than that based on recurrent neural networks (RNN) [3] and convolutional neural networks (CNN) [4]. GCN methods can construct a spatiotemporal topology graph of 3D positions of human skeleton joint nodes by regarding human joint nodes as vertices of a graph, treating natural topological connections between adjacent joint nodes as spatial edges of a graph and considering temporal correlation between adjacent frames as temporal edges. Then input the processed human skeleton topology graph sequence into the network for learning to finally achieve action classification. The GCN-based method has been proven to be an effective solution for achieving the task of human action recognition.

To further improve the performance of the model, they [5–8] focus on introducing adaptive graph residual masks to capture the relationships between different joints, that is, to extract more hidden information from the original human skeleton dataset, such as bone and velocity. In order to enhance the feature representation of every actions, they train this information through multiple network streams and fuse all the trained features together to obtain the score of each action and achieve the task of action classification. However, more information will cause information redundancy and model size doubling sacrificing model storage space and computational efficiency, which is extremely disadvantageous for model promotion in practical applications.

In response, SGN [9] achieves superior performance with a smaller model, however, it also has problems such as insufficient data mining and incomplete semantic utilization. Guided by literature [10,11], we consider combining channel attention with dilated convolution attention to enhance feature connections between frame dimensions and channels in the model. The main contributions of this paper can be summarized as follows:

- This paper introduces multiple hidden information of human skeletons after data preprocessing and effectively fuses them in the early stage of the model, enhancing feature representation of each information and obtaining a richer topology graph.
- In order to make full use of two semantic relationships, we integrate two semantic information into graph convolution modules by adjusting graph convolution layers and channel width effectively, solving defects in spatial-temporal separation processing.
- In the time module, we design a time multi-scale dilated convolution kernel attention (T-MDKA), to obtain a large receptive field by replacing large kernel convolution with dilated convolution, thereby simulating remote dependencies. In addition, we construct two branch time convolution blocks to more robustly learn the temporal features of actions.

2 Related Works

2.1 Attention Mechanism

The attention mechanism can be seen as simulating the degree of attention that people pay to a certain part when processing information by adjusting the size

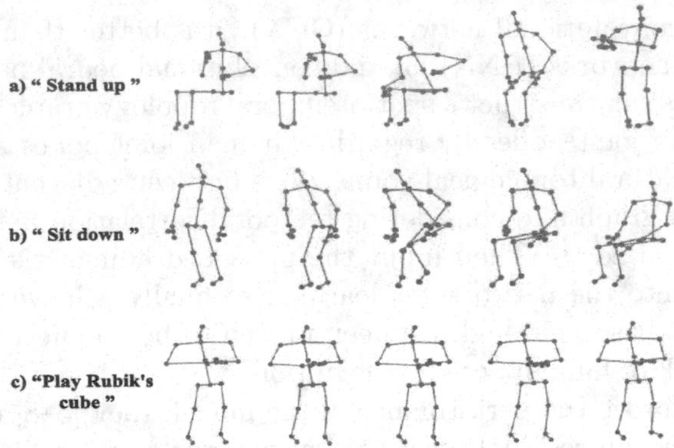

Fig. 1. Skeleton diagrams of 5 frames from three action sequences.

of the weights. It is currently widely used in various fields. [10–12]. SENet [10] proposes a squeeze-and-excitation block to learn global channel information, which enables the model to focus on more useful feature information. VAN [12] improves channel adaptability by using large kernels. To maximize the role of large convolution kernels, MAN [11] adopts the structure of transformer and introduces GSAU to replace MLP structure to obtain multi-scale remote modeling dependencies not only improving model representation ability, but also reducing model parameters and computational complexity.

2.2 Lightweight Models

In the image field [13] and object detection field [14], methods using depth-separable convolution and grouped convolution are proposed respectively to replace traditional convolution greatly reducing model parameters. Zhang et al. [9] based on graph convolutional neural networks introduce high-order semantic information to enhance feature expression ability achieving low parameters while maintaining high recognition accuracy. Cheng et al. [15] construct a lightweight network framework using dynamic displacement graph convolution instead of traditional convolution, In order to further simplify, they [16] use edge RELU distillation technology, which also improves model recognition performance. In addition, Song et al. [17] embed separable convolution layers into early multi-information fusion module. It makes the model's parameter size extremely small, making the model more lightweight.

3 Method

In this section, we will detail the composition of our proposed MDKA-GCN. Figure 2(a) is our overall model framework.

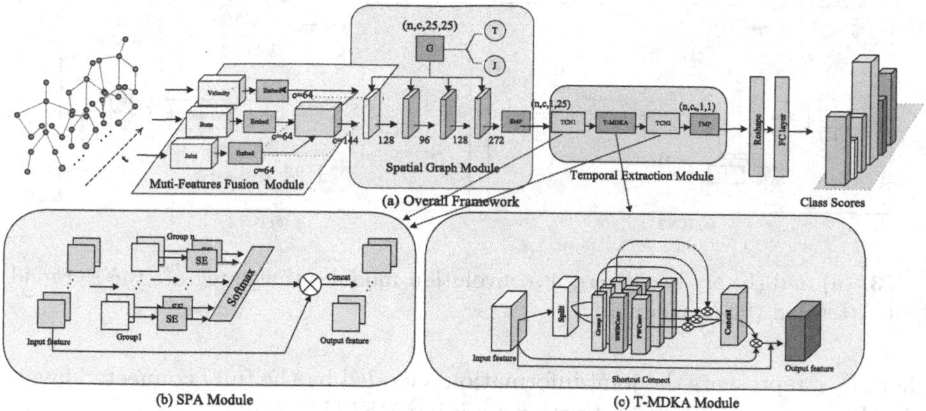

Fig. 2. (a) the overall framework of this paper's model includes three parts; (b) the pyramid partition attention module applied to TCN1 and TCN2; (c) the dilated convolution attention module proposed in this paper.

3.1 Multi-Branch Fusion Module

Earlier research [17] has shown that more skeleton topology graphs play a key role in model performance. In this work, we mine three types of input features from skeleton data: 1) joint stream, 2) velocity stream, 3) bone stream.

Specifically, this paper represents the skeleton sequence as a set of joint sets $\mathbf{S}_{t,k} = \{x_{t,k}|t = 1, 2, ..., T; k = 1, 2, ..., J\}$, where T represents the total number of time frames and J represents the total number of human joints.

Joint $\mathbf{S}_{t,k} \in \mathbb{R}^{C\times T\times V}$ is the original 3D coordinate provided by the datasets where channel C is equal to 3. Therefore, through formula $\mathrm{v}_{t,k} = \mathrm{x}_{t,k} - \mathrm{x}_{t,m}$, we can obtain the relative position of joints where $\mathrm{x}_{t,m}$ represents the position of the human skeleton's center of gravity. Considering that many subtle actions are concentrated on the hands such as "play Rubik's cube" in Fig. 1(c), we determine three central joints as the upper and lower spine and palm wrist joints of the action sequence.

Similarly joint velocity can also be easily defined by joint position that is the position change between adjacent frames represented as $\mathrm{v}_{t,k} = \mathrm{x}_{t,k} - \mathrm{x}_{t-1,k}$.

Like the definition of relative position, bone information can also be defined by the position difference between two adjacent joints on a skeleton represented as $\mathrm{b}_{t,k} = \mathrm{x}_{t,k} - \mathrm{x}_{t,i}$, where joint $\mathrm{x}_{t,k}$ represents a position away from the human body's center of gravity and joint $\mathrm{x}_{t,i}$ represents a position close to the human body's center of gravity adjacent to joint $\mathrm{x}_{t,k}$. Similar to formula of joint velocity, we can easily get bone velocity. Since acceleration information is crucial for capturing some small actions, we can obtains bone acceleration information from bone velocity information as input information. Finally, these input information are encoded through two fully connected layers (FC),

$$\mathrm{P}_{t,k} = \sigma(\mathrm{FC}(\sigma(\mathrm{FC}(\mathrm{x}_{t,k})))) \quad (1)$$

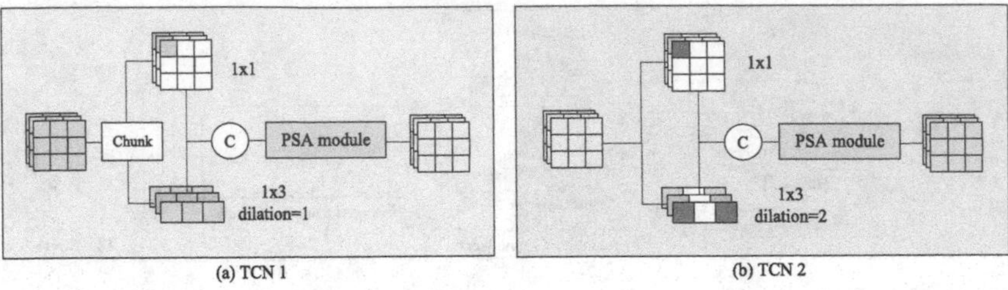

Fig. 3. (a) and (b) are two divergent convolution modules that integrate the Pyramid Split Attention (PSA) module.

where $\mathrm{P}_{t,k}$ represents the joint information encoded by the fully connected layer; σ in this paper represents activation function RELU.

In the feature fusion module, we obtain different channel attention weights through two layers of convolution operations, further enhancing feature representation of each information stream. Then fuse three information streams and input them into graph convolution modules,

$$\hat{\mathrm{P}}_{t,k} = Conv(\sigma(bn(\mathrm{p}_{t,k}))) \tag{2}$$

$$f_{\mathrm{in}} = Cat[\hat{\mathrm{P}}_{t,k}, \hat{\mathrm{V}}_{t,k}, \hat{\mathrm{B}}_{t,k}] \tag{3}$$

where $\hat{\mathrm{V}}_{t,k}$ and $\hat{\mathrm{B}}_{t,k}$ represents respectively the velocity information and bone information encoded by the fully connected layer; bn and Cat represent normalization function and concatenation operation respectively. $Conv$ represents pointwise convolution, which reduces the channel dimension to avoid generating a large number of parameters due to high dimensions after concatenation.

3.2 Semantic Information

In this work, we use a one-hot vector J'_k to represent the kth skeleton joint. Similarly use a one-hot vector T'_i to represent the ith frame index. This paper concatenates two semantic information in low dimensions, then convolves them through multi-layer perceptron (MLP). Finally, input them into each layer graph convolution,

$$\mathrm{G}_0 = Cat[\mathrm{J}'_k, \mathrm{T}'_i] \tag{4}$$

$$\mathrm{G}_j = \mathrm{MLP}(\mathrm{G}_{j-1}) \tag{5}$$

where J'_k and T'_i respectively refer to joint type and frame index semantic information, and $j = 1, 2, 3, 4$. The purpose of MLP is to increase channel dimensions to match input dimensions of each layer graph convolution.

3.3 Graph Convolution Module

The operation in this paper's graph convolution module is different from previous work [9], which often extracts spatial features through adjacency matrices composed of natural human joint nodes or deformations thereof. This paper considers the importance of two semantic information, inputs channel graphs fused with semantics into graph convolution modules,

$$f_{\text{out}} = \sigma(bn(Conv(f_{\text{in}} \otimes \mathrm{G}_j) + Conv(f_{\text{in}}))) \tag{6}$$

where f_{in} and f_{out} are respectively input and output of graph convolution, $\otimes$ represents matrix multiplication, the size of the convolutional kernel of $Conv$ is 1×1 with different training weights.

3.4 Time Convolution Module

Considering that time processing method of SGN only uses a fixed convolution operation which is not enough to distinguish some similar actions, such as Fig. 1(a) and Fig. 1(b). Therefore, we propose a time convolution module composed of two branch convolution blocks in Fig. 3 and multi-scale dilated attention in Fig. 2(c).

Branch Convolution Block. In order to obtain receptive fields at different scales while controlling model parameter volume and computational volume. Inspired by [18], We introduce the PSA module and design two types of pyramid split convolution modules in Fig. 2(b) for extracting multi-scale temporal features of action sequences,

$$F_i = Conv_i(\sigma(bn(chunk(\mathrm{X}_{\text{in}})))) \tag{7}$$

$$F = F_0 \oplus F_1 \tag{8}$$

where $chunk$ is the split operator, which divides the input information into two equal parts on the channel dimension. And $Conv_i$ means the convolution operation with kernel sizes of 1×1 and 1×3. $\oplus$ is the concat operator. Then, the SEweight module is used to obtain the attention weight from the input feature map with different scales,

$$Z_i = \mathrm{SEweight}(F_i) \tag{9}$$

$$Z = \mathrm{Softmax}(Z_0 \oplus Z_1) \tag{10}$$

$$Out = F \odot Z \tag{11}$$

where the $Softmax$ is used to obtain the re-calibrated weight, $\odot$ represents the dot product operation. Different from $TCN1$, $TCN2$ uses grouped convolutions and dilated convolutions with dilation rate 2 with kernel size 1×3.

Time Multi-scale Dilated Attention. Inspired by some large kernel works [8,11] without adding too much computational burden obtaining advantages of attention mechanisms in long-term modeling, we propose a time multi-scale dilated kernel attention and replace large kernel convolutions with dilated convolutions focusing on extracting temporal dependencies between action sequences. It is shown in Fig. 2(c), the formula is as follows:

$$x_i = Split(x) \tag{12}$$

$$DKA(x_i) = PWConv(Conv_{DWD}(x_i)) \tag{13}$$

$$MDKA(x) = Cat(DKA(x_i) \odot x_i) \tag{14}$$

where *Split* is the split operation, $i = 1, 2, 3, 4$; $Conv_{DWD}$ is a dilated separable convolution with a kernel size of 1×3, and the dilation rate can be $2, 3, 4$, *PWConv* represents a normal pointwise convolution.

4 Experiment

4.1 Dataset

NTU-RGB+D 60 Dataset [19]: One of the current mainstream skeleton-based action recognition datasets, containing 56880 skeleton sequences of 60 action categories captured simultaneously from 40 different subjects and 3 Microsoft Kinect V2 depth cameras. Each skeleton sequence contains three-dimensional spatial coordinates of 25 joints. The dataset provides two evaluation benchmarks: Cross-Subject (C-Sub) and Cross-View (C-View). C-Sub is completed by 40 subjects with half of the subjects used for training and the rest for testing. C-View selects samples captured by cameras 2 and 3 for training and the rest for testing.

NTU-RGB+D 120 Dataset [20]: This dataset is an expansion of the NTU RGB+D 60 dataset in terms of action categories and number of actors, containing 114480 action videos participated by 106 actors, with a total of 120 action categories including 82 daily life actions 12 medical conditions and 26 actions under two-person interaction. The dataset has two evaluation benchmarks: Cross-Subject (C-Sub120) and Cross-Setup (C-Set120). C-Sub120 divides this dataset into training set (63026 videos) and validation set(50919 videos) according to different actors in the video. C-Set120 divides the dataset according to the parity of video numbers. 54468 even-numbered videos are used as training sets, and 59477 odd-numbered videos are used as test sets.

4.2 Experimental Details

Similar to [9], the difference is that in order to facilitate the operation of channel graphs composed of two semantic information in the graph convolution module, we adjust the time dimension to 25. This paper sets the number of epochs in the model to 120 sets the batch size for each epoch to 64 sets, the initial learning rate to 0.001 and continues to decrease during iteration. When the number of iterations is 80 and 100 the learning rate drops tenfold. At the same time, this paper also uses Adam to optimize the model where weight decay is 0.0001.

4.3 Ablation Experiment

In this part, we mainly discuss the contributions of different components in this paper's model, which includes multi-branch fusion module, high-order semantic information time convolution module and necessity of attention module.

Table 1. Comparison of the accuracy of different input branches.

Input	Param(M)	C-Sub/%	C-View/%
Joint	0.60	88.4	94.4
Bone	0.60	89.8	94.8
Velocity	0.60	84.8	91.0
Joint and Bone	0.65	90.3	95.6
Joint and Velocity	0.65	90.4	95.8
Bone and Velocity	0.65	90.5	95.7
Baseline	0.69	**91.2**	**96.2**

Table 1 experimental results two conclusions are verified: first using information fusion of three input branches has obviously highest recognition accuracy on C-Sub and C-View; second this paper's information fusion method only increases about 1/7 of parameter volume while model performance has been significantly improved.

Table 2. Verification of the accuracy of two types of semantic information.

Input	Param(M)	C-Sub/%	C-View/%
w/o J	0.69	90.2	95.5
w/o T	0.69	90.5	95.3
w/o J and T	0.69	89.9	94.8
Baseline	0.69	**91.2**	**96.2**

Table 2 experimental results show the channel graph composed of two semantic information plays an important role in graph convolution operation. It is worth noting that when there is no channel graph the original graph convolution becomes ordinary pointwise convolution and the model's ability to aggregate different joint features will weaken resulting in a decline in model performance.

Table 3 verifies the effectiveness of the two branch convolution blocks and multi-scale dilated attention module. From the table it can be seen that the attention module proposed in this paper effectively improves model performance while adding a very small amount of parameters; and our reasonable combination of two branch convolution modules makes the model more balanced in terms of performance and parameter volume.

Table 3. Verification of the effectiveness of two divergent convolution modules and attention mechanisms.

Input	Param(M)	C-Sub/%	C-View/%
w/o Attention	0.70	90.7	96.0
w/o TCN1	0.62	90.9	95.8
w/o TCN2	0.55	90.8	95.4
TCN1 + TCN1	0.62	90.8	95.8
TCN2 + TCN2	0.77	90.8	96.0
Baseline	0.69	**91.2**	**96.2**

4.4 Comparison with State-of-the-Art

From Table 4, it can be seen that the best performance of our single-stream network MDKA-GCN (1s) on the two benchmarks is 91.2% and 96.2% respectively, while the establishment of multi-stream network enables the model to achieve better performance especially MDKA-GCN (4s) recognition accuracy on the two benchmarks respectively reach 92.1% and 96.8%, which is better than other SOTA models.

Table 4. Comparison of accuracy (%) with some recent SOTA methods.

Methods	Year	C-Sub/%	C-View/%	C-Sub120/%	C-Set120/%
ST-GCN [2]	2018	81.5	88.3	70.7	73.2
2s-AGCN [5]	2019	88.5	95.1	82.5	84.2
4s-Shift-GCN [15]	2020	90.7	96.5	85.9	87.6
SGN [9]	2020	89.0	94.5	79.2	81.5
MS-G3D [21]	2020	91.5	96.2	86.9	88.4
FGCN [22]	2021	90.2	96.3	85.4	87.4
CDGC [23]	2021	90.9	96.5	86.3	87.8
4s-Shift-GCN++ [16]	2021	90.5	96.3	85.6	87.2
Ta-CNN [4]	2022	90.7	95.1	85.7	87.3
4s-AGE-Ens [6]	2022	91.6	96.3	88.2	89.2
2s-ST-GCN++ [24]	2022	91.4	96.7	87.0	89.1
EffiencientB4 [17]	2022	**92.1**	96.1	**88.7**	88.9
SMotif-GCN [7]	2022	91.7	96.7	88.4	88.9
LKA-GCN [8]	2023	90.7	96.1	86.3	87.8
MDKA-GCN(1s)	–	91.2	96.2	86.8	88.3
MDKA-GCN(2s)	–	91.6	96.6	87.5	89.1
MDKA-GCN(4s)	–	**92.1**	**96.8**	87.9	**89.4**

Due to the randomness in the network during training, such as sample shuffling operations and frame extraction randomness, these random operations can cause incomplete feature learning in model training, resulting in unstable training results. To avoid this randomness and enhance the robustness of the network, we design a multi-stream network structure where each single-stream subnetwork structure is completely consistent. We fuse the output results of multiple single-stream sub-networks by adding them together and use them as the final output result of the multi-stream network.

In addition, Table 4 shows that on the C-Sub120 and C-Set120 compared with SGN [9], our single-stream method increases the accuracy by 7.6 percentage points and 6.8 percentage points respectively. In multi-stream networks (2s-AGCN [5], 4s-AGE-Ens [6], SMotif-GCN [7]), our multi-stream method MDKA-GCN(4s), although lower in accuracy on C-Sub120 than 4s-AGE-Ens and SMotif-GCN, reach the highest accuracy on C-Set120 benchmark.

Table 5. Comparisons with SOTA methods

Methods	Param.(M)	GFLOPS/%	C-View/%
Ta-CNN [4]	1.06	1.06	95.1
ST-GCN [2]	3.1	16.32	88.3
2s-AGCN [5]	6.94	37.32	95.1
4s-Shift-GCN [15]	2.76	10	96.5
SGN [9]	**0.69**	0.8	94.5
4s-Shift-GCN++ [16]	1.8	1.7	96.3
2s-ST-GCN++ [24]	1.39	2.8	96.7
EffiencientB4 [17]	1.1	4.05	96.1
MDKA-GCN(1s)	**0.69**	**0.69**	96.2
MDKA-GCN(2s)	1.38	1.38	96.6
MDKA-GCN(4s)	2.92	2.76	**96.8**

Comparison with Lightweight SOTA. To verify the overall performance of our model as shown in Table 5, we compare with SOTA methods in recent years in terms of accuracy model parameter volume and computational complexity. Compared to lightweight GCN method EfficientB4, our single-stream model has a smaller parameter size and computational complexity while achieving higher accuracy. We compare the model training process of MDKA-GCN and SGN on two datasets in Fig. 4. For fair comparison the hyperparameter settings and data preprocessing methods of the two models are kept consistent. Overall our method's overall performance has reached an advanced level and is more suitable for resource-limited mobile devices and practical application scenarios compared to most lightweight SOTA methods.

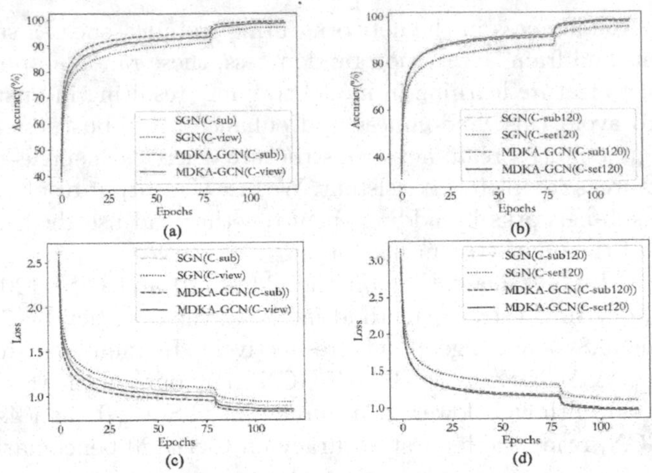

Fig. 4. comparison of the accuracy and convergence of MDKA-GCN and the baseline model SGN.

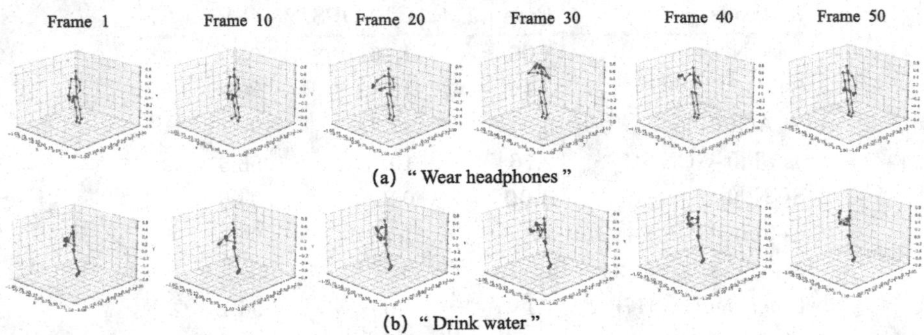

Fig. 5. Qualitative examples from NTU-RGB+D 60, Six frames are selected from each action.

5 Action Visualization

To more intuitively display the action process, this paper visualizes the skeleton diagrams of actions such as "wear headphones" and "drink water" by observing some similar or difficult-to-distinguish actions in Fig. 5, and selects a few frames from them. These actions are mainly completed by both hands and are extremely similar in spatial configuration and temporal dynamics, requiring long-term observation to distinguish.

6 Conclusion

In this paper, we introduce two semantic information into multiple graph convolution layers, and the model performance is improved while reducing model parameters. Our designed branch convolution block emphasizes significant motion features, the proposed time multi-scale dilated convolution attention module enlarges the receptive field and enriches the representation ability of various temporal features. We conduct extensive experiments on current mainstream action recognition datasets, whose results show that MDKA-GCN is more effective than most mainstream methods in terms of computational cost and performance with broader application prospects in the future.

Acknowledgement. This work is supported by the National Natural Science Foundation of China (61976006).

References

1. Carreira, J., Zisserman, A.: Quo vadis, action recognition? A new model and the kinetics dataset. In: Proceedings of the IEEE Conference on Computer Vision and Pattern Recognition, pp. 6299–6308 (2017)
2. Yan, S., Xiong, Y., Lin, D.: Spatial temporal graph convolutional networks for skeleton-based action recognition. In: Proceedings of the AAAI Conference on Artificial Intelligence, vol. 32 (2018)
3. Si, C., Chen, W., Wang, W., Wang, L., Tan, T.: An attention enhanced graph convolutional LSTM network for skeleton-based action recognition. In: Proceedings of the IEEE/CVF Conference on Computer Vision and Pattern Recognition, pp. 1227–1236 (2019)
4. Xu, K., Ye, F., Zhong, Q., Xie, D.: Topology-aware convolutional neural network for efficient skeleton-based action recognition. In: Proceedings of the AAAI Conference on Artificial Intelligence, vol. 36, pp. 2866–2874 (2022)
5. Shi, L., Zhang, Y., Cheng, J., Lu, H.: Two-stream adaptive graph convolutional networks for skeleton-based action recognition. In: Proceedings of the IEEE/CVF Conference on Computer Vision and Pattern Recognition, pp. 12026–12035 (2019)
6. Qin, Z., et al.: Fusing higher-order features in graph neural networks for skeleton-based action recognition. IEEE Trans. Neural Netw. Learn. Syst., 1–15 (2022)
7. Wen, Y.H., Gao, L., Fu, H., Zhang, F.L., Xia, S., Liu, Y.J.: Motif-GCNS with local and non-local temporal blocks for skeleton-based action recognition. IEEE Trans. Pattern Anal. Mach. Intell. **45**(2), 2009–2023 (2022)
8. Liu, Y., Zhang, H., Li, Y., He, K., Xu, D.: Skeleton-based human action recognition via large-kernel attention graph convolutional network. IEEE Trans. Visual Comput. Graphics **29**(5), 2575–2585 (2023)
9. Zhang, P., Lan, C., Zeng, W., Xing, J., Xue, J., Zheng, N.: Semantics-guided neural networks for efficient skeleton-based human action recognition. In: Proceedings of the IEEE/CVF Conference on Computer Vision and Pattern Recognition, pp. 1112–1121 (2020)
10. Hu, J., Shen, L., Sun, G.: Squeeze-and-excitation networks. In: Proceedings of the IEEE Conference on Computer Vision and Pattern Recognition, pp. 7132–7141 (2018)
11. Wang, Y., Li, Y., Wang, G., Liu, X.: Multi-scale attention network for single image super-resolution. arXiv preprint arXiv:2209.14145 (2022)
12. Guo, M.H., Lu, C.Z., Liu, Z.N., Cheng, M.M., Hu, S.M.: Visual attention network. arXiv preprint arXiv:2202.09741 (2022)

13. Howard, A.G., et al.: MobileNets: efficient convolutional neural networks for mobile vision applications. arXiv preprint arXiv:1704.04861 (2017)
14. Xie, S., Girshick, R., Dollár, P., Tu, Z., He, K.: Aggregated residual transformations for deep neural networks. In: Proceedings of the IEEE Conference on Computer Vision and Pattern Recognition, pp. 1492–1500 (2017)
15. Cheng, K., Zhang, Y., He, X., Chen, W., Cheng, J., Lu, H.: Skeleton-based action recognition with shift graph convolutional network. In: Proceedings of the IEEE/CVF Conference on Computer Vision and Pattern Recognition, pp. 183–192 (2020)
16. Cheng, K., Zhang, Y., He, X., Cheng, J., Lu, H.: Extremely lightweight skeleton-based action recognition with shiftGCN++. IEEE Trans. Image Process. **30**, 7333–7348 (2021)
17. Song, Y.F., Zhang, Z., Shan, C., Wang, L.: Constructing stronger and faster baselines for skeleton-based action recognition. IEEE Trans. Pattern Anal. Mach. Intell. **45**(2), 1474–1488 (2022)
18. Zhang, H., Zu, K., Lu, J., Zou, Y., Meng, D.: EPSANet: an efficient pyramid split attention block on convolutional neural network. arxiv 2021. arXiv preprint arXiv:2105.14447 (2021)
19. Shahroudy, A., Liu, J., Ng, T.T., Wang, G.: NTU RGB+ D: a large scale dataset for 3d human activity analysis. In: Proceedings of the IEEE Conference on Computer Vision and Pattern Recognition, pp. 1010–1019 (2016)
20. Liu, J., Shahroudy, A., Perez, M., Wang, G., Duan, L.Y., Kot, A.C.: NTU RGB+ D 120: a large-scale benchmark for 3d human activity understanding. IEEE Trans. Pattern Anal. Mach. Intell. **42**(10), 2684–2701 (2019)
21. Liu, Z., Zhang, H., Chen, Z., Wang, Z., Ouyang, W.: Disentangling and unifying graph convolutions for skeleton-based action recognition. In: Proceedings of the IEEE/CVF Conference on Computer Vision and Pattern Recognition, pp. 143–152 (2020)
22. Li, M., Chen, S., Chen, X., Zhang, Y., Wang, Y., Tian, Q.: Symbiotic graph neural networks for 3d skeleton-based human action recognition and motion prediction. IEEE Trans. Pattern Anal. Mach. Intell. **44**(6), 3316–3333 (2021)
23. Miao, S., Hou, Y., Gao, Z., Xu, M., Li, W.: A central difference graph convolutional operator for skeleton-based action recognition. IEEE Trans. Circuits Syst. Video Technol. **32**(7), 4893–4899 (2021)
24. Duan, H., Wang, J., Chen, K., Lin, D.: PYSKL: towards good practices for skeleton action recognition. In: Proceedings of the 30th ACM International Conference on Multimedia, pp. 7351–7354 (2022)

Auto-Learning-GCN: An Ingenious Framework for Skeleton-Based Action Recognition

Wentian Xin[1,2,3], Yi Liu[1,2,3], Ruyi Liu[1,2,3], Qiguang Miao[1,2,3(✉)], Cheng Shi[4], and Chi-Man Pun[5]

[1] School of Computer Science and Technology, Xidian University, Xi'an, China
{wtxin,ly330}@stu.xidian.edu.cn, {ruyiliu,qgmiao}@xidian.edu.cn
[2] Xi'an Key Laboratory of Big Data and Intelligent Vision, Xidian University, Xi'an, China
[3] Ministry of Education Key Laboratory of Collaborative Intelligence Systems, Xi'an, China
[4] School of Computer Science and Engineering, Xi'an University of Technology, Xi'an, China
C_shi@xaut.edu.cn
[5] Department of Computer and Information Science, University of Macau, Zhuhai, China
cmpun@umac.mo

Abstract. The Graph Convolutional Network (GCN) has garnered substantial interest over an extended period owing to its notable efficacy in addressing topological correlations, with particular achievements observed in skeleton-based action recognition. However, it is intriguing to note that the efficacy of the adaptive module within the existing model gradually diminishes as the learning capacity of the adjacency matrix improves. Inspired by this, AL-GCN constructs a comprehensively learnable group adjacency matrix in both spatial and temporal dimensions, thus forming an elegant and efficient GCN-based model. Firstly, the prevalent adaptive module is courageously removed, and the key points of the human body are astutely leveraged as motion compensation to provide dynamic correlation support. Secondly, a similar temporal adjacency matrix group is designed in the temporal domain to capture the global interaction characteristics more effectively. Subsequently, attention modules are purposefully crafted for both the temporal and spatial domains, which provide discriminative information between classes and improve learnability and interpretability. Finally, the Bayesian weight selection algorithm is utilized to achieve efficient and accurate fusion results for multi-stream data. On the NTU-60, NTU-120, and NW-UCLA, AL-GCN outperforms the state-of-the-art method, with up to **7.0%** improvement on the challenging UAV-Human, achieving superior performance in multiple settings, all while maintaining a lower computational cost. *Related code will be available on* Auto-Learning-GCN.

© The Author(s), under exclusive license to Springer Nature Singapore Pte Ltd. 2024
Q. Liu et al. (Eds.): PRCV 2023, LNCS 14425, pp. 29–42, 2024.
https://doi.org/10.1007/978-981-99-8429-9_3

Keywords: Skeleton-based Action Recognition · Graph Convolutional Network · Attention Mechanism · Model Optimization

1 Introduction

As a fundamental and important topic in computer vision, human action recognition is widely employed in automatic driving [1], video surveillance [2], and human-computer interaction [3], which provides reliable and human-centered action analysis results for real tasks. Different from the data of other modalities, the compact skeleton data provides the specific position and motion information of human joints, which facilitates the construction of spatio-temporal and motion features, avoids the interference of background environment, and can effectively improve the accuracy and robustness of action recognition in noisy scenes [4,5].

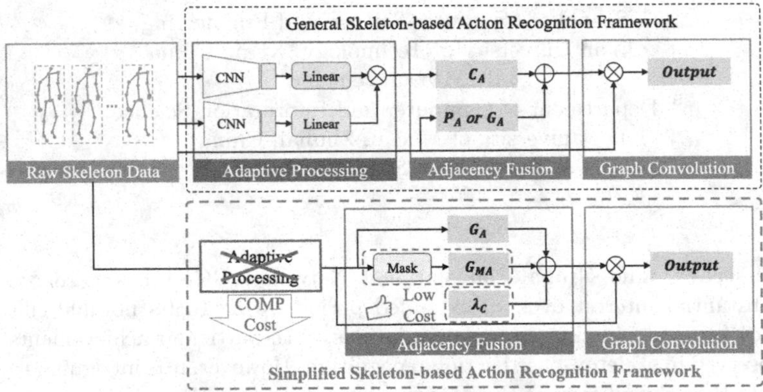

Fig. 1. In contrast to existing methods that rely on adaptive processing, our approach eliminates the adaptation module entirely. By leveraging the masked adjacency matrix group (G_{MA}) and the constant matrix (λ_C), we achieve significant performance improvements with minimal computational cost. G_A and P_A are the grouped and parameterized adjacency matrices, respectively.

Due to the processing advantages of the topology structure of GCN-based networks, related methods have gradually become mainstream. ST-GCN [6] is the first study to successfully process skeleton data applying spatio-temporal graph convolution. 2s-AGCN [7] introduced the concept of adaptation for the first time, which convolved the input and applied the self-attention mechanism to obtain the data-dependent adjacency matrix. Subsequent methods [8,9] have mostly continued this strategy, with the difference of adding linearity or additional information to improve the discrimination.

Obtaining an adaptive adjacency matrix through input convolution and subsequent processing has become a standard model approach. However, our experiments uncovered some interesting findings. Firstly, the adaptive module's positive role arises from the limited ability of the learnable adjacency matrix itself to capture associated features, which is often overlooked in most methods. Secondly, when the learnable adjacency parameter is sufficient, the adaptive module's role diminishes greatly and can be replaced by a small constant matrix. Finally, the importance of the attention mechanism is often underestimated. Unconventional changes in joint positions can disrupt the skeleton's information balance, making it necessary to employ an attention module to manage these significant features effectively.

We have significantly simplified and enhanced the existing spatio-temporal GCN-based framework, focusing on theoretical robustness and interpretability. Figure 1 shows our key improvements: (1) Spatial domain association learning. We have completely eliminated the adaptive module and replaced it with a self-learning adjacency matrix group. To compensate for motion, we utilize learnable key points and supplement them with a constant adjacency matrix, ensuring the fundamental association of joints. (2) Temporal domain global correlation. We introduce global correlation information by combining conventional multi-scale convolution with temporal graph convolution. For temporal graph convolution, we only utilize the self-learning adjacency matrix group. (3) Attention correction. We design specific attention correction modules for both spatial and temporal graph convolutions. In the spatial domain, we employ average pooling to balance the feature differences among individual joint components. In the temporal domain, we combine average and maximize pooling to accurately encode the action's start, end, and climax information, thereby adjusting the weight of the convolution.

Additionally, we introduce two universally applicable network enhancement algorithms. The first is the Jump Model, which extracts essential features prior to reducing the time series dimension and integrates them into the final output. The second is the multi-stream fusion weight adjustment algorithm. Utilizing Bayesian optimization, it enables rapid and precise adjustment of weights for different streams, resulting in improved fusion classification accuracy. In summary, our approach makes the following main contributions:

(1) An ingenious skeleton-based action recognition framework is proposed, called AL-GCN, which achieves efficient updates of skeleton spatio-temporal correlation features solely through the self-learning adjacency matrix group.
(2) We design attention correction modules for spatio-temporal graph convolutions. These modules address inter-group channel feature differences in space and enhance the model's ability to capture short-term action intervals and global action sequences.
(3) Universal network enhancement modules are proposed, which are suitable for multi-group spatio-temporal module combination and multi-stream fusion. These modules are beneficial to achieve higher recognition accuracy based on the existing model structure.
(4) Experiments validate our AL-GCN on four datasets across seven settings. In particular, AL-GCN improves by 3% 7% compared with SOTA on the challenging UAV-Human [23]. Ablation experiments are also conducted to demonstrate the effectiveness of individual modules.

2 Related Work

Adaptive Skeleton-Based Action Recognition. Previous research [10,11] primarily treated recognition as a sequence classification task, while employed CNNs to extract features from human skeleton data by transforming it into pseudo-images. However, these RNN-based or CNN-based methods tend to ignore the inherent spatial correlation among joints. In contrast, models utilizing Graph Convolutional Networks (GCNs) excel in capturing spatio-temporal information from skeleton sequences. The pioneering work of Yan *et al.* [6] introduced Spatio-temporal Graph Convolution. Subsequently, 2s-AGCN [7] proposed an adaptive structure to construct association topology graphs, and CTR-GCN [8] followed a similar strategy, performing linear group sum after convolution to derive the corresponding adaptive matrix. Other approaches have introduced additional information to the adaptation module. For instance, InfoGCN [9] modeled context-dependent inherent topology to enhance joint discrimination, Geng *et al.* [12] fused joint, bone, and angular features to improve the recognition accuracy of fine-grained actions, and HA-GCN [13] introduced a hybrid spatial attention mechanism to capture implicit connections between joints for better context learning. Additionally, some studies employ Transformer structures to build association topology. For example, STTR [14] directly substituted GCN or TCN with Transformers for spatial and temporal modeling, respectively. Notably, Transkeleton [15] unified spatial and temporal modeling within Transformers by temporally aggregating different perceptions and imposing physical connectivity constraints.

GCN-Based Temporal Processing. In the realm of temporal modeling, many approaches rely on stacking multiple temporal CNN layers, with few incorporating GCN-based methods. TE-GCN [16] introduced a novel approach called temporal augmented graph convolution, which constructed a correlation matrix for context learning in the temporal dimension. Another method, STF [17], employed an instance-dependent adjacency matrix to capture high-order spatio-temporal dynamics.

The distinguishing factor of our proposed AL-GCN lies in its simplicity and efficiency. Unlike previous works, our model does not bear the burden of redundant parameters in the adaptive module, while demonstrating the efficacy of using adjacency matrix groups in temporal graph convolutions.

3 Methodology

3.1 GCN-Based Skeleton Processing

The core of GCN, as an effective tool for dealing with such non-Euclidean data combinations, is the adjacency matrix, which enables the joints to have canonical topological association properties. Given the body joint sequence in 2D or 3D coordinates, the skeleton of the human body can be denoted as $\mathcal{G} = (\mathcal{V}, \mathcal{E})$, where $\mathcal{V} = (v_1, v_2, ..., v_N)$ represents the joint set of N vertices, $\mathcal{E}$ represents the bone set of the edges. In the adjacency matrix $A \in \mathbb{R}^{N \times N}$ (assuming $\mathcal{G}$ is an undirected graph), if the V_i and V_j have a skeleton directly connected, $A_{i,j} = 1$, otherwise, $A_{i,j} = 0$. If the skeleton sequence is represented by X and A, the layer-wise iteration and weights update can be formulated as $\mathcal{X}^{l+1} = \sigma(A\mathcal{X}^l W^l)$, where $W^l \in \mathbb{R}^{C_l \times C_{l+1}}$ represents the learnable matrix of the network at layer l. To make the adjacency matrix adaptive, some methods [7,8] convolve the input and use the self-attention method to obtain the adaptive relevance adjacency matrix, as $\mathcal{X}^{l+1} = \sigma(f(\mathcal{X}^l)^{\mathrm{T}} M^{\mathrm{T}} M f(\mathcal{X}^l))$, where M and $f(\cdot)$ represent learnable matrix and mapping operations, as shown in the upper part of Fig. 1.

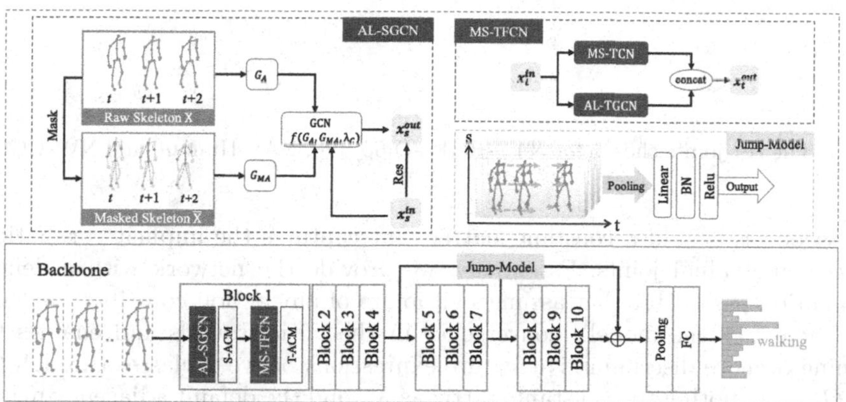

Fig. 2. Architecture Overview. We adopt a particularly concise framework consisting of Auto-Learning Spatial GCN (AL-SGCN) and Multi-scale Temporal Fusion Convolutional Network (MS-TFCN). Similar to CTR-GCN [9], MS-TCN consists of two dilated convolutional layers and a pooling layer. Different from AL-SGCN, AL-TGCN only rely on the learnable adjacency matrix group to obtain global temporal information.

3.2 The AL-GCN Module

The main structure of AL-GCN is illustrated in Fig. 2. To address the impact of adaptive processing on relevance learning, we introduce two lightweight modules as replacements:

The first is the Masked Skeleton Model, which consists of only seven key points. We observed that the majority of action takes place in the key joints of the human body, particularly in the limbs and torso, which represent the core dynamic information. Figure 3 displays the key joint points chosen for three different types of skeleton data. In particular, the skeleton data is often obtained from key-point prediction algorithms, which tend to produce coarse predictions for feet, hands, and heads, resulting in significant dynamic variations between frames. This noise hinders the network from learning general discriminative features effectively. Therefore, we remove these detailed joints in the branch network, allowing the abstracted features of the Masked Skeleton Model to be more universally applicable. Inspired by [19], the network adopts the grouping adjacency matrix strategy comprehensively. If the initial skeleton's adjacency matrix group is denoted as $G_A \in \mathbb{R}^{groups \times N' \times N'}$, the adjacency matrix group after the mask can be represented by the following equation:

$$Mask_{N \to N'}(G_A) = G_{MA} \in \mathbb{R}^{groups \times N' \times N'}, \tag{1}$$

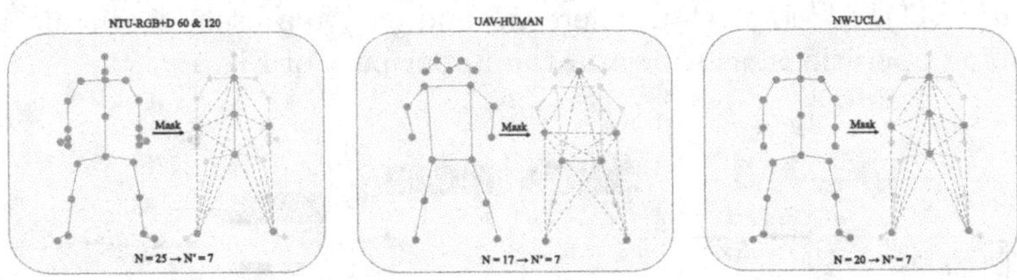

Fig. 3. The key joints chosen for NTU-RGB+D 60/120, UAV-Human, and NW-UCLA.

The second is to add a constant matrix that captures the implicit correlations between individual joints. Essentially, we provide the network with a default aggregation parameter that assumes all joints of any action contribute equally. This facilitates the model's convergence in the initial rounds and enables the learning of more discriminative feature expressions between classes through G_A and G_{MA}. Denoting the constant matrix as λ_c, and the default adjacency matrix is A, the fused adjacency matrices can be represented as follows:

$$G_{AF} = A + G_A + G_{MA} + \lambda_c, \tag{2}$$

We replaced the entire adaptation module of the existing network with these two simple changes, which greatly reduced the number of parameters of the network model and even improved the accuracy. Usually, multi-head mechanism is adopted to further expand the diversity of feature extraction. If the input of the spatial domain is X_s^{in}, then the output X_s^{out} can be expressed as follows:

$$X_s^{out} = MultiHead((Conv_{1 \times 1}(X_s^{in}))G_{AF})_{1 \sim H} + Res(X_s^{in}), \tag{3}$$

In the temporal processing phase, we incorporate the AL-TGCN module, where the graph convolutional network exclusively utilizes the learnable adjacency matrix group (G_{TA}). AL-TGCN captures global, long-range, and implicit associations, while MS-TCN captures local, discriminative, and short-term temporal features. These two modules, combined as MS-TFCN, form the temporal processing unit. If the input to the input space model is denoted as X_t^{in}, the corresponding output X_t^{out} can be expressed as follows:

$$\begin{aligned} X_t^{in} &\to Concat[X_t^{cnn}, X_t^{gcn}], \\ AL\text{-}TGCN(X_t^{gcn}) &= X_t^{gcn} G_{TA} + Res(X_t^{gcn}), \\ X_t^{out} &= concat[MS\text{-}TCN(X_t^{cnn}), AL\text{-}TGCN(X_t^{gcn})]. \end{aligned} \tag{4}$$

3.3 The Attention Correction and Jump Model

The temporal and spatial attention correction modules are depicted in Fig. 4. We employ an enhanced SE-Net [18] and have two key improvements. The first one is that the object of average pooling is changed according to different targets. For example, the target of spatial attention is the channel dimension, and the target of temporal attention is the time dimension. The second is to remove the *FC* layer, that is, to isolate the information interaction to maintain the feature purity. Specifically, we adopt channel dimension attention instead of joint dimension attention in the spatial domain to preserve discriminative information from the Masked Skeleton Model while enhancing feature coherence across channels. This approach helps the model fuse different channel features effectively and address group differences introduced by channel-balanced grouping networks.

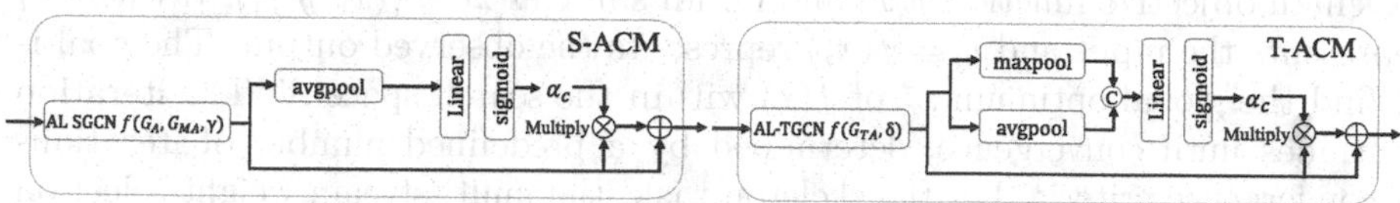

Fig. 4. The diagram of Spatial (L) and Temporal (R) Attention Correction Modules.

The Jump Model, shown in Fig. 2, is an effective residual information transfer model inspired by the jump connection structure in CNN. It utilizes an adaptive pyramid structure to transfer shallow features back to the final layer. This

Algorithm 1. Multi-stream Fusion with Bayesian Optimization

function OBJECTIVE(weights)
  right_num ← 0 ▷ Initialize the count of correct predictions
  total_num ← 0 ▷ Initialize the total number of predictions
  for $i \leftarrow 0$ **to** length(label) **do** ▷ Iterate over the labels
    $l, r_n \leftarrow$ label$[i]$, $s_n[i]$ ▷ Get the label and the corresponding stream value
    $r \leftarrow r_n \times$ weights$[n]$ ▷ Compute the weighted sum
    $r \leftarrow$ argmax(r) ▷ Find the index of the maximum value
    right_num ← right_num + int($r ==$ int(l)) ▷ Update the count of correct predictions
    total_num ← total_num + 1 ▷ Update the total number of predictions
  end for
  acc ← $\frac{\text{right\_num}}{\text{total\_num}}$
  return −acc ▷ We want to maximize accuracy, hence minimize −acc
end function
procedure FUSION
  Define search space for weights
  Result ← Perform Bayesian Optimization with the above function
  get optimal accuracy **from -Result**.accuracy
  get optimal weights **from Result**.weights
end procedure

approach helps mitigate the loss of crucial information during network iteration when the number of time dimensions is reduced.

3.4 Multi-stream Gaussian Weight Selection Algorithm

We validate the model under the widely used 3-stream fusion (3 s), 4-stream fusion (4 s), and 6-stream fusion (6 s), respectively. Following prior work [9], the input of multiple streams refers to $\tilde{\mathcal{X}}_k = (I - P^k)\mathcal{X}$, where $k = 1, 2, ..., K$, and K depends on different datasets[1]. In multi-stream fusion, fixed parameters limit the performance of each stream model and make it difficult to achieve optimal results. We introduce an efficient weight selection algorithm inspired by Gaussian Process Bayesian Optimization. This algorithm effectively finds the best solution by constructing a probabilistic model of the objective function. Given an objective function $f(x)$ and initial samples $\mathcal{D} = \{(x_i, y_i)\}_{i=1}^{N}$, where x_i represents the input and $y_i = f(x_i)$ represents the observed output. The goal is to find the global optimum x^* of $f(x)$ within the search space $\mathcal{X}$. The iteration continues until convergence, determined by a predefined number of iterations or convergence criteria. For the skeleton task, the multi-stream weight selection algorithm is shown in Algorithm 1. Please refer to our code for details.

4 Experimental Results and Analysis

4.1 Datasets

NTU-RGB+D 60. NTU RGB+D [20] is a 60-classes action recognition dataset completed by 40 volunteers, containing 56,880 skeletal action sequences. The 3D skeleton data includes the 3D positions of 25 main body joints in the human body. Two Benchmark evaluations named cross-subject (C-sub) and cross-view (C-view) are recommended. The testing set consists of 18,960 samples, including two 45-degree views on the left and right of the action.

NTU-RGB+D 120. NTU RGB+D 120 [21] is a 120-classes action recognition dataset completed by 106 volunteers, containing 113,945 skeletal action sequences, which is extended from NTU RGB+D 60. Two Benchmark evaluations named cross-subject (C-sub) and cross-set (C-set) are recommended.

Northwestern-UCLA. Northwestern-UCLA [22] is a 10-classes action recognition dataset, containing a total of 1494 video clips, which are shot by three Kinect cameras from different directions. We follow the evaluation method suggested by the author: training data from the first two cameras and test data from the other camera.

---

[1] According to the setting of [9], the six streams are S1: k = 1, w/o motion, S2: k = 2, w/o motion, S3: k = K, w/o motion, S4: k = 1, w/motion, S5: k = 2, w/motion, S6: k = K, w/motion, three-stream(3s) = S1+S2+S3; four streams(4 s) = S1+S2+S4+S5; six streams (6s) = S1+S2+S3+S4+S5+S6, where K = 8 in NTU 60/120, K = 6 in NW-UCLA/UAV-Human.

UAV-Human. UAV-Human [23] is a 155-classes action recognition dataset containing 22,476 video clips. The dataset was collected by a UAV in multiple urban and rural areas during the day and night. Action data are collected from 119 different subjects and 155 different activity categories at 45 different environmental locations. The authors suggest the following evaluation method: 89 subjects for training and 30 subjects for testing.

4.2 Implementation Details

All experiments are conducted on the Pytorch with two NVIDIA RTX 3090ti. We follow previous work [8] for data pre-processing. The batch size of NTU-60 is 64, and the others are 128, the training epoch is set to 90, and we use warm-up for the first 5 epochs. The weight decay is set to 0.0003 in NTU-60, and 0.0005 in the others. The initialized learning rate is set to 0.1 in NTU-60 & NTU 120 and 0.2 in NW-UCLA & UAV-Human, with a 10× reduction in rounds 35th, 55th, and 75th (only once in 50th for NW-UCLA & UAV-Human). The multi-stream fusion strategy [9] is adopted to further improve the performance.

Table 1. Classification accuracy comparison with state-of-the-art methods on different datasets. The best and second-best scores are shown in Red and Blue.

Method	Publisher	NTU-60		NTU-120		UAV		UCLA
		C-Sub	C-View	C-Sub	C-Set	CS-v1	CS-v2	Top-1
Auto-Learning-GCN-6s (**Ours**)		93.2	97.3	90.3	91.6	48.8	74.0	97.6
Auto-Learning-GCN-4s (**Ours**)		93.0	97.1	90.0	91.2	48.6	73.5	97.2
Auto-Learning-GCN-3s (**Ours**)		92.7	97.0	89.6	91.0	47.7	72.6	96.3
FR-GCN [24]	CVPR'2023	92.8	96.8	89.5	90.9	–	–	96.8
DST-HCN [25]	ICME'2023	92.3	96.8	88.8	90.7	–	–	96.6
TranSkeleton [15]	TCSVT'2023	92.8	97.0	89.4	90.5	–	–	–
Info-GCN [9]	CVPR'2022	93.0	97.1	89.8	91.2	–	–	97.0
EfficientGCN [26]	TPAMI'2022	92.1	96.1	88.7	88.9	–	–	–
FG-STFormer [27]	ACCV'2022	92.6	96.7	89.0	90.6	–	–	97.0
CTR-GCN [8]	ICCV'2021	92.4	96.8	88.9	90.6	43.4	–	96.5
ST-TR [14]	ICPR'2021	89.9	96.1	84.3	86.7	–	–	–
MS-G3D [28]	CVPR'2020	91.5	96.2	86.9	88.4	–	–	–
Shift-GCN [29]	CVPR'2020	90.7	96.5	85.9	87.6	38.0	67.0	–
2s-AGCN [7]	CVPR'2019	88.5	95.1	82.9	84.9	34.8	66.7	–
ST-GCN [6]	AAAI'2018	81.5	88.2	–	–	30.3	56.1	

4.3 Compared with the State-of-the-Art Methods

In Table 1, AL-GCN outperforms state-of-the-art methods on four datasets. Our algorithm achieves superior accuracy, particularly on the UAV-Human dataset, surpassing the latest published work by 5.4% and 7.0% respectively. Notably, we can only rely on 3 streams to achieve the accuracy of 4 streams of many methods, which proves that the model still has a strong ability to extract differential features after linear transformation of the data. During the final 6s fusion test, our method excels on NTU-RGBD-120, with C-sub accuracy exceeding 90% for the first time. Impressive results are also observed on the NW-UCLA dataset. Our efficient Bayesian fusion strategy enables us to explore multiple fusion accuracy points, surpassing the 97% baseline. While Transformer-based methods performs slightly worse than GCN-based, the accuracy gap is narrowing, demonstrating the potential of the Transformer in the skeleton-based HAR tasks.

4.4 Ablation Study

The Effectiveness of the Adaptive Module. We validate them on two typical methods with adaptive models [7,8]. It can be found from Table 2 that when the adaptive module is replaced by a constant matrix (λ_C), there is no obvious decline in accuracy, but the parameters are greatly reduced. This is consistent with our previous analysis, and the biggest role of the adaptive module is to model the global invisible correlation between joints, which can be completely replaced by the constant matrix. When the learnable matrix is replaced by the learnable matrix group (G_A), the accuracy will be greatly improved, without any computational cost.

Table 2. Ablation study on the proposed models. ✓ and ✗ denote adoption and deletion, respectively. All experiments are inferred on the NTU-60-Cview.

Methods	Joint	Bone	Param.	FLOPs	Methods	Joint	Bone	Param.	FLOPs
2s-AGCN [7]	94.31	94.05	3.47M	3.98G	CTR-GCN [8]	95.00	94.89	1.46M	1.97G
✗Adaptive	94.29	93.97	3.14M	3.56G	✗Adaptive	94.80	94.58	1.21M	1.50G
✓λ_C	94.27	94.33	3.14M	3.56G	✓λ_C	94.90	94.80	1.21M	1.50G
✓G_A	94.75	94.35	3.14M	3.56G	✓G_A	94.97	94.88	1.21M	1.50G
✓G_A+λ_C	**94.81**	**94.36**	3.14M	3.56G	✓G_A+λ_C	**95.17**	**94.90**	1.21M	1.50G

Table 3. Ablation study on the AL-SGCN (L) and the MS-TFCN (R). The L and R are inferred on the NTU-60-Cview with joint and NTU-60-Csub with bone, respectively.

Methods	Acc	Param.	FLOPs
Baseline	93.80	2.49M	3.11G
+ AL-SGCN	**95.50**	1.97M	2.44G
✗ λ_C	95.20	1.97M	2.44G
✗ G_A	95.07	1.97M	2.44G
✗ G_{MA}	95.39	1.97M	2.44G
✗ S-ACM	95.26	1.97M	2.43G

Methods	Acc	Param.	FLOPs
Baseline	88.65	4.99M	5.69G
+ MS-TCN	90.25	1.88M	2.34G
+ AL-TGCN	**91.27**	1.97M	2.44G
✓ λ_C	90.66	1.97M	2.44G
✗ T-ACM	90.98	1.97M	2.43G
✗ Jump Model	90.62	1.89M	2.37G

The Effectiveness of AL-SGCN. Table 3 (L) displays the experimental results. The baseline spatial processing method employs the ST-GC structure, while the baseline temporal processing method uses MS-TFCN to ensure a fair comparison. The results demonstrate a significant enhancement in the target model's performance, with an accuracy increase of 1.70%. Ablation experiments are conducted to assess the contribution of each component in the model. Our findings reveal that each module plays a vital role, with G_A exerting the most substantial influence on the final model performance. Removing G_A leads to a decrease in accuracy by 0.43%. Moreover, attention is indispensable and S-ACM can lead to 0.24% accuracy degradation.

The Effectiveness of MS-TFCN. Table 3 (R) presents the experimental results. The baseline temporal processing method utilizes the basic 2D-TCN structure, while the baseline spatial processing method employs AL-SGCN. The results demonstrate a significant enhancement in the target model's performance. MS-TCN contributes to a 1.60% improvement, and AL-TGCN further boosts performance by 1.02%. Interestingly, we observed a negative impact of the constant matrix on temporal graph convolution. This could be due to the predefined latent connections limiting the model's ability to discern critical time points accurately. Moreover, when we delete components T-ACM and Jump Model individually, the model's accuracy is reduced by 0.29% and 0.65%, respectively.

Hyperparameter Setting and Bayesian Weight Selection. We conducted separate analyses on the following hyperparameters: the constant matrix (λ_c), the number of groups (g_s) in AL-SGCN, and the number of groups (g_t) in MS-TFCN. The experimental results are displayed in Fig. 5 (L). It is evident that the optimal spatio-temporal grouping number is 8 for both, while the constant matrix is selected as 0.04. Next, we demonstrate the effectiveness of Bayesian weight selection in Table 4. Our proposed algorithm efficiently explores the optimal fusion solution within the weight range of 0.2 to 1.2 in a short amount of time. Compared to traditional manual weight settings, the accuracy is generally improved by 0.2% to 0.3%.

Table 4. Bayesian Optimization (Bayesian OPT) fusion results, on 4 different settings. The 6 streams (6s) are k = 8, k = 1, k = 8/motion, k = 1/motion, k = 2, k = 2/motion.

Datasets	Item	The fusion weights of 6 streams	ACC	Time
NTU-60-Cview	Manual Setup	[0.60, 0.60, 0.40, 0.40, 0.60, 0.40]	97.1	–
	Bayesian OPT (ours)	[1.20, 1.20, 0.72, 0.20, 1.00, 0.63]	**97.3**	37.99 s
NTU-60-Csub	Manual Setup	[0.60, 0.60, 0.40, 0.40, 0.60, 0.40]	93.0	–
	Bayesian OPT (ours)	[0.59, 0.95, 0.50, 0.20, 0.85, 0.67]	**93.2**	37.14 s
NTU-120-Cset	Manual Setup	[0.60, 0.60, 0.40, 0.40, 0.60, 0.40]	91.3	–
	Bayesian OPT (ours)	[0.20, 1.00, 0.44, 0.20, 0.87, 0.37]	**91.6**	68.21 s
NTU-120-Csub	Manual Setup	[0.60, 0.60, 0.40, 0.40, 0.60, 0.40]	90.0	–
	Bayesian OPT (ours)	[0.46, 1.20, 0.28, 0.52, 0.77, 0.20]	**90.3**	59.75 s

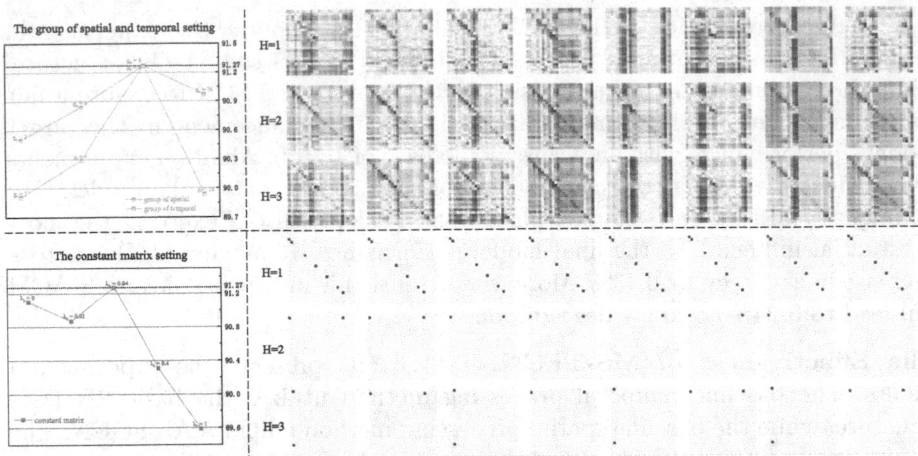

Fig. 5. The left is a table of multiple settings of λ_c and the number of groups, the top right is the visualization of G_A, and the bottom right is the visualization of G_{MA}, with $3(Head) \times 8(Group)$, respectively. The left tables are all inferred on the NTU-60-Csub with bone. The right visualization are inferred on the NTU-60-Cview with joint.

4.5 Visualization

Figure 5 (R) displays the visualization of the G_A and G_{MA} for the last layer network in the NTU-RGBD-60-Cview-Joint. When Heads are 3, and Groups are 8, there are a total of 48 parameterized matrices. Notably, the G_A exhibits significant variation in focus, indicating a reliance on auto-learning. Interestingly, the G_{MA} demonstrates distinct patterns. At H = 1, it solely concentrates on the selected seven joints, while at H = 2 and H=3, it focuses exclusively on the torso joints, suggesting autonomous selection. Based on these observations, we venture a bold hypothesis that the dynamics of the torso play a pivotal role in action recognition.

5 Conclusion

This paper proposes an elegant and efficient action recognition framework AL-GCN based on spatio-temporal skeleton, which aims to improve the performance of the algorithm with minimal computational cost by simplifying the traditional spatio-temporal paradigm to optimize the network structure. AL-GCN adopts a fully learnable adjacency matrix group in both time and space to automatically assign more detail weights between points. We introduce a general attention strategy and a jump strategy to further improve the feature discrimination ability of the model. In addition, we propose a Bayesian weight selection algorithm to efficiently handle the problem of weight selection for more and more streams. In an era of increasing admiration for complex networks, this paper makes the bold choice of "subtraction". It is worth noting that the code of the paper has been made public, hoping that more researchers will pay attention to our work.

Acknowledgement. The work was jointly supported by the National Key R&D Program of China under grant No. 2022ZD0117103, the National Natural Science Foundations of China under grant No. 62272364, the Teaching Reform Project of Shaanxi Higher Continuing Education under Grant No. 21XJZ004.

References

1. Huang, X., Wang, P., Cheng, X., et al.: The apolloscape open dataset for autonomous driving and its application. IEEE TPAMI **42**(10), 2702–2719 (2019)
2. Khan, M.A., et al.: Human action recognition using fusion of multiview and deep features: an application to video surveillance. MM Tools Appl., 1–27 (2020)
3. Kashef, M., et al.: Smart city as a smart service system: human-computer interaction and smart city surveillance systems. Comput. Hum. Behav. **124**, 106923 (2021)
4. Sun, Z., Ke, Q., Rahmani, H., et al.: Human action recognition from various data modalities: a review. IEEE TPAMI **45**, 3200–3225 (2022)
5. Xin, W., Liu, R., Liu, Y., et al.: Transformer for Skeleton-based action recognition: a review of recent advances. Neurocomputing **537**, 164–186 (2023)
6. Yan, S., Xiong, Y., Lin, D.: Spatial temporal graph convolutional networks for skeleton-based action recognition. In: AAAI, vol. 32, no. 1 (2018)
7. Shi, L., Zhang, Y., Cheng, J., Lu, H.: Two-stream adaptive graph convolutional networks for skeleton-based action recognition. In: CVPR, pp. 12026–12035 (2019)
8. Chen, Y., Zhang, Z., Yuan, C., et al.: Channel-wise topology refinement graph convolution for skeleton-based action recognition. In: ICCV, pp. 13359–13368 (2021)
9. Chi, H. G., Ha, M. H., Chi, S., et al.: InfoGCN: representation learning for human skeleton-based action recognition. In: CVPR, pp. 20186–20196 (2022)
10. Liu, J., Shahroudy, A., Xu, D., Wang, G.: Spatio-temporal LSTM with trust gates for 3D human action recognition. In: Leibe, B., Matas, J., Sebe, N., Welling, M. (eds.) ECCV 2016. LNCS, vol. 9907, pp. 816–833. Springer, Cham (2016). https://doi.org/10.1007/978-3-319-46487-9_50
11. Du, Y., Wang, W., Wang, L.: Hierarchical recurrent neural network for skeleton based action recognition. In: CVPR, pp. 1110–1118 (2015)
12. Geng, P., Lu, X., Hu, C., et al.: Focusing fine-grained action by self-attention-enhanced graph neural networks with contrastive learning. In: IEEE TCSVT (2023)
13. Xing, H., Burschka, D.: Skeletal human action recognition using hybrid attention based graph convolutional network. In: ICPR, pp. 3333–3340 (2022)
14. Plizzari, C., Cannici, M., Matteucci, M.: Spatial temporal transformer network for skeleton-based action recognition. In: ICPR, pp. 694–701 (2021)
15. Liu, H., Liu, Y., et al.: TranSkeleton: hierarchical spatial-temporal transformer for skeleton-based action recognition. In: IEEE TCSVT (2023)
16. Li, J., et al.: Temporal graph modeling for skeleton-based action recognition. In: arXiv:2012.08804 (2020)
17. Ke, L., Peng, K. C., Lyu, S.: Towards to-at spatio-temporal focus for skeleton-based action recognition. In: AAAI, pp. 1131–1139 (2022)
18. Hu, J., Shen, L., Sun, G.: Squeeze-and-excitation networks. In: CVPR, pp. 7132–7141 (2018)
19. Cheng, K., Zhang, Y., Cao, C., Shi, L., Cheng, J., Lu, H.: Decoupling GCN with DropGraph module for skeleton-based action recognition. In: Vedaldi, A., Bischof, H., Brox, T., Frahm, J.-M. (eds.) ECCV 2020. LNCS, vol. 12369, pp. 536–553. Springer, Cham (2020). https://doi.org/10.1007/978-3-030-58586-0_32

20. Shahroudy, A., Liu, J., et al.: NTU RGB+ D: a large scale dataset for 3d human activity analysis. In: CVPR, pp. 1010–1019 (2016)
21. Liu, J., Shahroudy, A., et al.: NTU RGB+ D 120: a large-scale benchmark for 3d human activity understanding. IEEE TPAMI **42**(10), 2684–2701 (2019)
22. Wang, J., et al.: Cross-view action modeling, learning, recognition. In: CVPR, pp. 2649–2656 (2014)
23. Li, T., Liu, J., Zhang, W., et al.: UAV-human: a large benchmark for human behavior understanding with unmanned aerial vehicles. In: CVPR, pp. 16266–16275 (2021)
24. Zhou, H., Liu, Q., Wang, Y.: Learning discriminative representations for skeleton based action recognition. In: CVPR, pp. 10608–10617 (2023)
25. Wang, S., Zhang, Y., et al.: Dynamic spatial-temporal hypergraph convolutional network for skeleton-based action recognition. In: arXiv:2302.08689 (2023)
26. Song, Y.F., Zhang, Z., et al.: Constructing stronger and faster baselines for skeleton-based action recognition. IEEE TPAMI **45**(2), 1474–1488 (2022)
27. Gao, Z., Wang, P., et al.: Focal and global spatial-temporal transformer for skeleton-based action recognition. In: ACCV, pp. 382–398 (2022)
28. Liu, Z., Zhang, H., et al.: Disentangling and unifying graph convolutions for skeleton-based action recognition. In: CVPR, pp. 143–152 (2020)
29. Cheng, K., Zhang, Y., et al.: Skeleton-based action recognition with shift graph convolutional network. In: CVPR, pp. 183–192 (2020)

Skeleton-Based Action Recognition with Combined Part-Wise Topology Graph Convolutional Networks

Xiaowei Zhu, Qian Huang(✉), Chang Li, Jingwen Cui, and Yingying Chen

School of Computer and Information, Hohai University, Nanjing, China
huangqian@hhu.edu.cn

Abstract. Graph Convolutional Network (GCN) has achieved promising performance in skeleton-based action recognition by modeling skeleton sequences as spatio-temporal graphs. However, most existing methods only focus on the overall characteristics of the skeleton, thus lacking fine-grained exploration of human body parts semantics. In this paper, we propose a novel Combined Part-wise Topology Graph Convolutional Network (CPT-GCN), including SPT-GC, TPT-GC, and STPT-GC modules, to refine the spatio-temporal topology from the spatial, temporal, and spatio-temporal perspectives, respectively. Specifically, SPT-GC aggregates spatial features by combining global topology and partial correlations. TPT-GC combines the overall motion trend and the motion details of parts to capture temporal dynamics. STPT-GC establishes a spatio-temporal dependency, focusing on exploiting the implicit spatio-temporal information in motions. Ultimately, the effectiveness of CPT-GCN is demonstrated through experiments on two large-scale datasets: NTU RGB+D 60 and NTU RGB+D 120.

Keywords: Skeleton action recognition · Graph convolutional network · Part-wise topology · Spatio-temporal correlation

1 Introduction

As a major research topic of computer vision technology, human action recognition plays an important role in applications such as video surveillance, human-computer interaction and abnormal behavior detection [1,3,8,14,27]. Skeleton data is a compact and expressive modality that has less data volume compared with RGB or depth modality, and is insensitive to complex backgrounds and dynamic camera perspectives. Therefore, skeleton-based human action recognition technology has received widespread attention [12,13,25,31,32].

Early deep learning-based action recognition methods manually construct human skeleton coordinates into vector sequences or pseudo-images, and feed them into a recurrent neural network (RNN) or convolutional neural network (CNN) to predict action results [5,7,10,36]. Kim et al. [11] used a one-dimensional residual CNN to identify skeleton sequences based on directly-concatenated joint coordinates. Li et al. [18] constructed an adaptive tree-structured RNN, and Si et al. [28] proposed a novel attention-enhanced graph

© The Author(s), under exclusive license to Springer Nature Singapore Pte Ltd. 2024
Q. Liu et al. (Eds.): PRCV 2023, LNCS 14425, pp. 43–59, 2024.
https://doi.org/10.1007/978-981-99-8429-9_4

convolutional LSTM network called AGC-LSTM for human action recognition from skeleton data. However, these methods ignore the important property of human skeleton as a topological structure, and it is difficult to capture the spatio-temporal dependencies between joints.

Graph Convolutional Network (GCN) can efficiently handle non-Euclidean data such as graphs, and it can generalize convolutions from images to graphs of arbitrary size and shape. In recent years, more and more skeleton action recognition models use GCN-based methods to extract spatio-temporal features [4,7,16,22,25,26,34]. Yan et al. [34] manually defined the human body topology, and Shi et al. [25] learned the human body topology dynamically through adaptive graph convolution. They all focus on graph convolution on the global human body topology while ignoring body part information. For many actions, such as clapping and throwing, the motion characteristics of parts are more important. Thakkar et al. [30] is the first to split the human skeleton into different parts for graph convolution. Wang et al. [33] proposed adaptive multi-part graph convolution to learn the spatial correlation between parts based on the self-attention mechanism. However, the topology of the human skeleton has not been fully utilized, and we construct the more refined local topology to extract more detailed features.

In this paper, we will further model the human skeleton topology from the three dimensions of spatial, temporal, and spatio-temporal based on human body parts. We then propose a novel network named Combined Part-wise Topology Graph Convolutional Networks (CPT-GCN), which focuses on exploring fine-grained features and capturing intrinsic spatio-temporal correlations. Specifically, we propose three modules, SPT-GC, TPT-GC and STPT-GC, to perform graph convolution based on locally refined topology. SPT-GC establishes specific global and local topologies in different channels, taking into account both global and local information to capture the spatial connections of joints in more detail. TPT-GC reasonably changes the receptive field of temporal convolution to extract the motion trend and motion details of the whole and part of the action. STPT-GC focuses on extracting the implicit spatio-temporal association information in the skeleton sequence, and establishes the part-enhanced spatio-temporal association topology. Combining the above three modules, our network dynamically aggregates high-dimensional features and achieves excellent performance on large-scale datasets.

Combining these efforts above, our main contributions are summarized as follows:

- Our proposed SPT-GC refines the spatial topology based on body parts by fusing global and local topology, which extracts more fine-grained spatial features.
- We propose the spatio-temporal module, including TPT-GC and STPT-GC, which establishes a specific temporal correlation topology and spatio-temporal correlation topology, and effectively extracts the temporal and spatio-temporal correlation of parts and joints.

- We propose a novel action recognition model CPT-GCN based on skeleton data. It accurately captures the relationship between and within parts, and effectively aggregates the spatial, temporal and spatio-temporal information of skeleton data.
- We conduct experiments on two widely-used datasets: NTU RGB+D [24] and NTU RGB+D 120 [19], on which our proposed method outperforms state-of-the-art approaches.

2 Related Work

2.1 Skeleton-Based Action Recognition

With the development of deep learning technology, deep learning methods have gradually rcplaced traditional manual feature methods. The mainstream methods can be divided into three categories according to the network architecture: convolutional neural network (CNN), recurrent neural network (RNN) and graph convolutional network. (GCN).

CNN-based method usually converts the skeleton data into a pseudo-image according to the manually designed conversion rules. RNN-based methods usually extract frame-level skeleton features, represent skeleton data as sequential data with predefined traversal rules [4,18]. However, human skelcton is a natural graphical structure, and GCN has obvious advantages in processing graph-structured data. Yan et al. was the first to use GCN to model human skeleton, proposing Spatio-temporal Graph Convolutional Network (ST-GCN). They build joint connection edgcs based on the natural connections of the human body,

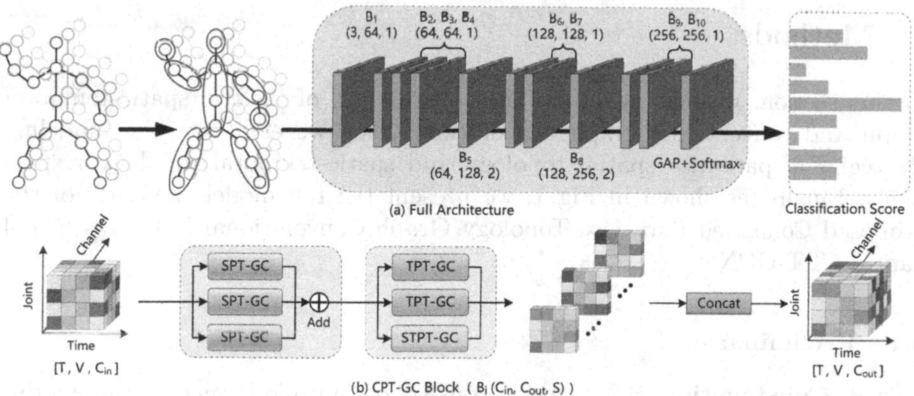

Fig. 1. The overview of the proposed CPT-GCN model. The entire combined part-wise topology graph convolutional block is represented as $B_i(C_{in}, C_{out}, S)$. C_{in}, C_{out} and S denote the number of input channels, the number of output channels and the stride, respectively. There are a total of 10 blocks. GAP represents the global average pooling.

and add temporal associations for the same joints in consecutive frames, constructing a skeletal spatio-temporal graph [34]. Shi et al. proposed an adaptive graph convolution network (AGCN), which uses the self-attention mechanism to change the topology of human skeletons and adaptively learns the connection between the original disconnected skeletons [25,26]. Liu et al. introduced a multi-scale graph topology to achieve multi-scale joint relationship modeling [21]. Cheng et al. proposed Shift-GCN [7], replacing the traditional convolution operator with the shift convolution operator, using shifted graph convolution. The CTR-GCN proposed by Chen et al. [5] designs channel-wise topology graphs to explore more possibilities for feature learning in different channels.

2.2 Partial Graph Convolution in Skeleton-Based Action Recognition

A complete action can be regarded as composed of different postures of human body parts. For example: in the process of clapping, the clapping of the palm plays a key role in the whole action, while the waving of the arm plays an auxiliary role. Previous studies [7,21,25,26,34] mostly learn the global features of actions based on the whole skeleton, ignoring the important contribution of local features to actions. Thakkar et al. [30] is the first to split the human skeleton into different parts for graph convolution, which effectively improves performance of recognition. Wang et al. [33] proposed adaptive multi-part graph convolution to learn the spatial correlation between parts based on the self-attention mechanism. Zhu et al. [38] focused on fusing global and local features from a spatial perspective, effectively aggregating multi-level joint features by constructing a topology based on bodyparts.

3 Methods

In this section, we first introduce the construction of skeletal spatio-temporal graph and conventional graph convolution. Then we elaborate the modeling strategies of part-wise spatial topology and spatio-temporal topology respectively. Finally, as shown in Fig. 1, we present the full model structure of the proposed Combined Part-wise Topology Graph Convolutional Networks model named CPT-GCN.

3.1 Preliminaries

Graph Construction. A full action consists of multiple frames containing different samples. We construct spatio-temporal skeleton graphs to describe the structured information between nodes along the spatial and temporal dimensions. The complete spatio-temporal skeleton graph is established based on the natural connections of the human body structure and the connection of consecutive frames, so it contains the connection edges between joints and the connection edges between frames. The graph is defined as $\mathcal{G} = (\mathcal{X}, \mathcal{V}, \mathcal{E})$. $\mathcal{X}$ denotes the feature set of vertices, which is represented as a matrix $X \in R^{C \times V \times T}$, there are V vertices, T frames and C channels. $\mathcal{V} = \{v_1, v_2, ..., v_V\}$ denotes the vertex set. $\mathcal{E}$ is the set of edges, reflecting the connection strength between vertices.

Graph Convolution. After the skeleton spatio-temporal feature map is constructed, we weight and sum the skeleton points in the input feature map with the features of their corresponding neighbor points to obtain the output feature map. The graph convolution implementation of feature maps can be intuitively formulated as:

$$f_{out} = \sum_{s}^{S} W_s \cdot f_{in} \cdot A_s \tag{1}$$

where f_{in} and f_{out} denote the input and output feature maps. S denotes the sampling area of the spatial dimension. A_s and W_s denote the adjacency matrix and weight function under the sampling area s.

3.2 Part-Wise Spatial Modeling

Almost any action is composed of sub-actions of different parts, and the difference mainly lies in the correlation between parts and the contribution of parts to the whole action. For example, clapping can be decomposed into the action of two palms and arms, and nodding can be regarded as the action of the head. Thus, optimizing the topology of skeleton based on human body parts can more accurately obtain the dependencies between joints.

Most of the previous studies explored the global features of the skeleton, and learned the spatial relationship of the skeleton through the natural connection of the human body or the attention mechanism [25,26,32,34], which will generate a lot of redundant information, and the spatial topology shared by each channel is also not optimal. Existing part-based models usually aim to extract features from body parts individually or only focus on discovering the importance of different body parts [29,35]. However, we take full account of inter-part dependencies and intra-part differences, and construct a refined part-wise topology for each channel.

Before performing GCN, body part correlations need to be modeled. Specifically, we divide the human body into 8 parts, which are head, body, two arms, two palms and two legs. The input features $X \in R^{C \times V \times T}$ is aggregated according to the proposed part division strategy, which is formulated as:

$$X_i^{part} = Concat(\{X_j \mid j \in L(i)\}) \quad i = 1, 2, ..., P \tag{2}$$

where P denotes the number of parts, $Concat(\cdot)$ denotes the splicing function, $L(i)$ denotes the set of joint numbers corresponding to the i_{th} part, and X_i^{part} denotes the feature of the i_{th} part after aggregation.

Parts Correlation Modeling. In order to obtain the best dependencies between parts, we propose the modeling strategie $\mathcal{M}(\cdot)$ to model the part dependencies.

Since each joint contributes to the corresponding body part, we perform an average pooling operation on the joints inside the part. In addition, in order to reduce the computation cost, we utilize linear transformations $\psi(\cdot)$ and $\varphi(\cdot)$ to reduce the feature dimension before the local topology modeling. $\mathcal{M}(\cdot)$ needs to calculate the distance of the channel dimension between different parts, and utilizes the nonlinear transformation of the distance to represent the correlation between parts, which is formulated as:

$$\mathcal{M}(i,j) = \sigma(\psi(AvgPool_{ST}(X_i^{part})) - \varphi(AvgPool_{ST}(X_j^{part}))) \quad (3)$$

where $AvgPool_{ST}(\cdot)$ denotes the average pooling in both spatial and temporal dimensions. $\mathcal{M}(i,j)$ is the modeling strategy, and its value denotes the correlation between parts i and j.

Part-wise Topology Modeling. The part correlation graph obtained by $\mathcal{M}(\cdot)$ represents the correlation between parts and cannot be directly applied to the human skeleton graph, so it needs to be mapped to joints relation graph through a mapping function. According to the relationship between the various parts obtained, the part correlation features are first connected into a whole vertex matrix, which is formulated as:

$$G_{part} = Concat(\{Concat(\{\mathcal{M}(i,j) \mid j = 1,2,...,P\}) \mid i = 1,2,...,P\}) \quad (4)$$

where G_{part} denotes the spliced inter-part relationship graph. It expresses different part correlations on each channel. But in fact, the joints within a part do not share weights, so the topology needs to be refined while mapping. We optimize the topology through learnable bias and linear transformation, which is formulated as:

$$G_s^{local} = \phi(\mathcal{R}(G_{part}) + B_0) \quad s = 1,2,...,S \quad (5)$$

where $\mathcal{R}(\cdot)$ denotes the mapping function, it maps the part association graph to the joint association graph. $\phi(\cdot)$ denotes the linear transformation function. B_0 denotes the positional bias of the channel and joint, which is a learnable parameter. G_s^{local} is the feature map based on body parts.

Spatial Part-wise Topology Graph Convolution(SPT-GC). Local topology captures both part correlations and intra-part differences. On this basis, a global topology is introduced to perform adaptive learning driven by data to capture the global spatial characteristics of actions. Our proposed CPT-GC is more flexible, which combines global and local topology to more accurately obtain the correlation between human skeletons. A gating mechanism α is introduced in the process of fusing the global graph and the individual refined graph to control the difference in the contribution of required parts and joints in different sampling regions. Finally, the graph convolution can be completed by performing Einstein summation of the part-wise topology and the input features in the spatial dimension.

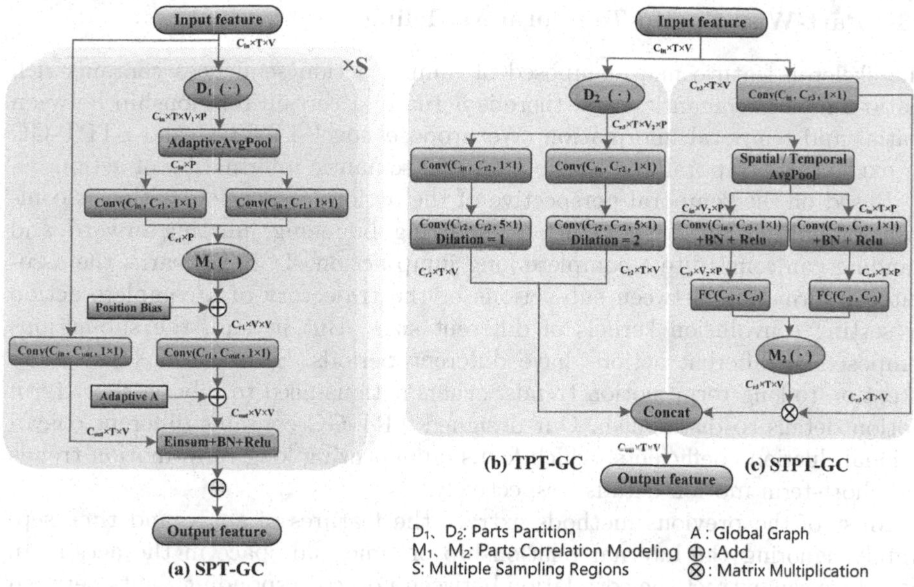

Fig. 2. Model architecture of CPT-GCN block. It consists of three modules: SPT-GC, TPT-GC, and STPT-GC. $D_1(\cdot)$, $D_2(\cdot)$ denote parts partition function. $M_1(\cdot)$, $M_2(\cdot)$ denote parts correlation modeling function. FC denotes the fully connected layer. BN denotes Batch Normalization. $Relu$ is the activation function.

GCN will dynamically update the global and local topology during the inference process to capture the features of the previously disconnected joints. Therefore, Eq. 1 is modified into the following form:

$$f_{out} = \sum_{s}^{S} W_s \cdot f_{in} \cdot (G_s^{global} + \alpha G_s^{local}) \tag{6}$$

where G_s^{global} is the global topology, which is initialized with the natural connection of the human skeleton, and changed by adaptively learning the correlation of actions.

The complete SPT-GC module is shown in Fig. 2 (a). We first divide the bone input feature X_{in} into parts, and then perform adaptive average pooling on the aggregated features. After that, they are respectively input to two convolutional layers with a convolution kernel of 1×1 for dimensionality reduction. After part-wise modeling, the associated topology graph of the part is obtained. Then it needs to be mapped to joint topology and fused with the global topology. In addition, multiple sampling regions S are set to learn semantic information at different levels.

3.3 Part-Wise Spatio-Temporal Modeling

The skeleton feature map composed of human action sequences contains rich spatio-temporal semantics, and there is actually a certain relationship between spatial and temporal information. We propose novel TPT-GC and STPT-GC for extracting temporal and spatio-temporal semantic information of action.

Based on the temporal perspective of the action, a complete action is composed of multiple sub-actions, such as squatting, bouncing, jumping forward, and standing can constitute a complete long jump action. TCN [2] learns the associated information between sub-actions or the trajectory of a complete action by setting convolution kernels of different sizes. But in fact, the sub-actions composed of different actions have different periods. Some actions pay more attention to long-term motion trends, others actions need to rely on short-term motion details to distinguish. Our designed TPT-GC contains different convolutional dilation coefficients, which focus on capturing long-term motion trends and short-term motion details, respectively.

Most of the previous methods extract the features of space and time separately, ignoring the internal relationship of time and space in the action. In fact, if we can extract the correlation between non-corresponding joints between frames, it will surely improve the accuracy of action recognition. Our proposed STPT-GC is used to capture spatio-temporal correlation features, and the effectiveness is verified in ablation experiments, as shown in Table 2.

In addition, the sub-actions that occur in different human body parts are also different. The arms and thighs may dominate the motion trend of this action, or the hands control the motion details of a certain action. It is obvious that adding part information helps to promote the learning of motion paterns. Therefore, we also introduced the concept of parts in the spatio-temporal modeling, and constructed the refined temporal and spatio-temporal topology respectively, achieving the part-enhanced effect.

Temporal Part-wise Topology Graph Convolution(TPT-GC). Inspired by Multi-scale Temporal Convolution [21], we design a part-based temporal modeling module for finer-grained extraction of joint motion trends and motion details. The part division strategy of Eq. 2 is used to aggregate the joint features of body parts. In order to reduce the computational complexity of the model, we utilize the $\psi(\cdot)$ linear transformation function to reduce the feature dimension. We set two convolution branches with different expansion coefficients in parallel to expand the neighborhood learned by graph convolution and extract semantic information at different levels of actions. The TPT-GC module is shown in Fig. 2 (b), which is formulated as:

$$f_{out}^{1}(i) = \sum_{k}^{K} W_1 \cdot \psi(f_{in}(i+k)) \quad i = 1, 2, ..., T \tag{7}$$

$$f_{out}^{2}(i) = \sum_{k}^{K} W_2 \cdot \psi(f_{in}(i+2k)) \quad i = 1, 2, ..., T \tag{8}$$

where f_{out}^1, f_{out}^2 denote the output feature obtained by the two branches. W_1, W_2 denote the weight corresponding to the convolution. K is the size of the convolution kernel in the time dimension.

Spatio-temporal Part-wise Topology Graph Convolution(STPT-GC). In order to obtain the inherent spatio-temporal correlation information of the action, we designed a novel spatio-temporal modeling module, which is also guided by the part information to establish the more refined spatio-temporal correlation topology. The STPT-GC module is shown in Fig. 2 (c). Specifically, STPT-GC relies on the spatio-temporal correlation topology to obtain spatio-temporal correlation information, and needs to construct a spatial correlation graph and a temporal correlation graph first. It uses the same part division strategy to aggregate joints features of the parts, and then aggregates the temporal and spatial information respectively through the average pooling operation. A linear transformation function is then used to reduce the temporal and spatial feature dimensions. It is formulated as:

$$G_{out}^S = W_S \cdot \sigma(\phi_1(AvgPool_S(f_{in})))) \tag{9}$$

$$G_{out}^T = W_T \cdot \sigma(\phi_2(AvgPool_T(f_{in}))) \tag{10}$$

where G_{out}^S and G_{out}^T denote the spatial and temporal correlation graphs, respectively. $AvgPool_S(\cdot)$ and $AvgPool_T(\cdot)$ denote the average pooling operation on the spatial and temporal dimensions, respectively. $\phi_1(\cdot)$ and $\phi_2(\cdot)$ denote the linear transformation function. $\sigma(\cdot)$ denotes the activation function. We add learnable parameters W_S and W_T to assist in learning the spatio-temporal features of actions, and then use the Kronecker product to model the spatio-temporal correlation topology. It is formulated as:

$$G_{out}^{ST} = \sigma(G_{out}^S \times G_{out}^T) \tag{11}$$

where G_{out}^S and G_{out}^T denote the spatial and temporal correlation graphs, respectively. G_{out}^{ST} denotes the obtained spatiotemporal correlation topology. Our proposed STPT-GC is parallel to TPT-GC. The output features of TPT-GC and STPT-GC are concatenated after spatio-temporal topological graph convolution. It is formulated as:

$$f_{out}^3 = W \cdot \phi_3(f_{in}) \cdot G_{out}^{ST} \tag{12}$$

$$f_{out} = Concat(f_{out}^{(i)}) \quad i = 1, 2, ..., N^{branch} \tag{13}$$

where f_{out}^3 denotes the output feature of the STPT-GC module. $\phi_3(\cdot)$ denotes the linear transformation function. $f_{out}^{(i)}$ denotes the output feature of the i_{th} branch. f_{out} denotes the output features after N^{branch} branches are cascaded. It can be understood that the first part of the channel represents the temporal characteristics of the action, and the latter part of the channel represents the spatiotemporal correlation characteristics of the action. The joints of each part can be restored to the original feature dimension through mapping and splicing strategies.

3.4 Model Architecture

We synthesize three modules of SPT-GC, TPT-GC and STPT-GC to construct a powerful graph convolutional network CPT-GCN for skeleton-based action recognition. The overall architecture is shown in Fig. 1 (a), which mainly consists of 10 basic blocks and a classification layer. The output channels of each block in the middle are 64, 64, 64, 64, 128, 128, 128, 256, 256, and 256. The residual network is connected between blocks [9], and finally perform global average pooling and softmax classification to obtain behavior prediction results.

Specifically, each individual block contains a spatial model and a spatio-temporal model, which are responsible for extracting spatial features and spatio-temporal joint features in skeleton information, respectively. As shown in Fig. 1 (b).

Spatial Modeling. The spatial model is mainly composed of SPT-GC modules, and three SPT-GCs are used in parallel to extract semantic information at different levels between parts and joints, as shown in Fig. 2 (a). For a single SPT-GC, first utilizes channel reduction rate r1 to compact representations, uses temporal and intra-part spatial pooling to aggregate features. After that, SPT-GC conducts pair-wise subtraction and activation, then fused with the global map. Finally, the graph convolution is completed to obtain the output feature map, as shown in Eq. 6.

Spatio-temporal Modeling. We demonstrate through ablation experiments that the spatio-temporal model with three branches has better performance. Among them, TPT-GC occupies two branches and STPT-GC occupies one branch, as shown in Fig. 2 (b) and (c).

TPT-GC first uses the channel reduction rate r2 to compress the channel information, and constructs two temporal convolutional layers of different scales to increase the receptive field, which are used to extract the motion trend and motion details of the action respectively.

STPT-GC aggregates temporal and spatial information through average pooling operations, and uses the channel reduction rate of r3 to reduce computational complexity. We use the Kronecker product to model spatio-temporal association topology. Finally, it performs a dot product of the compressed input features with the spatio-temporal correlation topology to complete the graph convolution, which can extract the spatio-temporal correlation information of the action.

4 Experiments

4.1 Datasets

NTU RGB+D. NTU RGB+D (NTU-60) [24] is currently the most widely used large-scale action recognition dataset, containing 60 action categories and 56,000 action clips. The clips were captured by three KinectV2 cameras with different perspectives and performed by 40 volunteers. Each sample contains one action and is guaranteed to have at most 2 subjects. The skeleton information consists of the 3D coordinates of 25 body joints and the corresponding action category labels. NTU-60 recommends two benchmarks [24]: Cross-View Evaluation (X-View) split according to different camera views and Cross-Subject Evaluation (X-Sub) split according to different subjects.

NTU RGB+D 120. NTU RGB+D 120 (NTU-120) [19] extends NTU-60 with a larger scale. It contains 120 action categories and 114,480 action clips. The clips were performed by 106 volunteers in 32 camera setups. NTU-120 also recommends two benchmarks [19]: the first is Cross-Subject Evaluation(X-Sub), which is the same cross-subject evaluation as NTU-60. The other is Cross-Setup Evaluation (X-Set), which splits training and test samples based on the parity of camera setup IDs.

4.2 Training Details

All experiments are conducted on one RTX 3070 TI GPU with the PyTorch deep learning framework. We use the stochastic gradient descent(SGD) with Nesterov momentum(0.9) as the optimizer and the cross-entropy as the loss function. Weight decay is 0.0004. The initial learning rate is set to 0.1 and a warmup strategy [9] is used in the first 5 epochs to make the training procedure more stable. The batch size is 32. The learning rate is divided by 10 at the 35_{th} epoch and 55_{th} epoch. The training process is ended at the 70_{th} epoch.Since the number of frames of the samples is not consistent, we uniformly downsample the frames to 64 frames. In addition, we adopt the data preprocessing strategy of [21] for the input skeleton features.

4.3 Ablation Studies

In this section, we use the X-Sub benchmark of the NTU-60 to verify the effectiveness of proposed modules in CPT-GCN.

Effectiveness of TPT-GC and STPT-GC. In order to test the performance of the space-time model proposed in Sect. 3.3 and obtain its optimal branch configuration, we conduct experiments on TPT-GC and STPT-GC with different branch numbers. We adopt ST-GCN [34] as the baseline method and replace the temporal module of the baseline model with the proposed spatio-temporal model. The specific ablation experiment configuration and results are shown in Table 1. The experimental results in the table show that the spatio-temporal model with two TPT-GCs and one STPT-GC branch configuration has better performance.

Model Configuration Exploration. As mentioned in Sect. 3.4, our proposed CPT-GCN contains three different modules, namely SPT-GC, TPT-GC and STPT-GC. We manually remove or only keep any kind of modules to test the parameter cost and model performance of different configurations of CPT-GCN. Additionally, we adopt ST-GCN [34] as the baseline method, which does not use any of these three modules.

Table 1. Comparison of the validation accuracy of spatio-temporal model with different settings.

Methods	Configuration	Acc(%)
Baseline	–	84.3
CPT-GCN (w/o SPT-GC)	TPT-GC + STPT-GC	87.5
	TPT-GC + 2STPT-GC	87.6
	TPT-GC + 3STPT-GC	86.6
	2TPT-GC + STPT-GC	**88.2**
	2TPT-GC + 2STPT-GC	88.0
	2TPT-GC + 3STPT-GC	87.1
	3TPT-GC + STPT-GC	87.7
	3TPT-GC + 2STPT-GC	87.4
	3TPT-GC + 3STPT-GC	86.9

Table 2. Comparison of the validation accuracy of CPT-GC with different settings.

Methods	SPT-GC	TPT-GC	STPT-GC	Param	Acc(%)
Baseline	–	–	–	1.27M	84.3
CPT-GCN	✓	✗	✗	2.30M	88.8
	✗	✓	✗	1.52M	87.5
	✗	✗	✓	1.45M	87.2
	✓	✓	✗	2.47M	89.1
	✓	✗	✓	2.40M	88.9
	✗	✓	✓	1.62M	88.2
	✓	✓	✓	2.57M	**89.5**

The specific ablation experiment configuration and results are shown in Table 2. The experimental results in the table show that although our proposed SPT-GC module introduces some additional parameters, it can effectively improve the performance of the model. The TPT-GC and STPT-GC modules have a significant effect on improving the performance of the model under the premise that a small number of parameters are required. The combination of the three modules of SPT-GC, TPT-GC and STPT-GC is the optimal configuration of this model. Under this configuration, CPT-GCN bring improvements of +5.2% over the baseline method on the X-Sub benchmark.

4.4 Comparison with the State-of-the-Art

Most state-of-the-art methods employ a multi-stream fusion framework to enrich semantic information. Our proposed method adopts the same strategy as [5,7,26] to generate four data modalities, namely joint, bone, joint motion and bone motion, and fuse the prediction scores of the four modalities.

We compare the final model with state-of-the-art skeleton-based action recognition methods on the NTU-60 and NTU-120 datasets. The results are shown in Tables 3 and 4. These methods for comparison include RNN-based methods [17,20,24], CNN-based methods [2,15,37] and GCN-based methods [6,7,16,21,25,34].

Table 3. Recognition accuracy comparison against state-of-the-art methods on the NTU RGB+D dataset.

Methods	X-Sub(%)	X-View(%)
Deep LSTM [24]	60.7	67.3
Ind-RNN [17]	81.8	88.0
TCN [2]	74.3	83.1
HCN [15]	86.5	91.1
SGN [37]	89.0	94.5
ST-GCN [34]	81.5	88.3
AS-GCN [16]	86.8	94.2
2 s-AGCN [25]	88.5	95.1
PT-GCN [38]	90.7	96.0
Shift-GCN [7]	90.7	96.5
MS-G3D [21]	91.5	96.2
MST-GCN [6]	91.5	**96.6**
CPT-GCN (Bone)	90.1	94.5
CPT-GCN (Joint+Bone)	91.9	96.2
CPT-GCN	**92.2**	96.5

Table 4. Recognition accuracy comparison against state-of-the-art methods on the NTU RGB+D 120 dataset.

Methods	X-Sub(%)	X-Set(%)
ST-LSTM [20]	55.7	57.9
SGN [37]	79.2	81.5
ST-GCN [34]	70.7	73.2
AS-GCN [16]	77.9	78.5
ST-Transformer [23]	82.7	84.7
2 s-AGCN [25]	82.9	84.9
PT-GCN [38]	85.0	87.3
Shift-GCN [7]	85.9	87.6
MS-G3D [21]	86.9	88.4
MST-GCN [6]	87.5	88.8
CPT-GCN	**88.9**	**89.8**

Our model achieves significant improvements of +1.4% and +1.0% over MST-GCN on the X-Sub and X-Set benchmark of NTU-120, respectively. Overall, CPT-GCN achieves better performance than other methods on both datasets, which demonstrates the superiority of our model.

5 Conclusion

In this work, we present a novel combined part-wise topology graph convolutional network (CPT-GCN) for skeleton-based action recognition. SPT-GC accurately learns the joint correlation of actions in a way that combines global topology and local topology. TPT-GC reasonably changes the receptive field of time convolution to extract the motion trend and motion details of the whole and part of the action. STPT-GC focuses on extracting the implicit spatio-temporal association information in the skeleton sequence, and establishes the part-enhanced spatio-temporal association topology. The combination of the three modules shows a powerful correlation modeling capability. We evaluate the proposed model on two large-scale datasets. The experimental results demonstrate that CPT-GCN has stronger performance than other graph convolutions, and the final model has excellent performance and generalization ability.

Acknowledgement. This work is supported by the Key Research and Development Program of China (No. 2022YFC3005401), the Key Research and Development Program of China, Yunnan Province (No. 202203AA080009), the Fundamental Research Funds for the Central Universities (No. B230205027), Postgraduate Research & Practice Innovation Program of Jiangsu Province(No. 422003261), the 14th Five-Year Plan for Educational Science of Jiangsu Province (No. D/2021/01/39), the Jiangsu Higher Education Reform Research Project (No. 2021JSJG143) and the 2022 Undergraduate Practice Teaching Reform Research Project of Hohai University.

References

1. Aggarwal, J.K., Ryoo, M.S.: Human activity analysis: a review. ACM Comput. Surv. (CSUR) **43**(3), 1–43 (2011)
2. Bai, S., Kolter, J.Z., Koltun, V.: An empirical evaluation of generic convolutional and recurrent networks for sequence modeling. arXiv preprint arXiv:1803.01271 (2018)
3. Carreira, J., Zisserman, A.: Quo vadis, action recognition? a new model and the kinetics dataset. In: Proceedings of the IEEE Conference on Computer Vision and Pattern Recognition, pp. 6299–6308 (2017)
4. Chen, T., et al.: Learning multi-granular spatio-temporal graph network for skeleton-based action recognition. In: Proceedings of the 29th ACM International Conference on Multimedia, pp. 4334–4342 (2021)
5. Chen, Y., Zhang, Z., Yuan, C., Li, B., Deng, Y., Hu, W.: Channel-wise topology refinement graph convolution for skeleton-based action recognition. In: Proceedings of the IEEE/CVF International Conference on Computer Vision, pp. 13359–13368 (2021)

6. Chen, Z., Li, S., Yang, B., Li, Q., Liu, H.: Multi-scale spatial temporal graph convolutional network for skeleton-based action recognition. In: Proceedings of the AAAI Conference on Artificial Intelligence, vol. 35, pp. 1113–1122 (2021)
7. Cheng, K., Zhang, Y., He, X., Chen, W., Cheng, J., Lu, H.: Skeleton-based action recognition with shift graph convolutional network. In: 2020 IEEE/CVF Conference on Computer Vision and Pattern Recognition (CVPR), pp. 180–189 (2020)
8. Feichtenhofer, C., Fan, H., Malik, J., He, K.: Slowfast networks for video recognition. In: Proceedings of the IEEE/CVF International Conference on Computer Vision, pp. 6202–6211 (2019)
9. He, K., Zhang, X., Ren, S., Sun, J.: Deep residual learning for image recognition. In: Proceedings of the IEEE Conference on Computer Vision and Pattern Recognition, pp. 770–778 (2016)
10. Ke, Q., Bennamoun, M., An, S., Sohel, F., Boussaid, F.: A new representation of skeleton sequences for 3d action recognition. In: Proceedings of the IEEE Conference on Computer Vision and Pattern Recognition, pp. 3288–3297 (2017)
11. Kim, T.S., Reiter, A.: Interpretable 3d human action analysis with temporal convolutional networks. In: 2017 IEEE Conference on Computer Vision and Pattern Recognition Workshops (CVPRW), pp. 1623–1631. IEEE (2017)
12. Li, C., Huang, Q., Li, X., Wu, Q.: Human action recognition based on multi-scale feature maps from depth video sequences. Multimedia Tools Appl. **80**, 32111–32130 (2021)
13. Li, C., Huang, Q., Li, X., Wu, Q.: A multi-scale human action recognition method based on laplacian pyramid depth motion images. In: Proceedings of the 2nd ACM International Conference on Multimedia in Asia, pp. 1–6 (2021)
14. Li, C., Huang, Q., Mao, Y.: DD-GCN: directed diffusion graph convolutional network for skeleton-based human action recognition. In: IEEE International Conference on Multimedia and Expo (ICME) (2023)
15. Li, C., Zhong, Q., Xie, D., Pu, S.: Co-occurrence feature learning from skeleton data for action recognition and detection with hierarchical aggregation. In: Proceedings of the 27th International Joint Conference on Artificial Intelligence, pp. 786–792 (2018)
16. Li, M., Chen, S., Chen, X., Zhang, Y., Wang, Y., Tian, Q.: Actional-structural graph convolutional networks for skeleton-based action recognition. In: Proceedings of the IEEE/CVF Conference on Computer Vision and Pattern Recognition, pp. 3595–3603 (2019)
17. Li, S., Li, W., Cook, C., Zhu, C., Gao, Y.: Independently recurrent neural network (INDRNN): building a longer and deeper RNN. In: Proceedings of the IEEE Conference on Computer Vision and Pattern Recognition, pp. 5457–5466 (2018)
18. Li, W., Wen, L., Chang, M.C., Nam Lim, S., Lyu, S.: Adaptive RNN tree for large-scale human action recognition. In: Proceedings of the IEEE International Conference on Computer Vision, pp. 1444–1452 (2017)
19. Liu, J., Shahroudy, A., Perez, M., Wang, G., Duan, L.Y., Kot, A.C.: NTU RGB+d 120: a large-scale benchmark for 3d human activity understanding. IEEE Trans. Pattern Anal. Mach. Intell. **42**(10), 2684–2701 (2019)
20. Liu, J., Shahroudy, A., Xu, D., Wang, G.: Spatio-temporal LSTM with trust gates for 3d human action recognition. In: Leibe, B., Matas, J., Sebe, N., Welling, M. (eds.) ECCV 2016. LNCS, vol. 9907, pp. 816–833. Springer, Cham (2016). https://doi.org/10.1007/978-3-319-46487-9_50
21. Liu, Z., Zhang, H., Chen, Z., Wang, Z., Ouyang, W.: Disentangling and unifying graph convolutions for skeleton-based action recognition. In: Proceedings of the

IEEE/CVF Conference on Computer Vision and Pattern Recognition, pp. 143–152 (2020)
22. Peng, W., Hong, X., Chen, H., Zhao, G.: Learning graph convolutional network for skeleton-based human action recognition by neural searching. In: Proceedings of the AAAI Conference on Artificial Intelligence, vol. 34, pp. 2669–2676 (2020)
23. Plizzari, C., Cannici, M., Matteucci, M.: Spatial temporal transformer network for skeleton-based action recognition. In: Del Bimbo, A., et al. (eds.) ICPR 2021. LNCS, vol. 12663, pp. 694–701. Springer, Cham (2021). https://doi.org/10.1007/978-3-030-68796-0_50
24. Shahroudy, A., Liu, J., Ng, T.T., Wang, G.: NTU RGB+ d: a large scale dataset for 3d human activity analysis. In: Proceedings of the IEEE Conference on Computer Vision and Pattern Recognition, pp. 1010–1019 (2016)
25. Shi, L., Zhang, Y., Cheng, J., Lu, H.: Two-stream adaptive graph convolutional networks for skeleton-based action recognition. In: Proceedings of the IEEE/CVF Conference on Computer Vision and Pattern Recognition, pp. 12026–12035 (2019)
26. Shi, L., Zhang, Y., Cheng, J., Lu, H.: Skeleton-based action recognition with multi-stream adaptive graph convolutional networks. IEEE Trans. Image Process. **29**, 9532–9545 (2020)
27. Shi, L., Zhang, Y., Cheng, J., Lu, H.: Action recognition via pose-based graph convolutional networks with intermediate dense supervision. Pattern Recogn. **121**, 108170 (2022)
28. Si, C., Chen, W., Wang, W., Wang, L., Tan, T.: An attention enhanced graph convolutional LSTM network for skeleton-based action recognition. In: proceedings of the IEEE/CVF Conference on Computer Vision and Pattern Recognition, pp. 1227–1236 (2019)
29. Song, Y.F., Zhang, Z., Shan, C., Wang, L.: Stronger, faster and more explainable: A graph convolutional baseline for skeleton-based action recognition. In: proceedings of the 28th ACM International Conference on Multimedia, pp. 1625–1633 (2020)
30. Thakkar, K., Narayanan, P.J.: Part-based graph convolutional network for action recognition. In: 29th British Machine Vision Conference, BMVC. p. Amazon et al. Microsoft; NVIDIA; SCANs; SCAPE. BMVA Press (2019)
31. Shi, L., Zhang, Y., Cheng, J., Lu, H.: Skeleton-based action recognition with directed graph neural networks. In: Proceedings of the IEEE/CVF Conference on Computer Vision and Pattern Recognition, pp. 7912–7921 (2019)
32. Shi, L., Zhang, Y., Cheng, J., Lu, H.: Decoupled spatial-temporal attention network for skeleton-based action-gesture recognition. In: Proceedings of the Asian Conference on Computer Vision (2020)
33. Wang, W., Xie, W., Tu, Z., Li, W., Jin, L.: Multi-part adaptive graph convolutional network for skeleton-based action recognition. In: 2022 International Joint Conference on Neural Networks (IJCNN), pp. 1–7 (2022)
34. Yan, S., Xiong, Y., Lin, D.: Spatial temporal graph convolutional networks for skeleton-based action recognition. In: Thirty-Second AAAI Conference on Artificial Intelligence (2018)
35. Zhang, H., et al.: Resnest: split-attention networks. In: Proceedings of the IEEE/CVF Conference on Computer Vision and Pattern Recognition, pp. 2736–2746 (2022)
36. Zhang, P., Lan, C., Xing, J., Zeng, W., Xue, J., Zheng, N.: View adaptive recurrent neural networks for high performance human action recognition from skeleton data. In: Proceedings of the IEEE International Conference on Computer Vision, pp. 2117–2126 (2017)

37. Zhang, P., Lan, C., Zeng, W., Xing, J., Xue, J., Zheng, N.: Semantics-guided neural networks for efficient skeleton-based human action recognition. In: proceedings of the IEEE/CVF Conference on Computer Vision and Pattern Recognition, pp. 1112–1121 (2020)
38. Zhu, X., Huang, Q., Li, C., Wang, L., Miao, Z.: Part-wise topology graph convolutional network for skeleton-based action recognition. In: Fang, L., Povey, D., Zhai, G., Mei, T., Wang, R. (eds.) Artificial Intelligence. CICAI 2022. LNCS, vol. 13604. Springer, Cham (2022). https://doi.org/10.1007/978-3-031-20497-5_26

Segmenting Key Clues to Induce Human-Object Interaction Detection

Mingliang Xue(✉), Siwei Wang, Bing Fu, Zhengyang Zhao, Tao Liu, and Lingfeng Lai

School of Computer Science and Engineering, Dalian Minzu University, Dalian 116650, China
xml@dlnu.edu.cn

Abstract. Two-stage HOI detectors have made great progress in training and inference, but still suffer from loss of original image features and ambiguous human-object relationships. To address the above issues, this paper proposes a network based on segmentation to extract key features to induce human-object interaction detection, which consists of two parts. First, a segmentation-based module is designed to extract fine-grained interaction features from original images, which is then refined by a feature learning encoder. Secondly, a graph-based module is proposed to encode the spatial relationships of detected human and object instances, which is to learn the interaction contexts from pair-wised human-object contexts. A transformer decoder is then utilized to equip the interaction features from the original images with the interaction contexts. The proposed method can directly learn fine-grained interaction features under the guidance of spatial relationships, achieving state-of-the-art performance on two standard benchmarks for HOI detection, HICO-DET and V-COCO.

Keywords: human-object interaction · object detection · graph network

1 Introduction

Human-object interaction (HOI) detection is the classification of interaction actions between humans and objects in an image while locating them. The detected HOI instances are represented in the form of triplets of humans, objects, and interaction actions. HOI detection can be categorized into one-stage methods and two-stage methods. The two-stage methods first use a target detector to detect humans and objects, then combine them for classification. Exemplified by FasterR-CNN [1], two-stage object detectors typically include a region proposal network, which explicitly encodes potential regions of interest in the form of bounding boxes. These bounding boxes can then be classified and further

This work was supported by the Research Foundation of Liaoning Educational Department (Grant No. LJKMZ20220400).

© The Author(s), under exclusive license to Springer Nature Singapore Pte Ltd. 2024
Q. Liu et al. (Eds.): PRCV 2023, LNCS 14425, pp. 60–71, 2024.
https://doi.org/10.1007/978-981-99-8429-9_5

refined via regression in a downstream network. In contrast, one-stage detectors, such as RetinaNet [2], retain the abstract feature representations of objects throughout the network and decode them into bounding boxes and classification scores. In previous studies, pre-trained object detectors were used to detect the bounding boxes of humans and objects. Then, these detected objects and humans are paired to classify interactions using a downstream classification network. These methods are consistent with the idea of two-stage methods. Due to high-quality object detection, two-stage methods have become the mainstream method. Meanwhile, due to the tremendous success of Transformer's decoder in using learnable queries for localization [3], one-stage HOI detectors have also undergone significant development. However, one-stage HOI detectors heavily rely on the transformer and can be difficult to train as a result. In contrast, two-stage HOI detectors do not require re-training of the target detection network, but shift their focus to the interaction classification in downstream tasks, resulting in more efficient use of memory and computing resources.

Although two-stage HOI detectors are relatively lightweight and flexible, they over-rely on the features of the target detector in downstream tasks, lacking key features of human and object in the image. The main contributions of this work are as follows:

1. A segmentation-based module (SBM) was proposed to extract key humans and objects that may have interactions in the image, providing strong semantic key features for interaction detection and leading to improved performance in HOI detection tasks.
2. More fine-grained features are required for accurate interaction classification, so the extracted key features of humans and objects are fused with the complete features map with background as input, and a key features learning encoder (KFEncoder) is proposed. The paired human-object feature representations are used as the query object and fine-grained features for attention calculation, and the interaction key features area will get a higher weight, thereby enhancing the correctness of the interaction classification.
3. A human-object pair spatial relationships learning graph-based module (GBM) is proposed, which uses feature representations from object detection to initialize nodes and incorporates spatial relationships into node calculations for asynchronous message transmission between humans and objects. Spatial relationships help disambiguate interactions by influencing node calculations.

2 Related Work

The transformer network was originally proposed for machine translation. Due to its expressive power, flexibility and global modeling ability through attention mechanisms, it has also been widely used in the field of computer vision. As in ViT [4], an image is evenly divided into a set of spatial patches, each of which is transformed linearly and encoded into a token. These approaches inspired further research in related areas. Carion et al. [3] used these queries as object queries, in place of the region recommendations of traditional target detectors, and combined them with bipartite matching losses to design a single-stage detection model, thereby stimulating more research on HOI detection [5–10]. To adapt the DETR model to HOI detection, Tamura et al. [6] and Zou et al. [7] added interactive detection heads after the detector to locate humans and objects while predicting actions. These one-stage detectors have achieved great success in advancing the state of the art, but require significant resources to train. However, they all require significant resources to train the models. Two-stage HOI detectors have also undergone significant development recently. Hou et al. explored few-shot learning by constructing object representations in features space and learning to transfer object affordance [11]. To explore spatial relations between human and object, Zhang et al. [12] proposed to fuse features of different modalities within a graphical model to produce more discriminative features. Despite providing more discriminative features, the image features of human and object were not utilized. Finally, Zhang et al. [13] propose the Unary-Pairwise Transformer (UPT), a two-stage detector that exploits unary and pairwise representations for HOIs, with the former preferentially increasing the scores of positive examples and the latter decreasing the scores of negative examples. Sanghyun Kim et al. [14] propose the multiplex relation network (MUREN) that performs rich context exchange between three decoder branches based on unary, pairwise, and ternary relations of human, object, and interaction tokens. In UPT, over-reliance on image features of target detection leads to the lack of fine-grained features of human and object in HOI classification, and it does not perform well in certain situations. In MUREN, the spatial relations between human and object are not fully utilized, resulting in negative pairs that affect the results. Therefore, our work is based on the two-stage method to solve the problem of the loss of some original image features and the fuzzy spatial relationships between humans and objects.

3 Approach

In this work, we have used a two-stage method as the foundation. We approach HOI detection as a problem of predicting sets of interactions. Figure 1 describes the proposed model. To enhance the model's efficiency, we use pre-trained DETR [3] as the target detector. The model obtains features such as bounding boxes, categories, and confidence scores for detected humans and objects using pre-trained DETR. ResNet extracts features from the key feature representations obtained using the SBM module, and these features are combined via a 1x1 convolution operation. The fusion highlights humans and objects, weakening the influence of background on interactive classification. The fused features will serve as inputs to the KFEncoder for learning global key features associated with humans and objects. The human and object features obtained by the target detector are initialized as human and object nodes, and a bipartite graph is constructed. Node features are updated by incorporating information about the relative positions of humans and objects into node updates and asynchronous message transmission. The updated human and object nodes are concatenated as input queries for the HOI interaction detector. The features learned by the KFEncoder are used as input keys and values for interaction detection. By using attention mechanisms, the model continuously enhances the query features. Finally, the model uses a linear layer to classify interactions.

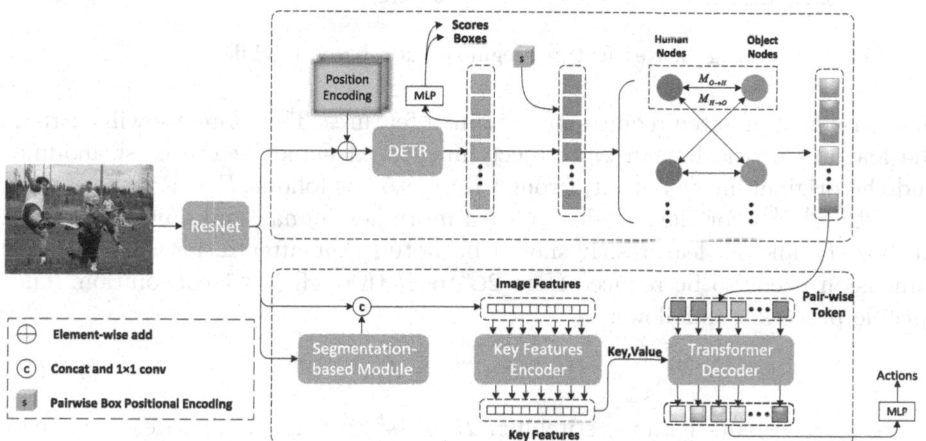

Fig. 1. Overview of our proposed network framework.

3.1 Key Features Segmentation-Based Module

In recent research, Alexander Kirillov et al. proposed the Segment Anything [15], which can segment everything in an image by pre-training on big data sets. We have been inspired by it. In human-object interaction detection, we also need to appropriately segment humans, objects and backgrounds in the image, especially in interaction classification. We hope that the model will pay more attention to key humans and objects. Therefore, Fig. 2 shows the SBM module. Specifically, we have fused and improved the target detector DETR and Segment Anything Model (SAM). The detection results are obtained through pre-trained DETR, and the correct results are first screened through non-maximum suppression and thresholding. This leaves a smaller set $\{d_i\}_{i=1}^{n}$, where a detection $d_i = (b_i, s_i, c_i, x_i)$ consists of the box coordinates $b_i \in \mathbb{R}^4$, the confidence score $s_i \in [0, 1]$, the object class $c_i \in \kappa$ for a set of object categories κ, and the object query or features $x_i \in \mathbb{R}^m$. Input the filtered bounding boxes and confidence scores into the SAM to obtain masks for each human and object box. The resulting masks are then passed through a standard transformer decoder to obtain the masked features image W.

3.2 Key Features Learning Encoder

The two-stage method for human-object interaction detection relies on the features of the target detector during the interaction detection stage, but interac-

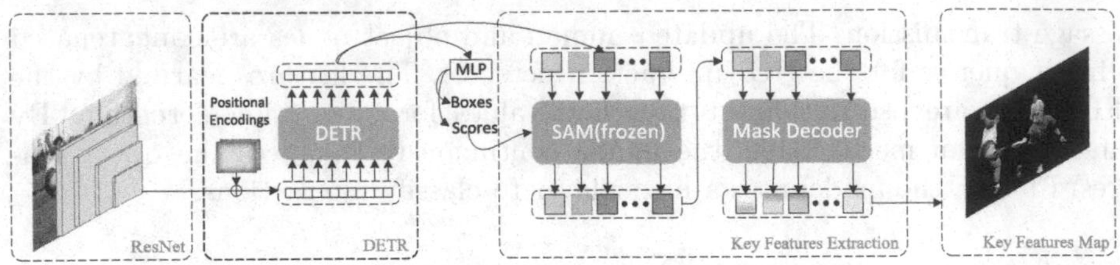

Fig. 2. Key features segmentation-based module.

tion classification often requires more refined features. Therefore, we will extract the features of the human and object images extracted by the mask module and the original image output through ResNet50 as follows $F_r \in \mathbb{R}^{H \times W \times C}$ and $F_s \in \mathbb{R}^{H \times W \times C}$. For the encoder to learn more key human and object features, we fuse F_r and F_s features. It should be noted that after fusion, the channel dimension needs to be reduced from $2C$ to C through 1×1 convolution. The specific process is as follows:

$$H = conv(F_s \oplus F_r) \tag{1}$$

where $\oplus$ represents features stitching, $H \in \mathbb{R}^{h \times w \times c}$ is the features after features fusion and convolution. The fused features H needs to reduce the channel dimension from c to d, and then input it to the key features learning encoder, and constantly expand the receptive field through self-attention to learn more features. For dimension reduction, we use the typical values $c = 2048, d = 512$, and the specific operational definition is:

$$H_i \in^{h \times w \times d}, H = (h_1, h_2, \cdots, h_{h \times w}) \tag{2}$$

$$K \in \mathbb{R}^{h \times w \times d}, K = (k_1, k_2, \cdots, k_{h \times w}) \tag{3}$$

$$K = KFEncoder(H_i) \tag{4}$$

where $\oplus$ represents features stitching, $H \in \mathbb{R}^{h \times w \times d}$ is the flattened ResNet features, $K \in \mathbb{R}^{h \times w \times d}$ is the fine-grained features learned (KFEncoder) from the key features learning encoder, which is a standard transformer encoder.

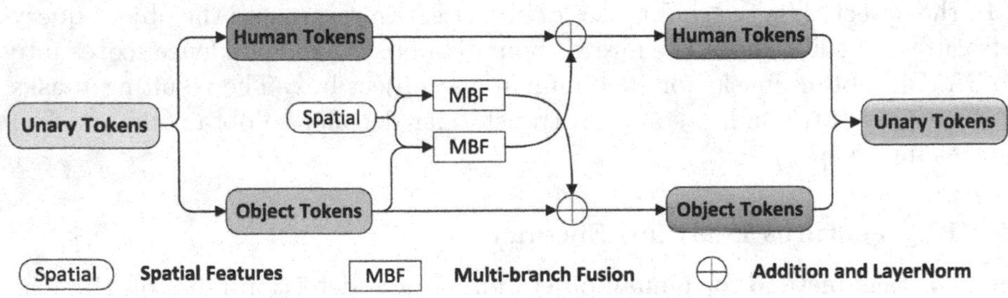

Fig. 3. Spatial relationships learning graph module.

3.3 Spatial Relationships Learning Graph-Based Module

We propose a spatial relationship learning graph module by using both appearance features and position features to jointly infer interactive actions. Figure 3 shows a spatial relationships learning graph module structure. We will represent the graph as $G = (H, O, \varepsilon)$, where H represents humans, O represents objects and ε represents edges of the graph. The improved graph network uses the features of DETR object detection to initialize the appearance features of human and object nodes, and the edge encodings are computed as handcrafted feature vectors. To describe the pairwise relationship between humans and objects, we drew inspiration from SCG [12]: coordinates of the bounding boxes, widths, heights, aspect ratios and areas, all normalised by the corresponding dimension of the image. Inspired by SCG, to describe the pairwise relationship between humans and objects, we also calculated the intersection of the bounding boxes of humans and objects, as well as the area ratio of the bounding boxes, and ultimately used the same positional encoding $P \oplus log(P + \epsilon)$. To initialize human and object nodes, we use the features hs of the target detector as the appearance features. We need to use a multilayer perceptron to map the dimensions to lower dimensions and obtain node encodings x_i^0, y_j^0 and time step $t = 0$, where $i \in \{1, \cdots, |H|\}, j \in \{1, \cdots, |O|\}$. Similarly, paired spatial relationships $s_{ij} \in \mathbb{R}^n$ are also mapped to the same dimension as appearance features through another multi-layer perceptron. When messaging between human and object nodes, the paired spatial relationships remain unchanged, and bidirectional messaging is defined as

$$x_i^{t+1} = LN\left(x_i^t + \sigma\left(\sum_{j=1}^{|O|} \alpha_{ij}^r M_{O\to H}(y_j^t, s_{ij})\right)\right) \tag{5}$$

$$y_j^{t+1} = LN\left(y_j^t + \sigma\left(\sum_{i=1}^{|H|} \alpha_{ij}^c M_{H\to O}(x_i^t, s_{ij})\right)\right) \tag{6}$$

where LN denotes the LayerNorm operation, σ is the activation function (RELU) and α is an adjacency weight between nodes. The key point is that to better model the relationship between humans and objects, the message function M is designed to be asynchronous, allowing nodes to send different messages based on the type of receiving node. The fusion strategy for human and object appearance features and spatial relationships is consistent with SCG, and a multi-branch fusion method is used for feature fusion. First, the dimension of appearance features and spatial relationships will be reduced to $c = 256$ dimension, and the point multiplication operation will be carried out for both features. The point multiplied features will be upgraded to the original dimension again through the activation function (RELU) to complete the fusion of appearance and spatial relationships. Feature fusion and message passing are defined as

$$M_{O\to H}(y_j^t, s_{ij}) = MBF_o(y_j^t, s_{ij}) \tag{7}$$

$$M_{H\to O}(x_i^t, s_{ij}) = MBF_h(x_i^t, s_{ij}) \tag{8}$$

where $M_{O\to H}$ denotes object-human transmission of messages, $M_{H\to O}$ is human-object transmission of messages. MBF_o and MBF_h denote feature fusion.

3.4 Training and Inference

To fully utilize the pre-trained object detector during the training process, we also included the predicted confidence scores of humans and objects in the final action scores of each human-object pair. So the final calculated score is expressed as

$$s_k = (s_i)^\lambda \cdot (s_j)^\lambda \cdot \sigma(\widetilde{s_k}) \tag{9}$$

where s_i is human confidence score, s_j is object confidence score, $\widetilde{s_k}$ is the action score of the human-object pair and σ is the sigmoid function. λ is a common practice used in the inference process to suppress overconfidence in target detection results. Similarly, we also use focal loss [2] for action classification to suppress the imbalance between positive and negative examples.

4 Experiments

In this section, we conducted experiments on the proposed SBM, KFEncoder, and GBM modules to achieve state-of-the-art performance on the HICO-DET [16] and V-COCO [17] datasets. We comprehensively analyzed the extracted key features of humans and objects and the impact of learning the location features of humans and objects through graphs. Our findings show that the features extracted through SBM effectively guide the encoder to focus more on key features of humans and objects, improving the scores of positive examples. By using graph networks, more attention is paid to the positions of humans and objects when paired, which helps suppress the score of negative examples. We also visualized the attention weights learned by the KFEncoder module and the extracted key feature maps. Our experiments were conducted on two NVIDIA RTX A6000 GPUs.

4.1 Implementation Details

Our approach uses the DETR model fine-tuned by UPT [13] on HICO-DET and V-COCO as our target detector to avoid the need for additional training. For the results of target detection, we discard all detection results with confidence scores below a threshold of 0.2. In order to maintain consistent and reliable training conditions, we reserve at least 3 humans and 3 objects, with up to 15 humans and 15 objects. For the feature dimensions of target detected humans and objects, we use $m = 256$, which is consistent with DETR [3]. At the same time, we use the same fusion strategy rate as SCG [12] to fuse features from the target detector and paired position encoded features, which is different from SCG [12]. The model trains 40 epochs, and the learning rate will decrease after every 10 epochs.

Table 1. Comparison of experimental results (mAP×100) between HICO-DET and V-COCO in the datasets.The best result has been highlighted.

Method	Backbone	Params	FPS	HICO-DET						V-COCO	
				Default Setting			Known Objects Setting				
				Full	Rare	Non-rare	Full	Rare	Non-rare	AP^{S1}_{role}	AP^{S2}_{role}
PPDM [18]	ResNet-101	-	14	21.94	13.97	24.32	24.81	17.09	27.12	-	-
VCL [19]	ResNet-50	-	-	23.63	17.21	25.55	25.98	19.12	28.03	48.3	-
IDN [20]	ResNet-50	-	-	24.58	20.33	25.86	27.89	23.64	29.16	53.3	60.3
HOTR [5]	ResNet-50	51	28	25.10	17.34	27.42	-	-	-	55.2	64.4
FCL [21]	ResNet-50	-	-	25.27	20.57	26.67	27.71	22.34	28.93	52.4	-
AS-Net [22]	ResNet-50	53	-	28.87	24.25	30.25	31.74	27.07	33.14	53.9	-
SCG [12]	ResNet-50-FPN	54	-	29.26	24.61	30.65	32.87	27.89	34.35	54.2	60.9
UPT [13]	ResNet-50	13	28	31.66	25.94	33.36	35.05	29.27	36.77	59.0	64.5
DOQ [23]	ResNet-50	41	-	33.28	29.19	34.50	-	-	-	**63.5**	-
IF [24]	ResNet-50	42	-	33.51	30.30	34.46	36.28	33.16	37.21	63.0	65.2
Ours	ResNet-50	10	33	33.72	29.76	34.81	36.28	31.65	37.78	59.4	65.2
HOI-Trans [7]	ResNet-101	41	26	26.61	19.15	28.84	29.13	20.98	31.57	52.9	-
QPIC [6]	ResNet-101	60	-	29.90	23.92	31.69	32.38	26.06	34.27	58.8	61.0
UPT [13]	ResNet-101	13	22	32.31	28.55	33.44	35.65	31.60	36.86	60.7	66.2
Ours	ResNet-101	10	25	34.37	30.56	35.39	37.88	33.90	38.86	61.5	67.4
UPT [13]	ResNet-101-DC5	13	15	32.62	28.62	33.81	36.08	31.41	37.47	61.1	67.1
Ours	ResNet-101-DC5	10	18	**34.92**	**31.67**	**35.85**	**38.79**	**35.43**	**39.60**	61.5	**67.5**

4.2 Implementation Results

Table 1 shows the comparison results of our model with existing methods on the HICO-DET and V-COCO datasets. Our model presents results on three different backbone networks with the same settings as UPT [13], indicating that our model has an average mAP value of 2 to 3% points higher than the advanced model UPT on the HICO-DET dataset. Improved by 5 to 6% points compared to the SCG [12]. Overall, our lightweight model outperforms suboptimal methods in almost all categories. Through experimental comparison, we have verified the effectiveness of our proposed key features extraction module, key features learning encoder. We will demonstrate the specific results through ablation experiments.

Table 2. Ablation study on HICO-DET.

Method	Full	Rare	Non-rare
Base	22.45	17.60	23.89
Base + KFEncoder	34.12	30.24	34.98
Base + KFEncoder + GBM	**34.37**	**30.56**	**35.39**

Table 3. Ablation study on V-COCO.

Method	AP^{S1}_{role}	AP^{S2}_{role}
Base	49.81	54.58
Base + KFEncoder	61.50	67.43
Base + KFEncoder + GBM	**61.52**	**67.54**

Fig. 4. To display the extracted key features of humans and objects in each image, we present a visual representation that includes both the original image and the corresponding key feature map. The left side of the display shows the original image, while the right side displays the extracted key features map.

Table 4. Scores for kicking a ball.

index	1	2	3	4	5	6
7	**0.7611**	0.3587	0.2742	0.0859	0.0464	0.0379

4.3 Ablation Study

We conducted ablation experiments on the HICO-DET and V-COCO datasets to verify the effectiveness of our proposed key features extraction module, fine-grained features learning encoder, and spatial relationships learning graph module. Adding the KFEncoder module and GBM module to the base model improved HOI interaction classification. Specifically, without the KFEncoder module, HOI interactive classification relies heavily on target detector features, leading to suboptimal interaction detection results. Furthermore, the absence of the GBM module causes incorrect pairing between humans and objects due to the lack of location features, which inhibits correct human-object pair interactions and reduces classification accuracy. Our findings confirm the importance of the KFEncoder and GBM modules. The results of the ablation experiments are shown in Tables 2 and 3.

Table 5. Scores for straddling a bicycle.

index	1	2
3	**0.9074**	0.0038
4	0.0008	**0.8101**

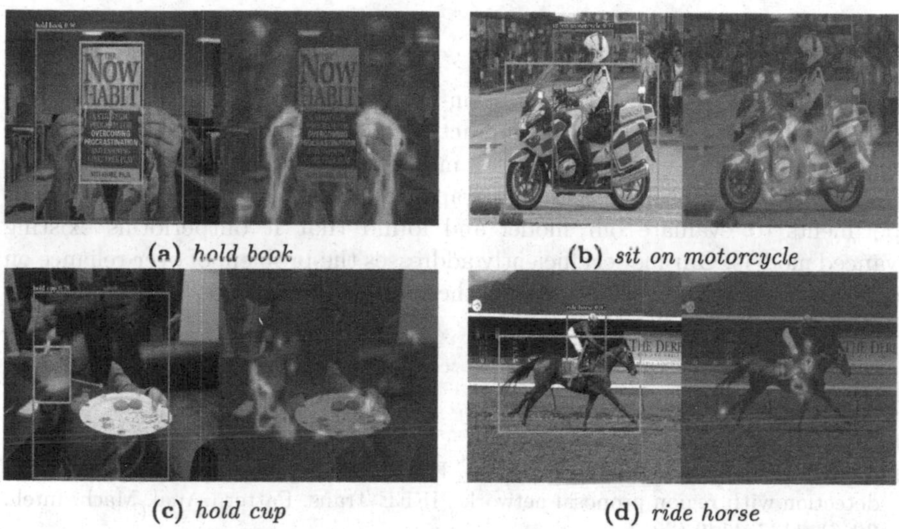

(a) *hold book* (b) *sit on motorcycle*

(c) *hold cup* (d) *ride horse*

Fig. 5. Qualitative results of detected HOIs. Interactive human–object pairs are connected by red lines, with the interaction scores overlaid above the human box. Pairs with scores lower than 0.2 are filtered out. (Color figure online)

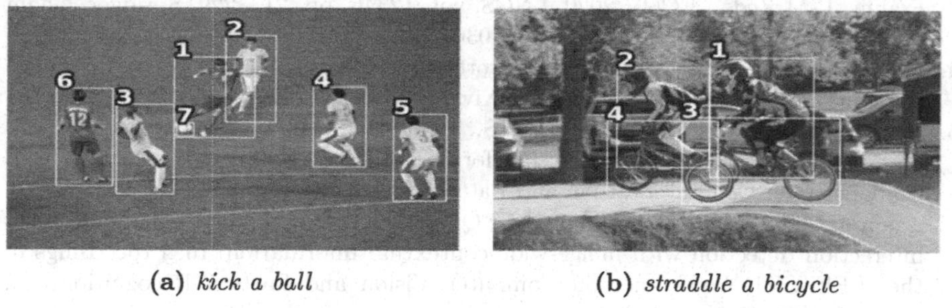

(a) *kick a ball* (b) *straddle a bicycle*

Fig. 6. Case study of using the proposed graph module to model human and object relationships in our model.

4.4 Qualitative Results

Figures 4 and 5 illustrate the key features extracted from various images and learned by the encoder, respectively. These key features enable the model to focus on interacting humans and objects, and the encoder continually learns the features that determine their interaction behavior. Figure 6 presents a successful case that uses the graph module to model human and object relationships in the model. By adding spatial relationships during node updates, positive human-ball pairs (1,7) have the highest scores in Fig. 6a. Similarly, in Fig. 6b, positive human-bicycle pairs (1,3) and (2,4) have the highest scores. The corresponding scores of human-object pairs are listed in Tables 4 and 5.

5 Conclusion

In this paper, we propose a segmentation-based network for extracting key features that facilitate human-object interaction detection. Our network comprises three modules: a key features extraction module, key features learning encoders, and a spatial relationships learning graph module. We conducted numerous experiments to evaluate our model and found that it outperforms existing advanced models. Our model efficiently addresses the problem of over-reliance on target detection features while ignoring the spatial relationships between humans and objects in interaction detection.

References

1. Ren, S., He, K., Girshick, R., Sun, J.: Faster R-CNN: towards real-time object detection with region proposal networks. IEEE Trans. Pattern Anal. Mach. Intell. **39**(6), 1137–1149 (2017)
2. Lin, T.-Y., Goyal, P., Girshick, R., He, K., Dollár, P.: Focal loss for dense object detection. IEEE Trans. Pattern Anal. Mach. Intell. **42**(2), 318–327 (2020)
3. Carion, N., Massa, F., Synnaeve, G., Usunier, N., Kirillov, A., Zagoruyko, S.: End-to-end object detection with transformers. In: Vedaldi, A., Bischof, H., Brox, T., Frahm, J.-M. (eds.) ECCV 2020. LNCS, vol. 12346, pp. 213–229. Springer, Cham (2020). https://doi.org/10.1007/978-3-030-58452-8_13
4. Dosovitskiy, A., et al.: An image is worth 16x16 words: transformers for image recognition at scale. arXiv preprint arXiv:2010.11929 (2020)
5. Kim, B., Lee, J., Kang, J., Kim, E.-S., Kim, H.J.: HOTR: end-to-end human-object interaction detection with transformers. In: Proceedings of the IEEE/CVF Conference on Computer Vision and Pattern Recognition, pp. 74–83 (2021)
6. Tamura, M., Ohashi, H., Yoshinaga, T.: QPIC: query-based pairwise human-object interaction detection with image-wide contextual information. In: Proceedings of the IEEE/CVF Conference on Computer Vision and Pattern Recognition, pp. 10410–10419 (2021)
7. Zou, C., et al.: End-to-end human object interaction detection with hoi transformer. In: Proceedings of the IEEE/CVF Conference on Computer Vision and Pattern Recognition, pp. 11825–11834 (2021)
8. Wang, H., Yao, M., Jiang, G., Mi, Z., Fu, X.: Graph-collaborated auto-encoder hashing for multiview binary clustering. IEEE Trans. Neural Netw. Learn. Syst. (2023)
9. Wang, H., Peng, J., Xianping, F.: Co-regularized multi-view sparse reconstruction embedding for dimension reduction. Neurocomputing **347**, 191–199 (2019)
10. Feng, L., Meng, X., Wang, H.: Multi-view locality low-rank embedding for dimension reduction. Knowl.-Based Syst. **191**, 105172 (2020)

11. Hou, Z., Yu, B., Qiao, Y., Peng, X., Tao, D.: Affordance transfer learning for human-object interaction detection. In: Proceedings of the IEEE/CVF Conference on Computer Vision and Pattern Recognition, pp. 495–504 (2021)
12. Zhang, F.Z., Campbell, D., Gould, S.: Spatially conditioned graphs for detecting human-object interactions. In: Proceedings of the IEEE/CVF International Conference on Computer Vision, pp. 13319–13327 (2021)
13. Zhang, F.Z., Campbell, D., Gould, S.: Efficient two-stage detection of human-object interactions with a novel unary-pairwise transformer. In: Proceedings of the IEEE/CVF Conference on Computer Vision and Pattern Recognition (CVPR), pp. 20104–20112 (2022)
14. Kim, S., Jung, D., Cho, M.: Relational context learning for human-object interaction detection, pp. 2925–2934 (2023)
15. Kirillov, A., et al.: Segment anything. arXiv preprint arXiv:2304.02643 (2023)
16. Chao, Y.-W., Liu, Y., Liu, X., Zeng, H., Deng, J.: Learning to detect human-object interactions. In: 2018 IEEE Winter Conference on Applications of Computer Vision (WACV), pp. 381–389. IEEE (2018)
17. Gupta, S., Malik, J.: Visual semantic role labeling. arXiv preprint arXiv:1505.04474 (2015)
18. Liao, Y., Liu, S., Wang, F., Chen, Y., Qian, C., Feng, J.: PPDM: parallel point detection and matching for real-time human-object interaction detection. In: Proceedings of the IEEE/CVF Conference on Computer Vision and Pattern Recognition, pp. 482–490 (2020)
19. Hou, Z., Peng, X., Qiao, Yu., Tao, D.: Visual compositional learning for human-object interaction detection. In: Vedaldi, A., Bischof, H., Brox, T., Frahm, J.-M. (eds.) ECCV 2020. LNCS, vol. 12360, pp. 584–600. Springer, Cham (2020). https://doi.org/10.1007/978-3-030-58555-6_35
20. Li, Y.-L., Liu, X., Xiaoqian, W., Li, Y., Cewu, L.: Hoi analysis: integrating and decomposing human-object interaction. Adv. Neural. Inf. Process. Syst. **33**, 5011–5022 (2020)
21. Hou, Z., Yu, B., Qiao, Y., Peng, X., Tao, D.: Detecting human-object interaction via fabricated compositional learning. In: Proceedings of the IEEE/CVF Conference on Computer Vision and Pattern Recognition, pp. 14646–14655 (2021)
22. Chen, M., Liao, Y., Liu, S., Chen, Z., Wang, F., Qian, C.: Reformulating hoi detection as adaptive set prediction. In: Proceedings of the IEEE/CVF Conference on Computer Vision and Pattern Recognition, pp. 9004–9013 (2021)
23. Qu, X., Ding, C., Li, X., Zhong, X., Tao, D.: Distillation using oracle queries for transformer-based human-object interaction detection. In: IEEE/CVF Conference on Computer Vision and Pattern Recognition, CVPR 2022, New Orleans, LA, USA, 18–24 June 2022, pp. 19536–19545. IEEE (2022)
24. Liu, X., Li, Y.L., Wu, X., Tai, Y.W., Lu, C., Tang, C.K.: Interactiveness field in human-object interactions. arXiv e-prints (2022)

Lightweight Multispectral Skeleton and Multi-stream Graph Attention Networks for Enhanced Action Prediction with Multiple Modalities

Teng Huang[1(✉)], Weiqing Kong[1], Jiaming Liang[1], Ziyu Ding[1], Hui Li[1], and Xi Zhang[2,3(✉)]

[1] Institute of Artificial Intelligence and Blockchain, Guangzhou University, Guangzhou, China
huangteng1220@gzhu.edu.cn

[2] School of Arts, Sun Yat-sen University, Guangzhou, China

[3] College of Music, University of Colorado Boulder, Boulder, USA
xizhangpiano@gmail.com

Abstract. Human action recognition methods often focus on extracting structural and temporal information from skeleton-based graphs. However, these approaches struggle with effectively capturing and processing extensive information during action transitions. To overcome this limitation, we propose LMS-GAT, a novel approach that facilitates information exchange through node concentration and diffusion across structural and temporal dimensions. By selectively suppressing and reinstating the representations of structural nodes for each specific action, and utilizing hierarchical shifted temporal windows for assessing temporal information, LMS-GAT addresses the challenge of dynamic changes in action recognition. Experimental evaluation on NTU RGB+D 60 and 120 datasets shows that LMS-GAT outperforms state-of-the-art methods in terms of prediction accuracy. This highlights the efficacy of our approach in capturing and recognizing human actions with improved performance.

Keywords: Human Action Recognition · Skeleton-based Graph · Graph Attention Networks

1 Introduction

Human action recognition (HAR) has ample practical applications in the real world [1–5]. For computer vision, skeleton-based HAR models are popular due to their robustness against cluttered backgrounds. Recognition of human actions

Supported by the National Natural Science Foundation of China under Grant 62002074 and 62072452; Supported by the Shenzhen Science and Technology Program JCYJ20200109115627045, in part by the Regional Joint Fund of Guangdong under Grant 2021B1515120011.

© The Author(s), under exclusive license to Springer Nature Singapore Pte Ltd. 2024
Q. Liu et al. (Eds.): PRCV 2023, LNCS 14425, pp. 72–83, 2024.
https://doi.org/10.1007/978-981-99-8429-9_6

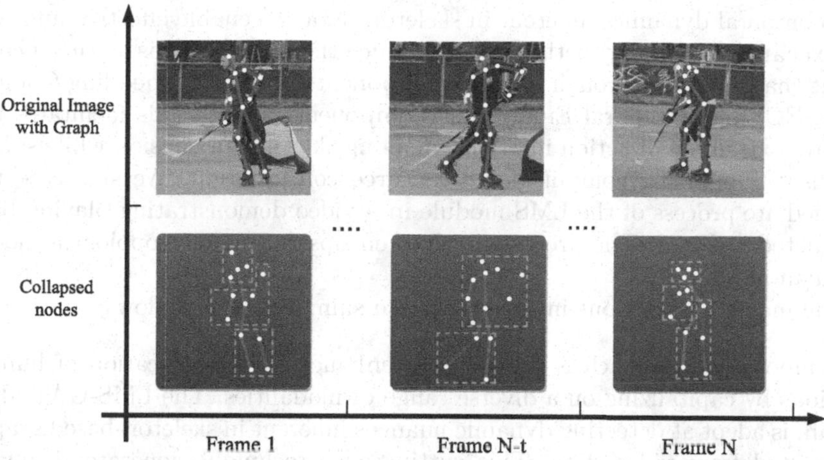

Fig. 1. The collapse strategy of our model. From left to right are the skeletal nodes of the character and the nodes collapsed result of different frames, which can effectively reduce calculation effort.

through skeleton-based analysis entails the ability to interpret a sequence of bodily postures or movements and identify the exact action being executed with precision. The latest approaches to action recognition using graph neural networks (GNN) work by converting skeletal data into a representation of non-Euclidean data. By doing so, these techniques enable the exploration of information flows that capture spatial and temporal details within the skeletal graph in an efficient manner [6].

The utility of GNN-based frameworks for action recognition offers a distinct advantage over other methods by enabling the integration of contextual information within the graph. This facilitates the aggregation of data from neighboring nodes, resulting in improved accuracy for recognizing actions by capturing contextual cues and global dependencies [7]. Furthermore, these approaches excel in capturing structural dependencies [8] and accounting for temporal dynamics [9], making them highly effective in action recognition tasks. One such example is the Spatio-Temporal Graph Convolutional Networks (ST-GCN) proposed by Yan et al. [10]. Combining spatial and temporal information by Using Graph Convolutional Networks (GCN), ST-GCN has demonstrated significant improvements in recognition accuracy. In addition, CTR-GCN [11] and Info-GCN [12] respectively use topology modeling and self-attention mechanism to obtain context information. However, these methods are difficult to deal with dynamic and changeable graph information, and it is challenging to effectively capture action data in the time dimension.

The current investigation puts forward a novel approach named LMS-GAT, which leverages dynamic GNN for HAR based on skeleton data. LMS-GAT is purposefully crafted to capture the dynamic intricacies of human behavior and extract a hierarchical representation that efficiently encapsulates the spatiotemporal dynamics inherent in skeleton data. To enable effective information exchange within our method, we introduce the LMS block as a crucial component that operates through two sub-components: Spatial Embedding Component (SEC) and Temporal Embedding Component (TEC). This facilitates the accurate extraction of action information along the time dimension while simultaneously minimizing computational resource consumption. We showcase the intermediate process of the LMS module in a video demonstrating playing ball, as depicted in Fig. 1. The process involves encapsulating key topological nodes of human motion.

The main contributions in this paper are summarized as follows:

- An innovative approach is proposed to enhance the identification of human actions by capitalizing on a diverse range of modalities. The LMS-GAT algorithm is adept at detecting dynamic nuances inherent in skeleton-based graphs extracted from video data. This cutting-edge technique generates a multi-layered depiction of the input data, thereby representing a paradigm shift in this domain of research.
- To capture the action-specific topologies, the LMS block uses an innovative technique of aggregating and feeding back neighboring node representations. This enhances information exchange and contextual understanding of actions in spatial-temporal dimensions, thus providing a more sophisticated comprehension of human actions.
- The performance of the LMS framework surpasses various baseline models for skeleton-based human action recognition on benchmark datasets such as NTU RGB+D 60, NTU RGB+D 120, and Northwestern-UCLA. As a result, it stands out as a state-of-the-art solution for human action recognition that holds immense potential for future research and development in this area.

2 Related Work

2.1 Skeleton-Based Action Recognition

Recently, GNNs have obtained immense popularity for skeleton-based action recognition tasks. ST-GCN emerged as a key player in this field by bringing Graph Convolutional Networks (GCN) into the picture for extracting spatial data in action recognition tasks. This approach was able to improve accuracy by taking into account the spatial and temporal aspects of the task at hand. However, GCN's limitations in efficiently aggregating information through adjacency matrices resulted in ST-GCN's inability to propagate information effectively for high-order neighboring nodes [13]. In pursuit of improved analysis of human actions, recent studies have explored various methods to fortify cooperative behavior features and enhance the extraction of dynamic temporal information in skeleton-based graphs. For example, AS-GCN [6] introduced an action link to reinforce such features in non-skeleton-connected joints during motion. On the other hand, MS-G3D [14] employed a disentangled multi-scale aggregation scheme to convey information effectively by reducing less significant interactions among nodes on the graph. Despite these approaches' localized focus, they failed to fully capture feature connections on long-range dependencies. To overcome this limitation, ST-TR [15] incorporated self-attention mechanisms to extract structural and temporal representations. However, their methodology did not account for the dynamic nature of temporal information, leading to errors when predicting diverse and coherent actions. In contrast, CTR-GCN and Info-GCN attempted to capture contextual information using context-dependent intrinsic topology modeling and self-attention-based graph convolution. While these methods showed progress, they still inadequately accounted for the dynamic fluctuations inherent in temporal information. This oversight often resulted in incomplete information capture during transitional actions when predicting diverse and coherent changes.

2.2 Dynamic Graph Neural Network

The utility of Recurrent Neural Network (RNN)-based Dynamic Graph Neural Networks (DGNNs) to incorporate temporal information into graph representations has been extensively explored [16,17]. Although these models are highly effective in capturing temporal dynamics over time, their practical implementation is impeded by high computational requirements due to the extensive graph data needed for training. Furthermore, they face scalability challenges when dealing with large temporal dimensions [13]. To alleviate these limitations, Transformer-based DGNNs provide practical strategies for managing temporal information along the temporal dimension. For instance, TGAT [18] incorporates temporal constraints into neighborhood aggregation techniques and integrates temporal topological representations onto continuous graph datasets via a dedicated temporal graph attention layer. Similarly, DySAT [19] uses a simultaneous consideration of structural and temporal information to generate dynamic representations on discrete graph datasets. Meanwhile, Sparse-Dyn [20] employs a sparse temporal transformer to efficiently propagate node representations among patches and relay nodes while harnessing the power of temporal dynamics. Drawing upon the advancements of DGNNs, we have designed the dynamic TEC block that leverages temporal representations in the context of dynamic action-driven skeleton-based graphs. The proposed block enables efficient propagation of temporal information while minimizing computational requirements, thus paving the way for more practical implementations of DGNNs.

3 Methods

Human action recognition in video data presents challenges such as long frame sequence length, resulting in excessive computation and difficulty for models to capture action transition information due to large pose changes during motion. In order to solve these problems, we introduce a lightweight spatio-temporal human action recognition approach, LMS-GAT, to effectively identify special human behavior from videos.

LMS-GAT operates on a representation matrix X: $T \times V \times C$, where T, V, and C indicate the number of video frames, joint points, and the dimension of the input graph G, respectively. First, to be able to better extract spatio-temporal features at multiple levels, the model utilizes a linear layer that boosts the X dimension to 64, thus combining the original data features into more effective features. Subsequent LMS-GAT loops use LMS Block to extract spatio-temporal map features and a 1D convolution with step size and a convolution kernel of 2 is used to downsample the time dimension in the course of the loop. Finally, the output of the features from the LMS Block is fed a linear layer to calculate the action classification probabilities.

In the process mentioned above, the crucial module is the LMS block, which consists of two pivotal components, Spatial Embedding Component (SEC) and Temporal Embedding Component (TEC). Since the human skeletal graph represents $G(V, E)$, with a single connectivity relationship, we use SEC, a special GAT, which uses the learnable parameter A initialized from the adjacency matrix to balance the node information aggregation and better extract the features of the single frame graph data. Following the SEC graph embedding, the model employs TEC to reduce the frame sequence length by dividing the time window into layers and collapsing nodes within these layers. By leveraging hierarchy and incorporating TEC, the LMS-GAT model overcomes challenges associated with long frame sequences and pose changes, resulting in enhanced behavior recognition performance. As shown in Fig. 2, this strategy enhances the model's capacity to capture information on action transitions has been enhanced while improving accuracy and reducing computational effort. More details of both components are discussed in Sect. 3.1 and 3.2.

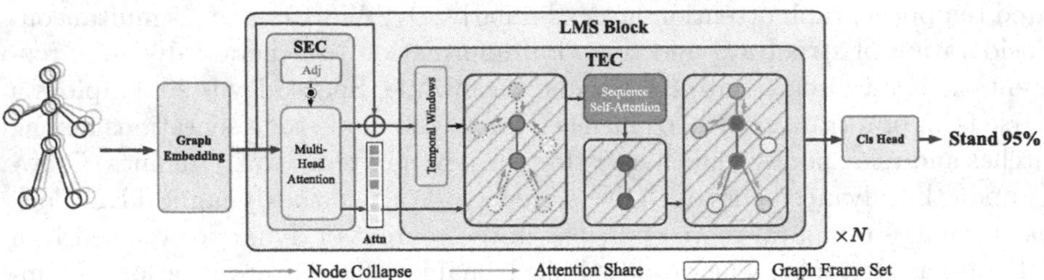

Fig. 2. The whole frame of LMS-GAT. The model is divided into three parts. First, it is the Graph Embedding composed of linear layers for feature enhancement, followed by the LMS Block, in which two sub-modules SEC and TEC the details are discussed in Sects. 3.1 and 3.2. Finally, it is the Classification Head composed of linear layers, which is used for action recognition using the spatio-temporal features extracted by LMS-GAT.

3.1 Spatial Embedding Component

The action video data consists of several frames of graph-structured data. To obtain the spatial information of a single frame, we used a modified GAT to extract features. Slightly different from the traditional GAT, because in behavior prediction, there may not be a physical connection between two joints with interaction, at this time the traditional GAT cannot aggregate the information of the more distant neighbor joints well due to the limitation of the fixed adjacency matrix, so we use Adjacent Matrix to initialize A learnable parameter instead of static Adjacent Matrix in traditional GAT for aggregate node information more efficiently. This allows better learning of the potential topological relationships between joints. The calculation process is as follows

$$\mathrm{H}_t = \mathrm{Softmax}\left(\frac{W_Q X_t \left(W_K X_t\right)^T}{\sqrt{D'}}\right) \cdot A \times X_t \tag{1}$$

where W_Q, W_K is learnable parameters, X is graph embedding of single frames, and A is learnable parameter that initialized from the adjacency matrix.

3.2 Temporal Embedding Component

Behavior recognition videos contain actions with significant posture changes, for which the use of static single-level embedding methods can limit the expressiveness of the model. Moreover, the long time series length also poses performance problems when the entire video frame sequence is used to compute temporal attention. To alleviate this situation, we propose a TEC module. The module reduces the calculated sequence length of calculation by splitting the time window to segment the time attention group and better capture contextual information. At the same time, a dynamic graph node collapse operation is performed within the split window, using the important nodes as the central condensed information, which not only reduces the number of nodes involved in the calculation but also dynamically adapts to the sudden change of actions in different time periods.

First, the model calculates temporal location codes for the input data and then combines this location information with the input data, which allows each frame to retain its location information when performing temporal attention calculations during the split window. Then, in the time dimension of the input data, we will perform two successive windowed self-attentive calculations on the feature sequence. Where the first operation divides the feature sequence into an even number of windows, the windows of the second operation will be shifted backward by half the window size with respect to the position of the first calculation.

Then we divide the feature matrix of the split time window obtained from the TW into k groups for several points using the Attn matrix of inter-node attention obtained from the SEC. By using Eqs. 2, we divide several groups by node importance and similarity, and then average pooling within the group is performed to compress the account of graph nodes. This method can effectively preserve the characteristics of each node while sharply reducing the computational load of the model, Which means that the model can achieve high-quality node feature compression by consuming only a very small amount of resources during computation.

$$Group = Hardmax\left(1/(1+\sqrt{\sum_{i=1}^{Dim}(CN_i - N_i)^2)}\right), CN \in Top_k\left(\sum_{m=1}^{M} SA_m\right) \tag{2}$$

Where the matrix *Group* is a one-hot matrix with dimensions of $(s \times V)$. A value of 1 at position (i, j) in the matrix indicates that the information of the jth joint point is condensed in the ith relay node. CN is nodes after collapse, M represents the number of nodes and SA_m is the attention value of the (m^{th}) node to the other nodes. This grouping of joints allows us to perform average pooling based on the grouping matrix *Group* to obtain the features of the relay nodes.

After calculating the temporal attention of the collapsed graph, we obtain the attention coefficient matrix containing rich temporal information for each node group. The model will recover the nodes according to their grouping at the time of node collapse while propagating the attention coefficient matrix to all nodes in the group, which facilitates the fusion of contextual feature information for multi-node frame sequences. The attention-sharing mechanism maintains the consistency of the input-output dimensions, allowing the module to overlay the model depth to enhance the model representation.

3.3 Action Prediction

Use the final output features of the framework for HAR. To exploit the spatio-temporal information extracted by the model, we average the both spatial and temporal dimensions of the features. These features are then sent into a linear classification head to calculate predicted probabilities for each class of action. LMS-GAT uses the classical cross-entropy loss as the loss function in Eq. 3.

$$L = \frac{1}{N}\sum_i L_i = -\frac{1}{N}\sum_i \sum_{c=1}^{M} y_{ic} \log\left(p_{ic}\right) \tag{3}$$

Where N is the batch size and M is the category account. The true label for the i-th sample indicates the class c, which is a one-hot vector $y_{ic} \in \mathbb{R}^C$, while whose predicted probability is denoted by $\hat{y}_{ic} \in [0, 1]$.

4 Experiments and Discussion

We conduct several experiments to test the classification accuracy of our proposed LMS-GAT on mainstream action recognition datasets. The results are then compared to state-of-the-art methods, while ablation studies are performed to verify the impact of each module.

Table 1. .

Methods	Mode	NTU-RGB+D 60		NTU-RGB+D 120		FLOPs
		X-Sub(%)	X-View(%)	X-Sub(%)	X-View(%)	
ST-TR	4 ensemble	89.9	96.1	82.7	84.7	-
Shift-GCN	4 ensemble	90.7	96.5	85.9	87.6	-
DC-GCN+ADG	4 ensemble	90.8	96.6	86.5	88.1	1.83G
MS-G3D	2 ensemble	91.5	96.2	86.9	88.4	5.22G
CTR-GCN	4 ensemble	92.4	96.8	88.9	90.6	1.97G
Info-GCN	4 ensemble	92.3	96.7	89.2	90.7	1.84G
LMS-GAT	4 ensemble	92.4	96.7	89.1	90.7	1.71G

4.1 NTU RGB+D Dataset

The NTU RGB+D dataset contains two versions (60 and 120) and is an important asset for conducting research on recognizing human actions based on skeletal data. The NTU RGB+D 60 dataset [21] offers 56,880 skeletal action sequences captured from multiple angles using three Microsoft Kinect-V2 depth sensors. It facilitates evaluation based on the X-Sub and X-View criteria, enabling model generalization across individuals and handling variations in camera viewpoints. This dataset provides a comprehensive assessment of model robustness to subject variations and viewpoint changes. In contrast, the NTU RGB+D 120 dataset [22] expands on the 60 dataset by incorporating 57,367 more skeleton sequences for an additional 60 action categories. It introduces 32 setups with diverse backgrounds and locations, broadening the range of evaluation scenarios. The dataset incorporates benchmark evaluations based on X-Sub and X-Setup standards, allowing researchers to assess model generalization across subjects and setups. With its diverse dataset and established assessment procedures, the NTU RGB+D 120 dataset plays a crucial role in advancing skeleton-based action recognition research, facilitating an exploration of various actions, scenarios, and improvements in model performance and generalization capabilities.

4.2 Experiments Setting

LMS-GAT model based on PyTorch [23] framework implementation, each GPU is allocated 128 batches size. The model utilizes the SGD optimizer [24] with a warm-up cosine scheduler. Trained on four NVIDIA RTX 3090 GPUs, some other settings are as follows: initial learning rate is 0.1, weight decay is 4×10^{-4}.

4.3 Evaluation of Human Action Recognition

This section shows the performance comparison of LMS-GAT with leading frameworks on two datasets: NTU RGB+D 60 & 120 with dual views, X-Sub, and X-View. In addition, the LMS- GAT exceeds the precision motion flow of two 4-stream integrations (including Joint Stream + Bone Stream (2S), and 4 Stream ensemble (4S),) as shown in Table 1.

With lower computation and lower FLOPs, it is 0.1% better than the second highest performing InfoGCN in terms of accuracy, and equal to X-Sub, X-View subsets of NTU RGB+D 60, respectively. In the X-Sub, X-View subset of NTU RGB+D 120, only a 0.1% drop, on par with InfoGCN.

4.4 Ablation Study

Table 2. 4-streams results of TEC block configurations on the NTU RGB+D 60 dataset. The modules definition: **B**: Baseline; **TW**: Temporal Window; **NC**: Node Collapse; The **bold** font score with the highest accuracy, while the green color indicates an accuracy improvement compared to the baseline.

Block Setting	Acc(%)	FLOPs
Baseline	91.5	2.22G
B + TW	91.8 +0.3	1.85G −0.37G
B + NC	91.9 +0.4	1.91G −0.31G
B + TW + NC (LMS-GAT)	**92.4** +0.9	**1.71G** −0.51G

This section presents experimental evidence of the effectiveness of dividing time windows and employing graph node collapse techniques for human behavior recognition tasks, as well as their impact on achieving model lightweights. In the baseline model, we do not perform the node collapse operation do not divide the time window, and compute the temporal attention for the full-time frame sequence length for all nodes.

As outlined in Table 2, after adding the time-slicing module, there was a 0.3 increase in accuracy for the baseline model, thanks to the segmentation time window mechanism, which enables the model to have a hierarchical feature extraction capability and better capture the action information at different time scales, while this mechanism reduces the length of the sequences involved in the computation and improves the FLOPs of the model to 1.85G. This is due to the fact that the node collapse operation allows the exchange of information between the nodes during the time-attentive computation, which optimization approach enhanced the model's representation, decreased the number of nodes involved in computations, and resulted in a reduction of approximately 0.37G FLOPs for the model.

4.5 Visualization

Action-Driven Node Distribution. In order to decrease the number of nodes used in the temporal self-attention operation of the model, a spatial node collapse method was employed, we use a spatial node collapse method. As shown in Fig. 3, the model will dynamically group the human skeletal nodes under different frames during the computation, and the white dashed line divides different groups of nodes, and the skeletal nodes within the groups tend to have similar feature distributions. This collapse process is dynamic and lightweight, and the number of groups in each frame will be obtained based on the feature distribution of the nodes. Nodes within the same group will be collapsed in the computation at the end of the grouping step, and the final action prediction and its corresponding confidence score are presented in the text located below the figure.

Fig. 3. Illustration of spatial feature aggregation method we proposed. From top to bottom are the original frame, the skeleton graph of the frame, the node aggregation method, and the prediction. We consider the nodes within circles as a group and aggregate all nodes by group.

5 Conclusion

We propose LMS-GAT, a lightweight spatio-temporal GAT model. The model is hierarchical in structure, and the node collapse operation at different levels makes it adaptively learn the topology of the dynamic discrete skeleton graph corresponding to the corresponding human actions, we also add a time window module to make the model further adaptive in the time dimension. Compared with traditional models using static skeletons, the dynamic skeleton feature of LMS-GAT allows better effective information acquisition during action transitions, especially for behaviors with large changes in posture. Moreover, the node collapse operation and time window division significantly reduce the computational effort of the model, which makes LMS-GAT also have significant advantages in terms of lightweight. Top results were achieved on two benchmark datasets.

References

1. Huang, T., Huang, J., Pang, Y., Yan, H.: Smart contract watermarking based on code obfuscation. Inf. Sci. **628**, 439–448 (2023)
2. Li, J., et al.: Efficient and secure outsourcing of differentially private data publishing with multiple evaluators. IEEE Trans. Dependable Secure Comput. **19**(01), 67–76 (2022)
3. Dong, C.-Z., Catbas, F.N.: A review of computer vision-based structural health monitoring at local and global levels. Struct. Health Monit. **20**(2), 692–743 (2021)
4. Senior, A., et al.: Enabling video privacy through computer vision. IEEE Secur. Priv. **3**(3), 50–57 (2005)
5. Kosch, T., Welsch, R., Chuang, L., Schmidt, A.: The placebo effect of artificial intelligence in human-computer interaction. ACM Trans. Comput.-Hum. Interact. **29**(6), 1–32 (2023)
6. Li, M., Chen, S., Chen, X., Zhang, Y., Wang, Y., Tian, Q.: Actional-structural graph convolutional networks for skeleton-based action recognition. In: Proceedings of the IEEE/CVF Conference on Computer Vision and Pattern Recognition, pp. 3595–3603 (2019)
7. Hao, X., Li, J., Guo, Y., Jiang, T., Yu, M.: Hypergraph neural network for skeleton-based action recognition. IEEE Trans. Image Process. **30**, 2263–2275 (2021)
8. Plizzari, C., Cannici, M., Matteucci, M.: Skeleton-based action recognition via spatial and temporal transformer networks. Comput. Vis. Image Underst. **208**, 103219 (2021)
9. Yang, C., Xu, Y., Shi, J., Dai, B., Zhou, B.: Temporal pyramid network for action recognition. In: Proceedings of the IEEE/CVF Conference on Computer Vision and Pattern Recognition, pp. 591–600 (2020)
10. Yan, S., Xiong, Y., Lin, D.: Spatial temporal graph convolutional networks for skeleton-based action recognition. In: Proceedings of the AAAI Conference on Artificial Intelligence, vol. 32, no. 1 (2018)
11. Chen, Y., Zhang, Z., Yuan, C., Li, B., Deng, Y., Hu, W.: Channel-wise topology refinement graph convolution for skeleton-based action recognition. In: Proceedings of the IEEE/CVF International Conference on Computer Vision, pp. 13359–13368 (2021)
12. Chi, H.-G., Ha, M. H., Chi, S., Lee, S.W., Huang, Q., Ramani, K.: InfoGCN: representation learning for human skeleton-based action recognition. In: Proceedings of the IEEE/CVF Conference on Computer Vision and Pattern Recognition, pp. 20186–20196 (2022)
13. Pang, Y., et al.: Graph decipher: a transparent dual-attention graph neural network to understand the message-passing mechanism for the node classification. Int. J. Intell. Syst. **37**(11), 8747–8769 (2022)
14. Liu, Z., Zhang, H., Chen, Z., Wang, Z., Ouyang, W.: Disentangling and unifying graph convolutions for skeleton-based action recognition. In: Proceedings of the IEEE/CVF Conference on Computer Vision and Pattern Recognition, pp. 143–152 (2020)
15. Plizzari, C., Cannici, M., Matteucci, M.: Spatial temporal transformer network for skeleton-based action recognition. In: Del Bimbo, A., et al. (eds.) ICPR 2021. LNCS, vol. 12663, pp. 694–701. Springer, Cham (2021). https://doi.org/10.1007/978-3-030-68796-0_50
16. Goyal, P., Chhetri, S.R., Canedo, A.: dyngraph2vec: capturing network dynamics using dynamic graph representation learning. Knowl.-Based Syst. **187**, 104816 (2020)

17. Hajiramezanali, E., Hasanzadeh, A., Narayanan, K., Duffield, N., Zhou, M., Qian, X.: Variational graph recurrent neural networks, arXiv preprint arXiv:1908.09710 (2019)
18. Xu, D., Ruan, C., Korpeoglu, E., Kumar, S., Achan, K.: Inductive representation learning on temporal graphs, arXiv preprint arXiv:2002.07962 (2020)
19. Sankar, A., Wu, Y., Gou, L., Zhang, W., Yang, H.: DySAT: deep neural representation learning on dynamic graphs via self-attention networks. In: Proceedings of the 13th International Conference on Web Search and Data Mining, pp. 519–527 (2020)
20. Pang, Y., et al.: Sparse-DYN: sparse dynamic graph multirepresentation learning via event-based sparse temporal attention network. Int. J. Intell. Syst. **37**(11), 8770–8789 (2022)
21. Shahroudy, A., Liu, J., Ng, T.T., Wang, G.: NTU RGB+ D: a large scale dataset for 3D human activity analysis. In: Proceedings of the IEEE Conference on Computer Vision and Pattern Recognition, pp. 1010–1019 (2016)
22. Liu, J., Shahroudy, A., Perez, M., Wang, G., Duan, L.-Y., Kot, A.C.: NTU RGB+ D 120: a large-scale benchmark for 3D human activity understanding. IEEE Trans. Pattern Anal. Mach. Intell. **42**(10), 2684–2701 (2019)
23. Paszke, A., et al.: Pytorch: an imperative style, high-performance deep learning library. In: Advances in Neural Information Processing Systems, vol. 32 (2019)
24. Rumelhart, D.E., Hinton, G.E., Williams, R.J.: Learning representations by back-propagating errors. Nature **323**(6088), 533–536 (1986)

Spatio-Temporal Self-supervision for Few-Shot Action Recognition

Wanchuan Yu[1], Hanyu Guo[1], Yan Yan[1], Jie Li[2], and Hanzi Wang[1](✉)

[1] Fujian Key Laboratory of Sensing and Computing for Smart City, School of Informatics, Xiamen University, Xiamen, China
{wanchuan,guohanyu}@stu.xmu.edu.cn, {yanyan,hanzi.wang}@xmu.edu.cn

[2] Video and Image Processing System Laboratory, School of Electronic Engineering, Xidian University, Xi'an, China
leejie@mail.xidian.edu.cn

Abstract. Few-shot action recognition aims to classify unseen action classes with limited labeled training samples. Most current works follow the metric learning technology to learn a good embedding and an appropriate comparison metric. Due to the limited labeled data, the generalization of embedding networks is limited when employing the meta-learning process with episodic tasks. In this paper, we aim to repurpose self-supervised learning to learn a more generalized few-shot embedding model. Specifically, a Spatio-Temporal Self-supervision (STS) framework for few-shot action recognition is proposed to generate self-supervision loss at the spatial and temporal levels as auxiliary losses. By this means, the proposed STS can provide a robust representation for few-shot action recognition. Furthermore, we propose a Spatio-Temporal Aggregation (STA) module that accounts for the spatial information relationship among all frames within a video sequence to achieve optimal video embedding. Experiments on several challenging few-shot action recognition benchmarks show the effectiveness of the proposed method in achieving state-of-the-art performance for few-shot action recognition.

Keywords: Few-shot learning · Action recognition · Self-supervised learning

1 Introduction

Deep learning has achieved remarkable success in the field of action recognition [10,13,18,22]. The main reason for the significant progress is the sufficiently large-scale labeled training data. However, the time-consuming and costly annotation process renders acquiring adequate data for network training an infrequent occurrence. Hence, recent research has placed greater attention on enhancing the generalization of the model to novel data with limited instances. Similar to the capacity of humans to transfer knowledge from only a few examples, few-shot learning (FSL) shows promise in mitigating data scarcity issues. While

© The Author(s), under exclusive license to Springer Nature Singapore Pte Ltd. 2024
Q. Liu et al. (Eds.): PRCV 2023, LNCS 14425, pp. 84–96, 2024.
https://doi.org/10.1007/978-981-99-8429-9_7

recent few-shot classification has made significant progress for images, progress in video classification has remained unsolved. Few-shot action recognition (FASR) is much more complicated due to the additional temporal dimension. Besides, video actions have different characteristics in terms of speed, duration, and occurrence scenarios.

To deal with the FASR problem, most existing metric-based FASR methods simply calculate the similarity between the embedding of a support class and a query video. Moreover, these methods mainly use frame-level embeddings [1,2], clip-level embeddings [12,16], or patch-level embeddings [24] for temporal alignment to obtain accurate video matching. Despite significant effects, these methods still need to address the key challenge of improving the generalization of the learned few-shot embedding model. Besides, unconstrained learning from training data will lead to the inductive bias of source classes and weaken the generalization performance of embeddings. ARN [23] has recently used spatial and temporal self-supervision to train a more robust encoder and attention. However, this method alone self-supervises support videos and query videos without fully capitalizing on their inherent connection, which is more suitable for few-shot learning. Therefore, using support and query videos together for self-supervision can help narrow the distances between the same categories and map instances of different categories to different clusters.

To address the above problems, a spatial-temporal self-supervision (STS) framework using self-supervised learning for few-shot action recognition is proposed. To be specific, we first propose a spatial cross self-supervision module (SCS) based on the spatial scale to enhance patch representations by establishing correlations between patches at different locations. This module effectively addresses the issue of key patches impacting correlation establishment in various video scenarios caused by displacement and indentation. Secondly, we develop a temporary cross self-supervision (TCS) module based on a temporary scale to fully enhance the temporality of the video, which can solve the problem of misclassification due to similar directionality of videos (e.g.,"moving something away from the camer" vs. "moving something towards the camer"). Moreover, the spatio-temporal aggregation (STA) module is utilized to aggregate video representations along the spatial and temporal dimensions, reducing the emphasis on a specific frame during temporal matching. Our model achieves competitive performance on several action recognition benchmarks: Something v2 (SSv2) [5], Kinetics [7], HMDB51 [8], and UCF101 [15].

2 Related Work

2.1 Few-Shot Action Recognition

CMN [25] proposes a memory network structure to obtain an optimal video representation and a multi-saliency embedding algorithm to encode a variable-length video sequence into a fixed-size representation. OTAM [2] proposes a dynamic time-warping algorithm to enhance long-term temporal ordering information by ordered temporal alignment. ARN [23] constructs a C3D encoder to capture short-term dependencies and leverage permutation-invariant pooling to learn discriminative action representations. The recent method TRX [12] compares the query samples to sub-sequences of all support samples with an attention mechanism to construct query-specific class prototypes for few-shot matching and achieves promising results. STRM [16] proposes spatio-temporal enrichment and temporal relationship modeling modules to measure query-class similarity. In contrast to previous works, our algorithm uses a self-supervised approach to construct separate spatial and temporal pretext tasks of the model, enhancing the generalization to novel classes of spatio-temporal modeling.

2.2 Self-supervised Learning (SSL)-Based Few-Shot Learning

The success of contrastive learning approaches like SimCLR [3] and MoCo [6] shows that the feature extraction network trained using self-supervised learning can have a robust representational capacity. Few-shot learning works [9,14] have achieved better results by combining the FSL framework with well-designed auxiliary self-supervised pretext tasks. This indicates that such methods can facilitate the transferability of learned feature representations. SLA [9] augments original labels through self-supervision of input transformation to relax invariant constraints during simultaneous learning of the original and self-supervised tasks. More recently, ESPT [14] proposes a new type of self-supervised pretext task for few-shot image classification that uses relations between local spatial features of multiple image samples in each episode to construct a supervision signal. However, most SSL augmented few-shot learning methods are not used in action recognition. Our algorithm is specifically designed for few-shot action recognition. By separately constructing self-supervised objective functions at the spatial and temporal levels, we combine them with the objective function of the original task to optimize the model parameters.

3 Method

3.1 Problem Definition

In the few-shot action recognition task, the goal is to classify an unlabeled video (query set) into one of the several classes represented by a limited number of labeled video samples (support set) that have not been seen during training. To this end, videos in a dataset are divided into two sets with disjoint classes: the meta-training set D_{train} and the meta-testing set D_{test}, i.e., $C_{train} \cap C_{test} = \emptyset$. Then, following previous work [2,23,25] using the episodic training strategy [17] to optimize the model with a meta-training set D_{train}. For each episode, we randomly sample N action classes each with K videos from D_{train} to construct the support set $S = \{(x_s, y_s), y_s \in C_{train}, s = 1, \ldots, n \times k\}$. And the query set $S = \{(x_q, y_q), y_q \in C_{train}, s = 1, \ldots, n \times p\}$ sampled from the rest of the videos of the N selected classes. To perform meta-learning, S and Q are completely disjoint, i.e., $S \cap Q = \emptyset$. Specifically, we use a large number of episodic tasks sampled from D_{train} for training to adapt to D_{test}. During the inference phase, episodic tasks are sampled on the D_{test} in a similar way as meta-training.

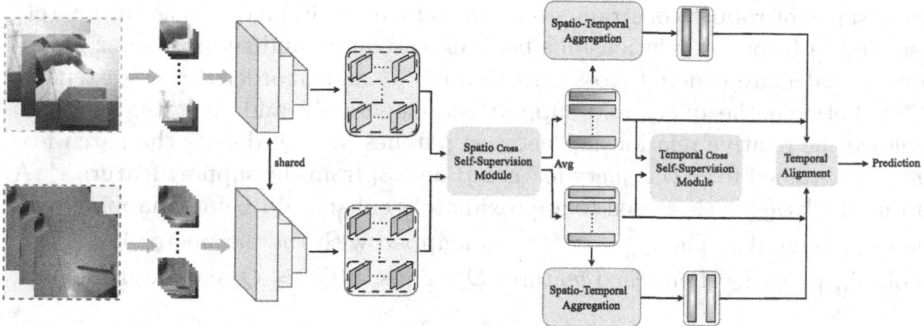

Fig. 1. Illustration of the proposed overall framework. Embedded video features are first fed into the spatial cross self-supervision (SCS) module to enhance patch representations. After that, the temporary cross self-supervision (TCS) module is used to boost the temporality of the video. Moreover, the spatiotemporal aggregation (STA) module incorporates the spatial relationships between frames to obtain optimal video embedding. Finally, we compute the similarity scores to make the final prediction.

3.2 Spatio-Temporal Self-supervision Framework

Overview. Figure 1 illustrates the overview of the proposed STS framework. The input support videos S and query videos Q with T frames are passed through an embedding network (e.g., ResNet-50) to extract support features $F_s = \{s_1, s_2, \ldots, s_{n\times k}\}$ and query features $F_q = \{q_1, q_2, \ldots, q_{n\times p}\}$, where $s_i = \{s_i^1, s_i^2, \cdots, s_i^T\}, q_i = \{q_i^1, q_i^2, \cdots, q_i^T\}$ and $s_i^j, q_i^j \in R^{P^2\times D}$. Each frame feature consists of $P \times P$ patch features with dimension D.

Spatial Cross Self-supervision Module. The position of the notable patches within a frame vary across different video scenes and motion postures. Thus, enhancing the correlation between associated patches is advantageous for capturing the precise frame-level appearance. Given the support feature F_s and query feature F_q, as illustrated in Fig. 2(a), we initially leverage self-attention to capture spatial relationships among patches within a frame. Let $x_i \in \mathbb{R}^{P^2\times D}$ denote the patch features s_i^j, q_i^j of a frame p_j ($j \in [1, T]$), where $P \times P$ is the number of patches. Then we use weights $W_q, W_k, W_v \in \mathbb{R}^{D\times D}$ to map patch features to x_i^q, x_i^k, and x_i^v, where $[x_i^q; x_i^k; x_i^v] = [W_q x_i; W_k x_i; W_v x_i]$ and D is the dimension of the input patch features. The attention matrix is computed by the dot-products between the query and key matrices. Then the value matrix and attention matrix are dot-products to reweight the correlations among all patches:

$$z_i = \lambda(\frac{x_i^q x_i^{k\top}}{D})x_i^v + x_i \tag{1}$$

where λ denotes the softmax function. Although this attention mechanism can establish correlations between patches, it cannot capture the relative positions and relationships of noteworthy patches in different video scenes. Meanwhile, the absence of robust constraints may introduce the induction bias, ultimately resulting in incorrect relationships between the query and support sets. Therefore, we calculate ordered cross-attention a^{space} and unordered cross-attention $\hat{a}^{space}$ between the query and support sets separately and then use $\hat{a}^{space}$ to enhance the relative relationship between patches. Let Q_i denote the patch features z_i obtained from the query feature q_i^j and S_i from the support feature s_i^j. A sub-network $\phi(\cdot)$ is then used to approximately enhance Q_i before mapping with the parameter W_q. The $Q_i^{\mathsf{T}} \in \mathbb{R}^{D \times P^2}$ is mapped with the parameter $W_p \in \mathbb{R}^{P^2}$ to obtain patch-level enriched features $\hat{Q}_i$. Thus, $\phi(Q_i) = \hat{Q}_i$ can be defined as:

$$\hat{Q}_i = \sigma(Q_i^{\mathsf{T}} W_p)^{\mathsf{T}} + Q_i \tag{2}$$

where σ denotes the ReLU non-linearity. After that, Q_i^q and S_j^k are mapped from weights $\hat{W}_q, \hat{W}_k \in \mathbb{R}^{D \times D}$, where $[Q_i^q; S_j^k] = [\hat{W}_q \hat{Q}_i; \hat{W}_k S_j]$. Let $o^s \in \mathbb{N}^{P^2}$ represent randomly shuffling the order of patches, then the ordered cross-attention $a_{(i,j)}^{space}$ and the unordered cross-attention $\hat{a}_{(i,j)}^{space}$ can be defined as:

$$a_{(i,j)}^{space} = Q_i^q S_j^k \tag{3}$$

$$\hat{a}_{(i,j)}^{space} = \nu(\phi(Q_i^q), o^s) S_j^k \tag{4}$$

where $\nu(x, o^s)$ is a function that shuffles the patches in x based on the order o^s. For instance, let o^s be the shuffle order $[2, 1, 3]$. In this case, the first patch of $\hat{Q}_i$ is obtained by enhancing the second patch of Q_i. Therefore, we can propose a spatial cross self-supervised loss to enhance the correlation between related patches in a frame, and the loss can be defined as:

$$\mathcal{L}_{spa}^{self} = \frac{1}{TN^2} \sum_{t=1}^{T} \sum_{k=1}^{N} \sum_{p=1}^{N} (1 - \hat{a}_{(k,p)}^{space-t}) \varepsilon(a_{(o^s(k),p)}^{space-t}) \tag{5}$$

where $\varepsilon(a) = 1$ if $a > \theta$, otherwise $\varepsilon(a) = -1$. The default setting for the judgment value θ is the P-th largest value in a. By employing ordered cross attention to guide unordered cross attention, we effectively enhance spatial connections between patches, thereby facilitating the utilization of a reweighted value embedding to generate more discriminative support class-specific embedding. The generation process can be defined as:

$$\alpha_i = \lambda(\frac{a_i}{D}) S_i^v \tag{6}$$

where S_i^v is mapped from weights $\hat{W}_v \in \mathbb{R}^{D \times D}$, the query features Q_i^v also uses the same weights. Let $[Q_i^v; S_i^v] = [\hat{W}_v Q_v; \hat{W}_v S_v]$. Then we calculate the distances between spatial patches point-to-point based on the ground truth label between query features and support class-specific features to define the cross-entropy loss $\mathcal{L}_{spa}$ at the spatial level.

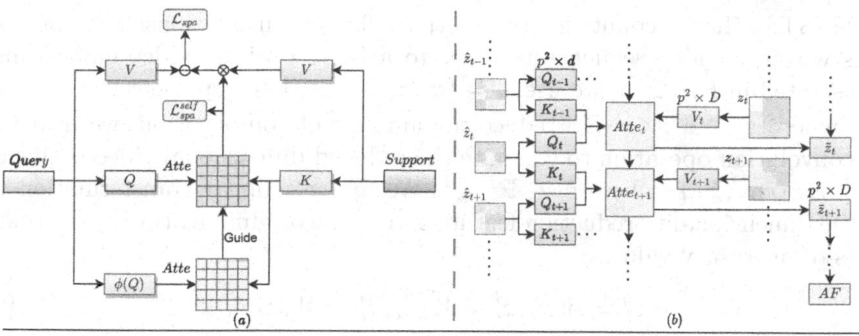

Fig. 2. (a) The spatial cross self-supervision (SCS) module, where the output self-supervision loss $\mathcal{L}_{spa}^{self}$ and the spatial cross-entropy loss $\mathcal{L}_{spa}$ are calculated by guiding the original cross-attention matrix using the disrupted cross-attention matrix. (b) The spatio-temporal aggregation (STA) module, where the output feature AF is aggregated by all the attention weights, which are computed by the cross-attention matrices for two adjacent frames in succession.

Temporal Cross Self-supervision Module. The temporary orders play an important role in the video, but using temporary matching directly does not reveal the difference, such as "moving something away from the camera" vs. "moving something towards the camera". Thus, learning about temporary orders is beneficial for processing more detailed temporal relationships. Similar to spatial processing, we use the Q_i^q, Q_i^v, S_i^k, and S_i^v for temporal cross self-supervision. Then spatial global-average pooling is applied to collapse the spatial dimension. Let $o^t \in \mathbb{N}^N$ represent randomly shuffling the order of video frames, and then the unordered cross-attention $\hat{a}_{(i,j)}^{time}$ can be defined as:

$$\hat{a}_{(i,j)}^{time} = \nu(\phi(\psi(Q_i^q)), o^t)\psi(S_j^k) \quad (7)$$

where $\nu(x, o^t)$ is a function that disrupts the video frames in x based on the order o^t, $\psi(\cdot)$ represents a spatial pooling function. Therefore, we can propose a temporal cross self-supervised loss to learn the temporary orders, and the loss can be defined as:

$$\mathcal{L}_{time}^{self} = \frac{1}{T^2}\sum_{t=1}^{T}\sum_{t'=1}^{T}(1 - \hat{a}_{(t,t')}^{time})\zeta((o^t(t), t')) \quad (8)$$

where $\zeta(i, j) = 1$ if $i > j$, otherwise $\zeta(i, j) = -1$. By supervising the order of videos, the directionality between frames can be enhanced to generate more discriminative temporal features.

Spatio-Temporal Aggregation Module. Many previous works [1,2] tend to ignore the long-term temporal relationships existing in the sequence of video. To address this limitation, our approach introduces a spatio-temporal aggregation module (STA) that accounts for the spatial information relationship among all frames within a video sequence in order to achieve optimal video embedding. As illustrated in Fig. 2(b), we use $Z = \{z_1, z_2, \ldots, z_T\}$ to represent the original input, where $z_i \in \mathbb{R}^{P^2 \times D}$. To reduce the amount of computation, we utilize a 1×1 convolution operation to compute the reduced dimensional video sequence $\hat{Z} = \{\hat{z}_1, \hat{z}_2, \ldots, \hat{z}_T\}$, where $\hat{z}_i \in \mathbb{R}^{P^2 \times d}$. We apply a linear transformation to both the dimensionality reduction feature $\hat{z}_t$ and the original feature z_t to obtain triplets of query-key-values:

$$z_t^q = \widetilde{W}_q \hat{z}_t, z_t^k = \widetilde{W}_k \hat{z}_t, z_t^v = \widetilde{W}_v z_t \tag{9}$$

where $\widetilde{W}_q$, $\widetilde{W}_k$, $\widetilde{W}_q$ denote the weight of the linear transformation layer. In a similar manner, other frame features are processed. The spatial relationship between adjacent frames is captured via the computation of the attention matrix of frame $t+1$ using the dot product between the query matrix of frame $t+1$ and the key matrix of frame t.

$$Atte_{t+1} = \frac{z_{t+1}^q {z_t^k}^\mathsf{T}}{d} \tag{10}$$

However, Eq. (10) only accounts for the interaction between two adjacent frames. To capture the spatial relationships of all frames effectively, we combine all the preceding attention matrices leading up to a specific frame t. This aggregation facilitates to compute the spatio-temporal characteristics of frame $t+1$:

$$\tilde{z}_{t+1} = \lambda(\sum_{i=1}^{t+1} Atte_i) z_{t+1}^v \tag{11}$$

where λ denotes the softmax function. Finally, the spatial-temporal aggregation feature $\tilde{Z}$ and the original feature Z are simply pooled in the spatio-temporal dimension and added together to obtain the final feature AF:

$$AF = \frac{\sum_{i=1}^{T} \psi(\tilde{z}_i)}{T} + \frac{\sum_{j=1}^{T} \psi(z_i)}{T} \tag{12}$$

where $\psi(\cdot)$ represents a spatial pooling function. Simply by calculating the Euclidean distance between the aggregated features AF, we can obtain the distance D_c. However, only using this strategy will fail to capture some fine-grained action information. To address this, we use the existing fine-grained distance function TRX [12] to calculate the fine-grained distance D_f. Finally, the global distance D_g can be expressed as follows:

$$D_g = D_c + D_f \tag{13}$$

Then we calculate the distances based on the ground truth label using the video-to-class distance D_g to define the cross-entropy loss $\mathcal{L}_g$ as the main loss. With ω_i as hyper-weights, our STS is trained using the joint formulation given by:

$$\mathcal{L} = \omega_1 \mathcal{L}_g + \omega_2 \mathcal{L}_{spa} + \omega_3 \mathcal{L}_{spa}^{self} + \omega_4 \mathcal{L}_{time}^{self} \tag{14}$$

4 Experiments

4.1 Experimental Settings

Datasets. We evaluate our method on four widely used datasets, including Something v2 (SSv2) [5], Kinetics [7], HMDB51 [8], and UCF101 [15] for few-shot action recognition. For Kinetics, we follow the splits in CMN [25] to select 100 action classes from Kinetics-400, which contains various activities in daily life and is rich in scene context. The 100 classes with 100 video clips per class are divided into 64, 12, and 24 for training, validation, and testing. For SSv2, we follow the two widely used splits denoted as SSv2† and SSv2*, proposed by [26] and [2] respectively. Both splits adopt 64, 24, and 12 non-overlapping action classes as the training set, validation set, and testing set. But compared with SSv2†, the training set of SSv2* uses approximately 10x videos per class. For UCF101, we use the splits from [23], which sample 70, 10, and 21 non-overlapping action classes as the training set, validation set and testing set. For HMDB51 with 51 classes with at least 101 video clips per class, we also use the split from [23] and select 31 training, 10 validation, and 10 testing classes.

Implementation Details. We follow the sparse sampling strategy described in TSN [19], which divides each input video into $N = 8$ segments and then randomly samples one frame in each segment. We resize the each frame scale into 224×224. Then we use ResNet-50 pretrained on ImageNet as the feature extractor. With $D = 2048$, an adaptive max pooling operation reduces the spatial resolution to P, where $P = 4$. During training, the weight of L_g, L_{spa}, L_{spa}^{self}, and L_{time}^{self} is set to 1, 0.5, 0.1, and 0.1, respectively. We train our model for 75,000 randomly sampled training episodes for SSv2* and SSv2† dataset with a learning rate of 1×10^{-4}. For the other three datasets, we set the learning rate to 1×10^{-3} and trained for 50,000 episodes. To evaluate few-shot performance on each benchmark, we randomly construct 10,000 episodes from the test set and report the average classification accuracy.

4.2 Comparison with State-of-the-Art Methods

In Table 1, we compare our method with state-of-the-art algorithms on Kinetics, SSV2†, SSV2*, UCF101, and HMDB51. On the five datasets, we conduct experiments under 5-way 5-shot settings.

Results on Kinetics. Table 1 shows that our model significantly outperforms all competing methods under 5-shot settings. For instance, our STS achieves new state-of-the-art results with 87.5%. Compared with current state-of-the-art methods, such as TRX [12] and STRM [16], our STS outperforms these methods by 1.7% and 0.9% under the 5-shot setting, respectively. This demonstrates that the attributes of the spatio-temporal self-supervision framework surpass these traditional spatio-temporal modeling methods.

Results on SSV2. We also evaluate the proposed STS on the SSV2 dataset, which is more complex in temporal reasoning. The gains of our STS on SSv2 are more evident, further demonstrating the advantages of our temporary self supervision on this dataset. Our method achieves +2.0% improvements on SSV2† compared with STRM [16] under the 5-shot settings. For the SSV2*, which includes more training data, our method achievements +1.1% improvement in the 5-shot settings. These results indicates that more training data can generate more discriminative embeddings through self -supervision.

Table 1. Comparison with state-of-the-art methods on Kinetics, SSV2†, SSV2*, UCF101, and HMDB51 in terms of 5-shot classification accuracy. "-" stands for the result is not available in published works. The best results are in bold.

Method	Kinetics	SSV2†	SSV2*	UCF101	HMDB51
MAML [4]	75.3	41.9	-	-	-
CMN [25]	78.9	-	-	-	-
TARN [1]	78.5	-	-	-	-
OTAM [2]	85.8	48.0	52.3	88.9	68.0
TRX [12]	85.9	**59.1**	64.6	96.1	75.6
MTFAN [21]	87.4	-	60.4	95.1	74.6
STRM [16]	86.7	55.3	68.1	96.9	77.3
HyRSM [20]	86.1	56.1	69.0	94.7	76.0
Nguyen *et al.* [11]	87.4	-	61.1	95.9	76.9
HCL [24]	85.8	55.4	64.9	93.9	76.3
Ours	**87.5**	57.3	**69.2**	**97.1**	**77.5**

Results on UCF101 and HMDB51. In order to further verify our STS, we also compare it with state-of-the-art methods on the UCF101 and HMDB51 datasets, whose data is simpler compared with Kinetics and SSv2. And the results are shown in Table 1. Our method improves over TRX [12] on HMDB51 e.g., +1.9% for the 5-shot settings. Similarly, STS improves over TRX on UCF101 e.g., +1.0% for the 5-shot settings.

4.3 Ablation Studies

Influence of the Different Training Losses in the STS Framework. To examine the effect of various training losses in our proposed STS, we performed ablation studies on the SSV2†, Kinetics, and HMDB51 datasets under the 5-way 5-shot setting. The results are shown in Table 2. The coarse-grained loss $\mathcal{L}_c$ primarily focuses on the similarity between the aggregated video embedding, while the fine-grained loss $\mathcal{L}_f$ mainly considers the similarity between the clip-level embedding, the total loss is given by $\mathcal{L}_g = \mathcal{L}_c + \mathcal{L}_f$. As evidenced in lines

1-3, adopting both coarse-grained and fine-grained classification losses without a doubt leads to better outcomes than just relying on $\mathcal{L}_c$ or $\mathcal{L}_f$. Since $\mathcal{L}_f$ is more suited to the 5-way 5-shot setup, it obtains considerably better results than $\mathcal{L}_c$. On the Kinetics and HMDB51 datasets, incorporating a distance loss $\mathcal{L}_{spa}$ between spatial patches point-to-point effectively enhances the model performance. Self-supervised learning of temporal and spatial features has demonstrated noticeable benefits, as shown in rows 5 and 6 of Table 2, respectively. Integrating self-supervision in both spatial and temporal domains can significantly improve the generality of the acquired embedding. Finally, we achieve the best results by combining $\mathcal{L}_g + L_{spa} + \mathcal{L}_{spa}^{self} + \mathcal{L}_{time}^{self}$ in row 6.

Table 2. Influence of various training losses in the STS framework on Kinetics, HMDB51, and SSV2† under the 5-way 5-shot setup. The best result are in bold.

	STS						Kinetics	HMDB51	SSV2†
	$\mathcal{L}_c$	$\mathcal{L}_f$	$\mathcal{L}_g$	$\mathcal{L}_{spa}$	$\mathcal{L}_{spa}^{self}$	$\mathcal{L}_{time}^{self}$	5-shot	5-shot	5-shot
1	✓	×	×	×	×	×	84.3	68.5	47.5
2	×	✓	×	×	×	×	85.0	75.4	55.1
3	×	×	✓	×	×	×	85.5	75.9	55.9
4	×	×	✓	✓	×	×	86.3	76.5	56.2
5	×	×	✓	✓	✓	×	87.1	77.2	56.5
6	×	×	✓	✓	✓	✓	**87.5**	**77.5**	**57.3**

Results on Using only the TRX Loss Under the 5-Way 1-Shot Setting. Table 3 shows the performance of our STS framework only based on TRX [12] alignment metrics in terms of 1-shot setting. TRX is designed for 5-shot, so its performance is not ideal under the 1-shot setting. To prove the validity of our model, we removed the coarse-grained loss $\mathcal{L}_c$ and kept only the fine-grained loss $\mathcal{L}_f$, i.e., the TRX loss, and compared the performance of TRX and its subsequent improvement STRM [16] under the 5-way 1-shot setting.

Table 3. Results on using only the TRX loss on Kinetics, SSV2†, SSV2*, UCF101, and HMDB51 under the 5-way 1-shot setup. The best results are in bold.

Method	Kinetics	SSV2†	SSV2*	UCF101	HMDB51
TRX [12]	63.6	36.0	42.0	78.2	53.1
STRM [16]	62.9	37.1	43.1	80.5	52.3
STS($\mathcal{L}_f$)	64.3	37.9	43.7	81.0	54.8
STS($\mathcal{L}_f$, $\mathcal{L}_c$)	**65.1**	**38.2**	**44.0**	**81.6**	**55.6**

5 Conclusions

In this paper, we propose a novel Spatio-Temporal Self-supervision (STS) framework for few-shot action recognition that consists of a spatial cross self-supervision (SCS) module, a temporary cross self-supervision (TCS) module, and a spatio-temporal aggregation (STA) module. The SCS and TCS modules for few-shot action recognition are proposed to generate self-supervision loss at the spatial and temporal levels as auxiliary losses to facilitate the transferability of learned feature representations. The STA module accounts for the spatial information relationship among all frames within a video sequence to achieve optimal video embedding. Extensive experiments on five commonly used benchmarks verify the effectiveness of our method and demonstrate that STS achieves state-of-the-art performance under the 5-shot setting.

Acknowledgements. This work was supported by the National Key Research and Development Program of China under Grant 2022ZD0160402, by the National Natural Science Foundation of China under Grant U21A20514, 62176195, and Grants 62372388, 62071404, and by the FuXiaQuan National Independent Innovation Demonstration Zone Collaborative Innovation Platform Project under Grant 3502ZCQXT2022008.

References

1. Bishay, M., Zoumpourlis, G., Patras, I.: Tarn: temporal attentive relation network for few-shot and zero-shot action recognition. arXiv preprint arXiv:1907.09021 (2019)
2. Cao, K., Ji, J., Cao, Z., Chang, C.Y., Niebles, J.C.: Few-shot video classification via temporal alignment. In: Proceedings of the IEEE/CVF Conference on Computer Vision and Pattern Recognition, pp. 10618–10627 (2020)
3. Chen, T., Kornblith, S., Norouzi, M., Hinton, G.: A simple framework for contrastive learning of visual representations. In: International Conference on Machine Learning, pp. 1597–1607. PMLR (2020)
4. Finn, C., Abbeel, P., Levine, S.: Model-agnostic meta-learning for fast adaptation of deep networks. In: International Conference on Machine Learning, pp. 1126–1135. PMLR (2017)
5. Goyal, R., et al.: The "something something" video database for learning and evaluating visual common sense. In: Proceedings of the IEEE International Conference on Computer Vision, pp. 5842–5850 (2017)
6. He, K., Fan, H., Wu, Y., Xie, S., Girshick, R.: Momentum contrast for unsupervised visual representation learning. In: Proceedings of the IEEE/CVF Conference on Computer Vision and Pattern Recognition, pp. 9729–9738 (2020)
7. Kay, W., et al.: The kinetics human action video dataset. arXiv preprint arXiv:1705.06950 (2017)
8. Kuehne, H., Jhuang, H., Garrote, E., Poggio, T., Serre, T.: HMDB: a large video database for human motion recognition. In: 2011 International Conference on Computer Vision, pp. 2556–2563. IEEE (2011)
9. Lee, H., Hwang, S.J., Shin, J.: Self-supervised label augmentation via input transformations. In: International Conference on Machine Learning, pp. 5714–5724. PMLR (2020)

10. Liu, Z., et al.: Video swin transformer. In: Proceedings of the IEEE/CVF Conference on Computer Vision and Pattern Recognition, pp. 3202–3211 (2022)
11. Nguyen, K.D., Tran, Q.H., Nguyen, K., Hua, B.S., Nguyen, R.: Inductive and transductive few-shot video classification via appearance and temporal alignments. In: Avidan, S., Brostow, G., Cissé, M., Farinella, G.M., Hassner, T. (eds.) ECCV 2022. LNCS, vol. 13680, pp. 471–487. Springer, Cham (2022). https://doi.org/10.1007/978-3-031-20044-1_27
12. Perrett, T., Masullo, A., Burghardt, T., Mirmehdi, M., Damen, D.: Temporal-relational crosstransformers for few-shot action recognition. In: Proceedings of the IEEE/CVF Conference on Computer Vision and Pattern Recognition, pp. 475–484 (2021)
13. Qiu, Z., Yao, T., Ngo, C.W., Tian, X., Mei, T.: Learning spatio-temporal representation with local and global diffusion. In: Proceedings of the IEEE/CVF Conference on Computer Vision and Pattern Recognition, pp. 12056–12065 (2019)
14. Rong, Y., Lu, X., Sun, Z., Chen, Y., Xiong, S.: ESPT: a self-supervised episodic spatial pretext task for improving few-shot learning. arXiv preprint arXiv:2304.13287 (2023)
15. Soomro, K., Zamir, A.R., Shah, M.: UCF101: a dataset of 101 human actions classes from videos in the wild. arXiv preprint arXiv:1212.0402 (2012)
16. Thatipelli, A., Narayan, S., Khan, S., Anwer, R.M., Khan, F.S., Ghanem, B.: Spatio-temporal relation modeling for few-shot action recognition. In: Proceedings of the IEEE/CVF Conference on Computer Vision and Pattern Recognition, pp. 19958–19967 (2022)
17. Vinyals, O., Blundell, C., Lillicrap, T., Wierstra, D., et al.: Matching networks for one shot learning. In: Advances in Neural Information Processing Systems, vol. 29 (2016)
18. Wang, L., Zhu, S., Li, Z., Fang, Z.: Complementary temporal classification activation maps in temporal action localization. In: Ma, H., et al. (eds.) PRCV 2021. LNCS, vol. 13020, pp. 373–384. Springer, Cham (2021). https://doi.org/10.1007/978-3-030-88007-1_31
19. Wang, L., et al.: Temporal segment networks: towards good practices for deep action recognition. In: Leibe, B., Matas, J., Sebe, N., Welling, M. (eds.) ECCV 2016. LNCS, vol. 9912, pp. 20–36. Springer, Cham (2016). https://doi.org/10.1007/978-3-319-46484-8_2
20. Wang, X., et al.: Hybrid relation guided set matching for few-shot action recognition. In: Proceedings of the IEEE/CVF Conference on Computer Vision and Pattern Recognition, pp. 19948–19957 (2022)
21. Wu, J., Zhang, T., Zhang, Z., Wu, F., Zhang, Y.: Motion-modulated temporal fragment alignment network for few-shot action recognition. In: Proceedings of the IEEE/CVF Conference on Computer Vision and Pattern Recognition, pp. 9151–9160 (2022)
22. Xie, S., Sun, C., Huang, J., Tu, Z., Murphy, K.: Rethinking spatiotemporal feature learning: Speed-accuracy trade-offs in video classification. In: Proceedings of the European Conference on Computer Vision (ECCV), pp. 305–321 (2018)
23. Zhang, H., Zhang, L., Qi, X., Li, H., Torr, P.H.S., Koniusz, P.: Few-shot action recognition with permutation-invariant attention. In: Vedaldi, A., Bischof, H., Brox, T., Frahm, J.-M. (eds.) ECCV 2020. LNCS, vol. 12350, pp. 525–542. Springer, Cham (2020). https://doi.org/10.1007/978-3-030-58558-7_31

24. Zheng, S., Chen, S., Jin, Q.: Few-shot action recognition with hierarchical matching and contrastive learning. In: Avidan, S., Brostow, G., Cissé, M., Farinella, G.M., Hassner, T. (eds.) ECCV 2022. LNCS, vol. 13664, pp. 297–313. Springer, Cham (2022). https://doi.org/10.1007/978-3-031-19772-7_18
25. Zhu, L., Yang, Y.: Compound memory networks for few-shot video classification. In: Proceedings of the European Conference on Computer Vision (ECCV), pp. 751–766 (2018)
26. Zhu, L., Yang, Y.: Label independent memory for semi-supervised few-shot video classification. IEEE Trans. Pattern Anal. Mach. Intell. **44**(1), 273–285 (2020)

A Fuzzy Error Based Fine-Tune Method for Spatio-Temporal Recognition Model

Jiulin Li, Mengyu Yang, Yang Liu, Gongli Xi, Lanshan Zhang, and Ye Tian(✉)

State Key Laboratory of Networking and Switching Technology,
Beijing University of Posts and Telecommunications, Beijing, China
{lijiulin,mengyuyang,lycs,kevinxgl,zls326,yetian}@bupt.edu.cn

Abstract. The spatio-temporal convolution model is widely recognized for its effectiveness in predicting action in various fields. This model typically uses video clips as input and employs multiple clips for inference, ultimately deriving a video-level prediction through an aggregation function. However, the model will give a high confidence prediction result, regardless of whether the input clips have sufficient spatio-temporal information to indicate its class or not. The inaccurate high confidence prediction errors can subsequently affect the accuracy of the video-level results. Although the current approach to mitigating this problem involves increasing the number of clips used, it fails to address this problem from its root causes. To solve this issue, we propose a fine-tuning framework based on Fuzzy error loss, aimed at further refining the well-trained spatio-temporal convolution model that relies on dense sampling. By giving a low confidence prediction output for clips with insufficient spatio-temporal information, our framework strives to enhance the accuracy of video-level motion recognition. We conducted extensive experiments on two motion recognition datasets, namely UCF101 and Kinetics-Sounds, to evaluate the effectiveness of our proposed framework. The results indicate a significant improvement in motion recognition accuracy at the video level on both data sets.

Keywords: Action recognition · Fine-tuning method · Spatio-temporal convolution

1 Introduction

The exponential growth of online video content necessitates the development of efficient and accurate systems for analyzing video content. In the realm of action recognition, the utilization of Spatio-temporal Convolutional Neural Network (3D CNN) models has emerged as aprominent research area [1–3]. Researchers have continuously proposed increasingly efficient and lightweight 3D CNN architectures in response to evolving demands [4–8]. Dense sampling and uniform sampling are two main sampling strategies, and using different sampling strategies will greatly affect the capability of the model. Traditionally, 3D CNN models

© The Author(s), under exclusive license to Springer Nature Singapore Pte Ltd. 2024
Q. Liu et al. (Eds.): PRCV 2023, LNCS 14425, pp. 97–108, 2024.
https://doi.org/10.1007/978-981-99-8429-9_8

have been designed to accommodate dense sampling methods during training. Also, recent studies [9] have revealed that the adoption of uniform sampling, as an alternative to dense sampling, can impact the upper limit of 3D CNN performance. Dense sampling involves extracting consecutive multi-frames from a single viewpoint within a video, with the input comprising short video clips that represent small segments. To effectively harness the spatio-temporal information embedded in videos, the classification results from multiple clips are uniformly combined and averaged to yield the final video-level classification outcome.

Therefore, for the spatio-temporal convolution models designed based on dense sampling strategy, the clip used in the inference process is only a short segment of the original video. Unfortunately, not all clips contain sufficient spatio-temporal information, leading to misjudgments by the classification model. However, the model often does not give an uncertain prediction result for misleading input, but gives a high confidence wrong answer. The stochastic distribution of video content further amplifies the limitations of aggregation-based classification models. Although increasing the number of clips used for inference has been attempted as a potential solution, it has not completely addressed this issue. Experimental investigations [10] into confidence separation also have revealed that improved performance does not solely rely on using a greater number of clips. Instead, the key lies in how to reduce the negative impact of the wrong predictions at the result aggregation stage. The inclusion of misleading results from erroneous clips in the result averaging process only serves to obfuscate the final classification outcome.

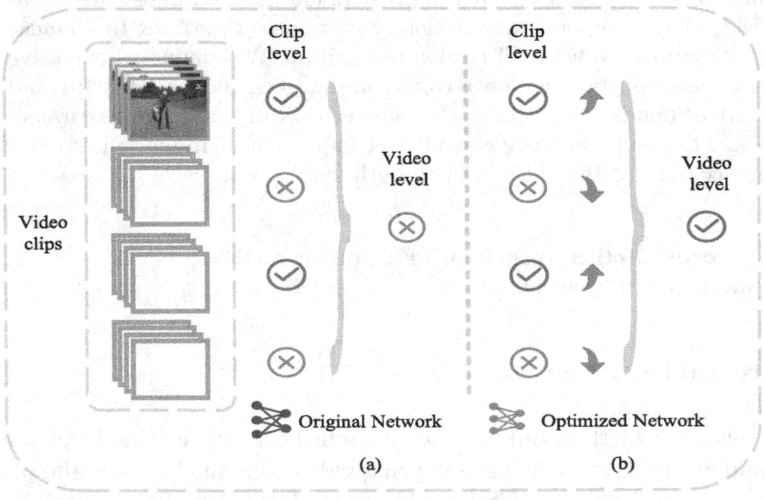

Fig. 1. An illustration of the fuzzy error fine-tuning effect

In practical scenarios, regardless of the availability of sufficient information in the input data, the classification model tends to produce high-probability results for specific classes. Figure 1(a) illustrates how individual clip inference results may not consistently contribute positively to the final classification outcome, underscoring the detrimental impact of misclassified clips on the model's performance.

In response to these challenges, we present a pioneering Fuzzy error fine-tuning framework designed to alleviate the influence of erroneous clips in video classification. This framework involves an iterative fine-tuning process applied to the model's classification header, leveraging a pre-trained spatio-temporal feature extractor as a foundation. As depicted in Fig. 1(b), the fine-tuned classification header mitigates the impact of misclassified clips while progressively amplifying the significance of accurate clips. Essentially, the 3D CNN block responsible for extracting spatio-temporal features is not tuned, and the classification head, explicitly tailored for interpreting these features, adopts a more ambiguous stance towards misleading inputs compared to its previous configuration. Consequently, when integrating multiple clips and averaging their results, clips with clearer information assume a relatively stronger influence, culminating in more precise video-level classification outcomes. The primary contributions of our research are as follows:

(1) We propose a novel approach for computing fuzzy target distribution and corresponding fuzzy error loss. This approach guides the model to deliver uncertain responses to misleading clips, deviating from the tendency to provide high-confidence definitive answers.
(2) We introduce the fuzzy error fine-tuning framework which effectively enhances the video-level classification capabilities of spatio-temporal convolution models based on dense sampling. By fine-tuning the model's classification header, this framework can be seamlessly applied to diverse pre-trained models.
(3) Through extensive experimentation, we empirically demonstrate the efficacy of our fine-tuning method in enhancing the classification accuracy of state-of-the-art methods across videos of varying lengths and content, including the UCF-101 and kinetics-sounds datasets.

2 Related Work

2.1 Spatio-Temporal (3D) Convolution Networks

Building upon the advancements in image classification models, researchers have naturally considered the utilization of spatio-temporal convolution to address the temporal dimension present in videos. A number of studies, such as [1–3,11,12] have progressively investigated and enhanced the accuracy and computational efficiency of the 3D-CNN method. Notably, the non-local operator [13] has leveraged attention mechanisms to effectively exploit temporal information within the time series. The Slow-fast network [4] has achieved improved recognition accuracy by employing a dual-flow model consisting of fast and slow flows. Furthermore, the X3D architecture [5], along with subsequent works [6,7] have further refined the structure of the 3D-CNN model, demonstrating remarkable achievements in terms of lightweight design and proficient extraction of spatio-temporal features.

2.2 Clips Selection and Features Aggregation

When confronted with a substantially larger volume of video data compared to image data, along with the inclusion of temporal sequences, it becomes impractical to directly employ all videos as the input for model inference. Numerous studies, such as [14] has delved into the exploration of utilizing subsets of videos for inference and devising sensible approaches to aggregate the complete time series information. Additionally, some research efforts [15–17] have focused on training samplers to select data with high informational content for model training. Within this realm, reinforcement learning methods [18] and multi-agent techniques [19] have been employed to extract and select pertinent features. Some work [20] has also considered the unification of local spatial information and global temporal information. In our methodology, we do not explicitly address the comprehensiveness of spatio-temporal information during the sampling stage. Instead, we opt to fine-tune the classification head of the model to mitigate the influence of insufficient input regarding spatio-temporal information on the resulting outcomes.

3 Proposed Method

Our objective is to enhance the robustness of a well-trained model through dense sampling, enabling it to provide nuanced responses when presented with inputs

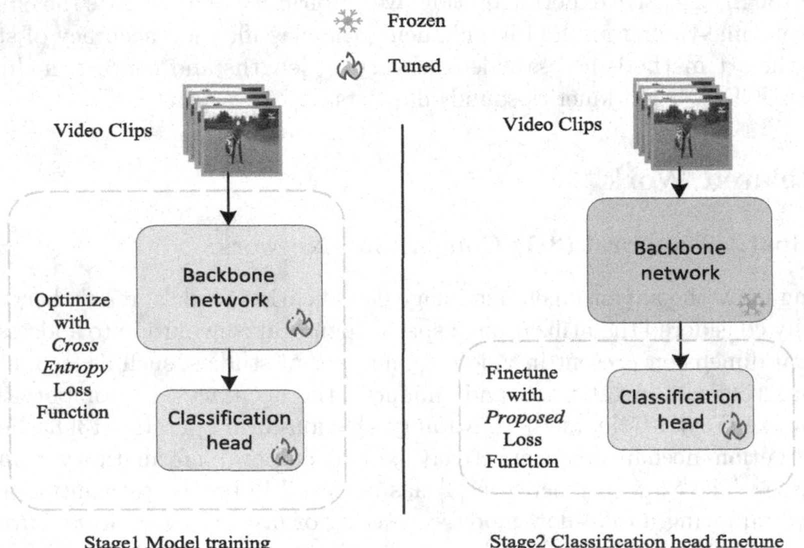

Fig. 2. Schematic diagram of model fine-tuning framework

that lack sufficient temporal and spatial characteristics or are misleading. Simultaneously, we aim to maintain its classification accuracy when provided with inputs containing ample information. The schematic overview of our approach is depicted in Fig. 2. Initially, the model is trained using standard methods, and once it nears convergence, we proceed to fine-tune it using our proposed technique. As the model acquires a strong capacity for extracting temporal features at this stage, we immobilize the parameters of the convolution module in the initial layers. Meanwhile, we continue to refine the classification head, responsible for interpreting spatio-temporal information, utilizing the loss function introduced below.

3.1 Problem Definition

The target of the action recognition task is to train a classifier F which use video data $V \in \mathbb{R}^{T*3*H*W}$ as the input and classify these videos into K classes. In many classical 3D CNN methods,a complete video is usually divided into N clips $\{C_1, C_2, \cdots, C_N\}$. One input of the classifier is only a small part of video frames, which are continuous M frames with size $H * W$ in a clip C_i.

The final video level prediction result is the average consideration of these N clips. For one clip input C_i, the classifier will generate the predictions of the probability of each class $p(C)$, which can be represented as Eq. 1. For the ground truth y, when the output of the model satisfies Eq. 2, the classification prediction is correct, otherwise it is wrong. However, not every clip has enough spatio-temporal information for the model to inference, and when the wrong classification result is taken into account in the average calculation, it may lead to the inaccurate video-level result. Our goal is to reduce the impact of these input data with ambiguous information to improve the classification accuracy of the model in video level.

Given a well-trained classifier F. Considering that the convolution module at the front of the classifier has been well trained, it has a strong ability to extract spatio-temporal features. Our method will lock the convolution module in the front and only retrain the classification head responsible for spatio-temporal feature classification by proposed fuzzy error loss function. It can give a fuzzy result instead of a definite wrong answer to the input data with insufficient or ambiguous spatio-temporal information.

In the fine-tuning process, most of the classifier model parameters will not be changed, and only a small part of the classifier header parameters will be tuned. This process is very parameter efficient, and it can be used in any trained model based on dense sampling to continue to improve its effect.

$$p(C) = [p_1(C), p_2(C), \cdots, p_K(C)] \tag{1}$$

$$y = argmax(p(C)) \tag{2}$$

3.2 Fuzzy Target

For a fully trained classifier F, when the input clip has insufficient spatio-temporal features, the model usually gives a wrong but definite classification answer. If we continue to use the ground-truth of the corresponding clip to optimize the model, it will usually lead to over-fitting of the model or the decline of the classification performance of the classifier. Based on the above assumptions, our method will construct a fuzzy target, whose purpose is to guide the model to give a result that the confidence of possible classes is close instead of a high confidence but wrong prediction result.

In order to achieve this goal, in the training stage, when the classifier outputs the wrong prediction result, we can adjust it to a fuzzy answer based on the prediction probability sequence it outputs as a new target rather than softmax, which is called fuzzy target $\widetilde{p}(C)$ as shown in Eq. 3.

$$\widetilde{p}(C) = [\widetilde{p}_1(C), \widetilde{p}_2(C), \cdots, \widetilde{p}_K(C)] \tag{3}$$

First of all, all probability items less than the threshold λ will be assigned to 0,and the intermediate quantity $p'(C)$ will be obtained, as shown in Eq. 4, Eq. 5.

$$p'(C) = [p'_1(C), p'_2(C), \cdots, p'_K(C)] \tag{4}$$

$$where\ p'_i(C) = \begin{cases} 0 & ,if\ p_i(C) < \lambda \\ p_i(C) & ,otherwise \end{cases} \tag{5}$$

Then an iterative smoothing formula is used to calculate the new target distribution, as shown in Eq. 6. The parameter μ controls the total number of iterations, and the parameter σ controls the smoothing strength of each iteration. In a smoothing calculation, every non-zero item of the input distribution will be closer to the L1 norm of the whole distribution, as shown in Eq. 7, thus achieving the purpose of smoothing operation. After the smoothed distribution $p''(C)$ is calculated, it is normalized to make the probability of all its terms sum to 1, as shown in Eq. 8, and the final target distribution $\widetilde{p}(C)$ is obtained.

$$p''(C) = \underbrace{F(F(\ldots F(p'(C))\ldots))}_{\mu \text{ iteration}} = F^{(\mu)}(p'(C)) \tag{6}$$

$$\forall p_i(C) \in p'(C), F(p_i(C)) = \begin{cases} \|p'(C)\|_1 * \sigma + p_i(C) * (1-\sigma) & ,if\ p_i(C) \neq 0 \\ p_i(C) & ,otherwise \end{cases} \tag{7}$$

$$\widetilde{p}_i(C) = \frac{p''(C)_i}{\sum_{j=1}^{n} p''(C)_j} \tag{8}$$

3.3 Fine Tune Loss Function

In order to maintain the correct classification results and weaken the influence of errors or fuzzy results, the loss function in the fine-tuning stage is divided into two cases, as shown in Eq. 9. When the prediction results are classified correctly, the cross entropy loss is used to continue to improve the correct direction, while when the classification results are wrong, we hope that the output result is a fuzzy answer, then the KL divergence loss with fuzzy target is used to fine-tune the model in another direction.

$$Loss(p(C), y) = \begin{cases} -\frac{1}{n}\sum_{j=i}^{n} y_i \log(P_i(C)) & ,if\ y = argmax(p(C)) \\ -\frac{1}{n}\sum_{i=1}^{n} \widetilde{p}_i(C) \log\left(\frac{P_i(C)}{\widetilde{p}_i(C)}\right) & ,otherwise \end{cases} \tag{9}$$

4 Experiment

4.1 Datasets and Implementation Details

Datasets. In order to evaluate the effectiveness of the proposed method, firstly, we employed the widely used benchmark dataset, UCF101 [21], which consists of 13,320 instances belonging to 101 categories. The average video length in this dataset is approximately 7 s. Secondly, we utilized the Kinetic-sounds dataset [22], which is a subset of the well-known benchmark dataset Kinetics [2]. Kinetic-sounds comprises 23,656 instances with an average video length of 10 s. We want to test our method on video datasets with different video length.

Training and Fine-Tuning Details. To assess the effectiveness of the proposed method, we employed the C3D [1] and X3D-S, X3D-L [5] models to test the efficiency of the proposed method. All the models are based on PyTorch [23] and were constructed based on the implementation details provided by Pyslowfast [24]. Specifically, for the C3D and X3D-L, we utilized an input size of 16 and the same resolution of 224×224 pixels with temporal stride 5. For the X3D-S, we utilized an input size of 13 frames and the resolution of 160×160 pixels with temporal stride 6. We perform scale augmentation by randomly resizing the shorter size of the training clip to between 182 and 228 pixels, and then randomly crop a video clip. We trained our models using SGD with momentum 0.9 and a weight decay of 0.0001 following the setting in X3D [5].

Upon convergence of the model, we proceed to freeze the parameters of all convolution blocks, while simultaneously reducing the learning rate to a range of 0.01–0.05 times the initial value. To fine-tune the classification header of the model, we conduct a series of experiments utilizing both the fuzzy error loss and cross entropy loss. The fine-tuning process is performed for a duration of 1 to 5 epochs. The fuzzy error loss function hyperparameter settings adopted for these experiments are as follows: threshold value λ set to 0.1, smoothing constant represented by σ set to 0.5, and the number of iterations denoted as μ set to 3. These choices were made to ensure an optimal balance between preserving model accuracy and incorporating the benefits of the fuzzy error loss during the fine-tuning phase.

Testing and Measurement. In accordance with the methodology employed by [4], we adopt the practice of sampling 10 clips evenly from each video for inference. Each clip is assigned a spatial resolution of 182×182 pixels. To obtain the final prediction, we calculate the average softmax scores of all clips. The accuracy metrics reported for the model on the dataset include top-1 and top-5 accuracy. The results of averaging with different numbers of clips K are also reported.

4.2 Performance Comparison

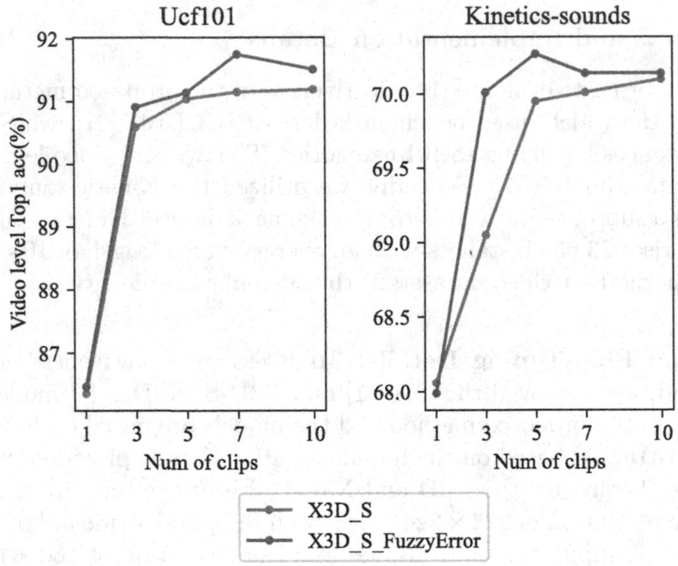

Fig. 3. Accuracy comparison on UCF101 and Kinetics-sounds. The top-1 accuracy is obtained by K-Center clip testing where the number of temporal clips $K \in \{1, 3, 5, 7, 10\}$ is shown in each curve

Performance on UCF101. Table 1 presents a comprehensive comparison of the inference results achieved by our model and other classical models on the UCF101 dataset. Notably, the model incorporating our proposed fine-tuning strategy exhibits significant improvements compared to the original model, outperforming several established models C3D [1], TSN [14] and the original models themselves. This outcome substantiates the efficacy of our proposed fine-tuning strategy.

Figure 3, the model fine-tuned by fuzzy error has a significant improvement when the clip number K used is 3, 5, 7. It can be reflected that the fine-tuning strategy has a positive effect on the results when the input spatio-temporal information is insufficient. When the used clip number is 10, the input spatio-temporal information is enough, so the promotion effect of fuzzy error loss is less.

Table 1. Comparison to state-of-the-arts on UCF-101 Split 1 set.

Model	GFLOPs	Params (M)	Top-1 (%)	Top-5 (%)
TSN	102.7	24.33	90.01	97.09
C3D	38.5	34.8	83.08	95.93
C3D(Fuzzy error)	38.5	34.8	**84.48**	**96.43**
X3D-S	1.96	3.76	91.19	98.02
X3D-S(Fuzzy error)	1.96	3.76	**91.51**	**98.89**
X3D-L	24.8	6.1	93.30	99.21
X3D-L(Fuzzy error)	24.8	6.1	**94.21**	**99.33**

Performance on Kinetics-Sounds. Table 2 illustrates the comparison of inference results obtained by our model and other classical models on the Kinetics-Sounds dataset. The model utilizing our proposed adjustment strategy exhibits superior performance compared to the UCF101 results. This substantial improvement can be attributed to the longer video length and the presence of redundant information in the Kinetics-Sounds dataset. Consequently, our proposed method effectively addresses these characteristics of the dataset.

As shown in Fig. 3, when clips number $K \in \{3, 5, 7\}$ is used, the accuracy of the model is greatly improved compared with the original model, which is similar to the performance on UCF101 dataset.

Table 2. Comparison to state-of-the-arts on Kinetics-sounds Test.

Model	GFLOPs	Params (M)	Top-1 (%)	Top-5 (%)
X3D-S	1.96	3.76	70.01	90.96
X3D-S(Fuzzy error)	1.96	3.76	**70.14**	**91.07**

4.3 Discussion

The aforementioned experimental results highlight the positive impact of fine-tuning with the Fuzzy error loss on enhancing the accuracy of model inference. Experiments show that the video-level accuracy of 3D convolution models such as C3D and X3D can be further optimized by the proposed method. However, as depicted in Fig. 3, this positive effect is less pronounced when the number of clips, denoted as K, is set to 1 or 10. Notably, significant improvements are observed when K takes on the values of 3, 5, and 7. This experimental finding provides compelling evidence for our problem hypothesis. When K equals 1, the video-level result is obtained solely from a single clip, eliminating the issue of aggregating different information across clips. Conversely, when K is set to 3, 5, and 7, there exists a disparity in the amount of spatio-temporal information captured by different clips. The results demonstrate that the fine-tuned model can effectively mitigate the influence of misleading clips and enhance the accuracy of the final averaged results. However, when K is set to 10, the increased number of clips diminishes the impact of misleading clip results. Consequently, the improvement achieved through the application of fuzzy error loss is less substantial compared to the model without fine-tuning.

5 Conclusion

This paper presents a novel fine-tuning framework for enhancing the video-level accuracy of dense sampling based spatio-temporal convolution video classification models. The proposed method aims to mitigate the impact of clip classification results characterized by substantial ambiguity. A fuzzy error loss function is introduced, which effectively reduces the influence of incorrect classification outcomes without significantly compromising the model's ability to accurately classify correct results. Importantly, this approach seamlessly integrates with existing dense sampling models, allowing for improved performance while incurring minimal retraining costs for the classification header. The efficacy of the proposed method is demonstrated through experiments conducted on two reputable datasets, namely ucf-101 and kinetics-sounds. The results show that this fine-tuning technique consistently enhances the performance of state-of-the-art models, particularly when multiple clips are utilized during inference.

Acknowledgement. This work was supported in part by the National Natural Science Foundation of China under Grant 62072048, and in part by Industry-University-Research Innovation Fund of Universities in China under Grant 2021ITA07005.

References

1. Tran, D., Bourdev, L., Fergus, R., Torresani, L., Paluri, M.: Learning spatiotemporal features with 3D convolutional networks. In: 2015 IEEE International Conference on Computer Vision (ICCV) (2015). https://doi.org/10.1109/iccv.2015.510

2. Hara, K., Kataoka, H., Satoh, Y.: Can spatiotemporal 3D CNNs retrace the history of 2D CNNs and imagenet? In: 2018 IEEE/CVF Conference on Computer Vision and Pattern Recognition (2018). https://doi.org/10.1109/cvpr.2018.00685
3. Tran, D., Wang, H., Torresani, L., Ray, J., LeCun, Y., Paluri, M.: A closer look at spatiotemporal convolutions for action recognition. In: 2018 IEEE/CVF Conference on Computer Vision and Pattern Recognition (2018). https://doi.org/10.1109/cvpr.2018.00675
4. Feichtenhofer, C., Fan, H., Malik, J., He, K.: Slowfast networks for video recognition. In: 2019 IEEE/CVF International Conference on Computer Vision (ICCV) (2019). https://doi.org/10.1109/iccv.2019.00630
5. Feichtenhofer, C.: X3D: expanding architectures for efficient video recognition. In: 2020 IEEE/CVF Conference on Computer Vision and Pattern Recognition (CVPR) (2020). https://doi.org/10.1109/cvpr42600.2020.00028
6. Jiang, Y., Gong, X., Wu, J., Shi, H., Yan, Z., Wang, Z.: Auto-X3D: ultra-efficient video understanding via finer-grained neural architecture search. In: 2022 IEEE/CVF Winter Conference on Applications of Computer Vision (WACV) (2022). https://doi.org/10.1109/wacv51458.2022.00241
7. Wang, J., et al.: Maximizing spatio-temporal entropy of deep 3D CNNs for efficient video recognition (2023)
8. Tan, Y., Hao, Y., Zhang, H., Wang, S., He, X.: Hierarchical hourglass convolutional network for efficient video classification (2022)
9. Chen, C.F.R., et al.: Deep analysis of CNN-based spatio-temporal representations for action recognition. In: 2021 IEEE/CVF Conference on Computer Vision and Pattern Recognition (CVPR) (2021). https://doi.org/10.1109/cvpr46437.2021.00610
10. Shalmani, S., Chiang, F., Zheng, R.: Efficient action recognition using confidence distillation (2021)
11. Carreira, J., Zisserman, A.: Quo vadis, action recognition? A new model and the kinetics dataset. In: 2017 IEEE Conference on Computer Vision and Pattern Recognition (CVPR) (2017). https://doi.org/10.1109/cvpr.2017.502
12. Xie, S., Sun, C., Huang, J., Tu, Z., Murphy, K.: Rethinking spatiotemporal feature learning for video understanding (2017)
13. Wang, X., Girshick, R., Gupta, A., He, K.: Non-local neural networks. In: 2018 IEEE/CVF Conference on Computer Vision and Pattern Recognition (2018). https://doi.org/10.1109/cvpr.2018.00813
14. Wang, L., et al.: Temporal segment networks: towards good practices for deep action recognition. In: Leibe, B., Matas, J., Sebe, N., Welling, M. (eds.) ECCV 2016. LNCS, vol. 9912, pp. 20–36. Springer, Cham (2016). https://doi.org/10.1007/978-3-319-46484-8_2
15. Wu, Z., Xiong, C., Ma, C.Y., Socher, R., Davis, L.S.: Adaframe: adaptive frame selection for fast video recognition. In: 2019 IEEE/CVF Conference on Computer Vision and Pattern Recognition (CVPR) (2019). https://doi.org/10.1109/cvpr.2019.00137
16. Alwassel, H., Caba Heilbron, F., Ghanem, B.: Action search: spotting actions in videos and its application to temporal action localization. In: Ferrari, V., Hebert, M., Sminchisescu, C., Weiss, Y. (eds.) ECCV 2018. LNCS, vol. 11213, pp. 253–269. Springer, Cham (2018). https://doi.org/10.1007/978-3-030-01240-3_16
17. Gao, R., Oh, T.H., Grauman, K., Torresani, L.: Listen to look: action recognition by previewing audio. In: 2020 IEEE/CVF Conference on Computer Vision and Pattern Recognition (CVPR) (2020). https://doi.org/10.1109/cvpr42600.2020.01047

18. Wang, Y., Chen, Z., Jiang, H., Song, S., Han, Y., Huang, G.: Adaptive focus for efficient video recognition. In: 2021 IEEE/CVF International Conference on Computer Vision (ICCV) (2021). https://doi.org/10.1109/iccv48922.2021.01594
19. Wu, W., He, D., Tan, X., Chen, S., Wen, S.: Multi-agent reinforcement learning based frame sampling for effective untrimmed video recognition. In: 2019 IEEE/CVF International Conference on Computer Vision (ICCV) (2019). https://doi.org/10.1109/iccv.2019.00632
20. Huang, H., Zhou, X., He, R.: Orthogonal transformer: an efficient vision transformer backbone with token orthogonalization (2022)
21. UCF101: a dataset of 101 human actions classes from videos in the wild
22. Arandjelovic, R., Zisserman, A.: Look, listen and learn. In: 2017 IEEE International Conference on Computer Vision (ICCV) (2017). https://doi.org/10.1109/iccv.2017.73
23. Paszke, A., et al.: Pytorch: an imperative style, high-performance deep learning library. In: Advances in Neural Information Processing Systems, vol. 32, pp. 8024–8035. Curran Associates, Inc. (2019). http://papers.neurips.cc/paper/9015-pytorch-an-imperative-style-high-performance-deep-learning-library.pdf
24. Fan, H., Li, Y., Xiong, B., Lo, W.Y., Feichtenhofer, C.: Pyslowfast (2020). https://github.com/facebookresearch/slowfast

Temporal-Channel Topology Enhanced Network for Skeleton-Based Action Recognition

Jinzhao Luo[1,2], Lu Zhou[1,2], Guibo Zhu[1,2,3](✉), Guojing Ge[1], Beiying Yang[1,2], and Jinqiao Wang[1,2,3,4](✉)

[1] Foundation Model Research Center, Institute of Automation, Chinese Academy of Sciences, Beijing 100190, China
{lu.zhou,gbzhu,guojing.ge,beiying.yang,jqwang}@nlpr.ia.ac.cn, luojinzhao2020@ia.ac.cn
[2] School of Artificial Intelligence, University of Chinese Academy of Sciences, Beijing 100049, China
[3] Wuhan AI Research, Wuhan 430073, China
[4] The Peng Cheng Laboratory, Shenzhen 518066, China

Abstract. Skeleton-based action recognition has become popular in recent years due to its efficiency and robustness. Most current methods adopt graph convolutional network (GCN) for topology modeling, but GCN-based methods are limited in long-distance correlation modeling and generalizability. In contrast, the potential of convolutional neural network (CNN) for topology modeling has not been fully explored. In this paper, we propose a novel CNN architecture, Temporal-Channel Topology Enhanced Network (TCTE-Net), to learn spatial and temporal topologies for skeleton-based action recognition. The TCTE-Net consists of two modules: the Temporal-Channel Focus module, which learns a temporal-channel focus matrix to identify the most important feature representations, and the Dynamic Channel Topology Attention module, which dynamically learns spatial topological features, and fuses them with an attention mechanism to model long-distance channel-wise topology. We conduct experiments on NTU RGB+D, NTU RGB+D 120, and FineGym datasets. TCTE-Net shows state-of-the-art performance compared to CNN-based methods and achieves superior performance compared to GCN-based methods. The code is available at https://github.com/aikuniverse/TCTE-Net.

Keywords: human skeleton · action recognition · topology modeling

1 Introduction

Action recognition is a crucial task with applications in various fields such as human-robot interaction and virtual reality. With the continuous development of depth sensors and pose estimators, obtaining high quality 3D skeletal data has

© The Author(s), under exclusive license to Springer Nature Singapore Pte Ltd. 2024
Q. Liu et al. (Eds.): PRCV 2023, LNCS 14425, pp. 109–119, 2024.
https://doi.org/10.1007/978-981-99-8429-9_9

become easier. As a result, skeleton-based action recognition received increasing attention in recent years, thanks to the compactness and robustness of human skeletal data against complicated backgrounds.

Graph Convolutional Networks (GCNs) [1–3] have become one of the most popular skeleton-based action recognition methods due to its ability to handle irregular topological information in skeletons [4,5]. Specifically, GCNs model skeleton sequences as spatio-temporal graph topologies. They use a handcrafted graph, which represents physically connected edges among human skeleton, to extract the spatial features representing the relationships between joint nodes in the human skeleton. ST-GCN [5], the first well-known GCN-based method, constructs spatial and temporal correlations in skeletal data via graph convolution. Subsequently, Li *et al.* [6] expands the receptive field based on the self-attention mechanisms to learn topology between joints, while Wang *et al.* [7] aggregates the spatio-temporal topological feature representations to improve the modeling capacity. However, GCN-based methods have limitations. Joint nodes in skeleton are treated equally, which means important nodes and edges cannot be identified [20]. Furthermore, GCNs struggle to model the complicated correlations between distant unnaturally connected joint nodes. However, for human action recognition, relationships between structurally distant joints are as important as between adjacent joints. Besides, GCN-based methods require complex network structure designs to fuse skeleton and other modalities [8]. Different from the previous GCN methods, in this paper, we attempt to alleviate the disadvantage of GCN methods that cannot identify the relationships between distant joint nodes.

Compared with GCN, Convolutional Neural Network(CNN) can model topological features more effectively with powerful local convolution characteristics and self-attention mechanism [9,10]. They can also be easily fused with other modalities [11]. Caetano *et al.* [12] converts the skeleton coordinates to a three channels pseudo image input, and then classifies the features extracted through the network. Such input cannot exploit the locality nature of convolution networks. Shi *et al.* [18] modeling spatial-temporal dependencies between joints without the requirement of knowing their positions or mutual connections by building the attention blocks. PoseC3D [11] generates 3D heatmap volumes from skeleton coordinates as input, and then classifies with a 3D-CNN. However, existing CNN-based methods do not utilize the natural topology of the bones. Processing in the temporal and spatial dimensions leads to the separation of spatio-temporal attributes of actions respectively, without forming a unified spatio-temporal feature representation. In addition, using the attention mechanism simply to assign different weights to each joint node cannot extract the relationship between joint nodes effectively.

In order to solve the above problems, in this paper, we propose a novel Temporal-Channel Topology Enhanced Network (TCTE-Net), which models the topological information of skeleton data effectively. Specifically, we propose a novel Temporal-Channel Focus (TCF) module, which emphasizes vital features to force the model focus on the critical joint nodes in action classification. And we propose a Dynamic Channel Topology Attention (DCTA) module, which can identify the relationships between distant joint nodes effectively and model the correlation between distant joint nodes dynamically.

Our contributions are summarized as follows:

1. We present a novel TCTE-Net for skeleton-based action recognition equipping with TCF and DCTA modules, which attempt to identify the critical joint nodes and relationships between distant joint nodes in action classification. The TCF module emphasizes the critical joint nodes with a focus matrix. DCTA learns distant channel-wise topology modeling based on the dynamic channel distance matrix and attention mechanism.
2. The extensive experiments verify the effectiveness of TCF and DCTA modules. The proposed TCTE-Net outperforms state-of-the-art CNN methods significantly and achieves remarkable performance compared to GCN-based methods on three skeleton-based action recognition datasets.

2 Proposed Method

In this section, we first propose the detailed architecture of TCTE-Net in Sect. 2.1. Then, the Temporal-Channel Focus Module and the Dynamic Channel Topology Attention Module are introduced in Sect. 2.2 and Sect. 2.3.

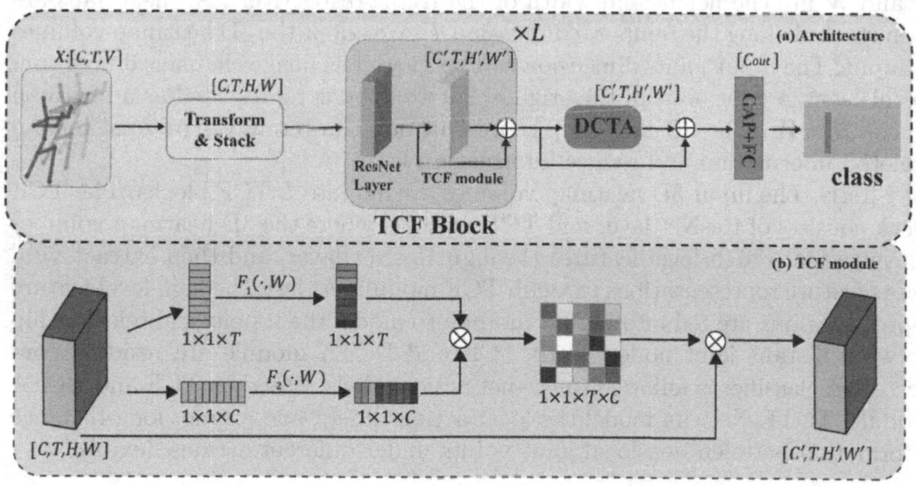

Fig. 1. (a) Pipeline of the proposed TCTE-Net, which consists of L TCF blocks and one DCTA module. We instantiate TCTE-Net with the SlowOnly backbone, where L is 3. (b) The detailed architecture of TCF module.

2.1 Network Architecture

Joint nodes in different body parts contribute to action classification differently. For example, in the case of a 'shaking hands' action, the weight of arm part is much higher than the body part. However, existing CNN-based methods are limited in identifying critical joint nodes, and CNNs cannot model the natural topology of the bones directly without an adjacency matrix, which is widely used in GCNs. To address these limitations, we propose the Temporal-Channel Topology Enhanced Network (TCTE-Net).

The TCTE-Net is illustrated in Fig. 1(a), which consists of three TCF blocks and one DCTA module. TCTE-Net adopts lightweight SlowOnly 3D-CNN [11, 13] as the backbone. Our approach focuses more on feature representation and topology modeling. The 2d skeleton coordinates extracted from the video frames stored as coordinate-triplets $(x,\ y,\ c)$, where $(x,\ y)$ represent the coordinates of the skeleton joints and c is the corresponding probability. In order to get the input of TCTE-Net, we first transform the 2d skeleton coordinates into a pseudo-image form. Specifically, a set of 2D skeleton coordinates $(x,\ y,\ c)$ is represented as a heatmap of size $K \times H \times W$ by composing K gaussian maps centered at joint:

$$J_{kij} = e^{-\frac{(i-x_k)^2+(j-y_k)^2}{2*\sigma^2}*c_k}, \tag{1}$$

where σ controls the variance of gaussian maps, $(x_k,\ y_k)$ and c_k are respectively the location and confidence score of the k-th joint, and K is the number of joints, H and W are the height and width of the frame respectively. All heatmaps are then stacked along the temporal dimension T. We adopt the 3D heatmap volumes as input. The input joints dimension can be viewed as image channels dimension. In this case, a joint node in the original 2D skeleton is represented as a heatmap of size $H \times W$. The 3D heat map representation can retain the original human skeleton information and reduce information loss.

Firstly, The input 3D heatmap volumes are fed into L TCF blocks. The TCF block consists of ResNet layer and TCF module, where the 3D heatmap volumes converted into high level features through ResNet layer, and then extract vital joints feature representations through TCF module. At last, the high level feature representations are fed into DCTA module to model the topological relationship between distant joint nodes. Both TCF and DCTA module are residual connected. A classifier is followed to predict action labels. Based on TCF and DCTA module, TCTE-Net can model the local actions finely, and extract long-distance information between non-local joint points under different actions flexibly.

2.2 Temporal-Channel Focus Module

To enhance the critical joint nodes, we propose TCF module, which is illustrated in Fig. 1(b). Before being fed into TCF module, the input features are transformed into high level representations $X \in \mathbb{R}^{C \times T \times H \times W}$. The features X are then fed into a two-stream structure of temporal and channel dimensions, which is implemented along the temporal dimension and channel dimension with a Global average pooling (GAP) layer and followed with a FC layer. The features X are downsampled and linearly transformed along the temporal and channel dimensions respectively. Through the above operations, we get the weighted vectors of joint node features in channel and temporal dimensions. The weighted vectors are then fused by element-wise multiplication. Activation function is applied to get the temporal-channel focus matrix.

In the early stages of the network, the heatmaps of different channels represent skeleton joint features of different body parts, which has different weights in classification. The proposed TCF module can distinguish critical local joint nodes, and extract the spatio-temporal local features that are more effective for action recognition adaptively. This makes the model pay more attention to the most discriminative local features, and addresses the difficulties of recognizing subtle and similar movements.

Finally, the joint node features X are strengthened by temporal channel focus matrix. The overall process of TCF can be formulated as:

$$X' = \sigma\big(F_1(GAP(X) * F_2(GAP(X))\big) * X, \tag{2}$$

where $X \in \mathbb{R}^{C\times T\times H\times W}$ and $X' \in \mathbb{R}^{C'\times T\times H'\times W'}$ are the input and output of the TCF module respectively. F_1 and F_2 mean FC layers. σ is activation function.

2.3 Dynamic Channel Topology Attention Module

To eliminate the weakness of CNN in modeling the irregular skeletal topology, we introduce the Dynamic Channel Topology Attention module (DCTA). As shown in Fig. 2, the self-attention matrix is used to extract global shared topology for all channels. Meanwhile, we learn specific relationships between joint nodes of different channels.

We utilize convolution and pooling operations on input feature X' to generate new feature representations Q, K, V, and reshape to $Q, K \in \mathbb{R}^{C'\times N}$ and $V \in \mathbb{R}^{C'\times T\times N}$, where $N = H' \times W'$. Then we calculate the spatial attention map $S \in \mathbb{R}^{N\times N}$:

$$S_{ji} = \frac{e^{q_i \cdot k_j}}{\sum_{i=1}^{N} e^{q_i \cdot k_j}}, \tag{3}$$

where S_{ji} measures the correlation between position q_i and k_j. CNN is capable of modeling the topology of joints implicitly [14]. Thus S are adopted to represent the global shared topological relationship between features of different joint nodes. Meanwhile, we calculate the channel-specific correlations between different features $M \in \mathbb{R}^{C'\times N\times N}$, which can be formulated as:

$$M(x_i, x_j) = \sigma\big(Q(x_i) - K(x_j)\big), \tag{4}$$

where $\sigma(\cdot)$ is activation function. $M(x_i, x_j)$ is dynamic channel distance matrix, which calculates distances between features x_i and x_j along channel dimension. Different channels represent different types of motion features in classification [1]. Therefore, $M(x_i, x_j)$ essentially models the topological relationship

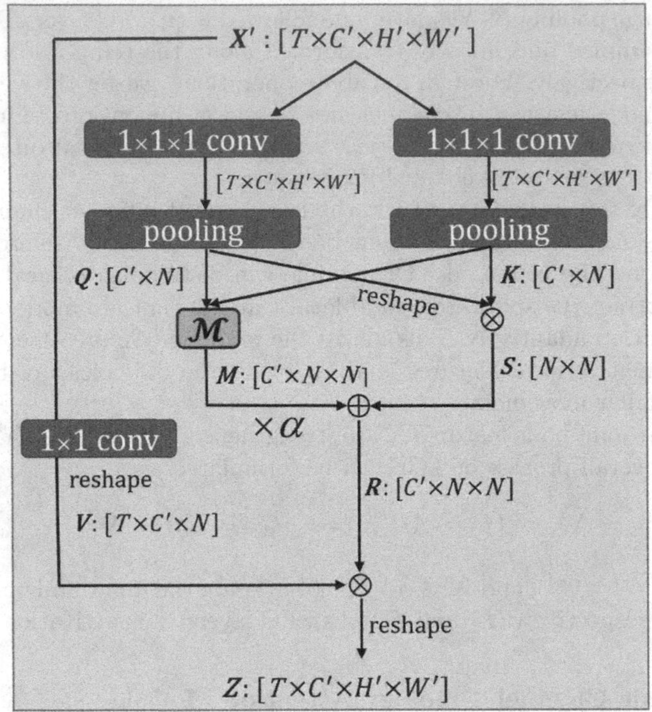

Fig. 2. The detailed architecture of DCTA module.

between joint nodes under different motion features. For distant unnaturally connected joint nodes, the dynamic channel distance matrix is able to capture their specific correlations under different motion features dynamically. The final topological relation $R \in \mathbb{R}^{C' \times N \times N}$ is formulated as:

$$R = S + \alpha \cdot M. \tag{5}$$

The dynamic channel distance matrix $M(x_i, x_j)$ is utilized to enhance the global shared topological representation S with a trainable scalar α. The addition is conducted in a broadcast way. Finally, we perform a matrix multiplication between R and V, and reshape the result to $\mathbb{R}^{C' \times T \times H' \times W'}$. The output of DCTA module is formulated as:

$$Z_j = X'_j + \sum_{i=1}^{N} r_{ji} v_i, \tag{6}$$

where r_{ji} and v_i represent the corresponding elements in the matrix respectively.

The Eq. (6) shows that the output feature Z is the sum of the final topological relation and original features, which models the long-range correlations between joint nodes dynamically.

3 Experiments

3.1 Datasets and Implementation Details

NTU RGB+D. NTU RGB+D is a large-scale human action recognition dataset. It contains more than 56k video samples of 60 human action classes performed by 40 distinct human subjects. Each sample is captured from different views by three Microsoft Kinect v2 cameras at the same time. The dataset has two benchmarks: Cross-subject (X-Sub), Cross-view (X-View), for which are split by action subjects, camera views in training and validation.

NTU RGB+D 120. NTU RGB+D 120 extends NTU RGB+D with 57k video samples of additional 60 action classes, which contains 113k samples over 120 human action classes performed by 106 human subjects. The authors recommend two benchmarks: Cross-subject (X-Sub) and Cross-setup (X-Set, split by camera setups).

FineGYM. FineGYM is a fine-grained action recognition dataset. It contains 29K videos of 99 fine-grained action classes collected from 300 professional gymnastics competitions. The characteristic of this dataset is that the behaviors are divided according to the level from fine-grained actions to coarse-grained events, and the human skeleton of gymnastics moves has a large deformation.

Implementation Details. TCTE-Net is implemented via Pytorch and trained with 8 RTX 2080 TI GPUs, where each GPU has 11 video clips in a mini-batch. The model is trained for 30 epochs with SGD optimizer. The initial learning rate is set to 0.1375 and decayed with Cosine Annealing scheduler [21]. The weight decay is set to 0.0003. For all datasets, we report the results of 10-clip testing.

3.2 Ablation Study

Table 1. Ablation study on NTU RGB+D. FM represents Focus Matrix.

Method	Param	X-Sub (%)	X-View (%)
Baseline	2.03M	93.3	96.2
+TCF (1)	+0.10M	93.5	96.3
+TCF (3)	+0.25M	93.7	96.5
+TCF (3) w/o FM	+0.25M	93.5	96.4
+DCTA	+0.30M	93.6	96.5
+ TCF (3) + DCTA	+0.53M	**93.8**	**96.6**

Effectiveness of TCF. Table 1 illustrates the performance gains brought about by TCF and DCTA on the NTU RGB+D dataset. We use SlowOnly as the baseline model and add the TCF module to it. When adding a TCF module, the accuracy increases by 0.2% compared to the baseline on the X-view benchmark. When increasing the number of TCF modules to 3, TCF boosts accuracy by 0.4% on X-view benchmark, with little parameter increase. Our results validating the effectiveness of TCF module. To verify the validity of the Focal Matrix in the TCF module, we then remove the Focus Matrix (FM) from TCF. The TCF without FM module parallels the weighted channel and temporal vectors. Compared to TCF, the performance of TCF without FM drops by 0.2%, indicating that the weighted focus matrix can identify classification-vital features efficiently, thereby assigning different attention weights to all spatio-temporal joint nodes.

Effectiveness of DCTA. By introducing the DCTA module, we further improve accuracy by 0.3%. DCTA dynamically models the relationship between long-distance joint points under different action characteristics based on the dynamic channel distance matrix, so that the model has a better recognition effect on actions with large deformation and fast movement speed. Our proposed TCTE-Net achieves an accuracy of 93.8% with the X-Sub benchmark, which improves the baseline accuracy by 0.5% with an efficient model.

3.3 Comparison with the State-of-the-Art

Table 2. Comparative Experiment of TCTE-Net and SOTA Model on NTU RGB+D Dataset.

Type	Method	X-Sub (%)	X-View (%)
CNN	DSTA-Net [18]	91.5	96.4
	Ta-CNN+ [14]	90.7	95.1
	PoseConv3D+ [11]	93.7	96.6
GCN	MS-G3D+ [3]	91.5	96.2
	STF [9]	92.5	96.9
	HD-GCN+ [20]	93.0	**97.0**
Ours	TCTE-Net	**93.8**	96.6

In the experimental results section, we evaluate the effectiveness of TCTE-Net on three benchmark datasets: NTU RGB+D, NTURGB+D 120, and FineGYM. Many state-of-the-art methods employ multi-stream fusion models [20,22,23], joint, bone. For a fair comparison, we compare our model with the state-of-the-art methods obtained by the best single models on each dataset, and our model significantly outperforms the other methods.

On the NTU RGB+D dataset, the results shown in Table 2 demonstrate that our model is effective. Although the performance of X-View benchmark is nearly saturated, our model still obtains remarkable performance. Moreover, TCTE-Net achieves an accuracy rate of 93.8% in the X-Sub benchmark test, outperforming other state-of-the-art single-model methods, and achieves a performance improvement of 0.1% compared with advanced multi-stream fusion methods. Compared with the state-of-the-art GCN model, TCTE-Net achieves 0.8% performance improvement on X-Sub benchmark test.

Table 3. Comparative Experiment of TCTE-Net and SOTA Model on NTU RGB+D 120 Dataset.

Type	Method	X-Sub (%)	X-View (%)
CNN	DSTA-Net [18]	86.6	89.0
	Ta-CNN+ [14]	85.7	87.3
	PoseConv3D [11]	86.0	89.6
GCN	Shift-GCN [24]	85.9	87.6
	InfoGCN [25]	85.1	86.3
	HD-GCN+ [20]	85.7	87.3
ours	TCTE-Net	**86.6**	**89.9**

On the challenging NTU RGB+D 120 dataset, our model achieves excellent performance, as shown in Table 3. Compared with the state-of-the-art single model, TCTE-Net achieves 0.6% and 0.3% performance improvements on the X-Sub and X-View benchmarks respectively, achieving comparable performance to the state-of-the-art multi-stream fusion methods, and outperforming the previous state-of-the-art CNN method. Compared with the most advanced GCN method, TCTE-Net achieves 0.7% and 2.3% performance improvements on the X-Sub and X-View benchmarks respectively, which verifies the effectiveness of the proposed method in the skeleton action recognition task.

Table 4. Comparative Experiment of TCTE-Net and SOTA Model on FineGYM Dataset.

Type	Mean Top-1 Accuracy (%)
MS-G3D+ [3]	92.6
PoseConv3D [11]	93.2
TCTE-Net	**93.8**

Furthermore, we evaluate TCTE-Net on the FineGYM dataset. The results shown in Table 4 demonstrate that our model achieves state-of-the-art performance on the FineGYM dataset. Our model obtains an accuracy of 93.8%, which outperforms the state-of-the-art GCN-based method by 1.2%. Notably, the GCN-based methods are weak in modeling non-connected joint relationships, while our model is able to capture long-range correlations of non-directly connected joints in the skeleton. Therefore, for FineGYM with large movement and deformation, TCTE-Net achieves higher performance than the GCN-based methods.

4 Conclusion

This paper proposes TCTE-Net, a novel framework for skeleton-based action recognition that addresses the limitations of CNN in modeling the irregular topology of the skeletal data. Through the proposed Temporal-Channel Focus module and Dynamic Channel Topology Attention module, we enhance the ability of TCTE-Net to identify critical joint nodes and model the correlation between joints under different motion features. Experiments on three benchmark datasets show that TCTE-Net outperforms the previous state-of-the-art models. Our work contributes to exploring the potential of CNNs for modeling skeletal data, and we hope that this will inspire further investigations in this direction.

Acknowledgements. This work was supported by National Key R&D Program of China under Grant No. 2021YFE0205700, National Natural Science Foundation of China (No. 62276260, 62076235, 62206283, 61976210, 62176254, 62002356, 62006230, 62002357, 62206290), sponsored by Zhejiang Lab (No. 2021KH0AB07).

References

1. Chen, Y., Zhang, Z., Yuan, C., Li, B., Deng, Y., Hu, W.: Channel-wise topology refinement graph convolution for skeleton-based action recognition. In: ICCV (2021)
2. Cheng, K., Zhang, Y., Cao, C., Shi, L., Cheng, J., Lu, H.: Decoupling GCN with DropGraph module for skeleton-based action recognition. In: Vedaldi, A., Bischof, H., Brox, T., Frahm, J.-M. (eds.) ECCV 2020. LNCS, vol. 12369, pp. 536–553. Springer, Cham (2020). https://doi.org/10.1007/978-3-030-58586-0_32
3. Liu, Z., Zhang, H., Chen, Z., Wang, Z., Ouyang, W.: Disentangling and unifying graph convolutions for skeleton-based action recognition. In: CVPR (2020)
4. Shi, L., Zhang, Y., Cheng, J., Lu, H.: Two-stream adaptive graph convolutional networks for skeleton-based action recognition. In: CVPR (2019)
5. Yan, S., Xiong, Y., Lin, D.: Spatial temporal graph convolutional networks for skeleton-based action recognition. In: AAAI (2018)
6. Li, B., Li, X., Zhang, Z., Wu, F.: Spatio-temporal graph routing for skeleton-based action recognition. In: AAAI (2019)
7. Wang, S., et al.: Skeleton-based Action Recognition via Temporal-Channel Aggregation. arXiv preprint arXiv: 2205.15936 (2022)
8. Das, S., Sharma, S., Dai, R., Brémond, F., Thonnat, M.: VPN: learning video-pose embedding for activities of daily living. In: Vedaldi, A., Bischof, H., Brox, T., Frahm, J.-M. (eds.) ECCV 2020. LNCS, vol. 12354, pp. 72–90. Springer, Cham (2020). https://doi.org/10.1007/978-3-030-58545-7_5

9. Vaswani, A., et al.: Attention is all you need. In: NIPS (2017)
10. Fu, J., et al.: Dual attention network for scene segmentation. In: CVPR (2019)
11. Duan, H., Zhao, Y., Chen, K., Lin, D., Dai, B.: Revisiting skeleton-based action recognition. In: CVPR (2022)
12. Caetano, C., Sena, J., Brémond, F., dos Santos, J.A., Schwartz, W.R.: SkeleMotion: a new representation of skeleton joint sequences based on motion information for 3D action recognition. In: AVSS (2019)
13. Feichtenhofer, C., Fan, H., Malik, J., He, K.: SlowFast networks for video recognition. In: ICCV (2019)
14. Xu, K., Ye, F., Zhong, Q., Xie, D.: Topology-aware convolutional neural network for efficient skeleton-based action recognition. In: AAAI (2022)
15. Shahroudy, A., Liu, J., Ng, T.T., Wang, G.: NTU RGB+D: a large-scale dataset for 3D human activity analysis. In: CVPR (2016)
16. Liu, J., Shahroudy, A., Perez, M., Wang, G., Duan, L.Y., Kot, A.C.: NTU RGB+D 120: a large-scale benchmark for 3D human activity understanding. TPAMI **42**(10), 2684–2701 (2019)
17. Shao, D., Zhao, Y., Dai, B., Lin, D.: FineGym: a hierarchical video dataset for fine-grained action understanding. In: CVPR (2020)
18. Shi, L., Zhang, Y., Cheng, J., Lu, H.: Decoupled spatial-temporal attention network for skeleton-based action recognition. In: ACCV (2020)
19. Ke, L., Peng, K.C., Lyu, S.: Towards to-a-t spatio-temporal focus for skeleton-based action recognition. In: AAAI (2022)
20. Lee, J., Lee, M., Lee, D., Lee, S.: Hierarchically Decomposed Graph Convolutional Networks for Skeleton-Based Action Recognition. arXiv preprint arXiv:2208.10741 (2022)
21. Loshchilov, I., Hutter, F.: SGDR: Stochastic Gradient De-scent with Warm Restarts. arXiv preprint arXiv:1608.03983 (2016)
22. Shi, L., Zhang, Y., Cheng, J., Lu, H.: Skeleton-based action recognition with multi-stream adaptive graph convolutional networks. TIP **29**, 9532–9545 (2020)
23. Song, Y.-F., Zhang, Z., Shan, C., Wang, L.: Constructing stronger and faster baselines for skeleton-based action recognition. TPAMI **45**(2), 1474–1488 (2022)
24. Cheng, K., Zhang, Y., He, X., Chen, W., Cheng, J., Lu, H.: Skeleton-based action recognition with shift graph convolutional network. In: CVPR (2020)
25. Chi, H., Ha, M.H., Chi, S., Lee, S.W., Huang, Q., Ramani, K.: InfoGCN: representation learning for human skeleton-based action recognition. In: CVPR (2022)

HFGCN-Based Action Recognition System for Figure Skating

Ying Zhou, Yana Zhang(✉), and Aiqiu Wu

Communication University of China, Beijing 100024, China
zynjenny@cuc.edu.cn

Abstract. With the rising popularity of winter sports, there is a growing interest in figure skating, and an action recognition algorithm for figure skating is increasingly needed. The figure skating action recognition algorithm is not only good for intelligent event understanding but also for objective competition judging. However, figure skating actions are difficult to distinguish by one frame, and the errors caused by the keypoint detection may be propagated in the action classification, and affect the performance of the figure skating action recognition algorithm. In this paper, a figure skating hierarchical dataset FSHD-10 and a Figure Skating Action Recognition System (FSARS) are established. The FSARS adopts a multi-stream structure and a decision fusion module to learn features at different dimensions. A Hierarchical Fine-Grained Graph Convolutional Neural Network (HFGCN) is also proposed in this paper. The HFGCN extracts the temporal features by a temporal modeling module and an attention module. The hierarchical classification adopted in HFGCN takes advantage of the hierarchy of figure skating actions and improves the precision of the action classification. The experimental results show that these improvements contribute to FSARS and make it achieve a final accuracy of 93.70% on the FSHD-10 dataset.

Keywords: Action Recognition · Keypoint Detection · Graph Convolutional Neural Network · Attention Mechanism · Figure Skating

1 Introduction

Human action recognition algorithms are widely applied in video surveillance, virtual reality, human-computer intelligent interaction, and other fields [1]. The existing action recognition algorithms are mainly focused on daily actions. Most of them performed well on the widely used dataset NTU RGB + D [2], which contains mostly daily actions in fixed scenes. Other algorithms focus on actions in specific scenes, such as drivers' actions while driving [3] and students' actions while learning [4].

With the growing interest in figure skating, an action recognition algorithm for figure skating is increasingly needed. A well-behaved action recognition system of figure skating could be used not only for a competition commentary assistant but also for a real-time action annotation in the live broadcasting of the figure skating competition. In addition, the algorithm could help umpires score objectively. However, unlike other sports such as

© The Author(s), under exclusive license to Springer Nature Singapore Pte Ltd. 2024
Q. Liu et al. (Eds.): PRCV 2023, LNCS 14425, pp. 120–130, 2024.
https://doi.org/10.1007/978-981-99-8429-9_10

basketball, actions in figure skating are complex in content and fast in action switching [5]. Many figure skating actions are similar, which is hard to tell the differences. For example, there is only a slight difference in the posture before takeoff between the loop jump and the salchow jump. Therefore, there is a need for an algorithm that can effectively handle action classification with complex action content, rapid action switching, and high similarity. This paper aims to establish a fine-grained action recognition system for figure skating. The main issue is how to learn fine-grained information about figure skating actions.

According to different data modalities, single modality-based action recognition algorithms can be mainly divided into video-based algorithms and skeleton-based algorithms. In the early days, most of the works focused on using RGB or gray-scale videos as input for human action recognition, due to their popularity and easy access [6]. One of the most classic methods for video-based action recognition is the iDT method proposed by Wang et al. [7]. After that, many studies have proposed some well-performing hand-designed features, such as spatial-temporal interesting point features proposed by Das et al. [8]. However, factors such as camera motion, object occlusion, and scene complexity, limit the performance of human action representation and recognition based on handcrafted features [9]. Therefore, deep learning has been widely used. Then, some video-based algorithms based on recurrent neural network [10] and convolutional neural network [11, 12] are proposed. Inspired by the success of Transformers [13] in Natural Language Processing (NLP), plenty of works [14, 15] employed Transformers for human action recognition from videos. Besides video data, skeleton joint information is also widely used in various action recognition tasks as the skeleton is more closely related to human motion than RGB frames or optical flow. Skeleton-based action recognition algorithms are lighter and more general than algorithms based on video. Researchers proposed algorithms based on RNN [16], LSTM [17], and CNN [18] for skeleton-based action recognition. Since few of them perform well in the learning of spatial-temporal co-occurrence features, Yan et al. [19] proposed a spatial-temporal graph convolution algorithm (ST-GCN). After that, researchers realized that graph convolutional networks have great advantages in modeling skeletons, so a lot of GCN-based algorithms [20–29] for action recognition have been proposed.

In summary, two main issues exist in the research of fine-grained action recognition algorithms for figure skating. The first issue is that errors made by the keypoint detection module may be propagated in the action recognition algorithms. The second issue is that figure skating actions are difficult to distinguish by a single frame. The above problems are concerned in this paper.

In this paper, contributions are made to build an action recognition system for figure skating, which are summarized as follows:

1) A figure skating hierarchical dataset (FSHD-10) is established for the research of action recognition of figure skating. The dataset organizes different kinds of figure skating actions hierarchically so that the algorithm can learn the features of different levels. According to the hierarchical structure, a new evaluation standard of figure skating action recognition algorithm is established, which is based on the proportion of different kinds of actions in international figure skating competitions.

2) A Hierarchical Fine-Grained Graph Convolutional Neural Network (HFGCN) is proposed to recognize figure skating actions that are similar in a frame. In order to identify figure skating actions better, this algorithm adopts hierarchical classification to assist the action recognition of subcategories through the common action features of major categories. ACTION-Net and temporal modeling modules are used to enhance the learning of temporal features and spatiotemporal co-occurrence features.
3) A Figure Skating Action Recognition System (FSARS) is designed in this paper. This system uses some data preprocessing methods to reduce noise in the FSHD-10 dataset. Multi-stream structure and decision fusion module are proposed to extract and fuse features at different dimensions of the data. By the HFGCN, this system could recognize the figure skating action more accurately.

The details will be described in the following sections.

2 Figure Skating Hierarchical Dataset

Experiments are done on a Figure Skating Hierarchical Dataset FSHD-10 [31]. The skeleton joints data in FSHD-10 are from the figure skating videos of the Olympic Winter Games Beijing 2022 after human keypoint detection and FSD-10[5] which is a public figure skating data set. FSHD-10 consists of 2300 samples. Considering that the Hollywood2 [32] dataset consists of 3669 clips and the FSD-10 dataset contains just 1484 clips, the inclusion of 2300 samples in FSHD-10 provides a substantial support for research. Each sample contains 2500 frames and each frame contains a human skeleton with 25 joints. The data in FSHD-10 are organized in a hierarchical structure. Each sample has two labels corresponding to different levels, one label to three major categories of actions and the other to ten subcategories of actions. The annotations are followed by the regulations of ISU [33], shown in Table 1.

Table 1. Classification standards of FSHD-10.

Major category	Subcategory	Abbreviation	Label
Jump	Toeloop jump	T	0
	Flip jump	F	1
	Salchow jump	Lz	2
	Loop jump	Lo	3
	Axel jump	A	4
	Lutz jump and Toeloop jump	Lz + T	5
Step sequence	Step sequence	Sq	6
Spin	Camel spin	CSp	7
	Upright spin	USp	8
	Sit Spin	SSp	9

The performance of the action recognition algorithm is always evaluated by classification accuracy. According to the regulations of the ISU, the ratio of jump, spin, and sequence in the competition is about 43:27:13. Therefore, the classification accuracy of ten subcategories acc_10 is calculated as:

$$acc_{10} = 0.5181 \times acc_{jump} + 0.3253 \times acc_{spin} + 0.1566 \times acc_{sequence} \quad (1)$$

in which, acc_{jump}, acc_{sp} and acc_{se} represent the accuracy of subcategories respectively. And the classification accuracy of three major categories is acc_3.

3 Figure Skating Action Recognition System

The framework of the proposed Figure Skating Action Recognition System (FSARS) is shown in Fig. 1. The human keypoint detection module is placed at the beginning of the system to obtain the joint data from figure skating videos. Next, the data preprocessing module removes the incorrect joint data and fills in the missing joint data. Multi-stream generation module transforms preprocessed joint data into three streams: bone stream, joint motion stream, and bone motion stream. After that, three generated streams are sent into HFGCN together with preprocessed joint data, the joint stream. Finally, the decision fusion module integrates the output data of HFGCN to obtain the type of figure skating action.

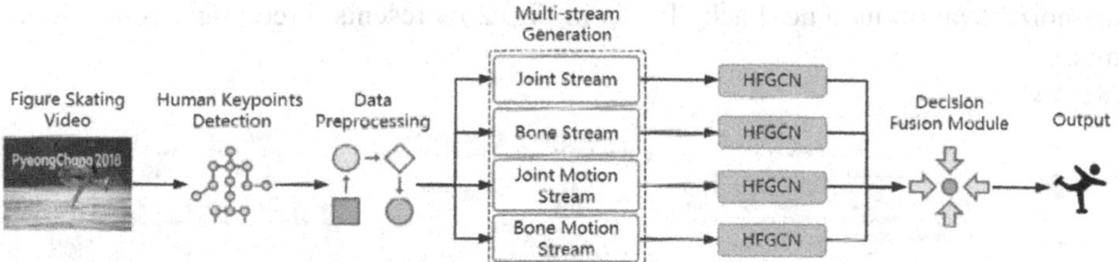

Fig. 1. Framework of FSARS.

3.1 Data Preprocessing

There are mainly two types of abnormal data in the output of the human keypoint detection module: the incorrect data detected from the audience instead of skaters, and the missing joint data not detected by the module. Since most incorrect data are presented as joints concentrated closely, the module distinguishes these incorrect data by detecting the joint coordinates at each frame and replacing them with origin coordinates. For the missing data, presented as origin coordinates, the module fills these data by linear interpolation.

3.2 Multi-stream Generation

To obtain more motion information, joint data are transformed into multi-stream data in this module. The joint data stream consists of the coordinates of each joint in the human body. As the joint data can only reflect the position information of the skeleton joint, the bone data stream is established by subtracting the joint data with a natural connection relationship to represent the length and direction information of the skeleton.

To enhance the learning of temporal information, joint motion stream and bone motion stream are also established. The joint motion stream is obtained by subtracting the joint data of two adjacent frames. Similarly, the bone motion matrix is obtained by subtracting the bone data of two adjacent frames.

These four data streams are trained and tested by HFGCN respectively, and the final result is obtained by adding the results of the four data streams in a certain proportion.

3.3 Hierarchical Fine-Grained Graph Convolutional Neural Network (HFGCN)

Structure of HFGCN. The hierarchical fine-grained graph convolutional neural network (HFGCN) as shown in Fig. 2, consists of ten graph convolution units and two ACTION-Net attention modules. A batch normalization (BN) layer is placed at the beginning of HFGCN to normalize the distribution of input data. After the BN layer, some graph convolution units and ACTION-Net attention modules are placed interleaved. At the end of the network is a hierarchical classification structure. Each graph convolution unit includes spatial graph convolution, temporal modeling module, and data normalization module. Each ellipsis in Fig. 2 represents three graph convolution units.

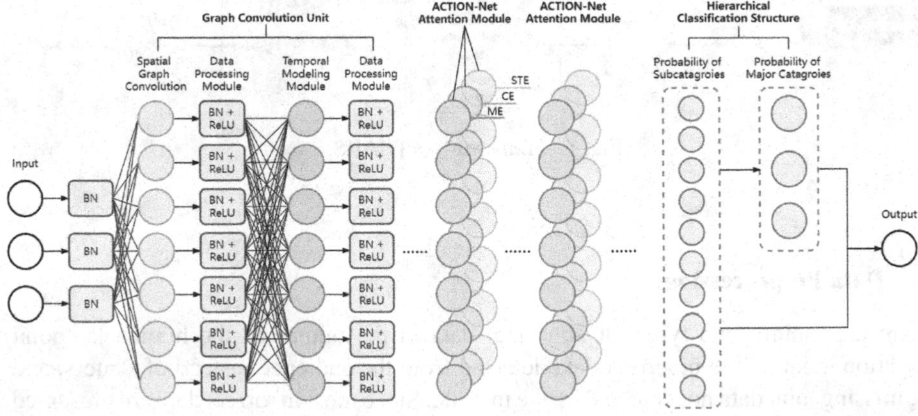

Fig. 2. Structure of the proposed hierarchical fine-grained graph convolutional neural network.

Hierarchical Classification Structure. Figure skating actions are divided into three major categories: jump, sequence, and spin. These three major categories of actions are subdivided into ten subcategories. The proposed algorithm also adopts the hierarchical classification structure to learn the features of both major categories and subcategories. The network calculates the probability that the sample belongs to the major category by adding the probability that the sample belongs to each subcategory.

The hierarchical classification structure is divided into three layers. The first layer calculates the accuracy and loss of each subcategory, the second layer calculates the probability that the sample belongs to a major category and calculates the accuracy and loss of each major category, and the third layer adds the output losses of the first and second layers.

Since figure skating competitions have different rules for major subcategories and subcategories, the proposed hierarchical classification structure matches the rules of the figure skating competition.

Graph Convolution Units. The graph convolution unit (shown in Fig. 3) consists of three parts: spatial graph convolutional module (Convs), data processing module (ModuleD), and temporal modeling module (ModuleT). This unit adopts residual linking to enhance model stability.

An adaptive graph convolutional layer [20] is introduced as Convs, which can learn the strength of the connection between joints. ModuleT and Convs are each followed by a ModuleD to process the output data. ModuleD normalizes the data distribution and alleviates over-fitting by the BN layer and rectified linear unit(ReLU).

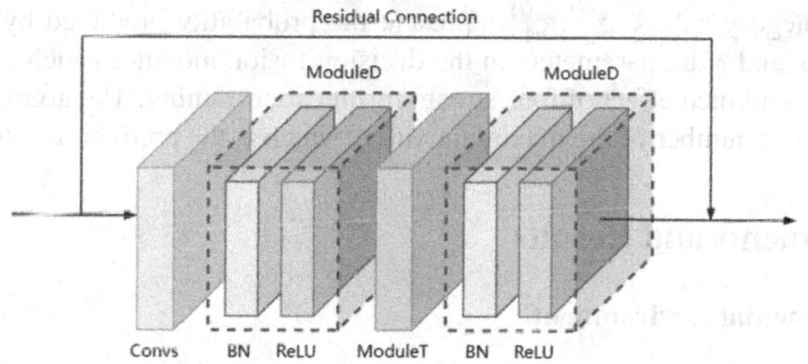

Fig. 3. Structure of graph convolution unit.

Temporal Modeling Module. To better learn figure skating motion features of different durations, a temporal modeling module [26] is set in the graph convolution unit. This module divides the input data into six branches in the channel by six convolution layers. The first five branches are subjected to convolution with four different dilations and a max pooling layer. To preserve the original features of the input data and enhance the model stability, the data of the sixth branch is not processed. After six branches are concatenated in the channel, the residual link structure is added. This module in graph convolution units helps to enhance the model's ability to learn temporal information.

ACTION-Net Attention Module. ACTION-Net attention module [30] includes Spatial-Temporal Excitation (STE), Channel Excitation (CE), and Motion Excitation (ME). STE extracts spatial-temporal features by convolution, CE models the channel interdependence, and ME learns data motion features by calculating the difference between two adjacent frames. The results of STE, CE, and ME are added as the result of the ACTION-Net attention module. This module is placed between the graph convolution units to help the model learn more critical joint information.

3.4 Decision Fusion Module

The decision fusion module is used to integrate the result of HFGCN and output the class of figure skating actions. The decision fusion module sums the probabilities predicted by the four data streams and predicts the category to which the sample belongs. The four-stream data of the sample are the joint data, the bone data, the joint motion data, and the bone motion data. Their contributions to the results can be appropriately distributed by changing the parameters in the decision fusion module. The decision fusion module is formulated as

$$r = argmax(z^{(0)}, \cdots, z^{(c)}), c = \begin{cases} 2, \textit{if} \text{ majorcategory} \\ 9, \textit{if} \text{ subcategory} \end{cases} \tag{2}$$

$$z^{(i)} = z_1^{(i)} + \alpha z_2^{(i)} + \delta\left(z_3^{(i)} + \omega z_4^{(i)}\right), i \in [0, c] \cap i \in Z^+ \tag{3}$$

where r is the prediction category. $z^{(i)}$ represents the probability that the sample belongs to the i-th category. $z_1^{(i)}$,$z_2^{(i)}$,$z_3^{(i)}$, $z_4^{(i)}$ represent the probability predicted by four data-streams. α, ω, and δ are parameters in the decision fusion module, which are adjusted based on the combined effect of four-stream results after training. The argmax function returns the serial number of the maximum value, which is the prediction category r.

4 Experiments and Results

4.1 Experimental Environment

The experimental environment is shown in Table 2.

Table 2. Experimental environment.

Category	Content
OS	Ubuntu 18.04.6 LTS (GNU/Linux 4.4.0–142-generic x86_64)
GPU	GeForce GTX 1080
Language	Python
Platform	Pytorch
ROM	32 GB

The same experimental parameters are made in each test, in which the training epoch is set to 50 and batch size is set to 16.

4.2 Experiment Results and Analysis

The first experiment done here is to test the contribution made by each module in HFGCN. A classic action recognition network 2s-AGCN [20] is set as a baseline.

As shown in Table 3, whether the data is preprocessed or not, 2s-AGCN only achieved an accuracy of less than 80%, indicating that 2s-AGCN does not perform well in the field of figure skating. The HFGCN improves the accuracy from 77.90% to 89.89% (+11.99%), mostly thanks to the module of hierarchical classification. The multi-stream structure improves the accuracy from 89.89% to 90.37%. Further, the ACTION-Net attention module and Temporal modeling module helped to increase the accuracy to 90.68% and 93.70%.

Table 3. The classification accuracy of ten subcategories. The table shows the classification accuracy of ten subcategories based on different models with or without such modules: Data Preprocessing (DP.), Multi-stream Structure (MS.), Temporal Modeling (TM.), and ACTION-Net (A.).

Model	DP	MS	acc_{10}
2s-AGCN (baseline)	/	/	32.09%
2s-AGCN	√	/	77.90%
HFGCN	√	/	89.89%
HFGCN	√	√	90.37%
HFGCN(with A.)	√	√	90.68%
HFGCN(with A.,TM.)	√	√	**93.70%**

The second experiment is conducted to verify the effectiveness of FSARS on FSHD-10, compared with action recognition algorithms: 2s-AGCN [20], MS-G3D [34], and CTR-GCN [26], which perform well in action recognition.

In Table 4, each algorithm has achieved more than 96% accuracy in the sequence subcategory. In the spin subcategory, FSARS has achieved an accuracy of 99.40%, much higher than that of the other three algorithms. In the jump subcategory, though FSARS has achieved an accuracy of 88.69% lower than that in the spin and sequence subcategory, 2s-AGCN, MS-G3D, and CTR-GCN just achieved an accuracy of 61.63%, 41.07%, and 83.93% respectively.No matter the distinction of major categories or the distinction of subcategories, FSARS gets the accuracy of 99.35% and 93.7% respectively, which are higher than other algorithms.

The accuracy of 2s-AGCN is 21.01% higher in major categories than in subcategories, the accuracy of MS-G3D is 31.81% higher, the accuracy of CTR-GCN is 8.53% higher, while the accuracy of FSARS is 5.65% higher. The accuracy differences between major categories and subcategories indicate that the hierarchical classification structure is valuable to closing the gap between the accuracy of major categories and subcategories, which improves the performance of the ten subcategories' figure skating action recognition.

Table 4. Accuracy of different action recognition algorithms on FSHD-10.

Algorithm	acc_3	acc_{jump}	$acc_{sequence}$	acc_{spin}	acc_{10}
2s-AGCN [20]	98.91%	61.63%	96.80%	95.21%	77.90%
MS-G3D [34]	96.62%	41.07%	97.60%	86.83%	64.81%
CTR-GCN [26]	98.52%	83.93%	**99.20%**	95.21%	89.99%
FSARS	**99.35%**	**88.69%**	98.40%	**99.40%**	**93.70%**

5 Conclusion

This paper focuses on figure skating action recognition and makes three main contributions: a figure skating hierarchical dataset FSHD-10, a hierarchical fine-grained graph convolutional neural network HFGCN, and a figure skating action recognition system FSARS. The experimental results on FSHD-10 show that the hierarchical classification structure, the ACTION-Net attention module, and the temporal modeling module contribute to HFGCN. In addition, FSARS has achieved an accuracy of 93.70% on the FSHD-10, which is higher than other methods. The research can help the public tell the different actions of figure skating, making it easier to understand figure skating. In addition, the research is helpful for action classification with complex action content, fast action switching, and high similarity. In the future, the research of figure skating action recognition could be devoted to making the model capture the slight variations in jumps. It is also worthy to be developed from single figure skating to pairs skating. Furthermore, figure skating action recognition is a significant part of fine-grained action recognition, whose research also contributes to improving other fine-grained action recognition algorithms.

Acknowledgment. This work was supported by the National Natural Science Foundation of China under Grant 61901422 and the National Key R&D Program of China 2021YFF0900700.

References

1. Li, R.F., Wang, L.L., Wang, K.: A survey of human body action recognition. Pattern Recognition Artificial Intelligence **27**(1), 35–48 (2014)
2. Shahroudy, A., Liu, J., Ng, T. T., Wang, G.: Ntu rgb+ d: A large scale dataset for 3d human activity analysis. In: Proceedings of the IEEE conference on computer vision and pattern recognition, pp. 1010–1019 (2016)
3. Duan, H., Zhang, C., Lv, Z.: Driving action recognition based on 3d convolution. In: 2021 17th International Conference on Computational Intelligence and Security (CIS), pp. 94–98 (2021)
4. Wu, D., Dang, D., Wang, J.: Recognition of students combining features of zernike moment and optical flow. In: 2016 2nd IEEE International Conference on Computer and Communications (ICCC), pp. 676–679 (2016)
5. Liu, S., et al.: Fsd-10: A fine-grained classification dataset for figure skating. Neurocomputing **413**, 360–367 (2020)

6. Sun, Z., Ke, Q., Rahmani, H., H., Bennamoun, M., Wang, G., Liu, J.: Human action recognition from various data modalities: A review. IEEE transactions on pattern analysis and machine intelligence 45(3), 3200–3225 (2022)
7. Wang, H., Klaser, A., Schmid, C., Liu, C.L.: Dense trajectories and motion boundary descriptors for action recognition. Int. J. Comput. Vision **103**(1), 60–79 (2013)
8. Dawn, D.D., Shaikh, S.H.: A comprehensive survey of human action recognition with spatio-temporal interest point (stip) detector. Vis. Comput. **32**(3), 289–306 (2016)
9. Zhang, H.B., et al.: A comprehensive survey of vision-based human action recognition methods. Sensors **19**(5), 1005 (2019)
10. Li, S., Li, W., Cook, C., Zhu, C., Gao, Y.: Independently recurrent neural network (indrnn): Building a longer and deeper rnn. In: Proceedings of the IEEE conference on computer vision and pattern recognition, pp. 5457–5466 (2018)
11. Tran, D., Bourdev, L., Fergus, R., Torresani, L., Paluri, M.: Learning spatiotemporal features with 3d convolutional networks. In: Proceedings of the IEEE international conference on computer vision, pp. 4489–4497 (2015)
12. Feichtenhofer, C., Pinz, A., Zisserman, A.: Convolutional two-stream network fusion for video action recognition. In: Proceedings of the IEEE conference on computer vision and pattern recognition, pp. 1933– 1941 (2016)
13. Vaswani, A., Shazeer, N., Parmar, N., Uszkoreit, J., Jones, L., Gomez, A. N., et al.: Attention is all you need. In: Advances in Neural Information Processing Systems 30, (2017)
14. Ryoo, M. S., Piergiovanni, A. J., Arnab, A., Dehghani, M., Angelova, A.: Tokenlearner: What can 8 learned tokens do for images and videos?. arXiv, arXiv:2106.11297 (2021)
15. Li, K., Wang, Y., Peng, G., Song, G., Liu, Y., Li, H., Qiao, Y.: UniFormer: Unified Transformer for Efficient Spatial-Temporal Representation Learning. In: International Conference on Learning Representations, (2022)
16. Du, Y., Wang, W., Wang, L.: Hierarchical recurrent neural network for skeleton-based action recognition. In: Proceedings of the IEEE conference on computer vision and pattern recognition, pp. 1110– 1118 (2015)
17. Liu, J., Wang, G., Duan, L.Y., Abdiyeva, K., Kot, A.C.: Skeletonbased human action recognition with global context-aware attention lstm networks. IEEE Trans. Image Process. **27**(4), 1586–1599 (2017)
18. Du, Y., Fu, Y., Wang, L.: Skeleton based action recognition with convolutional neural network. In: 2015 3rd IAPR Asian conference on pattern recognition (ACPR), pp. 579–583 (2015)
19. Yan, S., Xiong, Y., Lin, D.: Spatial temporal graph convolutional networks for skeleton-based action recognition. In: Thirty-second AAAI conference on artificial intelligence, pp. 7444–7452 (2018)
20. Shi, L., Zhang, Y., Cheng, J., Lu, H.: Two-stream adaptive graph convolutional networks for skeleton-based action recognition. In: Proceedings of the IEEE/CVF conference on computer vision and pattern recognition, pp. 12026–12035 (2019)
21. Shi, L., Zhang, Y., Cheng, J., Lu, H.: Skeleton-based action recognition with directed graph neural networks. In: Proceedings of the IEEE/CVF Conference on Computer Vision and Pattern Recognition, pp. 7912–7921 (2019)
22. Shi, L., Zhang, Y., Cheng, J., Lu, H.: Skeleton-based action recognition with multi-stream adaptive graph convolutional networks. IEEE Trans. Image Process. **29**, 9532–9545 (2020)
23. Tang, Y., Tian, Y., Lu, J., Li, P., Zhou, J.: Deep progressive reinforcement learning for skeleton-based action recognition. In: Proceedings of the IEEE conference on computer vision and pattern recognition, pp. 5323–5332 (2018)
24. Song, Y. F., Zhang, Z., Wang, L.: Richly activated graph convolutional network for action recognition with incomplete skeletons. In: 2019 IEEE International Conference on Image Processing (ICIP), pp. 1–5 (2019)

25. Obinata, Y., Yamamoto, T.: Temporal extension module for skeletonbased action recognition. In: 2020 25th International Conference on Pattern Recognition (ICPR), pp. 534–540 (2021)
26. Chen, Y., Zhang, Z., Yuan, C., Li, B., Deng, Y., Hu, W.: Channelwise topology refinement graph convolution for skeleton-based action recognition. In: Proceedings of the IEEE/CVF International Conference on Computer Vision, pp. 13359–13368 (2021)
27. Duan, H., Zhao, Y., Chen, K., Lin, D., Dai, B.: Revisiting skeleton-based action recognition. In: Proceedings of the IEEE/CVF Conference on Computer Vision and Pattern Recognition, pp. 2969–2978 (2022)
28. Liu, J., Wang, X., Wang, C., Gao, Y., Liu, M.: Temporal decoupling graph convolutional network for skeleton-based gesture recognition. IEEE Transactions on Multimedia, (2023)
29. Liu, M., Meng, F., Chen, C., Wu, S.: Novel motion patterns matter for practical skeleton-based action recognition. In: AAAI Conference on Artificial Intelligence, (2023)
30. Wang, Z., She, Q., Smolic, A.: Action-net: Multipath excitation for action recognition. In: Proceedings of the IEEE/CVF conference on computer vision and pattern recognition, pp. 13214–13223 (2021)
31. FSHD-10, https://github.com/YMediaLab/Auto-Montage, last accessed 2023/8/24
32. Marszalek, M., Laptev, I., Schmid, C.: Actions in context. In: Proceedings of the IEEE/CVF conference on computer vision and pattern recognition, pp. 2929–2936 (2009)
33. Special regulations technical rules single pair skating and ice dance 2021, https://www.isu.org/figure-skating/rules/fsk-regulations-rules/file, last accessed 2021
34. Liu, Z., Zhang, H., Chen, Z., Wang, Z., Ouyang, W.: Disentangling and unifying graph convolutions for skeleton-based action recognition. In: Proceedings of the IEEE/CVF conference on computer vision and pattern recognition, pp. 143–152 (2020)

Multi-modal Information Processing

Image Priors Assisted Pre-training for Point Cloud Shape Analysis

Zhengyu Li, Yao Wu, and Yanyun Qu(✉)

School of Informatics, Xiamen University, Xiamen, China
yyqu@xmu.edu.cn

Abstract. Self-Supervised Learning (SSL) is a viable technique to unleash the scalability and generalization of the network. Nevertheless, the representations learned by existing 3D SSL are still insufficient for point cloud shape analysis. Compared to point clouds, images have better strength in providing fine-grained semantic information. In this paper, we propose a framework of cross-modal SSL, enabling knowledge transfer across modalities. Specifically, two self-supervised tasks are meticulously designed. The former constructs instance-level consistency, which computes the similarity between objects to discriminate instances of different modalities. The latter constructs cluster-level consistency, which mines category coherence by grouping point clouds and corresponding images into the same semantic regions. By jointly learning feature representations and cluster assignments, the model incorporates the latent category information to reduce the intra-cluster variance as well as increase the inter-cluster variance. Extensive experimental results demonstrate the effectiveness on three representative datasets.

Keywords: Image Priors · Instance-Level · Cluster-Level

1 Introduction

3D geometric data such as point cloud provides a natural representation for the real world, which is crucial to a diversity of applications such as autonomous driving, robotics, and augmented/virtual reality. Recent advances in 3D neural networks tailored for irregular point clouds have achieved great success in classification and segmentation [5,7,15,20]. However, these methods are trained from scratch owing to the lack of plentiful labeled 3D datasets. This difficulty motivates a flux of research into 3D SSL. In recent years, some pioneering works [13,19,22] successfully bring the 2D SSL [4,10] paradigm into 3D, but they neglect the prior semantic information generated by the 2D pre-trained backbone. Afham *et.al* [1] first introduce cross-modal SSL to effectively utilize semantic information from images for point cloud shape analysis.

Supplementary Information The online version contains supplementary material available at https://doi.org/10.1007/978-981-99-8429-9_11.

© The Author(s), under exclusive license to Springer Nature Singapore Pte Ltd. 2024
Q. Liu et al. (Eds.): PRCV 2023, LNCS 14425, pp. 133–145, 2024.
https://doi.org/10.1007/978-981-99-8429-9_11

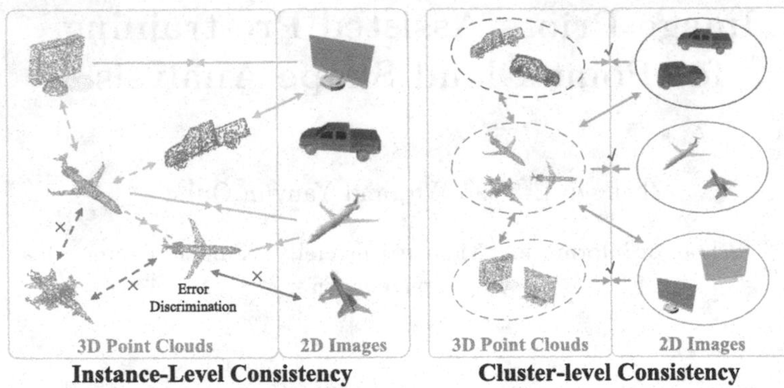

Fig. 1. An example of cross-modal SSL. The interactions between point clouds and images act on two levels: (i) instance-level consistency and (ii) cluster-level consistency.

Intuitively, point clouds and images are heterogeneous that treated as an independent problem. Although their low-level representations are drastically different, they can represent the same latent visual concept. Therefore, instance discrimination [10] can be applied to cross-modal learning for encouraging the 2D feature to be embedded close to the corresponding 3D feature [1]. Despite clear progress, existing cross-modal SSL methods only focus on instance-level consistency. As shown in Fig. 1(i), instance-level consistency maximizes the correlation between self-augmented samples but ignores latent category information, which restricts the representation ability of images and point clouds. Therefore, we hope that the latent category information can be incorporated to promote knowledge transfer across modalities.

Specifically, we propose a framework of cross-modal self-supervised learning for point cloud shape analysis, addressing the limitation of instance-level consistency and encouraging knowledge transfer across modalities. Firstly, each 3D and 2D sample is fed into the point cloud encoder and pre-trained image encoder to generate the 3D and 2D features. Afterward, these features are fed into the multi-task cross-modal SSL via the multi-task projector, which contains two self-supervised tasks, *i.e.*, instance-level consistency and cluster-level consistency. The former outputs a continuous feature vector through an instance projection head (IPH), whose similarity between objects is calculated to discriminate instances of different modalities. The latter outputs a cluster-assignment probability vector through a cluster projection head (CPH), which mines category coherence by grouping point clouds and corresponding images in the same semantic region. The backbone network serves as a medium for information interaction between the two heads, allowing the CPH to leverage discriminative fine-grained patterns captured by the IPH to extract correct coarse-grained cluster-level patterns. Via the cluster-level consistency, the model can effectively modulate the inter-cluster and intra-cluster similarities between different modalities as shown in Fig. 1(ii).

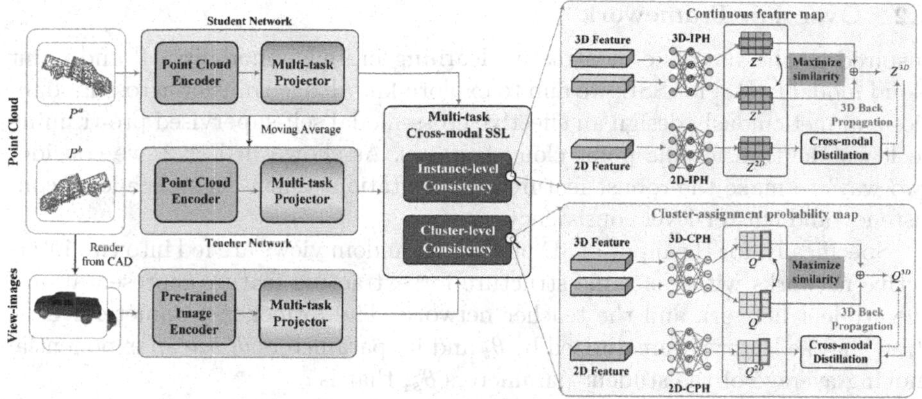

Fig. 2. An overview of our method. It consists of two major components in the multi-task cross-modal SSL: i) instance-level consistency and ii) cluster-level consistency. Parameters of the teacher network are updated by the moving average of the student network. The weights of the learned point cloud encoder of the student network are exploited as initialization for downstream tasks.

Our contributions can be summarized as follows:

- We propose a framework of cross-modal SSL for point cloud shape analysis from the perspective of knowledge transfer across modalities to improve self-supervised representation capabilities.
- We reveal that the latent category information of cluster-level consistency can be incorporated to leverage discriminative fine-grained patterns captured by the instance-level consistency to extract correct coarse-grained cluster-level patterns.
- Extensive experimental results demonstrate the effectiveness on three representative datasets, *i.e.*, ModelNet40, ScanObjectNN, and ShapeNetPart.

2 Proposed Method

2.1 Problem Setting

In this paper, we aim to pre-train a general initial weight for point cloud encoder without using labels. To this end, given a set of dataset $\mathcal{D} = \{(\mathbf{P}_k, \mathbf{I}_k)\}_{k=1}^{|\mathcal{D}|}$ with $\mathbf{P}_k \in \mathbb{R}^{N_P \times 3}$ and $\mathbf{I}_k \in \mathbb{R}^{H \times W \times 3}$, where $\mathbf{I}_k$ is the view image rendering from the 3D CAD model or perspective projecting the raw point cloud. Let $f_P : \mathbb{R}^{N_P \times 3} \rightarrow \mathbb{R}^{N_P \times D_P}$ be a point cloud encoder, that takes one point cloud as input and outputs one point-wise feature. Then, we leverage the availability of an image encoder $f_I : \mathbb{R}^{H \times W \times 3} \rightarrow \mathbb{R}^{H_I \times W_I \times D_I}$ pre-trained by a large-scale 2D image dataset (e.g. ImageNet [16]), that takes one image as input and outputs one image-wise feature.

2.2 Overview Framework

Inspired by the success of contrastive learning in image modality [4] and point cloud modality [13] for SSL, we aim to explore knowledge transfer across modalities and meticulously design an effective cross-modal self-supervised pre-training to learn the transferable point cloud features. As shown in Fig. 2, we consider two ways to make the robust feature representation, that is, instance-level consistency and cluster-level consistency.

Specifically, for the input of 3D modality, random views are fed into two interactive networks with the same structure for extracting feature representations: the student network and the teacher network. The former is parameterized by θ_s, while the latter parameterized by θ_t and its parameters θ_t are an exponential moving average of the student parameters θ_s, that is

$$\theta_t \leftarrow \lambda\theta_t + (1-\lambda)\theta_s, \tag{1}$$

where λ is the decay rate of the moving average. After feature extraction, 3D features are fed into the multi-task projector, which contains a learnable 3D instance projection head g_P^{IPH} and a learnable 3D cluster projection head g_P^{CPH}, gathering the positive pairs and dispersing the negatives by capturing the randomness and invariance across different views. To better exploit the 2D information, we transfer the knowledge from the image encoder to the point cloud encoder by using cross-modal distillation. Concretely, 2D features are generated from the pre-trained image encoder and then fed into the multi-task projector, which contains a learnable 2D instance projection head g_I^{IPH} and a learnable 2D cluster projection head g_I^{CPH}. It favors a solution where an image feature is more correlated to its corresponding point cloud than other instances and clusters.

2.3 Multi-task Cross-Modal SSL

Instance-Level Consistency. Invariance under transformations is crucial for the model to capture the features of 3D objects. To this end, we first design to utilize the representation discrepancy among different augmentations to formulate instance discrimination. Following [13], we augment the original input $\mathbf{P}$ into $\mathbf{P}^a$ and $\mathbf{P}^b$ by using spatial-temporal correlation. Both of them are fed into the point cloud encoder and output 3D feature matrices $Z^a, Z^b \in \mathbb{R}^{N\times M}$ via $g_P^{IPH}(f_P(\cdot))$. For specific samples z_i^a and z_i^b (the i-th row of Z^a and Z^b), the pair-wise similarity between instance pairs measured by cosine distance, that is

$$s_1(z_i, z_j) = \frac{(z_i)(z_j)^\top}{\|z_i\|\,\|z_j\|}, \tag{2}$$

where $i, j \in \{1, ..., N\}$ and N is the mini-batch size. To maximize pairwise similarities $s_1(\cdot,\cdot)$, following [4], we leverage NT-Xent loss for the intra-modal instance discrimination. The loss is in the form of

$$\ell_i^{a\to b} = -\log \frac{\exp(\frac{s_1(z_i^a, z_i^b)}{\tau_1})}{\sum\limits_{j=1, j\neq i}^{N}\left[\exp(\frac{s_1(z_i^a, z_j^a)}{\tau_1}) + \exp(\frac{s_1(z_i^a, z_j^b)}{\tau_1})\right]}, \tag{3}$$

where τ_1 is the instance-level temperature parameter to control the softness.

To better exploit the 2D knowledge for 3D pre-training, for each 3D object, we use multiple view-images $\mathbf{I}$ as additional data. In addition to that, we perform color jittering and random horizontal flip as image augmentation. Then, augmented images are fed into the pre-trained image encoder and output 2D feature matrix $Z^I \in \mathbb{R}^{N\times M}$ via $g_I^{IPH}(f_I(\cdot))$. Similar to Eq. (3), we conduct knowledge distillation for cross-modal instance discrimination by leveraging NT-Xent loss, which can be defined as

$$\ell_i^{I\to P} = -\log \frac{\exp(\frac{s_1(z_i^I, z_i^P)}{\tau_1})}{\sum_{j=1, j\neq i}^{N}\left[\exp(\frac{s_1(z_i^I, z_j^I)}{\tau_1}) + \exp(\frac{s_1(z_i^I, z_j^P)}{\tau_1})\right]}, \tag{4}$$

where z_i^P is the 3D feature vector of transformation mixing, *i.e.*, the i-th row of $Z_i^P = 0.5 \times (Z_i^a + Z_i^b)$.

For the losses $\ell_i^{a\to b}$ and $\ell_i^{I\to P}$, we aim to achieve knowledge distillation of instances by attracting a 3D feature vector to its self-augmentation and corresponding 2D feature while separating the 3D feature from other 2D/3D features at the same time. To this end, instance-level cross-modal (IC) loss is proposed to minimize the relative distance between the point cloud and the image representations. By traversing all instances, $\mathcal{L}_{IC}$ is then formulated as

$$\mathcal{L}_{IC} = \frac{1}{2N}\sum_{i=1}^{N}\left[\ell_i^{a\to b} + \ell_i^{b\to a} + \ell_i^{I\to P} + \ell_i^{P\to I}\right]. \tag{5}$$

Cluster-Level Consistency. Following the idea of "label as representation", when projecting a sample into a space whose dimensionality equals to the number of cluster. Therefore, we aim to learn a parametric projection head that maps a sample to a cluster assignment probability vector $q_i = (q_{i,1}, ..., q_{i,C})$ whose components $q_{i,c}$ characterize how likely point cloud belongs to the cluster $c \in \{1, ..., C\}$, C is the number of clusters. Intuitively, we obtain $Q^a, Q^b \in \mathbb{R}^{N\times C}$ as the representation of $\mathbf{P}$ via $g_P^{CPH}(f_P(\cdot))$ and use these matrices to capture the clustering information.

Since each sample belongs to only one cluster, the rows of Q^a and Q^b ideally tend to be one-hot vector. In this sense, specific samples q_i^a and q_i^b (the i-th column of Q^a and Q^b) can be seen as the representation of the c-th cluster and all columns should differ from each other. In other words, q_i^a and q_i^b could be understood as the distribution of two discrete cluster-assignment variables over C clusters. Then, the pair-wise similarity between cluster pairs can be defined as

$$s_2(q_i, q_j) = \frac{(q_i)^\top (q_j)}{\|q_i\| \, \|q_j\|}, \tag{6}$$

where $i, j \in \{1, ..., C\}$. To maximize pairwise similarities $s_2(\cdot,\cdot)$, without loss of generality, we propose ClusterNT-Xent loss, which is adopted to separate clusters for the intra-modal cluster discrimination, that is

$$\tilde{\ell}_i^{a\to b} = -\log \frac{\exp(\frac{s_2(q_i^a, q_i^b)}{\tau_2})}{\sum_{j=1,j\neq i}^{N}\left[\exp(\frac{s_2(q_i^a, q_j^a)}{\tau_2}) + \exp(\frac{s_2(q_i^a, q_j^b)}{\tau_2})\right]}, \tag{7}$$

where τ_2 is the cluster-level temperature parameter.

Similar to instance-level consistency, we conduct knowledge distillation for cross-modal cluster discrimination, which can be defined as

$$\tilde{\ell}_i^{I\to P} = -\log \frac{\exp(\frac{s_2(q_i^I, q_i^P)}{\tau_2})}{\sum_{j=1,j\neq i}^{N}\left[\exp(\frac{s_2(q_i^I, q_j^I)}{\tau_2}) + \exp(\frac{s_2(q_i^I, q_j^P)}{\tau_2})\right]}, \tag{8}$$

where q_i^P is the 3D probability vector of transformation mixing, *i.e.*, the i-th column of $Q_i^P = 0.5 \times (Q_i^a + Q_i^b)$. q_i^I is the i-th column of $Q^I \in \mathbb{R}^{N\times C}$ outputs from $g_I^{CPH}(f_I(\cdot))$.

For the loss $\tilde{\ell}_i^{a\to b}$ and $\tilde{\ell}_i^{I\to P}$, we aim to achieve knowledge distillation of clusters by grouping point clouds and corresponding images to the same semantic regions. It lowers the proportion of "false negatives" as we do not separate the feature likely to belong to one object. To this end, cluster-level cross-modal (CC) loss is proposed to minimize the relative distance between the point cloud and the image cluster. By traversing all clusters, $\mathcal{L}_{CC}$ is then formulated as

$$\begin{aligned}\mathcal{L}_{CC} = &\frac{1}{2C}\sum_{i=1}^{C}\left[\tilde{\ell}_i^{a\to b} + \tilde{\ell}_i^{b\to a} + \tilde{\ell}_i^{I\to P} + \tilde{\ell}_i^{P\to I}\right] \\ &- \mathcal{H}_{ab} - \mathcal{H}_{IP},\end{aligned} \tag{9}$$

where $\mathcal{H}_{ab}$ and $\mathcal{H}_{IP}$ are the information entropy of cluster assignment probabilities within a mini-batch following [12]. For instance, on the one hand, to learn cluster-level consistency of feature representation, it is encouraged to maximize the mutual information $I(Q^a, Q^b)$, which quantifies the amount of information shared by Q^a and Q^b. On the other hand, maximizing the entropy $\mathcal{H}_{ab}$ encourages the model to avoid a degenerate solution in which some instances fall into the same cluster.

2.4 Objective Function

Due to the limited representation capability of categorical probability vectors, model only trained by minimizing the loss $\mathcal{L}_{CC}$ in Eq. (9) is not able to discriminate objects in the same cluster. Because they may capture sub-optimal cluster-level patterns, which leads to unsatisfactory results. Therefore, we combine $\mathcal{L}_{IC}$ with $\mathcal{L}_{CC}$ into a unified framework, which lifts the instance-level consistency to the more compact cluster-level consistency. The overall objective function is:

$$\mathcal{L} = \mathcal{L}_{IC} + \alpha\mathcal{L}_{CC}, \tag{10}$$

Table 1. Quantitative results (Acc., %) of the linear classification. "Uni." denotes the uni-modality and "Multi." denotes the multi-modality. The best results are shown in boldface and the second best value is marked with an underline.

Method	Modality	ModelNet40	ScanObjectNN
PointNet + Jigsaw [17]	Uni.	87.3	55.2
PointNet + Rotation [14]		88.6	-
PointNet + OcCo [19]		88.7	69.5
PointNet + STRL [13]		88.3	74.2
PointNet + CrossPoint [1]	Multi.	<u>89.1</u>	<u>75.6</u>
PointNet + Ours		**89.3**	**77.0**
DGCNN + Jigsaw [17]	Uni.	90.6	59.5
DGCNN + Rotation [14]		90.8	-
DGCNN + OcCo [19]		89.2	78.3
DGCNN + Self-Contrast [6]		89.6	-
DGCNN + STRL [13]		90.9	77.9
DGCNN + CrossPoint [1]	Multi.	<u>91.2</u>	<u>81.7</u>
DGCNN + Ours		**91.5**	**84.9**

where α acts as a trade-off between instance-level consistency and cluster-level consistency across the training process.

3 Experiments and Analysis

3.1 Pre-training Setup

Dataset. For all experiments, we use ShapeNet [3] with rendered RGB images generated from [23] as the pre-training dataset, which has 43,783 CAD models and images. For a given point cloud, we randomly choose several 2D images captured from an arbitrary viewpoint.

Implementation Details. For fair comparisons, we choose PointNet [15] and DGCNN [20] as the point cloud encoder, and select standard ResNet-50 [11] as the pre-trained image encoder without modification. For the teacher network of the point cloud branch, exponential moving average parameter λ is set to 0.99. The instance projection head is implemented by a 2-layer MLP that consists of two successive linear layers of 256 output channels, with batch normalization (BN) and ReLU activation layer between them. The cluster projection head is implemented by the same 2-layer MLP of 13 output channels and attached to an extra SoftMax layer. Besides, the temperature parameter τ_1 and τ_2 is set to 0.1 and 0.5, respectively. More details are described in the supplementary material.

Table 2. Quantitative results (Acc., %) of the few-shot classification. We report the results with standard derivation over the 10 runs of four few-shot settings.

Method	5-way		10-way	
	10-shot	20-shot	10-shot	20-shot
	ModelNet40			
PointNet + Rand	52.0 ± 3.8	57.8 ± 4.9	46.6 ± 4.3	35.2 ± 4.8
PointNet + Jigsaw [17]	66.5 ± 2.5	69.2 ± 2.4	56.9 ± 2.5	66.5 ± 1.4
PointNet + cTree [18]	63.2 ± 3.4	68.9 ± 3.0	49.2 ± 1.9	50.1 ± 1.6
PointNet + OcCo [19]	89.7 ± 1.9	92.4 ± 1.6	83.9 ± 1.8	89.7 ± 1.5
PointNet + CrossPoint [1]	$\underline{90.9 \pm 4.8}$	$\underline{93.5 \pm 4.4}$	$\underline{84.6 \pm 4.7}$	$\mathbf{90.2 \pm 2.2}$
PointNet + Ours	$\mathbf{91.4 \pm 5.9}$	$\mathbf{93.7 \pm 4.8}$	$\mathbf{85.2 \pm 5.0}$	$\underline{90.0 \pm 5.3}$
DGCNN + Rand	31.6 ± 2.8	40.8 ± 4.6	19.9 ± 2.1	16.9 ± 1.5
DGCNN + Jigsaw [17]	34.3 ± 1.3	42.2 ± 3.5	26.0 ± 2.4	29.9 ± 2.6
DGCNN + cTree [18]	60.0 ± 2.8	65.7 ± 2.6	48.5 ± 1.8	53.0 ± 1.3
DGCNN + OcCo [19]	90.6 ± 2.8	92.5 ± 1.9	82.9 ± 1.3	86.5 ± 2.2
DGCNN + CrossPoint [1]	$\underline{92.5 \pm 3.0}$	$\underline{94.9 \pm 2.1}$	$\underline{83.6 \pm 5.3}$	$\underline{87.9 \pm 4.2}$
DGCNN + Ours	$\mathbf{93.5 \pm 4.4}$	$\mathbf{95.4 \pm 3.5}$	$\mathbf{88.6 \pm 3.4}$	$\mathbf{92.6 \pm 3.4}$
	ScanObjectNN			
PointNet + Rand	57.6 ± 2.5	61.4 ± 2.4	41.3 ± 1.3	43.8 ± 1.9
PointNet + Jigsaw [17]	58.6 ± 1.9	67.6 ± 2.1	53.6 ± 1.7	48.1 ± 1.9
PointNet + cTree [18]	59.6 ± 2.3	61.4 ± 1.4	53.0 ± 1.9	50.9 ± 2.1
PointNet + OcCo [19]	$\underline{70.4 \pm 3.3}$	72.2 ± 3.0	54.8 ± 1.3	61.8 ± 1.2
PointNet + CrossPoint [1]	68.2 ± 1.8	$\underline{73.3 \pm 2.9}$	$\underline{58.7 \pm 1.8}$	$\underline{64.6 \pm 1.2}$
PointNet + Ours	$\mathbf{72.9 \pm 7.3}$	$\mathbf{77.4 \pm 6.2}$	$\mathbf{62.2 \pm 4.6}$	$\mathbf{66.9 \pm 3.6}$
DGCNN + Rand	62.0 ± 5.6	67.8 ± 5.1	37.8 ± 4.3	41.8 ± 2.4
DGCNN + Jigsaw [17]	65.2 ± 3.8	72.2 ± 2.7	45.6 ± 3.1	48.2 ± 2.8
DGCNN + cTree [18]	68.4 ± 3.4	71.6 ± 2.9	42.4 ± 2.7	43.0 ± 3.0
DGCNN + OcCo [19]	72.4 ± 1.4	77.2 ± 1.4	57.0 ± 1.3	61.6 ± 1.2
DGCNN + CrossPoint [1]	$\underline{74.8 \pm 1.5}$	$\underline{79.0 \pm 1.2}$	$\underline{62.9 \pm 1.7}$	$\underline{73.9 \pm 2.2}$
DGCNN + Ours	$\mathbf{82.6 \pm 5.7}$	$\mathbf{88.3 \pm 4.6}$	$\mathbf{72.6 \pm 4.5}$	$\mathbf{79.5 \pm 3.9}$

3.2 Downstream Tasks

In this subsection, we report the experimental results on several downstream tasks, including classification and part segmentation. We conduct the comparison experiments with state-of-the-art (SOTA) methods, including uni-modality and multi-modality SSL methods. Besides the widely used benchmarks, we also learn the generalization and capacity of the model in few-shot settings.

3D Object Classification. We adopt the protocols presented in prior work [19] to demonstrate the generalization performance on ModelNet40 [21] and ScanObjectNN [2]. For evaluating the accuracy of our model in 3D object classification, we randomly sample 1,024 points from each object and freeze the point cloud encoder, fitting a simple linear Support Vector Machine (SVM) classifier on the top of the encoded global feature vectors of the train split. We report the test results based on the three runs in Table 1, which tabulates the linear evaluation results of classification accuracy. Significant gains of 7.0% and 3.2% in DGCNN, compared to the SOTA uni- and multi-modality methods tested on ScanObjectNN. These results show the effectiveness of our method, which can generalize to out-of-distribution data as well.

Table 3. Quantitative results (mIoU,%) of the 3D object part segmentation on ShapeNetPart using different labeled data. "Sup." indicates the model trained from scratch, while "Self-sup." indicates the model initialized with pre-training method.

Method	Task	Data Percentage		
		1%	5%	100%
DGCNN [20]	Sup.	-	-	85.2
SR-UNet [5]		-	-	84.7
DGCNN + Jigsaw [17]	Self-sup.	-	-	84.3
DGCNN + OcCo [19]		-	-	85.0
DGCNN + CrossPoint [1]		-	-	85.5
PointCapsNet [25]		67.0	70.0	-
Multi-Task [9]		68.2	77.7	-
DGCNN + Self-Contrast [6]		76.2	79.2	82.3
PointContrast [22]		74.0	79.9	85.1
DGCNN + Ours		**76.9**	**81.0**	**85.7**

Table 4. Ablation study on the effectiveness of components in our method.

$\mathcal{L}_{IC}$		$\mathcal{L}_{CC}$		Accuracy(%)	
$\ell^{a\to b}$	$\ell^{I\to P}$	$\tilde{\ell}^{a\to b}$	$\tilde{\ell}^{I\to P}$	ModelNet40	ScanObjectNN
✓				89.8	80.1
	✓			90.2	81.4
✓	✓			90.8	82.2
		✓	✓	90.2	81.6
✓	✓	✓		91.2	83.8
✓	✓	✓	✓	91.5	84.9

Few-Shot Object Classification. For fair comparisons, we follow previous work [1] to evaluate our model under the few-shot learning (FSL) setting, which aims to train a model that generalizes with limited data. Each FSL task is instantiated as the N-way K-shot 3D object classification, where the model is evaluated on N classes, and each class contains K samples. The model is trained on $N \times K$ samples (support set), and evaluate on the unseen 20 samples (query set) from each of N classes for testing. We evaluate the performance under four general settings. As shown in Table 2, our method achieves the best performance in the FSL task. Particularly, it obtains an absolute improvement of 7.8%, 9.3%, 9.7%, 5.6% in DGCNN over the CrossPoint [1] on the ScanObjectNN.

3D Object Part Segmentation. We perform 3D object part segmentation on ShapeNetPart [24]. We initially pre-train the DGCNN for part segmentation using our method on ShapeNet and fine-tune it in an end-to-end manner on ShapeNetPart. Each object is represented by 2,048 points. As shown in Table 3, we can observe that the pre-trained backbone using our method is superior to the randomly initialized DGCNN backbone with 0.5% mIoU (mean Intersection-over-Union) gains. Meanwhile, we also fine-tune the model with different proportion of annotations in semi-supervised settings. Our method outperforms all compared methods and achieves the largest performance gains in both settings. Furthermore, in order to illustrate the superiority of our approach more intuitively, we have conducted visual analyses in the supplementary materials and compared it with PointContast [22]. This further substantiates the effectiveness of our method.

Table 5. Ablation study on evaluating the pre-trained DGCNN model by linear evaluation under different spatial augmentations and temporal transformations.

Spatial Augmentations	Accuracy(%)
Full	84.9
Remove jitter	84.1
Remove dropout	84.0
Remove cutout	84.4
Remove jit. + drop. + cut.	83.1

(a) Spatial augmentation.

Temporal Transformations	Accuracy(%)
Full	84.9
Remove rotation	84.0
Remove translation	83.8
Remove scaling	84.2
Remove rot. + trans. + sca.	82.9

(b) Temporal transformations.

Table 6. Exploring the impact of different contrastive losses on model performance.

Backbone	Contrastive Loss	ModelNet40	ScanObjectNN
PointNet	MoCo-v1 [10]	89.0	76.5
	BYOL [8]	88.6	75.2
	SimCLR [4]	89.3	77.0
DGCNN	MoCo-v1 [10]	91.2	82.5
	BYOL [8]	90.8	80.1
	SimCLR [4]	91.5	84.9

3.3 Ablation Study

We conduct ablation studies to evaluate the effectiveness of the main components of our proposed method. In addition to that, we provide more details about the number of corresponding 2D images, the weight parameter α, model effectiveness, and visualize more qualitative results in the supplementary material.

Effectiveness of Components. We conduct an ablation study on the effectiveness of each component of our method in Table 4. Above all, we use intra-modal instance discrimination $\ell^{a\to b}$ as our baseline, achieving Acc. of 89.8%/80.1%. Then we leverage cross-modal instance discrimination $\ell^{I\to P}$ and jointly append $\ell^{a\to b}$, the variant of our method is improved in Acc. with the gain of 0.4%/1.3% and 1.0%/2.1%, respectively. The proposed $\mathcal{L}_{CC}$ consists of intra-modal cluster discrimination $\tilde{\ell}^{a\to b}$ and cross-modal cluster discrimination $\tilde{\ell}^{I\to P}$, which outperforms the baseline with the gain of 0.4%/1.5%, however, worse than $\mathcal{L}_{IC}$. Finally, our method combines $\mathcal{L}_{IC}$ with $\mathcal{L}_{CC}$ into a unified framework, which outperforms the baseline with the highest gain of 1.7%/4.8%.

Spatial-Temporal Correlation. We conduct a pilot study to understand the spatial-temporal correlation. As shown in Table 5, we disentangle the combinations by removing certain transformations and augmentations to generate training data on the DGCNN [20] backbone for the ShapeNet [3] dataset during the pre-training stage. The performance below reflects the linear evaluation results on ScanObjectNN [2]. Both of them introduce substantial performance gains.

Contrastive Loss. We exchange different contrastive loss functions such as InfoNCE in MoCo-v1 [10] and mean squared error in BYOL [8] to replace the NT-Xent loss in SimCLR [4]. The linear classification results are shown in the Table 6 and we can see that the NT-Xent loss is the best.

4 Conclusion

In this paper, we propose a framework of cross-modal SSL for point cloud shape analysis. We meticulously design two self-supervised tasks. The former constructs instance-level consistency and the latter constructs cluster-level consistency. With the joint optimization of two cross-modal SSL, the point cloud encoder of student network jointly learns representations and cluster assignments, incorporating the latent category information to reduce the intra-cluster variance as well as increase the inter-cluster variance. Extensive experimental results demonstrate the effectiveness on three datasets. We hope that our work could provide a new perspective on cross-modal SSL for point cloud shape analysis.

Acknowledgments. This work is supported by the National Key Research and Development Program of China No. 2020AAA0108301; National Natural Science Foundation of China under Grant No. 62176224; CCF-Lenovo Blue Ocean Research Fund.

References

1. Afham, M., Dissanayake, I., Dissanayake, D., Dharmasiri, A., Thilakarathna, K., Rodrigo, R.: Crosspoint: self-supervised cross-modal contrastive learning for 3D point cloud understanding. In: CVPR, pp. 9902–9912 (2022)
2. Angelina, M., Pham, Q.H., Hua, B.S., Nguyen, T., Yeung, S.K.: Revisiting point cloud classification: a new benchmark dataset and classification model on real-world data. In: ICCV (2019)
3. Chang, A.X., et al.: Shapenet: an information-rich 3D model repository. arXiv preprint arXiv:1512.03012 (2015)
4. Chen, T., Kornblith, S., Norouzi, M., Hinton, G.: A simple framework for contrastive learning of visual representations. In: ICML, pp. 1597–1607 (2020)
5. Choy, C.B., Gwak, J., Savarese, S.: 4D spatio-temporal convnets: minkowski convolutional neural networks. In: CVPR (2019)
6. Du, B.a., Gao, X., Hu, W., Li, X.: Self-contrastive learning with hard negative sampling for self-supervised point cloud learning. In: ACMMM (2021)
7. Engel, N., Belagiannis, V., Dietmayer, K.: Point transformer. In: ICCV (2021)
8. Grill, J.B., et al.: Bootstrap your own latent: a new approach to self-supervised learning. In: NIPS, pp. 21271–21284 (2020)
9. Hassani, K., Haley, M.: Unsupervised multi-task feature learning on point clouds. In: ICCV (2019)
10. He, K., Fan, H., Wu, Y., Xie, S., Girshick, R.: Momentum contrast for unsupervised visual representation learning. In: CVPR, pp. 9729–9738 (2020)
11. He, K., Zhang, X., Ren, S., Sun, J.: Deep residual learning for image recognition. In: CVPR (2016)
12. Hu, W., Miyato, T., Tokui, S., Matsumoto, E., Sugiyama, M.: Learning discrete representations via information maximizing self-augmented training. In: International Conference on Machine Learning, pp. 1558–1567. PMLR (2017)
13. Huang, S., Xie, Y., Zhu, S.C., Zhu, Y.: Spatio-temporal self-supervised representation learning for 3D point clouds. In: ICCV (2021)
14. Poursaeed, O., Jiang, T., Qiao, Q., Xu, N., Kim, V.G.: Self-supervised learning of point clouds via orientation estimation. In: 3DV (2020)
15. Qi, C., Su, H., Mo, K., Guibas, L.J.: Pointnet: deep learning on point sets for 3D classification and segmentation. In: CVPR (2017)
16. Russakovsky, O., et al.: Imagenet large scale visual recognition challenge. IJCV **115**(3), 211–252 (2015)
17. Sauder, J., Sievers, B.: Self-supervised deep learning on point clouds by reconstructing space. In: NIPS (2019)
18. Sharma, C., Kaul, M.: Self-supervised few-shot learning on point clouds. In: NIPS (2020)
19. Wang, H., Liu, Q., Yue, X., Lasenby, J., Kusner, M.J.: Unsupervised point cloud pre-training via occlusion completion. In: ICCV (2021)
20. Wang, Y., Sun, Y., Liu, Z., Sarma, S.E., Bronstein, M.M., Solomon, J.M.: Dynamic graph CNN for learning on point clouds. TOG **38**(5), 1–12 (2018)
21. Wu, Z., et al.: 3D shapenets: a deep representation for volumetric shapes. In: CVPR (2015)
22. Xie, S., Gu, J., Guo, D., Qi, C.R., Guibas, L., Litany, O.: PointContrast: unsupervised pre-training for 3D point cloud understanding. In: Vedaldi, A., Bischof, H., Brox, T., Frahm, J.-M. (eds.) ECCV 2020. LNCS, vol. 12348, pp. 574–591. Springer, Cham (2020). https://doi.org/10.1007/978-3-030-58580-8_34

23. Xu, Q., Wang, W., Ceylan, D., Mech, R., Neumann, U.: Deep implicit surface network for high-quality single-view 3D reconstruction. In: NIPS (2019)
24. Yi, L., et al.: A scalable active framework for region annotation in 3D shape collections. TOG **35**(7), 1–12 (2016)
25. Zhao, Y., Birdal, T., Deng, H., Tombari, F.: 3D point capsule networks. In: CVPR (2019)

AMM-GAN: Attribute-Matching Memory for Person Text-to-Image Generation

Wei Yue[1,2(✉)]

[1] Institute of Image Communication and Network Engineering, Shanghai Jiao Tong University, Shanghai 200240, China
[2] Shanghai Key Lab of Digital Media Processing and Transmission, Shanghai 200240, China
yuewei@sjtu.edu.cn

Abstract. Using GANs to generate realistic images from text descriptions has achieved great progress. However, it is difficult to represent attributes such as gender, clothing, hairstyle, and accessories completely and accurately, so generating full-body person images remains a challenge. In this paper, we propose the Attribute-Matching Memory Generative Adversarial Network(AMM-GAN) to generate high-quality full-body person images directly from the text. The AMM-GAN uses an attribute text feature extraction module to extract relevant attribute information from the text, and the encoder is dynamically updated to extract more accurate features. We also propose an attribute-matching memory (AMM) generator to refine the image by improving the attention to the whole attribute in a memory network composed of image and text information. Furthermore, the real-result-driven discriminators guide the network to produce more realistic and natural images and achieve better image-text matching. Experimental results demonstrate that our AMM-GAN achieves the best performance in FID and R-Precison metrics compared to existing models.

Keywords: Generative adversarial network · Text-to-image synthesis · Attribute matching · Person image generation

1 Introduction

Generative adversarial networks (GANs) [4] have exhibited remarkable efficacy in synthesizing highly realistic images. Conditional GAN [14] successfully overcomes modality barriers, allowing textual information extraction to generate corresponding images. Subsequent advancements in this field have witnessed rapid progress, with various approaches incorporating diverse modules [2,16,25,26] and designing novel network architectures [21,23,24] to enhance the alignment between text and images and improve the quality of generated outputs. Moreover, GANs have been found extensive application across a wide range of tasks, especially those related to person images [7,8,22].

© The Author(s), under exclusive license to Springer Nature Singapore Pte Ltd. 2024
Q. Liu et al. (Eds.): PRCV 2023, LNCS 14425, pp. 146–158, 2024.
https://doi.org/10.1007/978-981-99-8429-9_12

Person image generation tasks hold significant importance in domains such as virtual reality [15] and target tracking [1]. Generating person images from text is one of the important approaches. Some work [6,7,22] has focused on generating diverse person images by text guidance. They use different text-guided information to change the relative areas on the image to obtain realistic person images. In addition, there are also works [11] that generate the person images directly from the text without any reference information. They use advanced single-stage GAN to fuse text and image features and generate text-matching results. But they do not explore much about the matching degree of the generated images and text descriptions. Also, the generation quality is not remarkable due to the lack of utilization of text features at multiple granularities.

We hope to perform more challenging work to generate more photo-realistic and text-matching person images directly from the text. However, there are some problems in this work. First, the existing methods do not make sufficient use of textual information, especially in the case of more attribute information. Second, person images contain more appearance attributes, such as tops, pants, shoes, and backpacks. Each feature also has an extremely rich variety(jackets, undershirts, skirts, shorts, etc.). It is challenging for the network to learn these attributes accurately. Moreover, synthesizing smooth and natural person shapes requires the high capability of the network, and many methods have serious distortions in the generated results.

To solve these problems, we propose an Attribute-Matching Memory Generative Adversarial Network(AMM-GAN). Our AMM-GAN is a multi-stage network because one-stage network cannot fuse multi-granularity features conveniently. For the first problem, text descriptions are extracted by the encoder into features of different granularity, where attribute features consisting of phrases are purposely extracted to provide richer textual information. For the second problem, we propose a novel AMM generator. The generator uses a memory fusion approach and modifies the image regions with weights from attribute features to refine the image at the word and attribute level, allowing for more accurate image refinement operations and more sufficient feature fusion. The AMM generator and the simple attribute generator [17] are alternately stacked to form the multi-scale feature fusion generators. For the third problem, we design real-result-driven discriminators with gradient penalty [13], which can effectively influence the preference of the discriminator and so is used to guide it towards the true matching result. In order to get more accurate features, the encoder is also updated during the training process.

The main contributions of this paper are as follows:

1) We design a multi-stage GAN model for synthesizing person images directly from text descriptions. Attributed text features are used to enrich the learning capability of the network and use a strategy of updating the encoder to optimize the generative capability.
2) We design a multi-scale feature fusion generator consisting of an initial generator, a proposed AMM generator, and an attribute generator, which can fully fuse text and image features from multi-scale.

3) We use Real-result-driven Discriminators with gradient penalty to generate more realistic matching person images.
4) The experimental results show that our AMM-GAN achieves the best generation results with 39.5843 in FID and 0.8580 in R-Precision. Ablation studies demonstrate the effectiveness of our proposed module.

2 Related Work

2.1 Text-to-image Generative Adversarial Network

In the past few years, generative adversarial networks(GANs) [4] have shown promising results in image generation. Meanwhile, text-to-image generation has also achieved significant success. Firstly, conditional GAN [14] generates images by learning the corresponding features from text descriptions. AttnGAN [23] proposes a novel deep attentional multimodal similarity(DAMSM) model to compute the similarity between text and images and uses a multi-stage structure to generate more detailed images. The DAMSM model measures the matching loss of different modalities by computing their cosine similarities and driving the network to generate real images with consistent features. DMGAN [26] proposes a key-value memory network to implement word-level text modification of image sub-regions. Thus, the image can be consistent with the text content in detail. DAEGAN [17] introduces aspect-level text information. This method focuses more on the fine-grained synthesis of local features. They also devised a global-local image refinement process that matches the visual characteristics of the human eye. Alternating image synthesis with different granularity gives the generated images better global properties and more matching image details.

The models mentioned above are based on multi-stage architecture. Unlike these methods, DFGAN [21] proposes an end-to-end generation model. Specifically, realistic images can be generated using only a pair of generator and discriminator. By using an effective deep fusion block, the network can fuse text and image features and generate better-quality results. SSAGAN [10] adds a semantic space-aware module with mask prediction to further improve the generation quality.

2.2 GANs for Person Image

Current person image processing is mainly based on a text-guided modification of faces [8,22] and text-guided pedestrian pose changes [12]. InsetGAN [3] uses different pre-trained generators to fuse features in the latent space to get realistic person images. Liu et al. [11] performed more challenging work to generate human images directly from text descriptions. Unlike previous work that relies on reference images, this method generates a person with a corresponding appearance through textual information only and does not restrict to a fixed pose. They propose an identity-preserving network to ensure that highly identity-consistent images can be generated for people of the same identity. Then, the generated images of different identities are linearly interpolated using manifold mix-up to distinguish the boundaries of different identities. However, the previous method does not focus on the information of the phrases and the synthesis of attribute features is not comprehensive. We use attribute information as well as more adequate feature fusion methods to improve the quality of the results.

3 Method

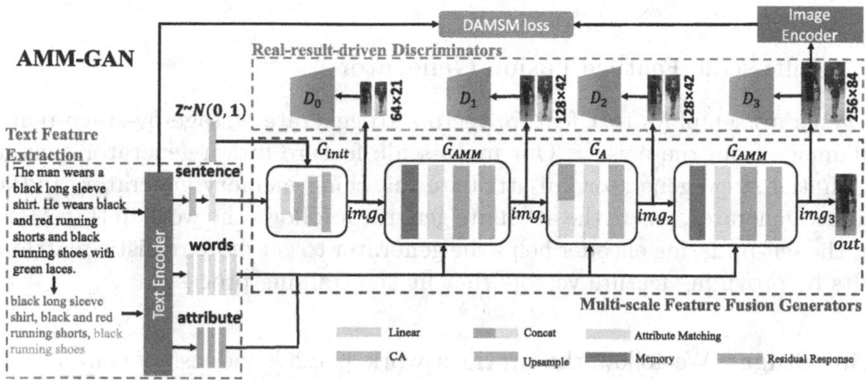

Fig. 1. The structure of AMM-GAN. The model consists of three parts: The Feature Extraction Module, Multi-scale Feature Fusion Generator (see Sect.3.2), and Real-result-driven Discriminator. DAMSM loss is used to update the encoder. G_{init}, G_{AMM}, and G_A are the initial generator, the AMM generator, and the attribute generator, respectively. m_i is the output image feature of the mth generator.

As shown in Fig. 1, our AMM-GAN consists of three main parts. Feature extraction module: features of different granularity are extracted from the input content, including text and images, respectively. Multi-scale Feature Fusion Generator: low-resolution initial images are refined from different granularity and different attention regions in the later stages, and the training process prompts the encoder to obtain more matching text and image feature vectors. Real-result-driven discriminator: a gradient penalty [13] is applied to the discriminator so that the discriminator can judge the real images better and improve the performance of the generation network.

3.1 Feature Extraction

We use the encoders from [23] to extract the feature vectors and retrain them. The text encoder is a bidirectional long short-term memory model [27]. It can extract global sentence features $s \in \mathbb{R}^{256}$ and word features $w \in \mathbb{R}^{256\times30}$ for coarse-grained and fine-grained feature fusion, respectively.

The image encoder is a pre-trained Inception-v3 [20] model on ImageNet [18]. It can also extract global image features $m \in \mathbb{R}^{256}$ and sub-regional image features $\bar{m} \in \mathbb{R}^{256\times17\times17}$ to correspond to coarse and fine granularity, respectively.

Following the idea of [17], we want the network to refine a certain attribute as a whole, such as “long black hair”, “white t-shirt”, etc. For different words describing the same attribute, the network fuses the features at the fine-grained level while maintaining the complete meaning of the attribute. Based on this idea, we use the NLTK tools to find phrases in text descriptions. Different from [17], we mainly select phrases of the “Adj.-Noun” type. The reason is that we hope the non-relevant words will not affect the attribute description. Then, the whole phrases are sent to the text encoder to get the attribute features $a \in \mathbb{R}^{256}$.

3.2 Multi-scale Feature Fusion Generator

The generator employs text feature vectors to generate a stage-by-stage refinement image from the noise z. Our multi-scale feature fusion generator consists of an initial stage generator, an attribute matching memory generator, and an attribute generator, where the attribute generator follows the work in [17]. Moreover, the self-updating encoder helps the generator to get more realistic matching results by providing feature vectors that fit the training process.

Initial Stage. We follow the previous work [23,26] and use a conventional generator to fuse sentence features s and random noise z to obtain the initial image img_0. The subsequent stages are refined on top of this. The img_0 contains global features that ensure the integrity of the whole person's appearance and the main attributes.

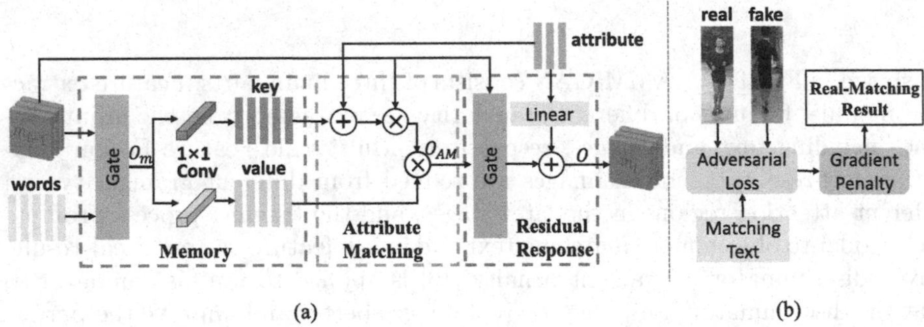

Fig. 2. (a). The main architecture of the proposed AMM. (b). The Real-result-driven Discriminator with a similar gradient penalty item.

Attribute Matching Memory. The one-stage process allows for end-to-end generation and maximizes the integrity of the image. However, for person images, we mainly care about the important attribute features that we expect to receive more attention. The one-stage network is difficult to achieve a rich fusion of information at different granularities, so we start from the best-performing multi-stage network [17] and design a generator that can fully fuse multiple levels of textual information.

As shown in Fig. 2(a), our AMM generator has three steps. the first memory module constructs a memory network [19] for the features of the two modalities by a gate:

$$O_m = G(m, w) \tag{1}$$

where m is the image feature and w is the word feature. The gate operation calculates the importance of each word to the image and maps the text and image information into the same feature space:

$$\begin{aligned} G(m, w) &= g * C(m) + (1 - g) * C(w) \\ g &= \sigma(C(m) + C(w)) \end{aligned} \tag{2}$$

where $C(\cdot)$ is $Conv_{1\times1}$ layer and σ is sigmoid function. The output O_m contains the correspondence between the image and the text. Then the *Key* used to calculate the weight and the memory content *Value* are obtained by $Conv_{1\times1}$ layer respectively.

In the Attribute Matching module, the attributes a are used to calculate the weight with the *Key*:

$$O_{AM} = ((K + a) * m) * V \tag{3}$$

where $(K+a)*m$ is the calculated weight that represents the regions to be refined obtained by multiplying the important words as well as the given attributes and *Value*. Finally, the output O is the result of the weight-guided image modification obtained by the gate operation and added to the attribute values $attr \in \mathbb{R}^{h\times w}$ after the linear layer operation, h and w are same with the size of O:

$$O = G(O_{AM}, w) + attr \tag{4}$$

Self-updating Encoder. For the more accurate generation of results, the encoder should be updated with the training process [10]. According to the general approach of gradient update, we initialize an optimizer for the encoder. We use the DAMSM loss [23] as the objective function.

To obtain the feature vectors of the final generated image directly, the image encoder is placed at the output stage so that the encoder can focus entirely on the quality of the generated image.

3.3 Real-Result-Driven Discriminator

In previous work [17,23,26], the discriminators were optimized by calculating the adversarial loss. Due to the inconsistent text feature vectors obtained by the encoder from different natural language descriptions of the same feature, the loss values will also be inconsistent. For this reason, discriminators often have difficulty judging text-image matching well.

The MAGP proposed in DFGAN [21] forces the discriminator to be more inclusive of the matched text-image pairs, i.e., a slight perturbation of the input vector does not affect the judgment significantly. We apply a similar gradient penalty to the discriminators in the multi-stage network (see Fig. 2(b)). For natural languages with similar meanings and corresponding feature vectors, this method can obtain better consistent generation results and improve text-image matching.

3.4 Objective Functions

The adversarial loss is used to evaluate the ability of the generator:

$$L_{G_i} = -\frac{1}{2}[\mathbb{E}_{\hat{x}\sim Pf_i} \log(D_i(\hat{x})) + \mathbb{E}_{\hat{x}\sim Pf_i} \log(D_i(\hat{x}), s)] \tag{5}$$

where x_i is the real image; $\hat{x}_i$ is the generated image; and s is the text embedding.

To further enhance the ability of the generator to process textual information, an attribute loss function is designed here:

$$L_A = -(\log P(A_i|M_i)P(M_i|A_i)) \tag{6}$$

where the $P(\cdot)$ is the posterior probability of the text to the matched image as mentioned in [23], A_i and M_i are the attribute features and the matched images, respectively. The DAMSM loss [23] is also added to constrain the generator to produce generation results corresponding to the text. CA loss [26] measures the Kullback-Leibler divergence between the standard Gaussian distribution and the training data distribution:

Finally, the loss of the generator is:

$$L_G = \sum_{i=1}^{n} L_{G_i} + L_{CA} + L_{DAMSM} + \lambda_A L_A \tag{7}$$

where λ_A is the weight for L_A.

For discriminators, real and fake images are used to jointly train their judgment ability. Meanwhile, a gradient penalty is applied to each discriminator to induce it to converge to the desired result, so the loss function of a single discriminator is:

$$\begin{aligned} L_{D_i} =& -\frac{1}{2}[\mathbb{E}_{x\sim P_r} \log(D_i(x)) + \mathbb{E}_{\hat{x}\sim Pf_i} \log(1 - D_i(\hat{x})) \\ &+ \mathbb{E}_{x\sim P_r} \log(D_i(x, s)) + \mathbb{E}_{\hat{x}\sim Pf_i} \log(1 - D_i(\hat{x}, s))] \\ &+ \lambda_k \mathbb{E}_{x\sim P_r}[(||\Delta_x D(x, s)||_2 + ||\Delta_x D(x, s)||_2)^p] \end{aligned} \tag{8}$$

where λ_k and p are hyperparameters for gradient penalty.

The total discriminator loss function is $L_D = \sum_{i=1}^{n} L_{D_i}$

4 Experiment

In this section, we first introduce the dataset, experimental details, and evaluation metrics. Then we validate the proposed AMM-GAN on the dataset CUHK-PEDES-ATTR and explore in detail the impact of the proposed module on the final results.

4.1 Dataset

There is no specific dataset used in the text-to-image generation task for person. We choose the CUHK-PEDES [9] used in pedestrian detection to verify the effectiveness of the method. In order to match the dataset to the current task, we obtained a subset CUHK-PEDES-ATTR by filtering out the parts of the original dataset with unobvious attribute features and severe image trailing.

Finally, the CUHK-PEDES-ATTR dataset contains 13,757 images, of which 9,052 are in the training set and 4,705 in the validation set. Each image has two text descriptions of 20 to 40 words in length. The number of non-repeated words in all descriptions is 4833.

4.2 Implementation

Our network is trained using AdamW optimizer with $\beta_1 = 0.5$, $\beta_2 = 0.999$, and weight_decay is set as 0.01. The rest of the parameters follow the settings in DAE-GAN [17]. The learning rate of the generator and discriminator is 0.0002. The smoothing parameters used to calculate the global sentence loss and word-level loss are $GAMMA1 = 4.0$, $GAMMA2 = 5.0$, $GAMMA3 = 10.0$ and the weighting parameter used to calculate the generator loss and discriminator loss is $LAMBDA = 5.0$. The hyperparameters are $\lambda_A = 0.2$, $\lambda_k = 2$, $p = 6$.

4.3 Evaluation Metrics

We use Fréchet Inception Distance (FID) [5] to evaluate the feature distance between the real and the generated images. A smaller FID means that the distribution of the two data is more similar, it also means the better quality of the generated images. Meanwhile, we use R-Precision [23] to evaluate the image-text matching. A larger value means that the generated image has a better image-text matching ability.

4.4 Quantitative Evaluation

We compare the proposed method with existing state-of-the-art models, including AttnGAN [23], ControlGAN [7], DMGAN [26], DFGAN [21], DAE-GAN [17], and SSAGAN [10]. We retrain these models on CUHK-PEDES-ATTR to obtain the experimental results.

The experimental results of all models are shown in Table 1. On the CUHK-PEDES-ATTR dataset, our method achieves the best FID and R-Precision. Compared with DMGAN, we reduce the FID by 1.9780 and achieve a 4.22% improvement in R-Precision. Compared with DAE-GAN, we improve by 0.54% in R-Precision and reduce the FID by 8.5055. Also, our method achieves better generation quality compared to the currently popular one-stage networks. Compared with DFGAN and SSA-GAN, our AMM-GAN reduces the FID by 6.3933 and 0.5110, respectively.

Table 1. FID and R-Precision of state of the art and our AMM-GAN on the CUHK-PEDES-ATTR dataset.

method		FID score ↓	R-Precision(%) ↑
Multi-Stage	AttnGAN [23] (CVPR 2018)	60.9694	70.30±1.36
	ControlGAN [7](NeurIPS 2019)	65.8544	69.24±0.95
	DMGAN [26](CVPR 2019)	41.5623	81.58±1.09
	DAE-GAN [17](ICCV 2021)	48.0898	85.26±1.08
Single-Stage	DFGAN [21](CVPR 2022)	45.9776	——
	SSAGAN [10](CVPR 2022)	40.0953	62.80±0.86
Multi-Stage	AMM-GAN(Ours)	**39.5843**	**85.80±0.75**

The experimental results prove the effectiveness and superiority of the method in this paper. The proposed method can generate results with both good quality and high text-image matching. Meanwhile, the method using one-stage network does not achieve a good text-image matching. This structure may not be sufficient to fully integrate textual information into the image.

4.5 Qualitative Evaluation

We selected DMGAN [26], DFGAN [21], SSAGAN [10], and the proposed AMM-GAN as a comparison. The results show in Fig. 3. Our model generates more realistic and natural results, while the character shape is more fluid. Our model is significantly better than the other two in terms of the portrayal of attributes. Both "skirts", "shirts", and "shorts" can be generated very well. The multi-scale feature fusion ensures that the detailed information is fully utilized. For example, in the second result (top right), "a blue shirt with white and blue design" is also generated accurately. The real-result-driven discriminator also ensures the authenticity of the generated results, such as the synthesis of the legs in result 1 (top left) and the representation of the entire person's pose in result 4 (bottom right).

4.6 Ablation Study

Our model incorporates the Multi-scale Feature Fusion Generator(MFF), Self-updating Encoder(SUE), and Real-result-driven Discriminator(RD). The previous section shows that our model has satisfactory performance. To further investigate the effectiveness and limitations of the improved approaches proposed in this paper, we design the ablation experiments to explore the performance of the three modules. The results of the ablation experiments are shown in Table 2.

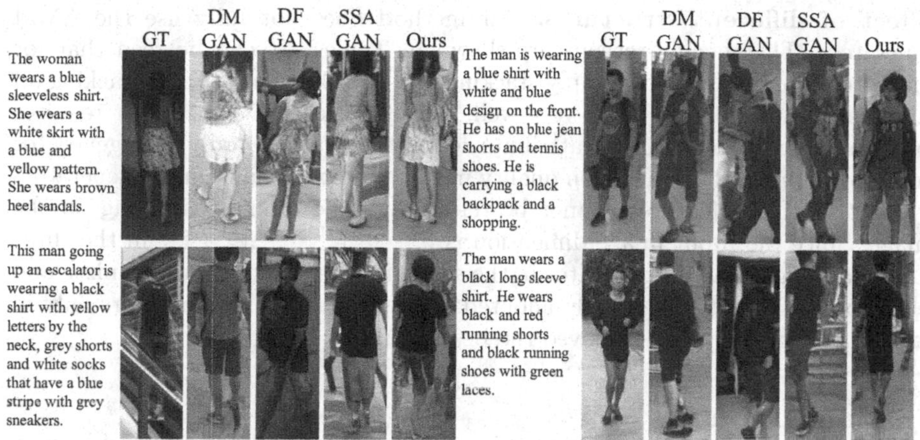

Fig. 3. Example results for text-to-image synthesis by DFGAN, DMGAN, SSAGAN, and AMM-GAN.

Effect of Proposed Modules. Compared with the baseline model, the use of MFF effectively improves the FID metric from 48.0898 to 42.8156, but with some decrease in R-Precision. This may be due to the complex structure used by the MFF module to refine the image quality using more textual features, so the series of operations on features results in matching deviations between feature vectors. The addition of SUE is useful in both generated image quality and text image matching. Among them, there is a significant improvement in the generation quality of the network when MFF and SUE are added together. Meanwhile, the effect of RD is shown in the improvement of the R-Precision index. It can be seen that the text image matching is better maintained in the group with RD, which diminishes the weakening of the MFF module in R-Precision. Finally, the three modules work together to achieve the best generation quality and the best text-image matching.

Table 2. The result of different modules of AMM-GAN on the CUHK-PEDES-ATTR datasets.

MFF	SUE	RD	FID↓	R-Precision(%)↑
—	—	—	48.0898	85.26±1.08
✓	—	—	42.8156	83.14±1.11
—	✓	✓	42.6471	84.77±1.04
✓	✓	—	41.9339	83.08±1.32
✓	—	✓	41.9709	84.00±1.20
✓	✓	✓	**39.5843**	**85.80±0.75**

Table 3. The performance of different architectures of AMM-GAN. G_i means i^{th} generator. ✓ means replace the ith generator with AMM.

G_1	G_2	G_3	FID↓	R-Precision(%)↑
✓	—	—	51.9456	85.77±0.67
✓	✓	—	44.0539	**85.90±0.92**
✓	✓	✓	46.3638	84.81±0.98
✓	—	✓	**39.5843**	85.80±0.75

Effect of different structures. Our method does not fully use the AMM generators. The specific reasons are shown in Table 3. It can be seen that for the latter three generators, simply stacking one AMM generator and stacking all AMM generators do not work very well. Because the AMM generator requires local image refinement using attribute features, replacing only one generator does not improve the generation quality much. Moreover, subsequent generators without attribute features cannot be refined in time after upsampling on the image. Since the input image dimension is $\mathbb{R}^{64\times21}$,$\mathbb{R}^{128\times42}$,$\mathbb{R}^{128\times42}$ in the three stages, adding an AMM generator in the second stage causes over-optimizes and leads to poor quality. We finally adopted to apply the AMM generators only in the upsampling step and achieved the best results.

5 Conclusion

In this paper, we propose a multi-stage generative adversarial network, AMM-GAN, which focuses on generating corresponding person images directly from text descriptions. We use attribute features to achieve full utilization of textual information, design a multi-scale feature fusion generator to achieve more effective image synthesis, and finally use a real-result-driven discriminator to obtain more realistic matching results. Experimental results show that our method outperforms existing methods in terms of both generation quality and text image matching.

Acknowledgements. This research was partly supported by grants of National Natural Science Foundation of China (NSFC, Grant No. 62171281), Science and Technology Commission of Shanghai Municipality (STCSM, Grant Nos. 20DZ1200203, 2021SHZDZX0102, 22DZ2229005).

References

1. Bai, Y., Zhang, Y., Ding, M., Ghanem, B.: Finding tiny faces in the wild with generative adversarial network. In: Proceedings of the IEEE Conference on Computer Vision and Pattern Recognition, pp. 21–30 (2018)
2. Cheng, J., Wu, F., Tian, Y., Wang, L., Tao, D.: RifeGAN: rich feature generation for text-to-image synthesis from prior knowledge. In: Proceedings of the IEEE/CVF Conference on Computer Vision and Pattern Recognition, pp. 10911–10920 (2020)
3. Frühstück, A., Singh, K.K., Shechtman, E., Mitra, N.J., Wonka, P., Lu, J.: Insetgan for full-body image generation. In: Proceedings of the IEEE/CVF Conference on Computer Vision and Pattern Recognition, pp. 7723–7732 (2022)
4. Goodfellow, I., et al.: Generative adversarial nets. In: Advances in Neural Information Processing Systems, vol. 27 (2014)
5. Heusel, M., Ramsauer, H., Unterthiner, T., Nessler, B., Hochreiter, S.: GANs trained by a two time-scale update rule converge to a local Nash equilibrium. Advances in Neural Information Processing Systems, vol. 30 (2017)
6. Jiang, Y., Yang, S., Qiu, H., Wu, W., Loy, C.C., Liu, Z.: Text2Human: text-driven controllable human image generation. ACM Trans. Gr. (TOG) **41**(4), 1–11 (2022)

7. Li, B., Qi, X., Lukasiewicz, T., Torr, P.: Controllable text-to-image generation. In: Advances in Neural Information Processing Systems, vol. 32 (2019)
8. Li, B., Qi, X., Lukasiewicz, T., Torr, P.H.: ManiGAN: text-guided image manipulation. In: Proceedings of the IEEE/CVF Conference on Computer Vision and Pattern Recognition, pp. 7880–7889 (2020)
9. Li, S., Xiao, T., Li, H., Zhou, B., Yue, D., Wang, X.: Person search with natural language description. In: Proceedings of the IEEE Conference on Computer Vision and Pattern Recognition, pp. 1970–1979 (2017)
10. Liao, W., Hu, K., Yang, M.Y., Rosenhahn, B.: Text to image generation with semantic-spatial aware GAN. In: Proceedings of the IEEE/CVF Conference on Computer Vision and Pattern Recognition, pp. 18187–18196 (2022)
11. Liu, D., Wu, L., Li, B., Ge, Z.: T-person-GAN: text-to-person image generation with identity-consistency and manifold mix-up (2023)
12. Ma, L., Jia, X., Sun, Q., Schiele, B., Tuytelaars, T., Van Gool, L.: Pose guided person image generation. In: Advances in Neural Information Processing Systems, vol. 30 (2017)
13. Mescheder, L., Geiger, A., Nowozin, S.: Which training methods for GANs do actually converge? In: International Conference on Machine Learning, pp. 3481–3490. PMLR (2018)
14. Mirza, M., Osindero, S.: Conditional generative adversarial nets. arXiv preprint arXiv:1411.1784 (2014)
15. Prajwal, K., Mukhopadhyay, R., Namboodiri, V.P., Jawahar, C.: A lip sync expert is all you need for speech to lip generation in the wild. In: Proceedings of the 28th ACM International Conference on Multimedia, pp. 484–492 (2020)
16. Qiao, T., Zhang, J., Xu, D., Tao, D.: MirrorGAN: learning text-to-image generation by redescription. In: Proceedings of the IEEE/CVF Conference on Computer Vision and Pattern Recognition, pp. 1505–1514 (2019)
17. Ruan, S., et al.: DAE-GAN: dynamic aspect-aware GAN for text-to-image synthesis. In: Proceedings of the IEEE/CVF International Conference on Computer Vision, pp. 13960–13969 (2021)
18. Russakovsky, O., et al.: Imagenet large scale visual recognition challenge. Int. J. Comput. Vision **115**(3), 211–252 (2015)
19. Sukhbaatar, S., Weston, J., Fergus, R., et al.: End-to-end memory networks. Advances in Neural Information Processing Systems, vol. 28 (2015)
20. Szegedy, C., Vanhoucke, V., Ioffe, S., Shlens, J., Wojna, Z.: Rethinking the inception architecture for computer vision. In: Proceedings of the IEEE Conference on Computer Vision and Pattern Recognition, pp. 2818–2826 (2016)
21. Tao, M., Tang, H., Wu, F., Jing, X.Y., Bao, B.K., Xu, C.: DF-GAN: a simple and effective baseline for text-to-image synthesis. In: Proceedings of the IEEE/CVF Conference on Computer Vision and Pattern Recognition, pp. 16515–16525 (2022)
22. Xia, W., Yang, Y., Xue, J.H., Wu, B.: TediGAN: text-guided diverse face image generation and manipulation. In: Proceedings of the IEEE/CVF Conference on Computer Vision and Pattern Recognition, pp. 2256–2265 (2021)
23. Xu, T., et al.: AttnGAN: fine-grained text to image generation with attentional generative adversarial networks. In: Proceedings of the IEEE Conference on Computer Vision and Pattern Recognition, pp. 1316–1324 (2018)
24. Zhang, H., et al.: StackGAN: text to photo-realistic image synthesis with stacked generative adversarial networks. In: Proceedings of the IEEE International Conference on Computer Vision, pp. 5907–5915 (2017)

25. Zhang, Z., Schomaker, L.: DTGAN: dual attention generative adversarial networks for text-to-image generation. In: 2021 International Joint Conference on Neural Networks (IJCNN), pp. 1–8. IEEE (2021)
26. Zhu, M., Pan, P., Chen, W., Yang, Y.: DM-GAN: dynamic memory generative adversarial networks for text-to-image synthesis. In: Proceedings of the IEEE/CVF Conference on Computer Vision and Pattern Recognition, pp. 5802–5810 (2019)
27. Zia, T., Zahid, U.: Long short-term memory recurrent neural network architectures for Urdu acoustic modeling. Int. J. Speech Technol. **22**(1), 21–30 (2019)

RecFormer: Recurrent Multi-modal Transformer with History-Aware Contrastive Learning for Visual Dialog

Liucun Lu[1], Jinghui Qin[2], Zequn Jie[3], Lin Ma[3], Liang Lin[4], and Xiaodan Liang[1,5](✉)

[1] Shenzhen Campus of Sun Yat-sen University, Guangzhou, China
lulc@mail2.sysu.edu.cn

[2] Guangdong University of Technology, Guangzhou, China

[3] Meituan, Beijing, China

[4] Sun Yat-sen University, Guangzhou, China
linliang@ieee.org

[5] DarkMatter AI Research, Shenzhen, China
xdliang328@gmail.com

Abstract. Recently, benefiting from the powerful representation ability learned from large-scale image-text pre-training, pre-trained vision-language models show significant improvements in visual dialog task. However, these works suffer from two main challenges: 1) how to incorporate the sequential nature of multi-turn dialog systems for better capturing temporal dependencies of visual dialog; 2) how to align the semantics among different modal-specific features for better multi-modal interactions and understandings. To address the above issues, we propose a recurrent multi-modal transformer (named **RecFormer**) to capture temporal dependencies between utterances via encoding dialog utterances and interacting with visual information turn by turn. Specifically, we equip a pre-trained transformer with a recurrent function that maintains cross-modal history encoding for the dialog agent. Thus, the dialog agent can make better predictions by considering temporal dependencies. Besides, we also propose history-aware contrastive learning as an auxiliary task to align visual features and dialog history features for improving visual dialog understanding. The experimental results demonstrate that our RecFormer can achieve new state-of-the-art performances on both VisDial v0.9 (72.52 MRR score and 60.47 R@1 on val split) and VisDial v1.0 (69.29 MRR score and 55.90 R@1 on test-std split) datasets.

Keywords: Visual dialog · Contrastive learning · Multi-modal learning

1 Introduction

The visual dialog task aims to ask a dialog agent to understand the multi-modal features from visual information (e.g. image) and textual information

© The Author(s), under exclusive license to Springer Nature Singapore Pte Ltd. 2024
Q. Liu et al. (Eds.): PRCV 2023, LNCS 14425, pp. 159–171, 2024.
https://doi.org/10.1007/978-981-99-8429-9_13

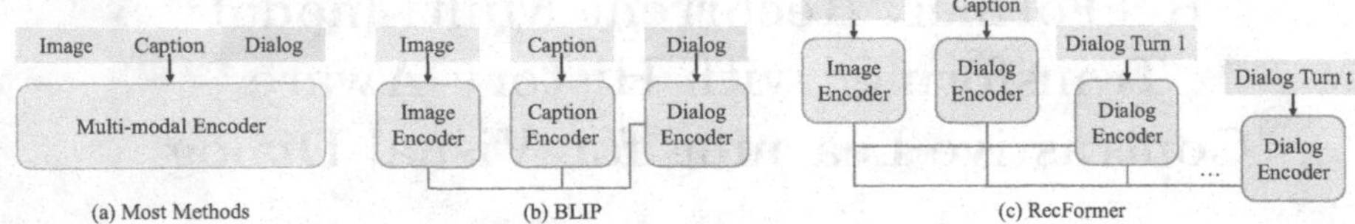

Fig. 1. Illustrations of different visual dialog methods. Compared with (a) most existing methods (e.g. ICMU [3], UTC [1], VD-BERT [22]) and (b) BLIP [12], our RecFormer (c) learns dialogs and interacts other modal information turn by turn in a recurrent way by sharing the dialog encoder and passing previous encodings.

(e.g. caption, sequential dialog history), and then answer questions related to visual information by selecting answers from lists of possible candidate options or generating answers. Therefore, it's a challenging vision-language task since the agent should be able to capture temporal dependencies inherent in multi-turn dialog for coherent dialog response while aligning the semantics among different modal-specific features for better inter-modal interactions and understandings. Due to its challenges, this task has received increasing attention [4,16] over the past few years with state-of-the-art obtaining new performance records.

This progress is largely powered by the adaptation of the more powerful transformer-based language models [1,12,15,22] which are learned from large-scale vision-language pre-trained on general image-text pairs. However, these works [1,12,15,22] only consider all utterances of dialog history equally important by concatenating them as the dialog context and interact directly with visual information by cross-attention [12] or multi-modal transformer [15], as shown in Fig. 1 (a) and (b). This means that these works hope to extract sufficient information for answering the current question with only one-time multi-modal interaction. But they ignore the temporal dependencies inherent in dialog history, which is a distinct characteristic of multi-turn dialog and crucial for the agent to predict response accurately. Besides, the semantic alignment between image and dialog history is important for agent to understand different modality consistently, but these encoders [12,15,22] are pre-trained only based on general image-caption pairs, lacking of alignment between image and dialog history.

To address above issues, we first propose a recurrent multi-modal transformer (called as **RecFormer**) to capture temporal dependencies of a dialog through encoding dialog utterances and interacting with visual information turn by turn. Specifically, we propose a multi-modal transformer with a recurrent function, named as **Recurrent Dialog Encoder**, to capture temporal dependencies better by encoding dialog history in a turn-by-turn fashion along with fusing other cross-modal encodings (e.g. image encoding, caption encoding, and multi-modal dialog encoding before current turn) via cross-attention mechanism as shown in Fig. 1 (c). In this way, the agent can capture temporal dependencies of dialog history better by encoding and fusing multi-modal information progressively so that it can pay more attention to the important information necessary to answer the current question and make better answer prediction. Besides, to align the semantics of different modalities in visual dialog, we further propose history-aware contrastive learning as an auxiliary task to align the features from different modalities for improving visual dialog understanding. More precisely, we enforce history-aware contrastive learning to align the semantics from different modalities by aligning the feature spaces between images and dialog histories with contrastive learning. We conduct extensive experiments on both VisDial v0.9 [7] and v1.0 datasets [7]. The experimental results show that our RecFormer achieves new state-of-the-art performances on both datasets, achieving 72.52 (0.87↑) MRR and 60.47 (0.83↑) Recall@1 on VisDial v0.9 validation set while achieving 69.29 (1.12↑) MRR and 55.90 (1.33↑) Recall@1 on VisDial v1.0 test-std set. Overall, the main contributions are summarized as follows:

- We propose a recurrent multi-modal Transformer, RecFormer, to capture the temporal dependencies of a dialog better by encoding dialog utterances and interacting with visual information in a turn-by-turn fashion.
- It is the first to propose history-aware contrastive learning to align semantics between turn-level dialog history and other modalities in visual dialog.
- The experiment results demonstrate that our RecFormer achieves new state-of-the-art performances on both VisDial v0.9 and VisDial v1.0.

2 Related Work

Vision-Language Pre-training. Inspired by the success of large-scale pre-trained language models like BERT [8] and GPT [19] , there are many recent works on extending BERT [8] to improve performance of downstream vision-language tasks by pre-training models on large-scale image-text pairs, such as VL-BERT [20], CLIP [18], and BLIP [12], etc. Most of these methods use general large-scale image- and alt-text pairs crawled from the web, lacking of alignment ability between image and dialog history. Besides, these methods are not designed for visual dialog, so it is suboptimal for adapting them into visual dialog task to capture temporal dependencies. In this work, building upon the vision-language pre-training, we propose RecFormer to capture temporal dependencies by encoding dialog context in a turn-by-turn fashion and history-aware contrastive learning to align different modalities at fine-tuning phase for visual dialog.

Visual Dialog. The Visual Dialog task was proposed by Das et al. [7]. Most previous works [10,11] focus on developing various attention mechanisms to model the interactions among entities including image, caption, question, and dialog history. To harvest powerful representation ability of pre-trained vision-language transformers [12,20], many recent works adapt various pre-trained vision-language models to achieve new performance records. VisDial-BERT [15] and VD-BERT [22] adopt BERT-based models [8] which is pre-trained on vision-language datasets and then fine-tuned on VisDial [7] dataset. UTC [1] proposes a contrastive learning-based framework to unify and facilitate both discriminative and generative tasks in visual dialog with a single model. Different from them, we capture temporal dependencies of a dialog better by encoding dialog context in a turn-by-turn fashion and align different modalities at fine-tuning phase so that the agent can pay more attention to the important information which is crucial for discriminating whether the answer matches current question.

3 Method

3.1 Preliminaries

As a step towards conversational visual AI, the Visual Dialog [7] task is as follows: given an image I, an optional caption C, a history H of a dialog consisting of a sequence of question-answer pairs, and a natural language follow-up question, the task for the machine is to answer the question in free-form natural language.

In this work, following the problem setting in prior work [7], we aim to learn a neural visual dialog answerer model that is given the followings as given: an image I and the 'ground-truth' dialog history (including the caption) $H = (\underbrace{C}_{H_0}, \underbrace{(Q_1, A_1)}_{H_1}, \underbrace{(Q_2, A_2)}_{H_2}, \ldots, \underbrace{(Q_{t-1}, A_{t-1})}_{H_{t-1}})$, the question Q_t, and a list of 100 candidate answers $\mathcal{A}_t = \left\{\mathcal{A}_t^{(1)}, \ldots, \mathcal{A}_t^{(100)}\right\}$, asked to return a sorting list, $\mathcal{A}_t$.

3.2 Model Architecture

An overview of RecFormer is presented in Fig. 2. It consists of two main components: Image Encoder and Recurrent Dialog Encoder. The image encoder extracts image features. Different from most of existing works that treat all dialog history as a long-form text and encode it once, our dialog encoder encodes dialog history and interacts with image features and antecedent dialog turn features from the previous turns via cross-attention mechanism in a turn-by-turn fashion.

Image Encoder. We use a 12-layer vision transformer ViT-B/16 [9] as the image encoder, and initialize it with weights pre-trained on bootstrapped datasets from [12]. Initially, an image I is resized to 384×384 and then split into multiple 16×16 patches as tokens. These tokens are fed into the image encoder to extract the visual feature $E_I = \{v_{cls}, v_1, \ldots, v_n\}$, where n is the number of 16×16 patches tokenized from the image I. Therefore, the image encoder can be expressed as follows:

$$E_I = ViT(I), \tag{1}$$

Recurrent Dialog Encoder. To capture temporal dependencies better in visual dialog and generate final multi-modal dialog feature E_{H_t} for scoring the matching degree between current question-answer pair H_t and dialog context consisting of an image I and a ground-truth dialog history (including the caption) $H = (H_0, H_1, H_2, \ldots, H_{t-1})$, we deploy a BERT-based uncased model [8],

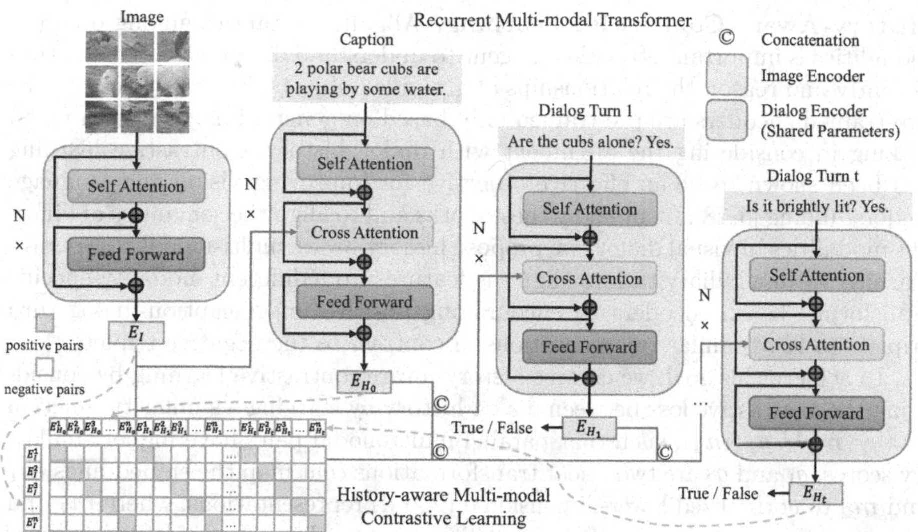

Fig. 2. The overview of RecFormer. It consists of two main components: image encoder and dialog encoder. The dialog encoder encodes the dialog context in a turn-by-turn fashion. Each dialog turn will be fused with the image features obtained from the image encoder and the antecedent dialog turn features via cross-attention mechanism. Current dialog turn feature will be passed to a linear layer to distinguish whether current dialog turn is matching with dialog context, image, and caption. Besides, the history-aware multi-modal contrastive loss is deployed to align different modal features involved by calculating the interactions between images, captions, and dialog turns. The loss will bring the positive samples, which are paired images, captions, and dialog turns, closer to each other and push away the distances between a sample and its negative samples.

which is initialized with weights pre-trained from BLIP [12], to encode current question-answer pair H_t and interact with context (I, H) by cross-attention mechanism [2]. For each question-answer pair, all the words are tokenized using the Word Piece tokenizer [24] to obtain a token sequence. Then, a task-specific *[Encode]* token is added at the beginning of the input sequence, and *[SEP]* is appended to the input sequence for marking the end of the input sequence. Specifically, for a multi-turn dialog, our recurrent dialog encoder encodes it in a turn-by-turn fashion to capture temporal dependencies. Therefore, our recurrent dialog encoder can be modeled in a recurrent way as follows:

$$E_{H_t} = BERT(H_t, [E_I, E_{H_0}, \ldots, E_{H_{t-1}}]), \quad (2)$$

where H_t is the input of the BERT encoder [8] and $[E_I, E_{H_0}, \ldots, E_{H_{t-1}}]$ is part of the input of cross-attention module in the recurrent dialog encoder. In this way, RecFormer encodes and interacts with multi-modal information in a turn-by-turn fashion and pays more attention to important information for understanding and reasoning current dialog turn, thus capturing temporal dependencies better.

History-Aware Contrastive Learning. Aligning semantics among different modalities is important for a dialog agent to understand different modalities consistently and reason the relationships of grounded objects. However, most of the pre-trained encoders are pre-trained only based on general image-caption pairs, lacking in considering the alignment with dialog history. Contrastive learning has been shown to be an effective objective for improving vision and language understanding [5,18]. Inspired by these works and to align the semantics of different modalities in visual dialog, we propose history-aware multi-modal contrastive learning as an auxiliary task to align the features from different modality-specific transformer-based encoders by encouraging positive image-caption-dialog turn triplets to have similar representations in contrast to the negative triplets.

To achieve this goal, we enforce history-aware contrastive learning by considering the contrastive loss between dialog history by learning a similarity function $\mathrm{s} = g_1(\boldsymbol{m_1})^\top g_2(\boldsymbol{m_2})$, such that parallel multi-modal pairs have higher similarity scores. g_1 and g_2 are two linear transformations that map the embeddings $\boldsymbol{m_1}$ and $\boldsymbol{m_2}$ to normalized lower-dimensional (256-d) representations, where $\boldsymbol{m_1}$ and $\boldsymbol{m_2}$ are two representations from two different encoders. Here, for dialog turn-image pairs, the $\boldsymbol{m_1}$ and $\boldsymbol{m_2}$ are $h_{[Encode]}$ and v_{cls}.

Therefore, the history-aware contrastive learning between dialog turn and image can be formulated as follows: For each dialog turn and image, we can calculate the softmax-normalized dialog turn-to-image similarity as:

$$\mathcal{L}_{d2i} = -\mathbb{E}_{p(H,I)}\left[\log \frac{exp(\mathrm{s}(H, I_+)/\tau)}{\sum_{k=1}^{K} exp\left(\mathrm{s}\left(H, \tilde{I}_k\right)/\tau\right)}\right] \tag{3}$$

where τ is a temperature parameter. $\tilde{I} = \left\{\tilde{I}_1, \tilde{I}_2, \ldots, \tilde{I}_K\right\}$ is a set of negative examples that are not matched to H. Similarly, the image-to-dialog turn similarity can be calculated as:

$$\mathcal{L}_{i2d} = -\mathbb{E}_{p(I,H)}\left[\log \frac{exp(\mathrm{s}(I, H_+)/\tau)}{\sum_{k=1}^{K} exp\left(\mathrm{s}\left(I, \tilde{H}_k\right)/\tau\right)}\right] \tag{4}$$

where τ is a temperature parameter. $\tilde{H} = \left\{\tilde{H}_1, \tilde{H}_2, \ldots, \tilde{H}_K\right\}$ is a set of negative examples that are not matched to I. Therefore, the dialog-image contrastive loss $\mathcal{L}_{hcl}$ can be expressed as:

$$\mathcal{L}_{hcl} = \lambda \frac{\mathcal{L}_{d2i} + \mathcal{L}_{i2d}}{2} \tag{5}$$

where λ is the loss weights for a trade-off between the main loss and $\mathcal{L}_{hcl}$.

3.3 Training Objectives

To train our RecFormer to achieve the above goal, we deploy a main loss $\mathcal{L}_{ce}$, which is a cross-entropy loss [25], to discriminate whether the answer is true or false for a question, given the entire dialog history, the image, and the caption. Besides, we also deploy the history-aware contrastive loss $\mathcal{L}_{hcl}$ to enforce history-aware multi-modal alignment for consistent understanding and reasoning. Therefore, the training objectives can be expressed as follows:

$$\mathcal{L} = \mathcal{L}_{ce} + \mathcal{L}_{hcl} \tag{6}$$

4 Experimental Setup

4.1 Dataset

The VisDial v0.9 [7] dataset was collected on COCO [13] images and captions. Each dialog lasts for 10 rounds for each image. The VisDial v0.9 contains 83k train dialogs and 40k validation dialogs. The VisDial v1.0 [7] dataset was released by extending the training dialogs with the combination of train and validation splits from VisDial v0.9. The validation and test splits were collected on Flickr images. Overall, VisDial v1.0 contains 123k train dialogs, 2k validation dialogs, and 8k test dialogs. Dialogs in v1.0 test split have a random length within 10.

4.2 Baselines

We compare RecFormer against a variety of single baselines in visual dialog task, including LF [7], MN [7], HCIAE [14], CoAtt [23], RvA [17], DVAN [11], ReDAN [10], VD-BERT [22], LTMI [16], VisDial-BERT [15], ICMU [3], UTC [1], BLIP [12], InstructBLIP [6], and AlignVD [4], which yield prominent performance on VisDial v0.9 dataset or VisDial v1.0 dataset.

4.3 Evaluation Metric

As in [7], we evaluate our RecFormer on VisDial v0.9 and v1.0 by using the ranking metrics including **Recall@K** ($K \in \{1, 5, 10\}$): the percentage of ground-truth answer option in top-k ranked responses, **Mean**: the average rank of ground-truth answer option, and **MRR** (Mean Reciprocal Rank): the reciprocal rank of ground-truth answer option. We also compute **NDCG** (Normalized Discounted Cumulative Gain), which penalizes the lower rank of answers with high relevance, to evaluate each answer candidate for v1.0.

4.4 Implementation Details

We use $BERT_{BASE}$ as the backbone of the recurrent dialog encoder, which consists of 12 Transformer [21] blocks, each with 12 attention heads and a hidden state dimension of 768. For image encoder, we deploy $ViT-B/16$ as the backbone, which also consists of 12 Transformer [21] blocks, each with 12 attention heads and a hidden state dimension of 768. To initialize our RecFormer, we load the base version of the weights of the image encoder and text encoder

Table 1. Performance on VisDial v0.9 validation set. ↑ indicates higher is better while ↓ indicates lower is better.

Method	MRR↑	R@1↑	R@5↑	R@10↑	Mean↓
LF [7]	58.07	43.82	74.68	84.07	5.78
MN [7]	59.65	45.55	76.22	85.37	5.46
HCIAE [14]	62.22	48.48	78.75	87.59	4.81
CoAtt [23]	63.98	50.29	80.71	88.81	4.47
RvA [17]	66.34	52.71	82.97	90.73	3.93
DVAN [11]	66.67	53.62	82.85	90.72	3.93
VD-BERT [22]	70.04	57.79	85.34	92.68	4.04
AlignVD [4]	71.65	59.64	**88.30**	**94.72**	**2.96**
RecFormer	**72.52**	**60.47**	87.67	94.18	2.98

Table 2. Results on VisDial v1.0 val set.

Method	NDCG↑	MRR↑	R@1↑	R@5↑	R@10↑	Mean↓
LTMI [16]	62.72	62.32	48.94	78.65	87.88	4.86
VisDial-BERT [15]	64.94	69.10	55.88	85.50	93.29	3.25
VD-BERT [22]	63.22	67.44	54.02	83.96	92.33	3.53
ICMU [3]	64.30	69.14	56.80	85.09	93.42	3.37
UTC [1]	63.22	68.58	55.48	85.38	93.20	3.28
BLIP [12]	-	69.41	56.44	85.90	93.30	3.20
InstructBLIP [6]	48.50	-	-	-	-	-
AlignVD [4]	**67.22**	70.45	57.64	**87.06**	**94.20**	**3.05**
RecFormer	65.42	**70.46**	**57.87**	86.37	93.70	3.14

Table 3. Results on VisDial v1.0 test-std.

Method	NDCG↑	MRR↑	R@1↑	R@5↑	R@10↑	Mean↓
ReDAN [10]	61.86	53.13	41.38	66.07	74.50	8.91
ReDAN+ [10]	64.47	53.73	42.45	64.68	75.68	6.63
LTMI [16]	60.92	60.65	47.00	77.03	87.75	4.90
VisDial-BERT [15]	63.87	67.50	53.85	84.68	93.25	3.32
VD-BERT [22]	59.96	65.44	51.63	82.23	90.68	3.90
ICMU [3]	61.30	66.82	53.50	83.05	92.05	3.59
UTC [1]	62.65	66.27	52.25	83.55	92.23	3.48
AlignVD [4]	**67.23**	68.17	54.57	85.65	93.38	3.23
RecFormer	64.87	**69.29**	**55.90**	**85.93**	**93.95**	**3.15**

of BLIP [12] as the initial weights of various encoders in our RecFormer. We set dropout probability to 0.1 and the learning rate with AdamW optimizer to 2e-5 with a weight decay of 0.05. A linear decay learning rate schedule with warmup is employed to train the model. The image resolution is set as 384×384. To train our RecFormer in memory-efficient setting, we follow the training way of BLIP [12] where the ground-truth answer and the other two negative answers sampled from the list of 100 candidate answers $\mathcal{A}_t$ excluding the ground-truth answer are used. We train our model with a batch size of 80 on 4 NVIDIA RTX 3090 with the help of gradient accumulation. The temperature parameter τ is set to 0.07. The weight λ of history-aware contrastive loss is set to 0.001.

5 Results and Analysis

5.1 Main Results

Results on VisDial v0.9 Validation Set. Table 1 shows discriminative results for our RecFormer and baselines on VisDial v0.9 val split. Our RecFormer achieves the best performance on the MRR and the Recall@1 while achieving the competitive performance on the Recall@5, Recall@10, and Mean. Compared to

AlignVD [4], our RecFormer results in 0.87% improvement in MRR and 0.83% improvement in Recall@1. We speculate other metrics can not achieve best due to the random selection on negative samples during training. Compared to another SOTA model VD-BERT [22], our RecFormer results in 2.48% improvement in MRR, 2.68% improvement in Recall@1, 2.33% improvement in Recall@5, 1.5% improvement in Recall@10, and 1.06 improvement in Mean. This validates the effectiveness of RecFormer using recurrent dialog encoder for capturing temporal dependencies and history-aware contrastive learning for multi-modal alignment.

Results on VisDial v1.0 Validation Set. As shown in Table 2, we also compare our results on VisDial v1.0 val split. Our RecFormer achieves the best performance on the MRR and the Recall@1 while achieving the competitive performance on the Recall@5, Recall@10, Mean, and NDCG. Besides, the training losses in all models for comparison target at improving the MRR and the Recall@1, leading that there's a certain amount of randomness on the Recall@5, the Recall@10, the Mean and the NDCG. Even so, our RecFormer still can achieve competitive results on those minor metrics (Recall@5, Recall@10, Mean, and NDCG) while achieving the best results on the two most important metrics MRR and Recall@1. This further indicates the effectiveness of RecFormer.

Results on VisDial v1.0 Test-Standard Set. To further validate the generalization of our RecFormer, we also report the discriminative results on VisDial v1.0 test-std split in Table 3. We can observe that compared with those baselines, our RecFormer achieves the best performance on MRR, Recall@1, Recall@5, Recall@10, and Mean. Specially, compared to AlignVD [4], our RecFormer results in 1.12% improvement in MRR, 1.33% improvement in Recall@1, 0.28% improvement in Recall@5, 0.57% improvement in Recall@10, and 0.08 improvement in Mean. This shows the strong generalization and effectiveness of our RecFormer.

Table 4. Ablation study on components of RecFormer on VisDial v1.0 validation set.

Model	NDCG↑	MRR↑	R@1↑	R@5↑	R@10↑	Mean↓
RecFormer	**65.42**	**70.46**	**57.87**	**86.37**	**93.70**	**3.14**
- $\mathcal{L}_{hcl}$	65.05	69.89	56.99	86.22	93.40	3.17
- recurrent	62.90	68.42	55.11	85.33	92.97	3.34

5.2 Ablation Study

Ablation Study on Components of RecFormer. To validate the effects of components of our RecFormer, we conduct experiments on VisDial v1.0 and show the results of val split in Table 4. We can observe that both our recurrent dialog encoder and history-aware contrastive loss are effective for visual dialog. Of these two, the recurrent dialog encoder improves the model more, showing that our recurrent mechanism is effective to capture temporal dependencies implicit in visual dialog. Besides, Recformer trained with $\mathcal{L}_{hcl}$ outperforms the model

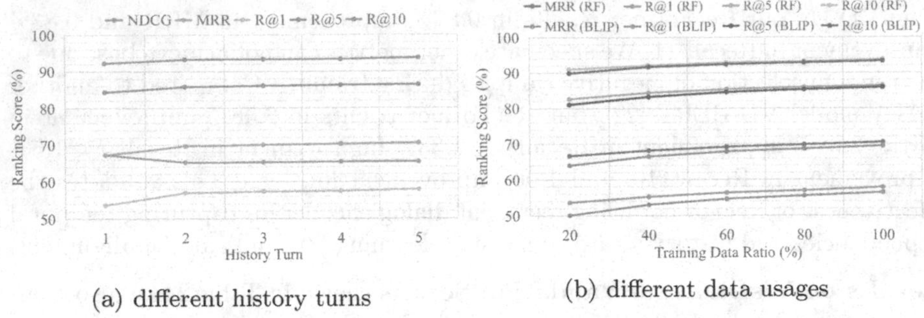

(a) different history turns (b) different data usages

Fig. 3. Ablation Study on VisDial v1.0 val set. RF stands for RecFormer.

without it, showing the effectiveness of our proposed history-aware contrastive learning to align the semantics among different modal-specific features for better multi-modal interactions and understandings. In summary, both recurrent dialog encoder and history-aware contrastive loss are effective for better visual dialog.

Ablation Study on Different Dialog History Turns. We investigate the effects of different dialog history turns for visual dialog. The results are shown in Fig. 3(a). Performance of our RecFormer continues to go up with increasing dialog history rounds on MRR, Recall@1, Recall@5, and Recall@10. The only exception is the NDCG score without history which is higher than others since the training objective is not NDCG-oriented and there is a certain contingency. However, the NDCG score of our RecFormer still continues to go up if the history round is more than 1. We believe these improvements are largely indicative of the RecFormer's ability to model temporal dependencies.

Ablation Study on Different Data Usages. We further study the versatility of our RecFormer on different data scales. We conduct experiments with RecFormer and BLIP [12] on different training data usages. The experimental results on different data usage are shown in Fig. 3(b). From the results, we can observe that our RecFormer outperforms BLIP [12] on various metrics and various data scales, especially on 20% data ratio. This shows that RecFormer is more versatile.

5.3 Attention Visualization

To explain the cross-modal learning of RecFormer, we visualize the attention weights of the cross-attention layers in Fig. 4. It can be observed that RecFormer tends to focus on the regions in the image that correspond to the content discussed in the current dialog turn. When referring to specific objects present in the image(e.g. Q_1, Q_3, Q_4), RecFormer can precisely locate the area in the image. When referring to objects that do not exist in the image(e.g. Q_2) or talking about the macro scene(e.g. Q_5, Q_8), the attention area of RecFormer is more open and scattered. As can be seen from the last two subgraphs, when answering

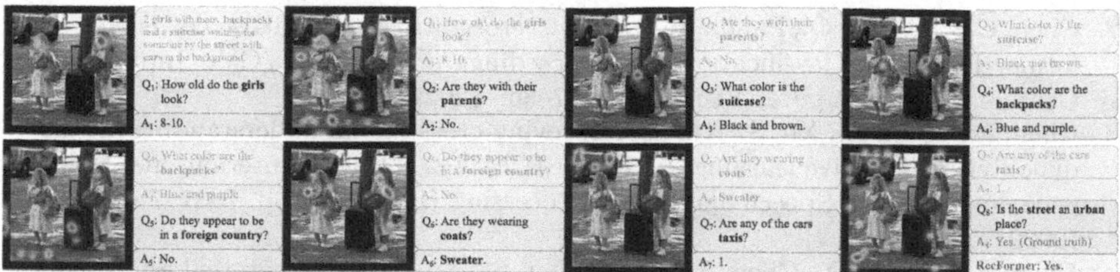

Fig. 4. Attention weight visualization of our RecFormer for each dialog turn. Inside the green box is the caption. The questions and answers for each turn are in blue boxes. Dark tones represent dialog history. Inside the yellow box is the predicted answer. (Color figure online)

the question Q_8, RecFormer can combine the content of "taxi" discussed before and focus on the corresponding area. So it can consistently predict the answer.

6 Conclusion

We propose a recurrent multi-modal transformer (named **RecFormer**) to capture temporal dependencies between utterances via encoding dialog utterances and interacting with visual information turn by turn. Specifically, we equip a pre-trained transformer with a recurrent function that maintains cross-modal history encoding for the dialog agent. Thus, the dialog agent can make better predictions by considering temporal dependencies. Besides, we also propose history-aware contrastive learning as an auxiliary task to align visual features and dialog history features for improving visual dialog understanding. The experimental results demonstrate that our RecFormer can achieve state-of-the-art performances on both the VisDial v0.9 and v1.0 datasets.

Acknowledgment. This work was supported in part by National Key R&D Program of China under Grant No. 2020AAA0109700, National Natural Science Foundation of China (NSFC) under Grant No. 62206314, Guangdong Outstanding Youth Fund (Grant No. 2021B1515020061), GuangDong Basic and Applied Basic Research Foundation under Grant No. 2022A1515011835, China Postdoctoral Science Foundation funded project under Grant No. 2021M703687, Shenzhen Science and Technology Program (Grant No. RCYX20200714114642083), Shenzhen Fundamental Research Program (Grant No. JCYJ20190807154211365), Shenzhen Science and Technology Program (Grant No. GJHZ20220913142600001), Nansha Key RD Program under Grant No. 2022ZD014, and Sun Yat-sen University under Grant No. 22lgqb38 and 76160-12220011. We thank MindSpore for the partial support of this work, which is a new deep learning computing framework (https://www.mindspore.cn/).

References

1. Chen, C., et al.: UTC: a unified transformer with inter-task contrastive learning for visual dialog. In: Proceedings of the IEEE/CVF Conference on Computer Vision and Pattern Recognition, pp. 18103–18112 (2022)

2. Chen, C.F.R., Fan, Q., Panda, R.: Crossvit: cross-attention multi-scale vision transformer for image classification. In: Proceedings of the IEEE/CVF International Conference on Computer Vision, pp. 357–366 (2021)
3. Chen, F., Chen, X., Xu, S., Xu, B.: Improving cross-modal understanding in visual dialog via contrastive learning. In: ICASSP 2022–2022 IEEE International Conference on Acoustics, Speech and Signal Processing (ICASSP), pp. 7937–7941. IEEE (2022)
4. Chen, F., Zhang, D., Chen, X., Shi, J., Xu, S., Xu, B.: Unsupervised and pseudo-supervised vision-language alignment in visual dialog. In: Proceedings of the 30th ACM International Conference on Multimedia, pp. 4142–4153 (2022)
5. Chen, T., Kornblith, S., Norouzi, M., Hinton, G.: A simple framework for contrastive learning of visual representations. In: International Conference on Machine Learning, pp. 1597–1607. PMLR (2020)
6. Dai, W., et al.: Instructblip: Towards general-purpose vision-language models with instruction tuning. arXiv preprint arXiv:2305.06500 (2023)
7. Das, A., et al.: Visual dialog. In: Proceedings of the IEEE Conference on Computer Vision and Pattern Recognition, pp. 326–335 (2017)
8. Devlin, J., Chang, M.W., Lee, K., Toutanova, K.: BERT: pre-training of deep bidirectional transformers for language understanding. arXiv preprint arXiv:1810.04805 (2018)
9. Dosovitskiy, A., et al.: An image is worth 16×16 words: transformers for image recognition at scale. arXiv preprint arXiv:2010.11929 (2020)
10. Gan, Z., Cheng, Y., Kholy, A.E., Li, L., Liu, J., Gao, J.: Multi-step reasoning via recurrent dual attention for visual dialog. arXiv preprint arXiv:1902.00579 (2019)
11. Guo, D., Wang, H., Wang, M.: Dual visual attention network for visual dialog. In: IJCAI, pp. 4989–4995 (2019)
12. Li, J., Li, D., Xiong, C., Hoi, S.: Blip: bootstrapping language-image pre-training for unified vision-language understanding and generation. In: International Conference on Machine Learning, pp. 12888–12900. PMLR (2022)
13. Lin, T.-Y., et al.: Microsoft COCO: common objects in context. In: Fleet, D., Pajdla, T., Schiele, B., Tuytelaars, T. (eds.) ECCV 2014. LNCS, vol. 8693, pp. 740–755. Springer, Cham (2014). https://doi.org/10.1007/978-3-319-10602-1_48
14. Lu, J., Kannan, A., Yang, J., Parikh, D., Batra, D.: Best of both worlds: transferring knowledge from discriminative learning to a generative visual dialog model. In: Advances in Neural Information Processing Systems, vol. 30 (2017)
15. Murahari, V., Batra, D., Parikh, D., Das, A.: Large-scale pretraining for visual dialog: a simple state-of-the-art baseline. In: Vedaldi, A., Bischof, H., Brox, T., Frahm, J.-M. (eds.) ECCV 2020. LNCS, vol. 12363, pp. 336–352. Springer, Cham (2020). https://doi.org/10.1007/978-3-030-58523-5_20
16. Nguyen, V.Q., Suganuma, M., Okatani, T.: Efficient attention mechanism for visual dialog that can handle all the interactions between multiple inputs. In: Computer Vision-ECCV 2020: 16th European Conference, Glasgow, UK, August 23–28, 2020, Proceedings, Part XXIV 16. pp. 223–240. Springer (2020)
17. Niu, Y., Zhang, H., Zhang, M., Zhang, J., Lu, Z., Wen, J.R.: Recursive visual attention in visual dialog. In: Proceedings of the IEEE/CVF Conference on Computer Vision and Pattern Recognition, pp. 6679–6688 (2019)
18. Radford, A., et al.: Learning transferable visual models from natural language supervision. In: International Conference on Machine Learning, pp. 8748–8763. PMLR (2021)
19. Radford, A., Narasimhan, K., Salimans, T., Sutskever, I., et al.: Improving language understanding by generative pre-training (2018)

20. Su, W., et al.: Vl-BERT: pre-training of generic visual-linguistic representations. arXiv preprint arXiv:1908.08530 (2019)
21. Vaswani, A., et al.: Attention is all you need. In: Advances in Neural Information Processing Systems, vol. 30 (2017)
22. Wang, Y., Joty, S., Lyu, M.R., King, I., Xiong, C., Hoi, S.C.: VD-BERT: a unified vision and dialog transformer with BERT. arXiv preprint arXiv:2004.13278 (2020)
23. Wu, Q., Wang, P., Shen, C., Reid, I., Van Den Hengel, A.: Are you talking to me? Reasoned visual dialog generation through adversarial learning. In: Proceedings of the IEEE Conference on Computer Vision and Pattern Recognition, pp. 6106–6115 (2018)
24. Wu, Y., et al.: Google's neural machine translation system: bridging the gap between human and machine translation. arXiv preprint arXiv:1609.08144 (2016)
25. Zhang, Z., Sabuncu, M.: Generalized cross entropy loss for training deep neural networks with noisy labels. In: Advances in Neural Information Processing Systems, vol. 31 (2018)

KV Inversion: KV Embeddings Learning for Text-Conditioned Real Image Action Editing

Jiancheng Huang[1,2], Yifan Liu[1], Jin Qin[1,2], and Shifeng Chen[1,2(✉)]

[1] ShenZhen Key Lab of Computer Vision and Pattern Recognition, Shenzhen Institute of Advanced Technology, Chinese Academy of Sciences, Shenzhen, China
{jc.huang,yf.liu2,shifeng.chen}@siat.ac.cn

[2] University of Chinese Academy of Sciences, Beijing, China

Abstract. Text-conditioned image editing is a recently emerged and highly practical task, and its potential is immeasurable. However, most of the concurrent methods are unable to perform action editing, i.e. they can not produce results that conform to the action semantics of the editing prompt and preserve the content of the original image. To solve the problem of action editing, we propose KV Inversion, a method that can achieve satisfactory reconstruction performance and action editing, which can solve two major problems: 1) the edited result can match the corresponding action, and 2) the edited object can retain the texture and identity of the original real image. In addition, our method does not require training the Stable Diffusion model itself, nor does it require scanning a large-scale dataset to perform time-consuming training.

Keywords: Real imgae editing · Diffusion model · Text-to-image generation · AIGC

1 Introduction

Along with the widespread adoption of diffusion model and muti-modal generative model, recent months have witnessed remarkable progress in text-to-image generation, which is mainly used for generation in many real-world scenarios, such as AI painting, commercial design, film making, etc. [25,31,32,34,47]. For example, Stable Diffusion [34] is able to generate diverse and high quality images according to user-provided text prompts. However, it is not enough just to generate a completely new image, and it will be more exciting to generate the desired

J. Huang and Y. Liu—Contributed equally to this work.

This work is supported by Key-Area Research and Development Program of Guangdong Province (2019B010155003), the Joint Lab of CAS-HK, and Shenzhen Science and Technology Innovation Commission (JCYJ20200109114835623, JSGG20220831105002004).

Supplementary Information The online version contains supplementary material available at https://doi.org/10.1007/978-981-99-8429-9_14.

© The Author(s), under exclusive license to Springer Nature Singapore Pte Ltd. 2024
Q. Liu et al. (Eds.): PRCV 2023, LNCS 14425, pp. 172–184, 2024.
https://doi.org/10.1007/978-981-99-8429-9_14

Fig. 1. Comparing with different concurrent image editing methods on the real natural world images, it is obvious that the object in the editing result of our KV Inversion can meet the edit prompt corresponding to the action editing, while retaining the original real image object.

image given a editing prompt, which we call text-conditioned image editing. There are many methods that use a pre-trained large-scale text-to-image model to implement text-conditioned image editing [12,25,27,42]. Text-conditioned image editing is a recently emerged and highly practical task, and its potential for successful application in comic book production, video editing, advertising material production is immeasurable.

Note that **synthetic image editing** and **real image editing** are two very different tasks. Synthetic image editing means that when an image is synthesised using a source prompt, an editing prompt can be used to create another image that is consistent with it, and the user cannot provide a real image in the process. Real image editing, on the other hand, means that the user can provide a real image and then use the edit prompt to edit that real image. We can divide this emerging field into 5 categories according to the content to be edited: 1) Object replacement, where the object in the original diagram is replaced with another object. 2) New objects creation, where a new object is added to the editing image and the rest remains unchanged. 3) Action editing, the object retains its original texture and identity, but the action becomes the new action given by the edit text. 4) Scene editing, where the object remains unchanged and the scene is turned into a given scene. 5) Style editing, the art style or colour style is changed.

A number of existing text-conditioned image editing methods [12,27,42] enable the above types of 1), 2), 4) and 5) to be edited. Most of them are able to perform tasks such as object replacement, new object creation and style transfer while keeping the overall structure and layout unchanged, and achieve satisfactory results. However, these methods are unable to perform type 3) edits as shown in Fig. 1. Some works aim to address this problem, such as Textual Inversion [10], Imagic [17], DreamBooth [36], Custorm Diffusion [19], ELITE [43], FastComposer [45] and MasaCtrl [5] are proposed to achieve the type 3) action editing. They can accomplish action editing while retaining the basic properties of the original image. However, Textual Inversion [10], Imagic [17], DreamBooth [36] and Custorm Diffusion [19] require finetune of the diffusion model itself as well as the text embedding on 5 images of the editing object, so the GPU memory and time cost required is too high. ELITE [43] requires training on a large dataset, which is even more expensive on time and GPU. MasaCtrl does not require training and finetuning, but its reconstruction performance on real images is unsatisfactory, making it difficult to editing.

In our method, we aim to solve the problem of action editing without using multiple images of the same object for finetuning (generally known as **Tuning-free** in this field), without training the diffusion model itself, and without training on a large dataset for a long time (generally known as **Training-free** in this field), and to propose a solution to the mentioned challenges in the above setting. The core problem is how to preserve the content of the original object when the action changes. Our KV Inversion is divided into 3 stages. The fundamental difference with the concurrnet works [6,12,27] is that KV Inversion directly learns Key and Value at the self-attention layer to preserve the content of the source image by these learnable parameters (KV embeddings), which we call upgrading the original self-attention to Content Preserving self-attention (CP-attn). Then, in the editing stage, we use the edit text prompt to introduce the action information and the learned KV embeddings to preserve the texture and identity of the object. We further control the timesteps of upgrading the CP-attn and the segmentation mask obtained by Segment Anything (SAM) [18], thus achieving a faster and more controllable editing.

Our main contributions are summarized as follows. 1) We propose a training-free text-conditoned image action editing method KV Inversion to solve the action editing problem. 2) We design an upgrade version of self-attention named content preserving self-attention, which can preserve the texture and identity of the original object and then be used to fill the editing image. 3) Comprehensive Experiments show that KV Inversion can achieve satisfactory performance in real image editing.

2 Background

2.1 Text-to-Image Generation and Editing

Text-to-image generation models [31,34,38,40,47] have experienced an unprecedented surge in popularity. Initially, most early image generation methods

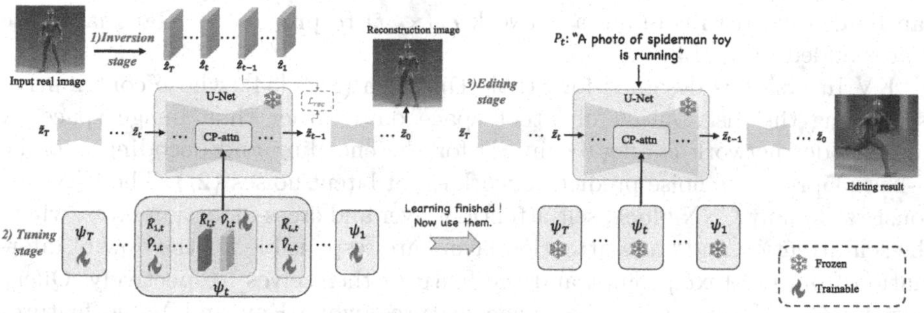

Fig. 2. Pipeline of the proposed KV Inversion, which can be divided into 3 stages, inversion stage, tuning stage and editing stage. Inversion stage is for getting supervision. Tunning stage is for learning KV embedding for content preserving and editing stage is using the learned content for consistency editing results.

relied on GANs [20,33,41,46,48,49,51] and conditioned on text descriptions. These models align text descriptions with synthesized image contents using multi-modal vision-language learning. However, more recently, diffusion models [8,10,13,14,26,28,40] have emerged as a dominant force, delivering exceptional generative power and achieving cutting-edge results in image quality and diversity. Notably, several text-to-image diffusion models, including DALL·E 2 [31], LDM [34], VQ-Diffusion [11], InstructPix2Pix [4] and GLIDE [25], have been developed to further enhance the synthesis process. The rise of diffusion-based models [10,28] has showcased the potential for manipulating given images without human intervention.

Editing images based on text descriptions is a complex task. Recently, there has been extensive research on text-conditioned editing using GANs [1–3,15,16,21,24,28,44]. The introduction of CLIP [29] has revolutionized image editing by allowing people to guide the editing process with text inputs. Diffusion models [31,37] exhibit inherent capabilities for image editing due to their strong ability to extract text features using CLIP. An innovative approach called VQGAN-CLIP [7] combines VQGAN [9] and CLIP [30] in an auto-regressive model, enabling the production of high-quality images and precise edits with diverse and controllable outcomes. Through training, the diffusion model generates subject-specific images aligned with textual descriptions, opening new possibilities for precise image generation.

2.2 Stable Diffusion Model

Diffusion models [13,26,39] utilize a forward process that gradually adds Gaussian noise to training data, and then recover the data distribution by a reverse process. The forward process follows the Markov chain that transforms a data sample $\mathbf{x}_0 \sim q(\mathbf{x}_0)$ into a sequence of noisy samples $\mathbf{x}_{1:T} = \mathbf{x}_1, \mathbf{x}_2, \cdots, \mathbf{x}_T$ in T steps. The above process can be reversed by $p_\theta(\mathbf{x}_{0:T})$ with learnable parameters θ, where $p_\theta(\mathbf{x}_{0:T})$ is implemented by a neural network. The optimization of it can be converted to training a network $\epsilon_\theta(\mathbf{x}_t, t)$ to predict the Gaussian noise vector added to $\mathbf{x}_t$ [13].

KV Inversion is designed for Stable Diffusion (SD) [34], whose contribution is learning the distribution of latent space data rather than image space. A autoencoder network is adopted in SD for the encoding and decoding and the U-Net [35] is as the noise predict network ϵ_θ for latent noises $\{\hat{\mathbf{z}}_t\}$. The U-Net ϵ_θ consists of many CNN block, self-attention layer and cross-attention layer, where the self-attention and cross-attention layers are responsible for the flow of information from the text prompt and the features themselves, respectively. Query features Q is derived from the image features, while Key and Value features K, V are derived from the text embedding (cross-attention) or the image itself (self-attention). There are several phenomena: 1) The attention map of cross-attention determines the structure of the image (the position of the appearing objects) [6,12,42]. 2) K and V of the cross-attention layer change the texture and detail of generated image. 3) K, V and the attention map in self-attention have a dramatic effect on the generated result in terms of content.

3 KV Inversion: Training-Free KV Embeddings Learning

3.1 Task Setting and Reason of Existing Problem

Given a real rather than a synthetic source image I_s and a corresponding text prompt P_s (which can be semantic text or empty), the goal of our task is to synthesize a new ideal image I_t with a pretrain stable diffusion model that matches the target editing text prompt P_t. This editing result image I_t should meet the following requirements: 1) I_t semantically matches the text prompt of P_t, e.g., it can satisfy that the corresponding object is performing the corresponding action. 2) The object inside I_t should be consistent with I_s in terms of content. For example, given a real image (corresponding to I_s) with a cat standing still, we edit the text prompt P_t to "a running cat", and then generate a new image with the cat running, while the other contents of the image remain basically unchanged.

This task is very difficult, especially when used on real images, and most of the current stble diffusion based image editing methods do not allow for action editing on real images while maintaining good reconstruction performance [12, 42]. A naive baseline method for action editing is to directly use P_t to synthesize a new image $\bar{I}_t$ that matches the semantics of the actions and objects in P_t. However, even using $\hat{\mathbf{z}}_T$ as the starting point of the reverse process obtained from DDIM inversion of the original real image I_s, the scene and object generated in $\bar{I}_t$ are often very different from the original real image I_s [12].

The core problem mentioned above is that the scene and object generated in $\bar{I}_t$ are often very different from the original real image I_s. The primary cause of this phenomenon is that during editing, and the Key K and Value V of cross-attention in the U-Net are derived from the target editing text prompt P_t, which brings new content features different from those of the original real image I_s, so that the editing result has changed dramatically in content. Therefore, our core idea is to first implement a better inversion method to preserve the contents in the source image I_s in the feature space. Then, we utilize these preserved content and combine it with the action layout generated by the target prompt P_t, and finally synthesize the desired editing image I_t.

3.2 KV Inversion Overview

To realise the above core idea, we propose KV Inversion, which can be divided into 3 stages, inversion stage, tuning stage and editing stage.

1) Inversion Stage: Getting $\mathbf{z}_t$ for Supervision. Specifically, as illustrated in Fig. 2, for obtaining the trace of $\hat{\mathbf{z}}_t$ in the reserve process, we first perform DDIM inversion [8,39] to synthesize a series of latent noises $\{\hat{\mathbf{z}}_t\}$ including $\hat{\mathbf{z}}_T$. However, directly using $\hat{\mathbf{z}}_T$ as the starting point of the reverse process can't reconstruct the original real image I_s well since the U-Net can't predict the noise accurately enough which resulting in the next stage for better content preserving.

2) Tuning Stage: Learning KV Embeddings for Better Content Preserving. For better reconstruct I_s, we need more parameters to learn the content of I_s. Since we find that the contents (texture and identity) are mainly controled in the self-attention layer, we choose to learn the K and V embeddings in the self-attention layer. Here we use $\hat{K}_{l,t}$ and $\hat{V}_{l,t}$ to denote the learnable Key and Value embeddings in the l-th self-attention layer during step t of reverse process. Then we define our proposed Content Preserving self-attention (CP-attn) as follows. For instance, given the l-th self-attention layer during step t of reverse process, we have:

$$\bar{K}_{l,t} = \lambda^k_{l,t} K_{l,t} + \gamma^k_{l,t} \hat{K}_{l,t}, \quad \bar{V}_{l,t} = \lambda^v_{l,t} V_{l,t} + \gamma^v_{l,t} \hat{V}_{l,t}, \tag{1}$$

$$\text{Attention}(Q_{l,t}, \bar{K}_{l,t}, \bar{V}_{l,t}) = \text{Softmax}(\frac{Q_{l,t}\bar{K}^T_{l,t}}{\sqrt{d}})\bar{V}_{l,t}, \tag{2}$$

where $(K_{l,t}, V_{l,t})$ and $(\bar{K}_{l,t}, \bar{V}_{l,t})$ denote the original and the new (Key, Value), respectively. Besides, $(\lambda^k_{l,t}, \lambda^v_{l,t})$ and $(\gamma^k_{l,t}, \gamma^v_{l,t})$ denote the learnable weights for the original and the learnable (Key, Value) embeddings.

Figure 2 shows that our framework consists of the above learnable parameters $(\hat{K}_{l,t}, \hat{V}_{l,t}, \lambda^k_{l,t}, \lambda^v_{l,t}, \gamma^k_{l,t}, \gamma^v_{l,t})$ and the original diffusion networks, and we denote all the above learnable parameters with ψ and the learnable parameters for step t with ψ_t.

In this tuning stage, we first use $\widetilde{\mathbf{z}}_T = \hat{\mathbf{z}}_T$ as the begining of reverse process $(T \rightarrow 1)$. For each specific timestep t, we train the corresponding parameters ψ_t. We get the output of U-Net $\epsilon_\theta(\widetilde{\mathbf{z}}_t, t, \mathbf{c}_s, \psi_t)$ and use DDIM sampling to produce the next noise sample $\widetilde{\mathbf{z}}_{t-1}$, where $\mathbf{c}_s$ is the text embedding of source prompt P_s. Since the latent noise $\hat{\mathbf{z}}_{t-1}$ represents a reasonable trace whose corresponding result $\hat{\mathbf{z}}_0$ is close to the real image I_s, we use $\hat{\mathbf{z}}_{t-1}$ as supervision to train our learnable parameters ψ_t to output a more precise noise $\widetilde{\mathbf{z}}_{t-1}$, which is closer to

Algorithm 1. The 3 Stages of KV Inversion

Require: the latent of the original real image $\mathbf{z}_0$ and the source prompt embeddings $\mathbf{c}_s$. The initialization learnable parameters $\psi = \{\psi_t,\ t = 1, ..., T\}$.

1) Inversion stage: Set guidance scale $w = 7.5$ of stable diffusion model, in this stage we utilize DDIM inversion to obtain the trace of noises $\{\hat{\mathbf{z}}_t, t = 1, ..., T\}$.

2) Tuning stage: Set guidance scale $w = 7.5$ and the begining of reverse process $\widetilde{\mathbf{z}}_T = \hat{\mathbf{z}}_T$;

for $t = T, T-1, \ldots, 1$ **do**
  while *not converge* **do**
    obtain $\widetilde{\mathbf{z}}_{t-1}$ by $\epsilon_\theta(\widetilde{\mathbf{z}}_t, t, \mathbf{c}_s, \psi_t)$;
    $\psi_t \leftarrow \psi_t - \eta \nabla_{\psi_t} \mathcal{L}_{rec}$;
  end
  obtain final $\widetilde{\mathbf{z}}_{t-1}$ by learned ψ_t as Eq. 4;
end
Return learned parameters ψ

3) Editing stage: Set guidance scale $w = 7.5$ and the begining of reverse process $\bar{\mathbf{z}}_T = \hat{\mathbf{z}}_T$. Given a editing prompt P_t, then get its embedding $\mathbf{c}_t$ and an uncondition embedding $\mathbf{c}_u$;

for $t = T, T-1, \ldots, 1$ **do**
  $\epsilon_c = \epsilon_\theta(\bar{\mathbf{z}}_t, t, \mathbf{c}_t, \psi_t)$;
  $\epsilon_u = \epsilon_\theta(\bar{\mathbf{z}}_t, t, \mathbf{c}_u, \psi_t)$;
  $\epsilon_t = \epsilon_u + w(\epsilon_c - \epsilon_u)$;
  $\bar{\mathbf{z}}_{t-1} = \text{Sample}(\bar{\mathbf{z}}_t, \epsilon)$;
end
$I_t = \text{Decode}(\bar{\mathbf{z}}_0)$;
Return Editing result I_t

the noise representation $\hat{\mathbf{z}}_{t-1}$ than baseline DDIM inversion and other inversion methods [22]. The loss function is

$$\mathcal{L}_{rec} = \min_{\psi_t} \left\| \hat{\mathbf{z}}_{t-1} - \widetilde{\mathbf{z}}_{t-1} \right\|^2. \tag{3}$$

After training the ψ_t, we use the trained ψ_t and obtain a better $\widetilde{\mathbf{z}}_{t-1}$ by following:

$$\widetilde{\mathbf{z}}_{t-1} = \sqrt{\frac{\alpha_{t-1}}{\alpha_t}} \widetilde{\mathbf{z}}_t + \left(\sqrt{\frac{1}{\alpha_{t-1}} - 1} - \sqrt{\frac{1}{\alpha_t} - 1} \right) \cdot \epsilon_\theta(\widetilde{\mathbf{z}}_t, t, \mathbf{c}_s, \psi_t). \tag{4}$$

With the learning procedure above, the well-optimized ψ_t actually preserves the content of the source real image I_s well at timestep t of reverse process, which is important for the next editing stage. The algorithm of the inversion and tuning stages is provided in Algorithm 1.

3) Edit Stage: Using Learned KV Embeddings for Better Content Consistency. The overall architecture of the proposed pipeline to perform editing is shown in Fig. 2. During each denoising step t of generating the target editing image I_t, we also upgrade the self-attention layer in the U-Net to our

Fig. 3. The intermediate predicted $\mathbf{x}_0$ during the sampling process. KV Inversion can ensure the performance of both reconstruction and editing.

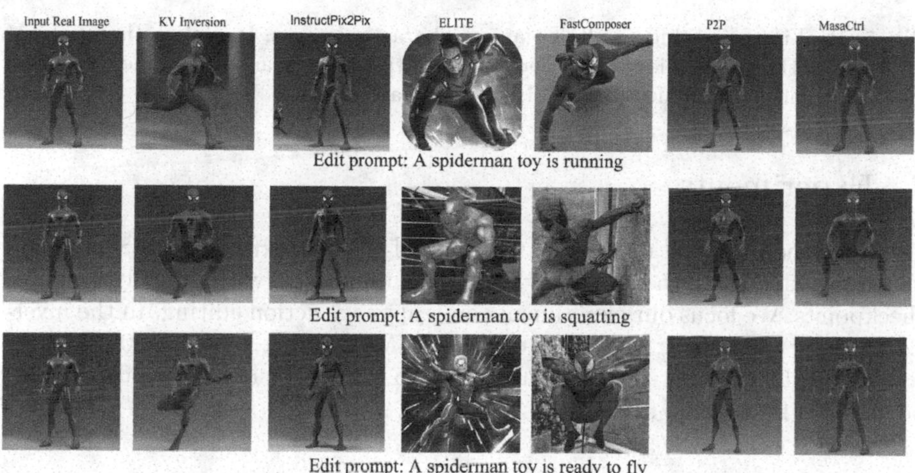

Fig. 4. Real image editing results of different editing methods on real human shape images. It is obvious that our KV Inversion can achieve very good action editing performance on humanoid objects, and the semantics of the actions corresponding to its limbs are almost imperceptible to editing.

CP-attn layer with the learned parameters ψ_t. Specifically, we keep the Query features $Q_{l,t}$ unchanged and use the learned $(\bar{K}_{l,t}, \bar{V}_{l,t})$ from the corresponding place of the tuning stage as the new Key and Value features here. Note that $(\bar{K}_{l,t}, \bar{V}_{l,t})$ preserve the content of the source image I_s. Thus, we can perform attention according to Eq. 2 to transfer the contents from I_s to the editing features as shown in Fig. 3. We provide the detailed implementation of the method inside the supplementary material.

Fig. 5. Real image editing results on anime character images. It is clear that our KV inversion method can achieve very good action editing performance on anime characters and their limbs correspond to natural action semantics.

4 Experiments

We apply the proposed method to the state-of-the-art text-to-image Stable Diffusion [34] model and the anime-style model Anything-V3 with publicly available checkpoints. We focus our experiments on real image action editing. In the inversion, tuning and editing stages, we utilize DDIM sampling [39] with 50 denoising steps, and the classifier-free guidance is set to 7.5. Other hyperparameters may be changed for specific models.

4.1 Comparisons with Other Concurrent Works

On Pretrained Stable Model. We mainly compare the proposed KV Inversion to the concurrent text-conditioned image editing methods, including Custorm Diffusion [19], InstructPix2Pix [4], MasaCtrl [5], P2P [12], PnP [42], ELITE [43], FastComposer [45]. We use their codes and checkpoints to produce the editing results.

The editng results are shown in Figs. 1 and 4. Our KV Inversion achieves better results in real image editing. As can be seen, action editing for real images is still a very challenging task. Most methods are unable to solve both 1) the resulting edits are not sufficiently similar or even completely different from the source image, and 2) the resulting edits do not match the semantics of the edit prompt in terms of actions. We see that the proposed KV Inversion solves both of these problems relatively well, and these results demonstrate the effectiveness of the proposed method. Also, unlike methods such as Custorm Diffusion [19], which require fine-tuning of the stable diffusion network itself, our method does not require fine-tuning of the network itself, but only learning some

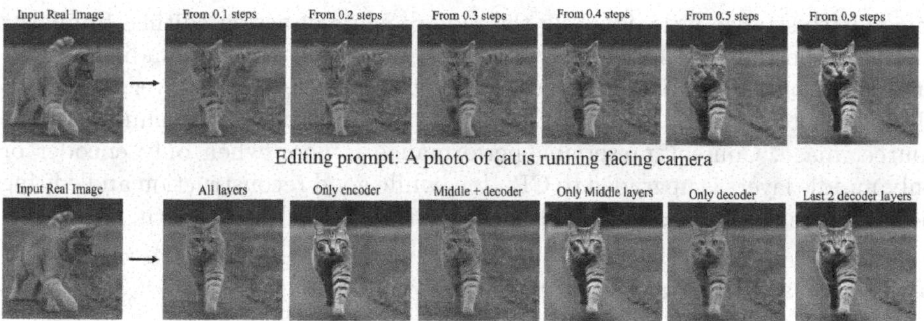

Fig. 6. Ablation study of upgrading self-attention layer to CP-attn in different reverse timesteps (top) and U-Net places (bottom).

parameters corresponding to the image. In addition, unlike methods such as FastComposer [45] and ELITE [43], which require training on large datasets, we do not need to collect and use large datasets, nor do we need to spend graphics resources and time training on them. The reasons for the failure of the existing methods can be explained by the fact that P2P [12] uses the attention map of the source image in the generation of the edited image, thus replicating the original spatial integrity. MasaCtrl [5] generates the corresponding action edits, but the reconstruction performance on the real image is still not good enough.

On Preatrained Anything-V3 Model. We continue to test the effectiveness of our method on the editing of images in the animation domain, specifically using Anything-V3. Figure 5 shows the editing results of our method and a number of other methods, including P2P, ELITE, MasaCtrl, InstructPix2Pix, PnP, FastComposer. Note in particular that our method is focused on real image input, and these results in the illustration are edited using the input image provided by the user, rather than given a prompt and then edited during generation, which are two completely different tasks. The proposed KV Inversion is a further demonstration of the generality of the proposed method by retaining the textures and identities of the animated objects in the source image while obtaining the corresponding editing actions.

4.2 Ablation Study

The effectiveness of the proposed KV Inversion method and the upgrading of self-attention with CP-attn can be demonstrated on top of the comparison results of action editing of real images.

In order to analyse why the method works, we control the specific parameters for upgrading self-attention to CP-attn, such as the step and position in the U-Net, and observe the effect of the reconstruction and the effect of the editing. From Fig. 6 (top), we see that using CP-attn in too few steps (From 0.9 steps) leads to poor performance of the reconstructed image, resulting in an edited result that does not look like the original. Conversely, using CP-attn in too many steps (From 0.1 steps) leads to the edited result too similar to the source image with strange pose, ignoring the requirements of action editing. Similarly, in Fig. 6 (bottom) we observe a similar phenomenon when upgrading to CP-attn is performed in different layers of U-Net (Encoder, Middle layer, Decoder). Upgrading to CP-attn in all layers makes the edit texture too similar to the source image. Content preserving performance is poor when only encoder or only middle layers is upgraded to CP-attn, while good reconstruction and editing performance is obtained when only decoder is upgraded to CP-attn.

5 Limitations and Conclusion

Our method also has some obvious problems, the reconstruction performance of our method is good enough, so the problem is not in the tuning stage, but in the editing stage, that is, if the action semantics from the editing prompt that produces an action that is seriously incompatible with the original real image, the editing result will fail. This problem can be solved by having some user provided additional information (such as skeleton joint maps, depth maps, sketch images, segmentation masks, etc.) to control the action [23,50], but this is not in line with our original intention of action editing, because providing such additional information is very tedious for the user.

In general, we propose KV Inversion, a method that can achieve highly reconstructed effects and action editing, which can solve two major problems: 1) the edited result can match the corresponding action, and 2) the edited object can keep the texture and identity of the original real image. In addition, our method does not require training U-Net itself, nor does it require scanning large datasets for time-consuming training.

References

1. Abdal, R., Qin, Y., Wonka, P.: Image2stylegan++: how to edit the embedded images? In: Proceedings of the IEEE/CVF Conference on Computer Vision and Pattern Recognition, pp. 8296–8305 (2020)
2. Abdal, R., Zhu, P., Mitra, N.J., Wonka, P.: StyleFlow: attribute-conditioned exploration of StyleGAN-generated images using conditional continuous normalizing flows. ACM Trans. Graph. **40**(3), 21:1-21:21 (2021)
3. Brock, A., Donahue, J., Simonyan, K.: Large scale GAN training for high fidelity natural image synthesis. In: 7th International Conference on Learning Representations, ICLR 2019, New Orleans, LA, USA, May 6–9, 2019. OpenReview.net (2019)
4. Brooks, T., Holynski, A., Efros, A.A.: Instructpix2pix: learning to follow image editing instructions. arXiv preprint arXiv:2211.09800 (2022)
5. Cao, M., Wang, X., Qi, Z., Shan, Y., Qie, X., Zheng, Y.: MasaCtrl: tuning-free mutual self-attention control for consistent image synthesis and editing. arXiv preprint arXiv:2304.08465 (2023)
6. Chefer, H., Alaluf, Y., Vinker, Y., Wolf, L., Cohen-Or, D.: Attend-and-excite: Attention-based semantic guidance for text-to-image diffusion models. arXiv preprint arXiv:2301.13826 (2023)
7. Crowson, K., et al.: VQGAN-CLIP: open domain image generation and editing with natural language guidance. In: Avidan, S., Brostow, G., Cissé, M., Farinella, G.M., Hassner, T. (eds.) Computer Vision. ECCV 2022. LNCS, vol. 13697, pp. pp. 88–105. Springer, Cham (2022). https://doi.org/10.1007/978-3-031-19836-6_6

8. Dhariwal, P., Nichol, A.: Diffusion models beat GANs on image synthesis. NeurIPS **34**, 8780–8794 (2021)
9. Esser, P., Rombach, R., Ommer, B.: Taming transformers for high-resolution image synthesis. In: CVPR, pp. 12873–12883 (2021)
10. Gal, R., et al.: An image is worth one word: Personalizing text-to-image generation using textual inversion. arXiv preprint arXiv:2208.01618 (2022)
11. Gu, S., et al.: Vector quantized diffusion model for text-to-image synthesis. In: CVPR, pp. 10696–10706 (2022)
12. Hertz, A., Mokady, R., Tenenbaum, J., Aberman, K., Pritch, Y., Cohen-Or, D.: Prompt-to-prompt image editing with cross attention control. arXiv preprint arXiv:2208.01626 (2022)
13. Ho, J., Jain, A., Abbeel, P.: Denoising diffusion probabilistic models. NeurIPS **33**, 6840–6851 (2020)
14. Huang, Y., Huang, J., Liu, J., Dong, Y., Lv, J., Chen, S.: WaveDM: wavelet-based diffusion models for image restoration. arXiv preprint arXiv:2305.13819 (2023)
15. Karras, T., et al.: Alias-free generative adversarial networks. Adv. Neural. Inf. Process. Syst. **34**, 852–863 (2021)
16. Karras, T., Laine, S., Aittala, M., Hellsten, J., Lehtinen, J., Aila, T.: Analyzing and improving the image quality of StyleGAN. In: Proceedings of the IEEE/CVF Conference on Computer Vision and Pattern Recognition, pp. 8110–8119 (2019)
17. Kawar, B., et al.: Imagic: text-based real image editing with diffusion models. arXiv preprint arXiv:2210.09276 (2022)
18. Kirillov, A., et al.: Segment anything. arXiv preprint arXiv:2304.02643 (2023)
19. Kumari, N., Zhang, B., Zhang, R., Shechtman, E., Zhu, J.Y.: Multi-concept customization of text-to-image diffusion. In: Proceedings of the IEEE/CVF Conference on Computer Vision and Pattern Recognition (CVPR) (2023)
20. Li, B., Qi, X., Lukasiewicz, T., Torr, P.: Controllable text-to-image generation. In: NeurIPS, vol. 32 (2019)
21. Li, B., Qi, X., Lukasiewicz, T., Torr, P.H.: ManiGAN: text-guided image manipulation. In: CVPR, pp. 7880–7889 (2020)
22. Mokady, R., Hertz, A., Aberman, K., Pritch, Y., Cohen-Or, D.: Null-text inversion for editing real images using guided diffusion models. arXiv preprint arXiv:2211.09794 (2022)
23. Mou, C., et al.: T2i-adapter: learning adapters to dig out more controllable ability for text-to-image diffusion models. arXiv preprint arXiv:2302.08453 (2023)
24. Nam, S., Kim, Y., Kim, S.J.: Text-adaptive generative adversarial networks: manipulating images with natural language. In: NeurIPS, vol. 31 (2018)
25. Nichol, A., et al.: Glide: towards photorealistic image generation and editing with text-guided diffusion models. arXiv preprint arXiv:2112.10741 (2021)
26. Nichol, A.Q., Dhariwal, P.: Improved denoising diffusion probabilistic models. In: ICML, pp. 8162–8171. PMLR (2021)
27. Parmar, G., Singh, K.K., Zhang, R., Li, Y., Lu, J., Zhu, J.Y.: Zero-shot image-to-image translation. arXiv preprint arXiv:2302.03027 (2023)
28. Patashnik, O., Wu, Z., Shechtman, E., Cohen-Or, D., Lischinski, D.: Styleclip: text-driven manipulation of StyleGAN imagery. In: ICCV, pp. 2085–2094 (2021)
29. Radford, A., et al.: Learning transferable visual models from natural language supervision. In: International Conference on Machine Learning, pp. 8748–8763. PMLR (2021)
30. Radford, A., et al.: Learning transferable visual models from natural language supervision. In: ICML, pp. 8748–8763. PMLR (2021)

31. Ramesh, A., Dhariwal, P., Nichol, A., Chu, C., Chen, M.: Hierarchical text-conditional image generation with clip Latents. arXiv preprint arXiv:2204.06125 (2022)
32. Ramesh, A., et al.: Zero-shot text-to-image generation. In: ICML, pp. 8821–8831. PMLR (2021)
33. Reed, S., Akata, Z., Yan, X., Logeswaran, L., Schiele, B., Lee, H.: Generative adversarial text to image synthesis. In: ICML, pp. 1060–1069. PMLR (2016)
34. Rombach, R., Blattmann, A., Lorenz, D., Esser, P., Ommer, B.: High-resolution image synthesis with latent diffusion models. In: CVPR, pp. 10684–10695 (2022)
35. Ronneberger, O., Fischer, P., Brox, T.: U-net: convolutional networks for biomedical image segmentation. In: Navab, N., Hornegger, J., Wells, W.M., Frangi, A.F. (eds.) MICCAI 2015. LNCS, vol. 9351, pp. 234–241. Springer, Cham (2015). https://doi.org/10.1007/978-3-319-24574-4_28
36. Ruiz, N., Li, Y., Jampani, V., Pritch, Y., Rubinstein, M., Aberman, K.: Dreambooth: fine tuning text-to-image diffusion models for subject-driven generation. arXiv preprint arXiv:2208.12242 (2022)
37. Saharia, C., et al.: Photorealistic text-to-image diffusion models with deep language understanding. arXiv preprint arXiv:2205.11487 (2022)
38. Saharia, C., et al.: Photorealistic text-to-image diffusion models with deep language understanding. In: Advances in Neural Information Processing Systems (2022)
39. Song, J., Meng, C., Ermon, S.: Denoising diffusion implicit models. arXiv preprint arXiv:2010.02502 (2020)
40. Song, Y., Ermon, S.: Generative modeling by estimating gradients of the data distribution. NeurIPS **32**, 1–13 (2019)
41. Tao, M., Tang, H., Wu, F., Jing, X.Y., Bao, B.K., Xu, C.: DF-GAN: a simple and effective baseline for text-to-image synthesis. In: CVPR, pp. 16515–16525 (2022)
42. Tumanyan, N., Geyer, M., Bagon, S., Dekel, T.: Plug-and-play diffusion features for text-driven image-to-image translation. arXiv preprint arXiv:2211.12572 (2022)
43. Wei, Y., Zhang, Y., Ji, Z., Bai, J., Zhang, L., Zuo, W.: Elite: Encoding visual concepts into textual embeddings for customized text-to-image generation (2023)
44. Xia, W., Yang, Y., Xue, J.H., Wu, B.: TediGAN: text-guided diverse face image generation and manipulation. In: CVPR, pp. 2256–2265 (2021)
45. Xiao, G., Yin, T., Freeman, W.T., Durand, F., Han, S.: FastComposer: tuning-free multi-subject image generation with localized attention. arXiv preprint arXiv:2305.10431 (2023)
46. Xu, T., Zhang, P., Huang, Q., Zhang, H., Gan, Z., Huang, X., He, X.: AttnGAN: fine-grained text to image generation with attentional generative adversarial networks. In: CVPR, pp. 1316–1324 (2018)
47. Yu, J., et al.: Scaling autoregressive models for content-rich text-to-image generation. arXiv preprint arXiv:2206.10789 (2022)
48. Zhang, H., Koh, J.Y., Baldridge, J., Lee, H., Yang, Y.: Cross-modal contrastive learning for text-to-image generation. In: CVPR, pp. 833–842 (2021)
49. Zhang, H., et al.: StackGAN++: realistic image synthesis with stacked generative adversarial networks. IEEE TPAMI **41**(8), 1947–1962 (2018)
50. Zhang, L., Agrawala, M.: Adding conditional control to text-to-image diffusion models. arXiv preprint arXiv:2302.05543 (2023)
51. Zhu, M., Pan, P., Chen, W., Yang, Y.: DM-GAN: dynamic memory generative adversarial networks for text-to-image synthesis. In: CVPR, pp. 5802–5810 (2019)

Enhancing Text-Image Person Retrieval Through Nuances Varied Sample

Jiaer Xia, Haozhe Yang, Yan Zhang, and Pingyang Dai(✉)

Key Laboratory of Multimedia Trusted Perception and Efficient Computing, Ministry of Education of China, School of Informatics, Xiamen University, 361005 Xiamen, People's Republic of China

pydai@xmu.edu.cn

Abstract. Text-image person retrieval is a task that involves searching for a specific individual based on a corresponding textual description. However, a key challenge in this task is achieving modal alignment while conducting fine-grained retrieval. Current methods utilize classification and metric losses to enhance discrimination and alignment. Nevertheless, the substantial dissimilarities between samples often impede the network's capacity to learn discriminative fine-grained information. To tackle this issue and enable the network to focus on intricate details, we introduce the Nuanced Variation Module (NVM). This module generates artificially difficult negative samples, which serve as a guide for directing the network's attention towards discerning nuances. The incorporation of NVM-constructed hard-negative samples enhances the alignment loss and facilitates the network's attentiveness to details. Additionally, we leverage the image text matching task to explicitly augment the network's fine-grained ability. By adopting our NVM method, the network can extract an ample amount of fine-grained features, thereby mitigating the interference caused by challenging negative samples. Extensive experiments demonstrate that our proposed method achieves competitive performance compared to state-of-the-art approaches on publicly available datasets.

Keywords: Text-image person retrieval · Text-based person re-identification

1 Introduction

Text-to-image person retrieval combines elements of image-text retrieval and person re-identification tasks. Instead of using an image as the query for person re-identification, it utilizes a textual description of the target individual. In real-world surveillance scenarios, obtaining a text description of a target person is often easier compared to acquiring an image of the individual. This aspect has led to increased attention and research on the task of text-to-image person retrieval in the field of surveillance retrieval.

© The Author(s), under exclusive license to Springer Nature Singapore Pte Ltd. 2024
Q. Liu et al. (Eds.): PRCV 2023, LNCS 14425, pp. 185–196, 2024.
https://doi.org/10.1007/978-981-99-8429-9_15

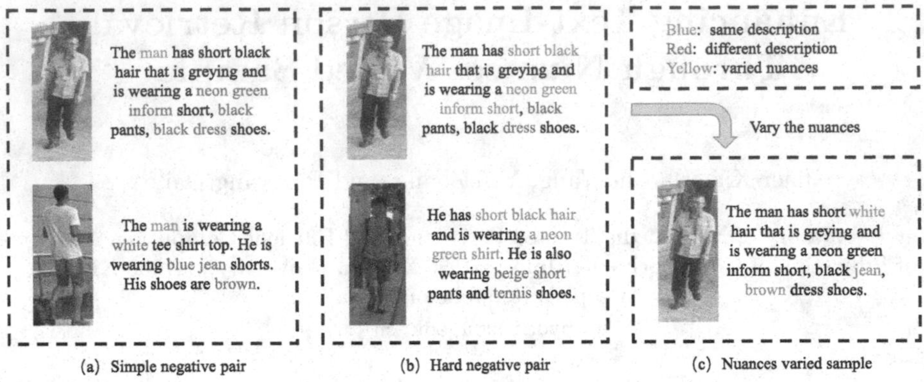

Fig. 1. Illustration for different difficulty sample pairs.

Text-to-image person retrieval encounters challenges found in both image-text retrieval and person re-identification tasks, including significant inter-modality differences and intra-modality diversity. The inter-modality difference arises from the inherent heterogeneity between images and text. Aligning the feature representations of both modalities becomes crucial to enable retrieval across these two modalities. Intra-modality diversity stems from variations in viewpoint, illumination, pose, and other factors encountered in person re-identification tasks. To address these challenges in text-to-image person retrieval, the primary approaches involve extracting discriminative features using encoders for each modality and subsequently aligning the features of both modalities. Commonly, some methods [13,29,30] employ cross-modality alignment loss to minimize the discrepancy between modalities. However, all alignment losses are confined to a single batch, and due to the enormous diversity of the dataset, the samples within a batch are more likely to have very significant differences. As shown in Fig. 1(a), these samples with significant differences can be viewed as simple pairs. Obviously, using such simple samples to align will make the network ignore some more fine-grained information and only need to focus on some significant differences, which is harmful to the robustness of the network. It can be seen from Fig. 1(b) that for hard sample pairs, there will be a lot of similarities, while with only very fine-grained differences, the network learned from easy samples will have difficulty dealing with such hard sample pairs.

To address this issue, we propose the Nuanced Variation Module (NVM) to intentionally introduce nuanced variations in the text descriptions, creating challenging sample pairs that encourage the network to focus on fine-grained information. Our approach involves continuously collecting nouns and adjectives from the text and utilizing them as an attribute library. We randomly replace nouns or adjectives in each sample with entries from the attribute library, generating artificially constructed hard samples. As shown in Fig. 1(c), the transformed samples exhibit similar characteristics to the challenging samples, featuring numerous shared descriptions with only slight differences.

To make effective use of these constructed hard samples, we not only incorporate them into the standard alignment loss but also employ an image text matching task. This additional task explicitly compels the network to prioritize distinctive fine-grained variations, enabling discrimination between positive samples and artificially constructed hard negative samples.

The main contributions can be summarized as follows:

- We propose a Nuanced Variation Module(NVM) to construct more difficult sample pairs for alignment loss and improve the fine-grained ability of the network.
- We introduce the Image Text Matching (ITM) technique for the first time in the context of text-to-image person retrieval. This approach explicitly enables the network to extract more detailed features.
- Through extensive experiments conducted on well-established benchmark datasets, including CUHK-PEDES [12] and RSTPReid [31], we demonstrate that NVM outperforms state-of-the-art methods with significant improvements.

2 Relataed Work

2.1 Text-Image Retrieval

Text-image retrieval has gained significant attention in recent years due to its potential applications in various domains, including multimedia information retrieval and content-based image retrieval. The emergence of deep learning has revolutionized text-image retrieval by enabling end-to-end learning of multimodal representations. Convolutional neural networks [8] and Transformer [20] have been successfully employed to extract visual and textual features, respectively. Joint embedding techniques, such as siamese networks and triplet loss, have been proposed to map the text and image modalities into a shared latent space, facilitating cross-modal similarity measurement [4,6,10,21]. For example, Frome *et al.* [4] present a new deep visual-semantic embedding model to leverage unlabeled data to improve the semantic representation while mapping the two modalities into the same feature space. Additionally, attention mechanisms, such as cross-modal attention and self-attention, have been incorporated to capture fine-grained correspondences between textual and visual elements. Niu *et al.* [14] introduced an innovative approach for feature alignment across different modalities by employing a cross-modal fusion mechanism and effectively captures and integrates features at various levels.

2.2 Text-Image Person Retrieval

Text-image person retrieval is an active research area that aims to bridge the semantic gap between textual descriptions and visual representations of people. In recent years, several approaches have been proposed to tackle this challenging task [2,18,23,29]. One popular direction is the use of deep learning models, such

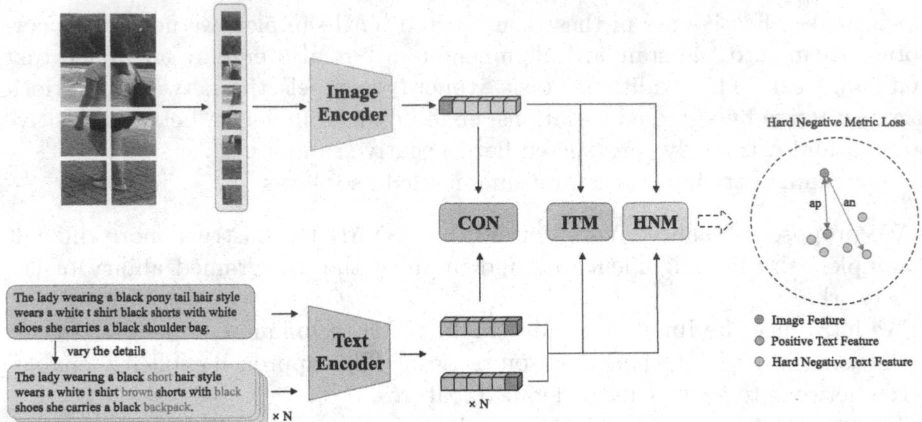

Fig. 2. Overview of the proposed NVM framework. The model adopts a Dual-Encoder structure, image and text features are extracted from separate encoders, and contains three loss modules, namely contrastive loss (CON), Image-Text Matching loss (ITM), and Hard Negative Metric loss (HNM).

as the CLIP framework, which leverages pre-trained models for image and text encoding. CLIP-based methods [9,27] have shown promising results by aligning the two modalities and capturing fine-grained features. However, these methods focus on the local information in a single sample in an explicit or implicit way, and cannot well mine the fine-grained differences between hard negative samples. Different from the methods above, NVM artificially constructs hard sample pairs to encourage the network to focus on subtle differences.

3 Method

In this section, we introduce the proposed Nuanced Variation Module (NVM). The architecture of NVM is depicted in Fig. 2.

3.1 Feature Extraction and Alignment

The crux of text-image person retrieval lies in extracting high-quality image and text features and aligning them effectively. To accomplish this, we employ a dual encoder structure, which allows us to extract image and text features independently. This approach enables us to capture fine-grained details in both modalities while leveraging the benefits of pre-training with CLIP [15] to improve alignment.

Image Feature Extraction. In the text-image person retrieval process, an input image $I \in \mathbb{R}^{N\times H\times W\times C}$ is processed using the image encoder of CLIP to obtain its image embedding. Initially, I is divided into patches of the same size, and these patches are then transformed into patch embeddings using a fixed patch size. To incorporate positional information and include an additional [CLS] token, a learnable class embedding token denoted as x_{cls} is appended to the input sequences. Moreover, a learnable position embedding is added to the transformer to incorporate spatial information into the transformer. Lastly, the [CLS] token, representing a global feature of the image, is projected into the joint space of the text-image through a linear projection. This results in the final image feature representation denoted as $f^I \in \mathbb{R}^{N \times C}$.

Text Feature Extraction. To generate the text representation from an input text T, we employ the CLIP text encoder. Similar to CLIP, we utilize the lower-cased byte pair encoding (BPE) method [17] with a vocabulary size of 49152 to tokenize the input text description. The tokenization process involves encapsulating the text description with [SOS] (start of sequence) and [EOS] (end of sequence) tokens. Next, the tokenized text sequence is fed into a transformer, where masked self-attention is employed to capture correlations between each token. This allows for the modeling of contextual relationships within the text. Finally, the [EOS] token is projected into the joint image-text space using a linear projection. This step yields the global representation of the text, and the final text feature is denoted as $f^T \in \mathbb{R}^{N \times C}$.

Feature Alignment. To align the image and text features effectively, we utilize the CLIP pre-trained model. To retain the valuable features acquired during CLIP pre-training, we employ the contrastive learning loss, which is also utilized in CLIP pre-training, to align the features in the text-image person retrieval task. The formulation of the contrastive loss can be summarized as follows:

$$\mathcal{L}_{con}^{i \to t} = \sum_{i=0}^{N} -\log \frac{\exp\left(\overline{f}_i^I \cdot \overline{f}_i^T / \tau\right)}{\sum_{k=0}^{N} \exp\left(\overline{f}_i^I \cdot \overline{f}_k^T / \tau\right)}, \tag{1}$$

$$\mathcal{L}_{con}^{t \to i} = \sum_{i=0}^{N} -\log \frac{\exp\left(\overline{f}_i^T \cdot \overline{f}_i^I / \tau\right)}{\sum_{k=0}^{N} \exp\left(\overline{f}_i^T \cdot \overline{f}_k^I / \tau\right)}, \tag{2}$$

$$\mathcal{L}_{con} = (\mathcal{L}_{con}^{i \to t} + \mathcal{L}_{con}^{t \to i})/N. \tag{3}$$

Here, $\overline{f}^I$ and $\overline{f}^T$ represent the normalized image and text features, respectively. N denotes the batch size, and τ is a temperature parameter used to control the smoothness of the probability distribution.

3.2 Nuanced Variation Module

To enhance the network's ability to distinguish subtle differences, we introduce a strategy of actively modifying certain details in the text descriptions to construct hard negative samples. This variation introduces challenging instances that require the network to capture fine-grained distinctions.

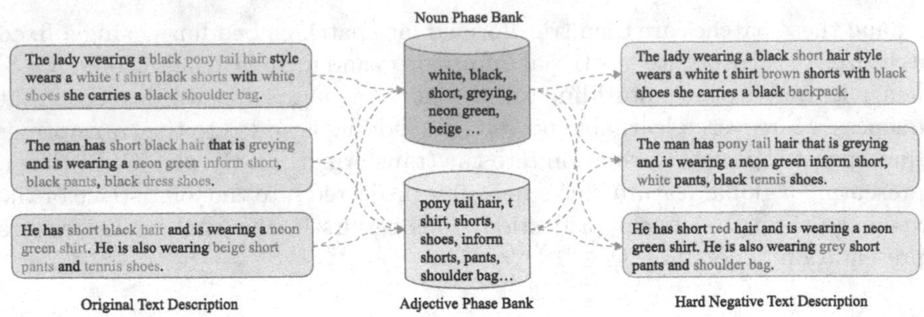

Fig. 3. An illustration of the generation of hard-negative samples.

Additionally, we employ various losses to explicitly encourage the network to differentiate between positive samples and the artificially constructed hard negative samples. By doing so, we aim to enhance the network's capacity to extract fine-grained features. These losses act as guiding signals, providing explicit supervision for the network to learn and discriminate between nuanced differences in the data.

To construct hard negative samples, we employ Noun Phrase Extraction and Adjective Phrase Extraction techniques to extract noun phrases and adjective phrases from the text. These extracted phrases are then continuously stored in a memory bank for future reference. For each text instance, we randomly select noun phrases or adjective phrases and replace them with alternative words sourced from the memory bank. This substitution process ensures that the replaced phrases exhibit variations while preserving the syntactic structure of the original text. A visual representation of this process can be found in Fig. 3.

3.3 Image Text Matching Loss

The purpose of image-text matching is to determine whether a given sample pair is a positive sample or a negative sample. By identifying positive samples, we can improve the alignment of matched instances and reduce the interference caused by negative samples.

To obtain the representation features of text-image sample pairs, we employ a cross-attention layer to enable interaction between the features of both modalities. Subsequently, these features are passed through two Transformer layers for further processing. The resulting output from the [CLS] token is considered as the joint feature representation of the image-text pair. Finally, a fully-connected (FC) layer is used to classify the joint feature and obtain the prediction probability.

The loss function for image-text matching (ITM) can be summarized as follows:

$$\mathcal{L}_{\text{itm}} = \mathbb{E}_{(I,T)\sim D}\text{H}\left(\boldsymbol{y}^{\text{itm}}, \boldsymbol{p}^{\text{itm}}(f^I, f^T)\right), \tag{4}$$

Here, $\boldsymbol{p}^{\text{itm}}$ represents the probability of a positive pair, and y^{itm} denotes the ground-truth label. The positive samples utilized for ITM are the original image-text pairs, while the hard-negative samples generated by the Nuanced Variation Module (NVM) serve as the negative samples. The ITM loss measures the discrepancy between predicted probabilities and the ground truth labels for both positive and negative sample pairs.

3.4 Hard Negative Metric Loss

The use of Hard Negative Metric Loss hopes to use hard negative samples to perform harder metric learning so that the model can extract subtle differences from the hard samples, so as to push the distance between the hard negative samples.

The commonly used Triplet Loss is used to optimize the distance between the positive and hard-negative sample pairs. At the same time, contrary to the previous hard sample mining strategy, we mine the simplest negative sample pairs for optimization. This is because the constructed negative sample pairs are hard enough in themselves. If we continue to mine the most difficult samples, the network may collapse, so we use the simplest negative samples to make the network better optimized. The Hard Negative Metric Loss is defined as:

$$\mathcal{L}_{hnm} = \left[\alpha + \left(\left\|f^I - f_p^T\right\|_2^2 - \left\|f^I - \hat{f_n^{T'}}\right\|_2^2\right)\right]_+, \tag{5}$$

$$\hat{f_n^{T'}} = argmin(\left\|f^I - f_n^{T'}\right\|_2^2). \tag{6}$$

The whole loss function of Nuanced Variation Module can be summarized as:

$$\mathcal{L}_{NVM} = \mathcal{L}_{con} + \lambda_1\mathcal{L}_{itm} + \lambda_2\mathcal{L}_{hnm}, \tag{7}$$

where the λ_1 and λ_2 is the hyper-parameter coefficient of $\mathcal{L}_{itm}$ and $\mathcal{L}_{hnm}$.

4 Experiment

4.1 Datasets and Evaluation Setting

To validate the effectiveness of our proposed method, we perform extensive experiments on publicly available text-to-image person retrieval datasets, including CUHK-PEDES [12] and RSTPReid [31].

CUHK-PEDES [12] is the first dataset designed for text-image person retrieval task, and it is also the most widely used large-scale dataset. It includes 40,206 images of 13,003 persons, and each image has 2 descriptions totaling 80,412 text descriptions, each of which has an average length of not less than 23 words. In accordance with the official data split, the training set comprises 11,003 unique identities, with 34,054 corresponding images and 68,108 textual descriptions. The validation set and test set consist of 3,078 and 3,074 images, accompanied by 6,158 and 6,156 textual descriptions, respectively. Both the validation and test sets include 1,000 distinct identities each.

RSTPReid [31] consists of 20,505 images belonging to 4,101 unique identities captured by 15 different cameras. Each identity is associated with 5 images taken from different cameras, and each image is annotated with 2 textual descriptions. Based on the official data split, the training set contains 3,701 identities, the validation set has 200 identities, and the test set also consists of 200 identities.

Evaluation Metrics. The evaluation metrics used in this study include the Rank-k metrics as the primary evaluation criteria. The Rank-k metric measures the probability of finding at least one matching person image within the top-k candidate list when a textual description is used as a query. Additionally, for a comprehensive evaluation, additional retrieval criteria, mean Average Precision (mAP), are employed. Higher values for Rank-k and mAP indicate better performance in the retrieval task.

Implementation Details. The NVM comprises three main components: a pre-trained image encoder (CLIP-ViT-B/16), a pre-trained text encoder (CLIP text Transformer), and a Transformer for obtaining ITM features, which is randomly initialized. During training, various data augmentation techniques are applied to the image data, including random horizontal flipping, random crop with padding, and random erasing. All input images are resized to a size of 384×128 pixels. The maximum length of the textual token sequence is set to 77 tokens. The model is trained using the SGD optimizer for 80 epochs, with the initial learning rate of $1 \times 10^{-}6$ and a cosine learning rate decay. The learning rate is gradually increased from $1 \times 10^{-}7$ to $1 \times 10^{-}6$ over the first 10 warm-up epochs. The hyper-parameter λ_1 and λ_2 in Eq. (7) are set to 0.5 and 0.1, respectively.

4.2 Comparison with State-of-the-Art Methods

Result on CUHK-PEDES Datasets. Table 1 presents a comparison of our proposed Nuanced Variation Module (NVM) with existing state-of-the-art (SOTA) methods on the CUHK-PEDES dataset. The compared methods are categorized into Global-based and Part-based approaches.

From the results in Table 1, we observe that in order to achieve better performance, most methods adopt Part-based techniques to obtain more fine-grained feature representations. However, our proposed NVM, which utilizes only global features, achieves competitive results. Specifically, on the CUHK-PEDES dataset, NVM achieves a rank-1 accuracy of 71.09% and an mAP of 65.10%. These results represent a significant improvement over the Global-based SOTA method IVT [19], with a rank-1 accuracy improvement of 5.5%. Moreover, NVM outperforms the Local-based SOTA method CFine [27] by +1.5% in rank-1 accuracy.

Overall, these results demonstrate the effectiveness of our proposed NVM in text-to-image person retrieval, even when using only global features. The performance improvements over both Global-based and Local-based SOTA methods highlight the advantages of NVM in extracting fine-grained features and aligning image-text representations.

Table 1. Performance comparisons with state-of-the-art methods on CUHK-PEDES dataset.

Method	Match Type	Ref	Rank-1	Rank-5	Rank-10	mAP
CMPM/C [29]	Local	ECCV18	49.37	-	79.27	-
TIMAM [16]	Global	ICCV19	54.51	77.56	79.27	-
ViTAA [22]	Local	ECCV20	54.92	75.18	82.90	51.60
NAFS [5]	Local	arXiv21	59.36	79.13	86.00	54.07
DSSL [31]	Local	MM21	59.98	80.41	87.56	-
SSAN [2]	Local	arXiv21	61.37	80.15	86.73	-
LapsCore [26]	Local	ICCV21	63.40	-	87.80	-
ISANet [28]	Local	arXiv22	63.92	82.15	87.69	-
LBUL [25]	Local	MM22	64.04	82.66	87.22	-
TextReID [7]	Global	BMVC21	64.08	81.73	88.19	60.08
SAF [11]	Local	ICASSP22	64.13	82.62	88.40	-
TIPCB [1]	Local	Neuro22	64.26	83.19	89.10	-
CAIBC [24]	Local	MM22	64.43	82.87	88.37	-
AXM-Net [3]	Local	AAAI22	64.44	80.52	86.77	58.73
LGUR [18]	Local	MM22	65.25	83.12	89.00	-
IVT [19]	Global	ECCVW22	65.59	83.11	89.21	-
CFine [27]	Local	arXiv22	69.57	85.93	91.15	-
NVM(Ours)	Global	-	**71.09**	**87.77**	**92.41**	**65.10**

Table 2. Performance comparisons with state-of-the-art methods on RSTPReid dataset.

Method	Match Type	Rank-1	Rank-10
DSSL [31]	Global	39.05	73.95
SSAN [2]	Local	43.50	77.15
LBUL [25]	Local	45.55	77.85
IVT [19]	Global	46.70	78.80
CFine [27]	Local	50.55	81.60
NVM(Ours)	Global	**58.20**	**86.90**

Table 3. Ablation study of each proposed module in NVM on CUHK-PEDES dataset.

Method	CUHK-PEDES	
	Rank-1	Rank-10
CLIP+CMPM	59.7	86.7
baseline	67.2	91.3
+HNM	69.0	91.3
+ITM	69.1	91.7
NVM(full)	**71.1**	**92.4**

Result on RSTPReid Datasets. We also conducted experiments on the RSTPReid dataset to evaluate the performance of our proposed method. Similar to the CUHK-PEDES dataset, we categorized the methods into global and local types based on the features used.

As shown in Table 2, our proposed method NVM achieves impressive results on the RSTPReid dataset. With a rank-1 accuracy of 58.2%, NVM outperforms the Global-based SOTA method IVT [19] by a significant margin of 11.5%. Additionally, NVM surpasses the Local-based SOTA method CFine [27] by +7.7% in rank-1 accuracy.

4.3 Ablation Study

In this section, we conducted ablation studies on the CUHK-PEDES dataset to analyze the impact of each module of the proposed NVM method.

For the baseline method, we used CLIP as the backbone network and trained it using the simple Contrastive Loss described in Sect. 3.1.

The results of the ablation studies are presented in Table 3. Firstly, we justified our choice of the baseline loss by comparing it with the CMPM loss used in CLIP. The results showed that using the CMPM loss, which was commonly used in the past, did not yield the best performance when applied to CLIP. In fact, the performance decreased from 67.2% to 59.7% in Rank-1 accuracy and from 91.3% to 86.7% in Rank-10 accuracy compared to our baseline with the simple contrastive learning loss.

In the module ablation experiments, we observed that training with the Hard Negative Metric (HNM) Loss significantly improved the model performance. It resulted in a performance increase of +1.8% in Rank-1 accuracy compared to the baseline. Additionally, Image-Text Matching showed an even more efficient improvement effect, with performance increasing from 67.2% to 69.1% in Rank-1 accuracy and from 91.3% to 91.7% in Rank-10 accuracy beyond the baseline. Finally, the best performance was achieved by combining both methods, demonstrating the effectiveness of the proposed HNM and Image-Text Matching modules.

5 Conclusion

In conclusion, the proposed Nuanced Variation Module (NVM) addresses the challenges of text-image person retrieval through several key contributions. By generating hard negative samples and incorporating them into the training process, the NVM module guides the network's attention towards subtle differences and improves alignment loss. Additionally, leveraging the image-text matching task enhances the network's ability to extract fine-grained features, allowing it to capture more detailed information from both modalities. The effectiveness of NVM is demonstrated through extensive experiments conducted on the CUHK-PEDES and RSTPReid datasets, two widely used benchmarks for text-image person retrieval. The results of these experiments showcase the competitive performance of our approach, surpassing state-of-the-art methods in terms of rank-1 accuracy and mean Average Precision (mAP).

Acknowledgement. This work was supported by National Key R&D Program of China (No. 2022ZD0 118202), the National Science Fund for Distinguished Young Scholars (No. 620256 03), the National Natural Science Foundation of China (No. U21B2037, No. U22B2051, No. 62176222, No. 62176223, No. 62176226, No. 62072386, No. 620723 87, No. 62072389, No. 62002305 and No. 62272401), and the Natural Science Foundation of Fujian Province of China (No. 2021J01002, No. 2022J06001).

References

1. Chen, Y., Zhang, G., Lu, Y., Wang, Z., Zheng, Y.: TIPCB: a simple but effective part-based convolutional baseline for text-based person search. Neurocomputing **494**, 171–181 (2022)
2. Ding, Z., Ding, C., Shao, Z., Tao, D.: Semantically self-aligned network for text-to-image part-aware person re-identification. arXiv preprint arXiv:2107.12666 (2021)
3. Farooq, A., Awais, M., Kittler, J., Khalid, S.S.: AXM-net: implicit cross-modal feature alignment for person re-identification. Pattern Recogn. **36**(4), 4477–4485 (2022)
4. Frome, A., et al.: Devise: a deep visual-semantic embedding model. Adv. Neural. Inf. Process. Syst. **26**, 1–9 (2013)
5. Gao, C., et al.: Contextual non-local alignment over full-scale representation for text-based person search. arXiv preprint arXiv:2101.03036 (2021)
6. Gu, J., Cai, J., Joty, S.R., Niu, L., Wang, G.: Look, imagine and match: Improving textual-visual cross-modal retrieval with generative models. In: Proceedings of the IEEE Conference on Computer Vision and Pattern Recognition, pp. 7181–7189 (2018)
7. Han, X., He, S., Zhang, L., Xiang, T.: Text-based person search with limited data. arXiv preprint arXiv:2110.10807 (2021)
8. He, K., Zhang, X., Ren, S., Sun, J.: Deep residual learning for image recognition. In: Proceedings of the IEEE Conference on Computer Vision and Pattern Recognition, pp. 770–778 (2016)
9. Jiang, D., Ye, M.: Cross-modal implicit relation reasoning and aligning for text-to-image person retrieval. In: Proceedings of the IEEE/CVF Conference on Computer Vision and Pattern Recognition, pp. 2787–2797 (2023)
10. Klein, B., Lev, G., Sadeh, G., Wolf, L.: Associating neural word embeddings with deep image representations using fisher vectors. In: Proceedings of the IEEE Conference on Computer Vision and Pattern Recognition, pp. 4437–4446 (2015)
11. Li, S., Cao, M., Zhang, M.: Learning semanticaligned feature representation for text-based person search. In: ICASSP 2022-2022 IEEE International Conference on Acoustics, Speech and Signal Processing (ICASSP), pp. 2724–2728. IEEE (2022)
12. Li, S., Xiao, T., Li, H., Zhou, B., Yue, D., Wang, X.: Person search with natural language description. In: Proceedings of the IEEE Conference on Computer Vision and Pattern Recognition, pp. 1970–1979 (2017)
13. Lv, X., Zhao, C., Chen, W.: A novel hard mining center-triplet loss for person re-identification. In: Lin, Z., et al. (eds.) PRCV 2019. LNCS, vol. 11859, pp. 199–210. Springer, Cham (2019). https://doi.org/10.1007/978-3-030-31726-3_17
14. Niu, K., Huang, Y., Ouyang, W., Wang, L.: Improving description-based person re-identification by multi-granularity image-text alignments. IEEE Trans. Image Process. **29**, 5542–5556 (2020)
15. Radford, A., et al.: Learning transferable visual models from natural language supervision. In: International Conference on Machine Learning, pp. 8748–8763. PMLR (2021)
16. Sarafianos, N., Xu, X., Kakadiaris, I.A.: Adversarial representation learning for text-to-image matching. In: Proceedings of the IEEE/CVF International Conference on Computer Vision, pp. 5814–5824 (2019)
17. Sennrich, R., Haddow, B., Birch, A.: Neural machine translation of rare words with subword units. arXiv preprint arXiv:1508.07909 (2015)

18. Shao, Z., Zhang, X., Fang, M., Lin, Z., Wang, J., Ding, C.: Learning granularity-unified representations for text-to-image person re-identification. arXiv preprint arXiv:2207.07802 (2022)
19. Shu, X., et al.: See finer, see more: implicit modality alignment for text-based person retrieval. arXiv preprint arXiv:2208.08608 (2022)
20. Vaswani, A., et al.: Attention is all you need. In: Advances in Neural Information Processing Systems, vol. 30 (2017)
21. Wang, L., Li, Y., Lazebnik, S.: Learning deep structure-preserving image-text embeddings. In: Proceedings of the IEEE Conference on Computer Vision and Pattern Recognition, pp. 5005–5013 (2016)
22. Wang, Z., Fang, Z., Wang, J., Yang, Y.: ViTAA: visual-textual attributes alignment in person search by natural language. In: Vedaldi, A., Bischof, H., Brox, T., Frahm, J.-M. (eds.) ECCV 2020. LNCS, vol. 12357, pp. 402–420. Springer, Cham (2020). https://doi.org/10.1007/978-3-030-58610-2_24
23. Wang, Z., Xue, J., Zhu, A., Li, Y., Zhang, M., Zhong, C.: Amen: Adversarial multi-space embedding network for text-based person re-identification. In: Pattern Recognition and Computer Vision: 4th Chinese Conference, PRCV 2021, Beijing, China, October 29-November 1, 2021, Proceedings, Part II 4. pp. 462–473. Springer (2021)
24. Wang, Z., Zhu, A., Xue, J., Wan, X., Liu, C., Wang, T., Li, Y.: Caibc: Capturing all-round information beyond color for text-based person retrieval. arXiv preprint arXiv:2209.05773 (2022)
25. Wang, Z., Zhu, A., Xue, J., Wan, X., Liu, C., Wang, T., Li, Y.: Look before you leap: Improving text-based person retrieval by learning a consistent cross-modal common manifold. In: Proceedings of the 30th ACM International Conference on Multimedia. pp. 1984–1992 (2022)
26. Wu, Y., Yan, Z., Han, X., Li, G., Zou, C., Cui, S.: LapScore: language-guided person search via color reasoning. In: Proceedings of the IEEE/CVF International Conference on Computer Vision, pp. 1624–1633 (2021)
27. Yan, S., Dong, N., Zhang, L., Tang, J.: Clip-driven fine-grained text-image person re-identification. arXiv preprint arXiv:2210.10276 (2022)
28. Yan, S., Tang, H., Zhang, L., Tang, J.: Image-specific information suppression and implicit local alignment for text-based person search. arXiv preprint arXiv:2208.14365 (2022)
29. Zhang, Ying, Lu, Huchuan: Deep cross-modal projection learning for image-text matching. In: Ferrari, Vittorio, Hebert, Martial, Sminchisescu, Cristian, Weiss, Yair (eds.) ECCV 2018. LNCS, vol. 11205, pp. 707–723. Springer, Cham (2018). https://doi.org/10.1007/978-3-030-01246-5_42
30. Zheng, Z., Zheng, L., Garrett, M., Yang, Y., Xu, M., Shen, Y.D.: Dual-path convolutional image-text embeddings with instance loss. ACM Trans. Multimed. Comput. Commun. Appl. (TOMM) **16**(2), 1–23 (2020)
31. Zhu, A., et al.: DSSL: deep surroundings-person separation learning for text-based person retrieval. In: Proceedings of the 29th ACM International Conference on Multimedia, pp. 209–217 (2021)

Unsupervised Prototype Adapter for Vision-Language Models

Yi Zhang[1,2], Ce Zhang[2], Xueting Hu[2], and Zhihai He[2,3](✉)

[1] Harbin Institute of Technology, Harbin, China
[2] Southern University of Science and Technology, Shenzhen, China
{zhangyi2021,zhangc2019,huxt2022}@mail.sustech.edu.cn, hezh@sustech.edu.cn
[3] Pengcheng Laboratory, Shenzhen, China

Abstract. Recently, large-scale pre-trained vision-language models (*e.g.* CLIP and ALIGN) have demonstrated remarkable effectiveness in acquiring transferable visual representations. To leverage the valuable knowledge encoded within these models for downstream tasks, several fine-tuning approaches, including prompt tuning methods and adapter-based methods, have been developed to adapt vision-language models effectively with supervision. However, these methods rely on the availability of annotated samples, which can be labor-intensive and time-consuming to acquire, thus limiting scalability. To address this issue, in this work, we design an unsupervised fine-tuning approach for vision-language models called Unsupervised Prototype Adapter (UP-Adapter). Specifically, for the unannotated target datasets, we leverage the text-image aligning capability of CLIP to automatically select the most confident samples for each class. Utilizing these selected samples, we generate class prototypes, which serve as the initialization for the learnable prototype model. After fine-tuning, the prototype model prediction is combined with the original CLIP's prediction by a residual connection to perform downstream recognition tasks. Our extensive experimental results on image recognition and domain generalization show that the proposed unsupervised method outperforms 8-shot CoOp, 8-shot Tip-Adapter, and also the state-of-the-art UPL method by large margins.

Keywords: Vision-Language Models · Contrastive Language-Image Pre-training · Unsupervised Learning · Image Recognition

1 Introduction

Transferring knowledge to unseen scenarios is a capability natural to humans yet challenging for deep learning models to reproduce [35,42]. During the past decade, researchers have focused on designing sophisticated architectures and advanced algorithms to train visual recognition systems to predict accurately

© The Author(s), under exclusive license to Springer Nature Singapore Pte Ltd. 2024
Q. Liu et al. (Eds.): PRCV 2023, LNCS 14425, pp. 197–209, 2024.
https://doi.org/10.1007/978-981-99-8429-9_16

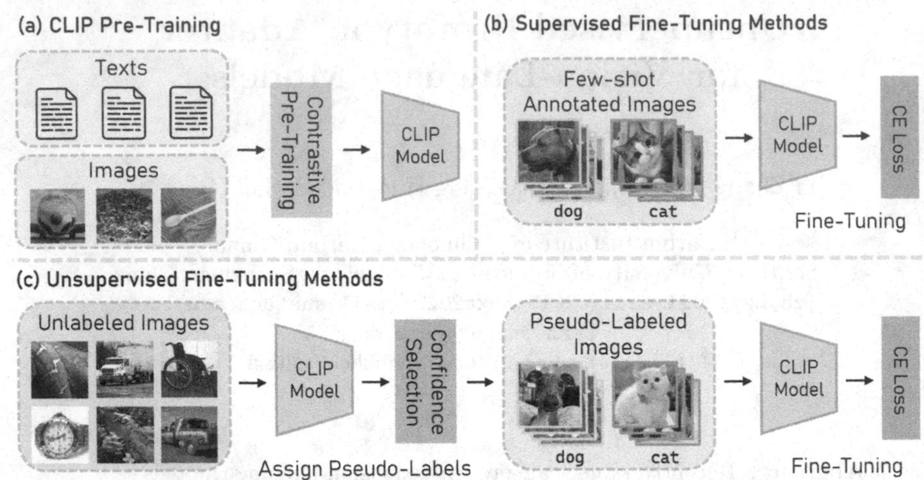

Fig. 1. (a) CLIP pre-training process. (b) Typical supervised fine-tuning methods, *e.g.*, CoOp [44] and Tip-Adapter [40]. (c) Our proposed unsupervised fine-tuning method.

for a limited set of visual categories. However, this approach often fails to perform consistently on unseen classes unless the classifier is retrained on a more comprehensive dataset.

Recently, pioneering CLIP [24] has provided a new paradigm for generic visual recognition: pre-training, then fine-tuning. Under this paradigm, large-scale vision-language models are established based on pre-training on image-text pairs available on the Internet [14,24,36,39], then practitioners fine-tune these powerful models to downstream tasks through supervised learning on annotated samples [8,40,44]. Illustrations of the CLIP pre-training process and fine-tuning process are shown in Fig. 1 (a) and (b). Those fine-tuned vision-language models have shown impressive performance on various vision tasks, such as image recognition [8,24,44], object detection [5,28], and image captioning [16,38].

Although those supervised fine-tuning approaches, such as CoOp [44] and Tip-Adapter [40] have shown robust adaptation capabilities, these methods still require annotated samples, which are labor-intensive and time-consuming to obtain and limit the scalability. In this work, we provide an unsupervised fine-tuning approach as shown in Fig. 1 (c) to address this important issue. In our approach, we make use of the text-image aligning capability of CLIP to generate pseudo-labels for unlabeled samples. We then carefully select samples based on confidence scores for each class and generate class prototypes from these selections. These prototypes are utilized to initialize the prototype model, which can be significantly boosted by fine-tuning with a few training epochs. The prediction of the prototype model is integrated with the original CLIP's prediction to predict the final labels of images. Our extensive experimentation in image recognition and domain generalization demonstrates significant performance improvements of our unsupervised method UP-Adapter compared to 8-shot CoOp, 8-shot Tip-Adapter, and the state-of-the-art unsupervised UPL method.

Our method has the following three *unique novelties*: (1) Leveraging the visual-textual correlation capability of CLIP, we address the challenge of requiring a substantial amount of annotated data. (2) We have concurrently addressed the challenge of mitigating the substantial computational costs. For ImageNet, UPL needs 200 epochs and 15 h to achieve an accuracy of 61.09%, while our method achieves an accuracy of 63.58% with only 20 epochs and 3 min. (3) We are the first to explore the potential of the adapter-based fine-tuning method in the unsupervised setting. Our method works directly on the image feature, which can guarantee the preservation of prior knowledge in pre-trained CLIP.

2 Related Work

2.1 Large-Scale Pre-trained Vision-Language Models

Over the past few years, vision-language pre-trained models have emerged as a powerful approach to tackle various tasks that involve the interaction between visual and textual information [24]. These models leverage large-scale datasets to learn representations that capture the semantic understanding of both images and their associated textual descriptions. For example, CLIP [24] employs contrastive learning between the embeddings of 400 million carefully selected image-text pairs. Researchers have demonstrated that large-scale pre-trained vision-language models have great potential in visual representation learning and transfer learning, therefore can be utilized in several tasks such as image retrieval [6,17], visual question answering [45], *etc.*

2.2 Adaptation Methods for Vision-Language Models

To adapt vision-language models to various downstream tasks, researchers have developed a number of *supervised* and *unsupervised* approaches.

Recent *supervised* methods typically try to adapt the models using few-shot samples to inherit the well-learned prior knowledge of vision-language models to downstream tasks. These methods can be classified into two major directions: prompt tuning methods and adapter-based methods. **Prompt tuning methods** focus on refining prompts and incorporating learnable context to extract task-relevant information from the encoded knowledge [43,44]. A notable example is CoOp [44], which optimizes the prompt context using learnable vectors. CoCoOp [43] builds upon CoOp by generating vectors conditioned on each image, addressing the challenge of generalizing to unseen classes. TPT [19] enables the learning of adaptive prompts on the fly with just a single test sample. **Adapter-based methods** focus on tuning the text and image feature representations directly [40,41]. For example, CLIP-Adapter [8] designs a feature adapter to enhance traditional fine-tuning results. Building upon this, Tip-Adapter (Training-free CLIP-Adapter) [40] achieves improved performance by establishing a key-value cache model from few-shot samples to perform fine-tuning.

Another line of research delves into the *unsupervised* setting where the labels for the target images are unavailable. For instance, UPL [13] proposes to generate pseudo-labels for the unannotated images to convert this problem to a supervised problem. Some approaches also suppose the target images are inaccessible, and propose to adapt the vision-language models with only the class names (called "name-only transfer"). For instance, VisDesc [20] queries large language models for concept descriptors for image classification, CHiLS [22] performs inference in the subclass space, while SuS-X [32] constructs a support set using text-to-image generation model or large-scale vision-language dataset.

2.3 Self-training with Pseudo-Labeling

Semi-supervised learning has been extensively studied over several decades. Its primary objective is to develop a predictive model using a small set of annotated samples and a large set of unlabeled samples [33]. One notable approach within semi-supervised learning is self-training, which was initially proposed by Scudder [27]. Self-training approaches typically leverage a model trained on annotated samples to generate pseudo-labels for unlabeled samples. In the realm of fine-tuning vision-language models, the concept of self-training was initially introduced by UPL [13]. UPL specifically utilizes the pre-trained CLIP model directly for pseudo-labeling and employs these pseudo-labeled samples to optimize the prompt representations. In this work, we aim to further explore the potential of self-training in fine-tuning vision-language models.

3 Method

3.1 Background

CLIP has two parallel encoders: an image encoder and a text encoder. For the image encoder, ResNet [9] or Vision Transformer (ViT) [4] are commonly utilized to extract the visual features, while for the text encoder, we typically use a Transformer-based text. During the training phase, CLIP utilizes a contrastive loss function to facilitate the similarity between the features of the images and text in the embedding space. After training, the two modalities are aligned within a shared embedding space.

A CLIP model is denoted as $\{E_t, E_v\}$, where E_t refers to the text encoder, and E_v refers to the image encoder. In the image recognition task, a single test image X_{test} belonging to a specific class y is provided, where $X_{test} \in \mathbb{R}^{C\times H\times W}$ and $y \in \mathbb{R}^K$ for a K-class classification problem. In zero-shot CLIP, each y_i in the set $Y = \{y_1, y_2, \cdots, y_K\}$ is concatenated with a pre-defined prompt such as $\rho =$ "a photo of," to create textual inputs for different classes, denoted as $\{\rho; y_i\}$. Text features, $\{t_1, t_2, \cdots, t_K\}$, are generated by the text encoder E_t, where $t_i = E_t(\rho; y_i)$. Subsequently, we compute a cosine similarity score using each text feature t_i is combined with the image feature $v = E_v(X_{test})$, which can be used to predict the probability of X_{test} belonging to class y_i.

$$\operatorname{sim}(t_i, v) = \frac{t_i \cdot v}{\|t_i\|\|v\|}. \tag{1}$$

The prediction probability on X_{test} can be denoted by

$$p(y_i|X_{test}) = \frac{\exp\left(\mathrm{sim}\left(t_i, v\right)/\tau\right)}{\sum_{j=1}^{K}\exp\left(\mathrm{sim}\left(t_j, v\right)/\tau\right)}, \tag{2}$$

where τ refers to the temperature hyperparameter of the softmax function. The final predicted label $\hat{y}$ can be written by:

$$\hat{y} = \underset{y_i}{\mathrm{argmax}}\, p_{y_i}. \tag{3}$$

3.2 Unsupervised Prototype Adapter

In Fig. 2, we present an overview of our proposed UP-Adapter method. The proposed UP-Adapter consists of 3 stages: pseudo-label generation, class prototype estimation, and prototype adapter.

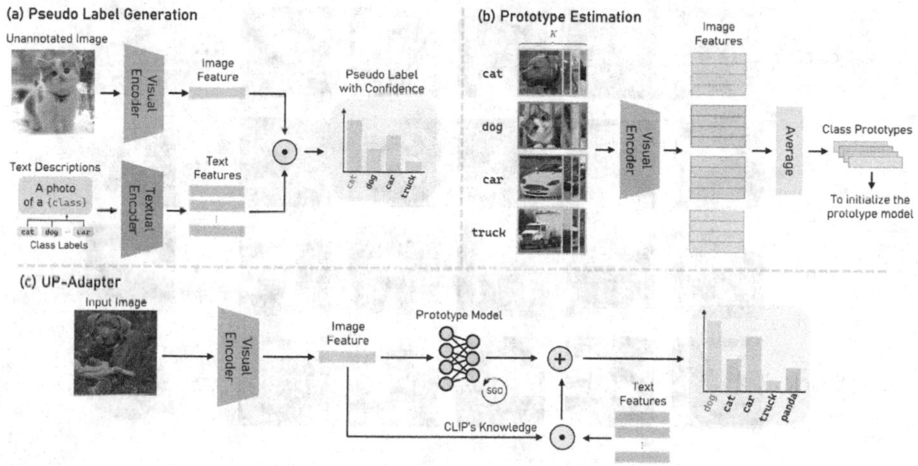

Fig. 2. An overview of our proposed UP-Adapter method. (a) shows the pseudo-label generation process, for every class in the target dataset, we choose the most confident top-K samples for subsequent optimization. (b) is the class prototype estimation process, for every class, we generate the features of 16 selected samples by image encoder, then average the features as the class prototype. (c) presents the overall process of our proposed UP-Adapter, we fuse the predictions of original pre-trained CLIP and the prototype model for final classification.

Pseudo-Label Generation. Utilizing the pre-trained CLIP model, we can leverage Eqs. (2) and (3) to derive pseudo-labels for unlabeled samples in the target dataset. To prevent the model from being overwhelmed by the abundance of samples from certain categories, we propose to select the most confident top-K samples per class using Eqs. (2) and (3) for subsequent optimization. Given an unlabeled training set, for each class, we choose the most confident top-K samples. Specifically, we compute the similarities of all the images with the class-specific text descriptions and select the top-K similar samples. In our experiments, we empirically set $K = 16$. We present an illustration of our selected top-K images in Fig. 3.

Class Prototype Estimation. We can obtain the prototype by averaging the $L2$ normalized features of these K images. For each class c in a C-class recognition problem, we generate the visual features f_k by CLIP's visual encoder E_v, then take the average of them to obtain the class prototype

$$P_c = \frac{1}{K}\sum_{k=1}^{K} f_k. \tag{4}$$

Here, P_c represents the class prototype for the particular class c. The prototype set for all the classes can be denoted as $P \triangleq \{P_c\}_{c=1}^{C}$.

Fig. 3. Top-K pseudo-labeled images with the highest confidence level for 4 classes in ImageNet: `cat`, `dog`, `car`, and `truck`.

Prototype Adapter. During the inference stage, the $L2$ normalized test image feature $v_{te} \in \mathbb{R}^{1\times C}$ generated by the visual encoder E_v, is used to calculate the affinities [26] between the test image and prototype, denoted as

$$H = \exp\left(-\eta\left(1 - \mathrm{sim}(v_{te}, P^\top)\right)\right). \tag{5}$$

Here, we use an exponential function to convert the similarities into non-negative values. η is another hyperparameter to control its sharpness. $\mathrm{sim}\,(v_{te}, P)$ stands for the cosine similarity between test feature v_{te} and prototype P. We denote it as:

$$\mathrm{sim}\,(v_{te}, P) = v_{te}P^\top. \tag{6}$$

Since both v_{te} and P are $L2$-normalized, $v_{te}P^\top$ stands for their cosine similarity shown in Eq. (5), the affinities between the test image and prototype can be denoted as:

$$H = \exp\left(-\eta(1 - v_{te}P^\top)\right). \quad (7)$$

Therefore, the prediction from the prototypes can be denoted as H. Besides the knowledge from the prototypes, the prior knowledge of pre-trained CLIP is calculated by $v_{te}f_t^\top$, where f_t is the text features generated from textual encoder E_t. By combining the predictions of CLIP and the prototype adapter, the output logits of the test image by training-free UP-Adapter can be written as:

$$\begin{aligned} \text{logits} &= \beta H + v_{te}f_t^\top \\ &= \beta \exp\left(-\eta(1 - v_{te}P^\top)\right) + v_{te}f_t^\top, \end{aligned} \quad (8)$$

where β denotes the residual ratio.

Even though the prediction from the prototype adapter can be achieved without training, there is still room for improving its performance. Therefore, we design a prototype model with only one linear layer for better adaptation, which can be denoted as

$$Q(x) = \exp\left(-\eta(1 - Wx)\right). \quad (9)$$

Similarly, the exponential function is adopted to convert the similarities into non-negative values. As shown in Fig. 2, we initialize the network with prototype P, which sets the model at a high starting point. Therefore, given the image feature v extracted by E_v, the logits of the UP-adapter with training can be denoted as

$$\begin{aligned} \text{logits} &= \beta Q + v f_t^\top \\ &= \beta \exp\left(-\eta(1 - vW^\top)\right) + v f_t^\top. \end{aligned} \quad (10)$$

During training, W is updated by gradient descent. We use cross-entropy loss as the loss function.

4 Experiments

4.1 Image Recognition

Datasets. Following prior fine-tuning methods for CLIP [40], we perform a few-shot assessment on 11 commonly used datasets for image classification. These datasets encompass various categories such as generic object classification (ImageNet [25], Caltech101 [7]), fine-grained object classification (OxfordPets [23], StandfordCars [15], Flowers102 [21], Food-101 [1], FGCV Aircraft [18]), texture classification (DTD [2]), remote sensing recognition (EuroSAT [10]), scene recognition (SUN397 [37]) and action recognition (UCF101 [29]). These datasets provide a comprehensive benchmark to evaluate the few-shot learning performance of each method.

Baselines. We compare our method with four baseline methods: zero-shot CLIP [24], CoOp [44], Tip-Adapter [40] and UPL [13]. Therein, CoOp [44] and Tip-Adapter [40] are supervised prompt tuning method and adapter-based method, respectively. UPL [13] is one of the first approaches in fine-tuning vision-language models in an unsupervised setting. To ensure a fair comparison, their results are taken directly from their original paper. We also report CoOp's accuracies with the best-performance setting.

Implementation Details. For pseudo-label generation, we utilize ViT-B/16 as its image encoder, while we use ResNet-50 as its image encoder in the training and inference period, we exploit Transformer as its text encoder for all the processes. We adopt prompt ensembling defined in [24] and follow the data preprocessing methods defined in CLIP [24] for all the datasets. Our model is trained for 30 epochs for ImageNet and 20 epochs for the other 10 datasets. We conduct all the experiments on one NVIDIA RTX 3090 GPU.

Results. Table 1 presents a comprehensive comparison of our proposed method with state-of-the-art techniques across 11 datasets. The results clearly demonstrate that our approach yields substantial performance enhancements when compared to zero-shot CLIP [24]. Moreover, our method outperforms both the 4-shot supervised CoOp [44] and Tip-Adapter [40] methods across all datasets, and surpasses the 8-shot versions in 10 out of 11 datasets. In comparison to another unsupervised method UPL [13], our approach consistently outperforms it in the majority of cases, resulting in an average accuracy improvement of +2.35%, which is quite significant.

Table 1. Performance comparisons of our proposed methods and other state-of-the-art methods on 11 datasets. We compare our unsupervised approach with: 1) original CLIP with prompt engineering [24]; 2) supervised methods including CoOp [44] and Tip-Adapter [40]; 3) previous unsupervised UPL [13] method.

Methods	CLIP	**Ours**	UPL	CoOp			Tip-Adapter		
Settings	-	**Unsupervised**	**Unsupervised**	2-shot	4-shot	8-shot	2-shot	4-shot	8-shot
ImageNet	60.34	**63.58**	61.09	57.13	59.72	61.52	60.96	60.98	61.45
Caltech101	86.09	**91.65**	91.40	87.76	89.67	90.14	89.25	89.41	89.94
DTD	41.61	**59.06**	55.08	47.48	54.19	58.65	49.76	54.14	57.33
EuroSAT	38.23	**72.49**	71.04	59.98	62.17	68.73	61.10	65.30	66.89
FGVCAircraft	16.92	**25.04**	21.75	20.36	22.10	24.99	21.25	21.54	24.48
Food101	77.33	**78.06**	77.93	72.92	73.74	76.28	77.58	77.60	77.79
Flowers102	66.06	86.19	76.65	76.58	84.59	**88.27**	76.82	81.53	85.95
OxfordPets	85.83	**89.66**	89.51	84.53	87.11	87.71	87.38	87.67	87.87
SUN397	60.18	**68.52**	66.42	61.35	65.08	67.47	62.82	64.32	65.57
StandfordCars	55.64	68.31	**70.97**	59.49	61.92	65.25	59.86	62.03	63.35
UCF101	62.70	**75.35**	70.18	65.06	68.26	71.67	66.59	67.51	69.10
Average	59.18	**70.72**	68.37	62.97	66.23	69.15	64.85	66.55	68.16

4.2 Domain Generalization

Experiment Setup. It is critical for the machine learning model to have strong robustness to distribution shifts [30,31]. To demonstrate the generalization capabilities of our proposed method, We evaluate the performance on the domain generalization task. We follow prior methods to train our model on 16-shot ImageNet [3] and test on four ImageNet variant datasets (ImageNet-V2 [25], ImageNet-Sketch [34]), ImageNet-A [12], and ImageNet-R [11].

Baselines. We include six previous methods for comparison: zero-shot CLIP [24], linear probe CLIP [24], CoOp [44], CoCoOp [43], and TPT [19]. All of these state-of-the-art methods are supervised methods. For a fair comparison, we directly include the results of other baselines from their original paper.

Results. Table 2 presents the performance comparison results on domain generalization tasks (from ImageNet to ImageNet-V2/-Sketch/-A/-R). Our proposed unsupervised UP-Adapter outperforms the supervised baselines in accuracy, which suggests remarkable robustness to distribution shifts of our method.

Table 2. Comparison with other methods on domain generalization task. The best results are in **bold**.

Method	Unsupervised?	Source	Target				
		ImageNet	-V2	-Sketch	-A	-R	OOD Average
Zero-Shot CLIP [24]	-	60.33	53.27	35.44	21.65	56.00	41.59
Linear Probe CLIP [24]	-	56.13	45.61	19.13	12.74	34.86	28.09
CoOp [44]	✗	63.33	55.40	34.67	23.06	56.60	42.43
CoCoOp [43]	✗	62.81	**55.72**	34.48	23.32	57.74	42.82
TPT [19]	✗	60.74	54.70	35.09	26.67	59.11	43.89
Ours	✓	**63.58**	54.90	**35.56**	**27.65**	**59.36**	**44.37**

4.3 Ablation Study

Effects of Different Algorithm Components. To better demonstrate the effectiveness of our UP-Adapter, we have conducted an ablation study on the ImageNet [3] dataset to study the effects of different algorithm components. The results are shown in Table 3. As we can see from the table, directly using the class prototypes is not fully effective. The initialization of the prototype is also crucial, which will enhance the performance by 5.54%.

Table 3. Effects of different algorithm components. We report the average accuracies on 11 image recognition datasets.

Methods	Accuracy
Zero-shot CLIP [24]	59.18
Prototype adapter only	66.70
CLIP + class prototypes	67.31
CLIP + prototype adapter (w/o initialization)	65.18
CLIP + prototype adapter (w/ initialization)	70.72

Number of Selected Images for Each Class K. In our approach, we carefully choose the top-K most confident samples for the prototype estimation process. In our experiments, we empirically set $K = 16$. In Table 4, we show an ablation study on the number of selected images for each class K. We report the average accuracies over 11 datasets on image recognition tasks using $K = 4, 8, 16, 32$ to investigate the impact of K. This ablation study suggests that our setting of $K = 16$ yields the optimal performance.

Visual Backbones for Pseduo-Labeling. Table 4 provides a summary of average accuracies across 11 datasets using different visual backbones, including ResNets and ViTs. Notably, employing more advanced backbones results in more accurately generated pseudo-labels, which in turn leads to improved overall performance. When using the ViT-B/16 backbone, we achieve the best average accuracy of 70.72%.

Table 4. Parametric analysis on values of K and visual backbones.

Number of Selected Images K				Visual Backbones for Pseduo-Labeling			
$K = 4$	$K = 8$	$K = 16$	$K = 32$	ResNet-50	ResNet-101	ViT-B/32	ViT-B/16
67.03	68.97	**70.72**	70.21	65.44	67.26	68.81	**70.72**

5 Conclusion

In this work, we propose an unsupervised fine-tuning approach for vision-language models to tackle the challenge of requiring a substantial amount of annotated data. We utilize CLIP's capabilities to generate pseudo-labels for unlabeled samples, and we carefully choose the most confident samples for each class. Using these selected samples, we create prototypes specific to each class, which serve as initializations for an adapter module. This initialized adapter module can be further fine-tuned to handle recognition tasks effectively. Extensive experimentation in the domains of few-shot image recognition and domain generalization reveals that our proposed unsupervised method outperforms the state-of-the-art unsupervised UPL method by substantial margins.

References

1. Bossard, L., Guillaumin, M., Van Gool, L.: Food-101 – mining discriminative components with random forests. In: Fleet, D., Pajdla, T., Schiele, B., Tuytelaars, T. (eds.) ECCV 2014. LNCS, vol. 8694, pp. 446–461. Springer, Cham (2014). https://doi.org/10.1007/978-3-319-10599-4_29
2. Cimpoi, M., Maji, S., Kokkinos, I., Mohamed, S., Vedaldi, A.: Describing textures in the wild. In: Proceedings of the IEEE/CVF Conference on Computer Vision and Pattern Recognition, pp. 3606–3613 (2014)

3. Deng, J., Dong, W., Socher, R., Li, L.J., Li, K., Fei-Fei, L.: Imagenet: a large-scale hierarchical image database. In: Proceedings of the IEEE/CVF Conference on Computer Vision and Pattern Recognition, pp. 248–255 (2009)
4. Dosovitskiy, A., et al.: An image is worth 16×16 words: transformers for image recognition at scale. In: International Conference on Learning Representations (2020)
5. Du, Y., Wei, F., Zhang, Z., Shi, M., Gao, Y., Li, G.: Learning to prompt for open-vocabulary object detection with vision-language model. In: Proceedings of the IEEE/CVF Conference on Computer Vision and Pattern Recognition, pp. 14084–14093 (2022)
6. Duan, J., et al.: Multi-modal alignment using representation codebook. In: Proceedings of the IEEE/CVF Conference on Computer Vision and Pattern Recognition, pp. 15651–15660 (2022)
7. Fei-Fei, L., Fergus, R., Perona, P.: Learning generative visual models from few training examples: an incremental Bayesian approach tested on 101 object categories. In: Proceedings of the IEEE/CVF Conference on Computer Vision and Pattern Recognition Workshops, pp. 178–178 (2004)
8. Gao, P., et al.: Clip-adapter: better vision-language models with feature adapters. arXiv preprint arXiv:2110.04544 (2021)
9. He, K., Zhang, X., Ren, S., Sun, J.: Deep residual learning for image recognition. In: Proceedings of the IEEE/CVF Conference on Computer Vision and Pattern Recognition, pp. 770–778 (2016)
10. Helber, P., Bischke, B., Dengel, A., Borth, D.: EuroSAT: a novel dataset and deep learning benchmark for land use and land cover classification. IEEE J. Sel. Top. Appl. Earth Observ. Rem. Sens. **12**(7), 2217–2226 (2019)
11. Hendrycks, D., et al.: The many faces of robustness: a critical analysis of out-of-distribution generalization. In: Proceedings of the IEEE/CVF International Conference on Computer Vision, pp. 8340–8349 (2021)
12. Hendrycks, D., Zhao, K., Basart, S., Steinhardt, J., Song, D.: Natural adversarial examples. In: Proceedings of the IEEE/CVF Conference on Computer Vision and Pattern Recognition, pp. 15262–15271 (2021)
13. Huang, T., Chu, J., Wei, F.: Unsupervised prompt learning for vision-language models. arXiv preprint arXiv:2204.03649 (2022)
14. Huang, X., et al.: Idea: increasing text diversity via online multi-label recognition for vision-language pre-training. In: Proceedings of the 30th ACM International Conference on Multimedia, pp. 4573–4583 (2022)
15. Krause, J., Stark, M., Deng, J., Fei-Fei, L.: 3d object representations for fine-grained categorization. In: Proceedings of the IEEE/CVF International Conference on Computer Vision Workshops, pp. 554–561 (2013)
16. Li, L.H., et al.: Grounded language-image pre-training. In: Proceedings of the IEEE/CVF Conference on Computer Vision and Pattern Recognition, pp. 10965–10975 (2022)
17. Lu, J., Batra, D., Parikh, D., Lee, S.: VilBERT: pretraining task-agnostic visiolinguistic representations for vision-and-language tasks. In: Advances in Neural Information Processing Systems, vol. 32 (2019)
18. Maji, S., Rahtu, E., Kannala, J., Blaschko, M., Vedaldi, A.: Fine-grained visual classification of aircraft. arXiv preprint arXiv:1306.5151 (2013)
19. Manli, S., et al.: Test-time prompt tuning for zero-shot generalization in vision-language models. In: Advances in Neural Information Processing Systems (2022)
20. Menon, S., Vondrick, C.: Visual classification via description from large language models. In: International Conference on Learning Representations (2023)

21. Nilsback, M.E., Zisserman, A.: Automated flower classification over a large number of classes. In: Indian Conference on Computer Vision, Graphics and Image Processing (2008)
22. Novack, Z., McAuley, J., Lipton, Z.C., Garg, S.: Chils: zero-shot image classification with hierarchical label sets. In: International Conference on Machine Learning, pp. 26342–26362. PMLR (2023)
23. Parkhi, O.M., Vedaldi, A., Zisserman, A., Jawahar, C.: Cats and dogs. In: Proceedings of the IEEE/CVF Conference on Computer Vision and Pattern Recognition, pp. 3498–3505 (2012)
24. Radford, A., et al.: Learning transferable visual models from natural language supervision. In: International Conference on Machine Learning (2021)
25. Recht, B., Roelofs, R., Schmidt, L., Shankar, V.: Do imagenet classifiers generalize to imagenet? In: International Conference on Machine Learning (2019)
26. Ru, L., Zhan, Y., Yu, B., Du, B.: Learning affinity from attention: end-to-end weakly-supervised semantic segmentation with transformers. In: Proceedings of the IEEE/CVF Conference on Computer Vision and Pattern Recognition, pp. 16846–16855 (2022)
27. Scudder, H.: Probability of error of some adaptive pattern-recognition machines. IEEE Trans. Inf. Theory **11**(3), 363–371 (1965)
28. Shi, H., Hayat, M., Wu, Y., Cai, J.: Proposalclip: unsupervised open-category object proposal generation via exploiting clip cues. In: Proceedings of the IEEE/CVF Conference on Computer Vision and Pattern Recognition, pp. 9611–9620 (2022)
29. Soomro, K., Zamir, A.R., Shah, M.: Ucf101: a dataset of 101 human actions classes from videos in the wild. arXiv preprint arXiv:1212.0402 (2012)
30. Tang, Y., Guo, Q., He, Z.: Cross-inferential networks for source-free unsupervised domain adaptation. arXiv preprint arXiv:2306.16957 (2023)
31. Tang, Y., et al.: Neuro-modulated Hebbian learning for fully test-time adaptation. arXiv preprint arXiv:2303.00914 (2023)
32. Udandarao, V., Gupta, A., Albanie, S.: SUS-X: training-free name-only transfer of vision-language models. In: Proceedings of the IEEE/CVF International Conference on Computer Vision (2023)
33. Van Engelen, J.E., Hoos, H.H.: A survey on semi-supervised learning. Mach. Learn. **109**(2), 373–440 (2020)
34. Wang, H., Ge, S., Lipton, Z., Xing, E.P.: Learning robust global representations by penalizing local predictive power. In: Advances in Neural Information Processing Systems, vol. 32 (2019)
35. Wang, Y., Yao, Q., Kwok, J.T., Ni, L.M.: Generalizing from a few examples: a survey on few-shot learning. ACM Comput. Surv. **53**(3), 1–34 (2020)
36. Wang, Z., Yu, J., Yu, A.W., Dai, Z., Tsvetkov, Y., Cao, Y.: SimVLM: simple visual language model pretraining with weak supervision. In: International Conference on Learning Representations (2022)
37. Xiao, J., Hays, J., Ehinger, K.A., Oliva, A., Torralba, A.: Sun database: large-scale scene recognition from abbey to zoo. In: Proceedings of the IEEE/CVF Conference on Computer Vision and Pattern Recognition, pp. 3485–3492 (2010)
38. Yao, Y., Zhang, A., Zhang, Z., Liu, Z., Chua, T.S., Sun, M.: CPT: COLORFUL prompt tuning for pre-trained vision-language models. arXiv preprint arXiv:2109.11797 (2021)
39. Yu, J., Wang, Z., Vasudevan, V., Yeung, L., Seyedhosseini, M., Wu, Y.: COCA: Contrastive captioners are image-text foundation models. Transactions on Machine Learning Research (2022)

40. Zhang, R., et al.: Tip-adapter: training-free adaption of CLIP for few-shot classification. In: Avidan, S., Brostow, G., Cissé, M., Farinella, G.M., Hassner, T. (eds.) Computer Vision. ECCV 2022. LNCS, vol. 13695. Springer, Cham (2022). https://doi.org/10.1007/978-3-031-19833-5_29
41. Zhang, Y., Zhang, C., Tang, Y., He, Z.: Cross-modal concept learning and inference for vision-language models. arXiv preprint arXiv:2307.15460 (2023)
42. Zhou, K., Liu, Z., Qiao, Y., Xiang, T., Loy, C.C.: Domain generalization: a survey. In: IEEE Transactions on Pattern Analysis and Machine Intelligence (2022)
43. Zhou, K., Yang, J., Loy, C.C., Liu, Z.: Conditional prompt learning for vision-language models. In: Proceedings of the IEEE/CVF Conference on Computer Vision and Pattern Recognition, pp. 16816–16825 (2022)
44. Zhou, K., Yang, J., Loy, C.C., Liu, Z.: Learning to prompt for vision-language models. Int. J. Comput. Vision **130**(9), 2337–2348 (2022)
45. Zhou, M., Yu, L., Singh, A., Wang, M., Yu, Z., Zhang, N.: Unsupervised vision-and-language pre-training via retrieval-based multi-granular alignment. In: Proceedings of the IEEE/CVF Conference on Computer Vision and Pattern Recognition, pp. 16485–16494 (2022)

Multimodal Causal Relations Enhanced CLIP for Image-to-Text Retrieval

Wenjun Feng, Dazhen Lin, and Donglin Cao(✉)

Department of Artificial Intelligence, Xiamen University, Xiamen, Fujian, China
another@xmu.edu.cn

Abstract. Traditional image-to-text retrieval models learn joint representations by aligning multimodal features, typically learning the weak correlation between image and text data which can introduce noise during modality alignment. To solve this problem, we propose a Multimodal Causal CLIP (MMC-CLIP) network that integrates causal semantic relationships into CLIP for image-to-text retrieval task. Firstly, we employ the Multimodal Causal Discovery (MCD) method, which models the causal relationships of causal variables in both image and text data to construct the multimodal causal graph. Subsequently, we seamlessly integrate the causal nodes extracted from the multimodal causal graph as learnable prompts within the CLIP model, giving rise to the novel framework of Multimodal Causal CLIP. By integrating causal semantics into CLIP, MMC-CLIP effectively enhances the correlation between causal variables in image and text modalities, leading to an improved alignment capability for multimodal image-text data. We demonstrate the superiority and generalization of the proposed method by outperforming all strong baselines in image-to-text retrieval task on the Flickr30K and MSCOCO datasets.

Keywords: Multimodal causal discovery · Image-to-text retrieval · CLIP

1 Introduction

With the continuous development of deep learning, multimodal processing has emerged as one of the most promising research topics. Particularly, there has been a significant amount of work focused on the joint modeling of image-text data and numerous remarkable achievements have been made in tasks such as image caption [2], visual question answering [18], and image-text retrieval [4, 13]. Among these tasks, image-text retrieval plays a crucial role in the retrieval of relevant text or images from a vast amount of data, making it essential in recommendation systems [8] and search engines [29].

D. Cao — is the corresponding author.

© The Author(s), under exclusive license to Springer Nature Singapore Pte Ltd. 2024
Q. Liu et al. (Eds.): PRCV 2023, LNCS 14425, pp. 210–221, 2024.
https://doi.org/10.1007/978-981-99-8429-9_17

GT Captions

(1) The players of the baseball team are standing on the field , with many people watching from the stands.
(2) Baseball players are playing on a field in a stadium filled with people.
(3) A team plays baseball at a large crowded stadium.
(4) A crowd cheers on a baseball team.
(5) Game is playing in the stadium.

Top-5 Recall Captions of CLIP

(1) The players of the baseball team are standing on the field , with many people watching from the stands.
(2) Baseball players are playing on a field in a stadium filled with people.
(3) A team plays baseball at a large crowded stadium.
(4) A baseball player in a white outfit swings to hit a baseball. ✕
(5) Man in white suit throws first pitch at baseball game. ✕

Fig. 1. A bad case of CLIP on Flickr30K dataset. On the left are the ground truth captions, and on the right are the captions recalled by the CLIP model. The text highlighted in red represents the incorrectly retrieved captions. (Color figure online)

When provided with an image and the task of describing it in natural language, it is often observed that the resulting caption may not comprehensively capture all the entities depicted in the image. Indeed, the content of the image and text do not always exhibit a strong correlation. The weak correlation between modalities in multimodal data can introduce noise during the modal alignment process, potentially impacting the model's performance. Figure 1 shows a bad case of CLIP [27] on the Flickr30K [33] dataset. In this sample, two captions from the annotations fail to sufficiently describe the entity information depicted in the image, resulting in a weak correlation between the image and text. Therefore, during the image-to-text retrieval process, CLIP erroneously retrieves captions that exhibit greater alignment with the image rather than the ground truth captions.

Causal discovery is promising in tackling the above weak correlation issues between image and text caused by data distribution. Establishing causal relationships among multimodal variables offers an effective means to counteract this weak correlation. Notably, causal discovery has found applications across various domains, including visual tasks [30], language comprehension [25], and recommendation systems [31]. However, the aforementioned methods have not specifically focused on modeling causal relationships in the context of multimodal image-text data. To tackle the challenge of weak correlation in image-text data, we propose a multimodal causal discovery method MCD. MCD employs the FGES (Fast Greedy Equivalence Search) [28] algorithm to jointly model causal relationships among the causal variables in both image and text data, resulting in a multimodal causal graph. Subsequently, we consider the nodes within the multimodal causal graph as learnable prompts and integrate them into the CLIP framework obtaining MMC-CLIP. This integration aims to enhance the correlation between the image and text data, thereby strengthening the capability of the model to align multimodal data.

To sum up, in this paper, we propose the following contributions:

- To the best of our knowledge, MCD is the first method to jointly model the causal relationships among causal variables in both image and text data to construct the multimodal causal graph. We utilize causal graphs in the image-to-text retrieval task to assist the model in mitigating the weak correlation between multimodal data.

- We introduce a novel approach, MMC-CLIP, which incorporates the nodes of the multimodal causal graph as learnable prompts within the CLIP framework. This augmentation enhances the model's ability to align multimodal data.
- The proposed MMC-CLIP outperforms all strong baselines in the image-to-text retrieval task across both the Flickr30K and MSCOCO [20] datasets. This outcome highlights the superiority and generalizability of our approach. Furthermore, we observed that the model incorporating multimodal causal graph also exhibits the capability for few-shot learning.

2 Related Works

Image-to-text retrieval aims to retrieve relevant textual captions for a given image query. [4,13] focus on learning a joint multimodal embedding space to align the semantics of both image and text. As multimodal pre-training models have advanced, leveraging pre-training on unlabeled data followed by fine-tuning for the image-to-text retrieval task has resulted in the emergence of several state-of-the-art (SOTA) models, such as [10,19,23]. However, these pre-training models still rely on ROI features extracted by object detectors [7] to handle visual features, which can impact the inference speed of the models. Recently, contrastive image-text pre-training has emerged as a new paradigm for multimodal representation learning. Works such as [16,17,27] have employed image-text matching as a pre-training task, enabling models to learn the ability to align image and text data during the pre-training stage.

However, these approaches frequently presume a strong level of correspondence and correlation between images and text, often overlooking the existence of weakly correlated instances within the data. This limitation can impact the model's ability to effectively align the modalities. To address the issue of weak correlation, we propose MMC-CLIP, which integrate the multimodal causal relationships from multimodal causal graph into the CLIP framework. This integration is designed to compensate for the weak correlation observed in image-text data and enhance the model's capability for modality alignment.

In recent years, causal discovery has been widely applied in various domains of deep learning such as natural language processing (NLP) and computer vision. In NLP, [12,21] have introduced causal discovery methods to text classification tasks to mitigate the impact of confounding factors in text. [5,25] utilize causal graphs to generate counterfactual samples, aiming to alleviate model biases introduced by data distribution. In computer vision, [30,32] also leverages counterfactual samples to enhance the robustness of models on imbalanced data. While these works mainly focus on causal discovery in either image or text unimodal data, providing limited assistance for the task of image-to-text retrieval, which involves multimodal data. In this paper, we propose multimodal causal discovery method (MCD) that jointly models the multimodal causal variables from both image and text data to construct a multimodal causal graph. MCD goes beyond capturing the causal relationships between modalities and also captures the causal relationships within each individual modality, resulting in a comprehensive multimodal causal graph.

3 Method

3.1 Overview

Figure 2 shows our proposed MCD and MMC-CLIP framework. Firstly, the MCD method extracts causal variables from both image and text, utilizing the FGES [28] algorithm to construct the multimodal causal graph $\mathcal{G}$. Secondly, we use a 2-layer GCN [11] to model $\mathcal{G}$ and inject graph nodes into CLIP as learnable prompts, resulting in MMC-CLIP. The image prompts are generated by applying a coupling function to the text prompts, with the aim of enhancing the correlation between these text and image prompts. Importantly, only these prompts are fine-tuned while all other parameters of CLIP remain frozen. Finally, a triplet loss is employed to align the multimodal features for the image-to-text retrieval task.

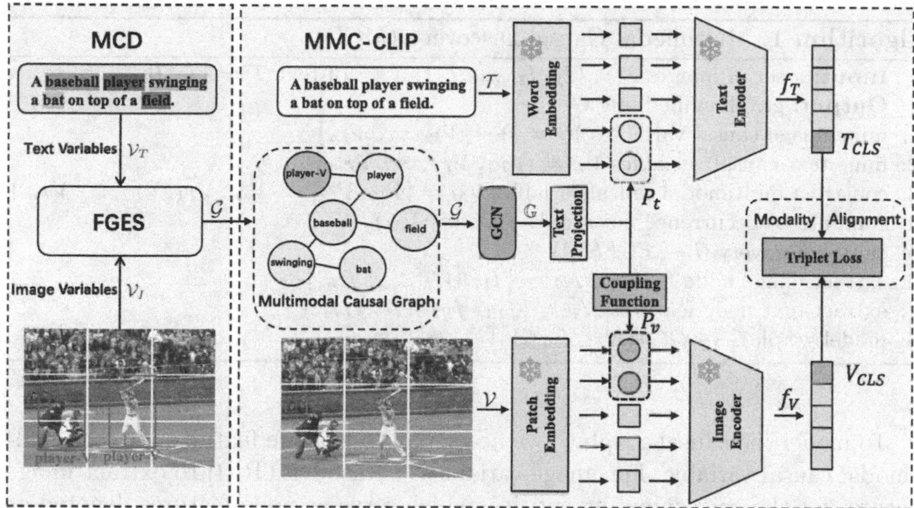

Fig. 2. Overview of our MCD and MMC-CLIP. On the left side, our proposed method, Multimodal Causal Discovery (MCD), combines causal variables from multimodal image-text data to construct a multimodal causal graph $\mathcal{G}$ using the FGES algorithm. Then, we utilize GCN to model the graph $\mathcal{G}$, considering it as learnable prompts that are injected into frozen CLIP, resulting in MMC-CLIP. Finally, a triplet loss is utilized to align the multimodal features.

3.2 MCD: Multimodal Causal Discovery

Algorithm 1 illustrates the workflow of MCD. We utilize MCD to construct a multimodal causal graph using the MSCOCO multimodal dataset. Causal entities are extracted from both images and texts in the dataset. We extract causal entities from the images and texts in the dataset using Faster R-CNN [7] and

NLTK library [22], respectively. These extracted entities are utilized as causal variables for the construction of the multimodal causal graph. Then we merge the causal variables from images and texts into multimodal causal variables $\mathcal{V}_M$ and compute the co-occurrence matrix $\mathcal{A}$ based on the frequency of occurrence of multimodal causal variables. We employ FGES (Fast Greedy Equivalence Search) [28], an efficient causal discovery algorithm, to explore the causal relationships among the multimodal variables and construct the multimodal causal graph $\mathcal{G}$. FGES is a score-based method that minimizes the required computation by leveraging the properties of scoring functions and storing previous intermediate evaluations. This optimization leads to a notable reduction in computation time, rendering FGES particularly suitable for modeling high-dimensional multimodal causal variables within our context,

$$\mathcal{G} = FGES(\mathcal{A}). \tag{1}$$

Algorithm 1. Multimodal Causal Discovery (MCD)

Input: a set of images $\mathcal{I} = \{I_1, I_2, \ldots, I_n\}$, a set of text $\mathcal{T} = \{T_1, T_2, \ldots, T_m\}$
Output: graph embedding $\mathbb{G}$

mine image causal variables $\mathcal{V}_I = \{\mathcal{V}_{I1}, \mathcal{V}_{I2}, \ldots \mathcal{V}_{In}\}$
mine text causal variables $\mathcal{V}_T = \{\mathcal{V}_{T1}, \mathcal{V}_{T2}, \ldots \mathcal{V}_{Tm}\}$
construct multimodal causal variables $\mathcal{V}_M = \{\mathcal{V}_{I1}, \mathcal{V}_{I2}, \ldots \mathcal{V}_{In}, \mathcal{V}_{T1}, \mathcal{V}_{T2}, \ldots \mathcal{V}_{Tm}\}$
compute co-occurrence matrix $\mathcal{A} \leftarrow Count(\mathcal{V}_M)$
causal discovery $\mathcal{G} \leftarrow FGES(\mathcal{A})$
extract image node features $\mathcal{N}_I = \{f_{I1}, f_{I2}, \ldots, f_{In}\}$
extract text node features $\mathcal{N}_T = \{f_{T1}, f_{T2}, \ldots, f_{Tm}\}$
model graph $\mathbb{G} \leftarrow GCN(\mathcal{N}_I, \mathcal{N}_T, \mathcal{G})$

To model the causal graph $\mathcal{G}$, we need to extract node features for each multimodal causal variable. For image variables, we use DETR [1] to extract image features for the corresponding entities, which serve as node features denoted as $\mathcal{N}_I$. For text variables, we use GloVe [26] to learn word vectors as node features $\mathcal{N}_T$. Finally, we utilize a 2-layer GCN to model the nodes within the causal graph and obtain the final graph embedding $\mathbb{G}$.

$$\mathbb{G} = GCN(\mathcal{N}_I, \mathcal{N}_T, \mathcal{G}). \tag{2}$$

3.3 MMC-CLIP

CLIP, a contrastive multimodal pre-trained model, is commonly employed as a baseline for the image-to-text retrieval task. However, it encounters challenges in handling weak correlation within multimodal data. To address this limitation, we introduce MMC-CLIP, a novel framework that integrates the multimodal causal graph $\mathcal{G}$ into CLIP. By leveraging multimodal causal relationships, MMC-CLIP aims to improve the correlation between different modalities in the data.

The multimodal causal graph $\mathbb{G}$, modeled by GCN, consists of image nodes $\mathcal{N}_v$ and text nodes $\mathcal{N}_t$. Each node in $\mathbb{G}$ aggregates information from its neighboring nodes, capturing rich causal relationships. Drawing inspiration from the commonly used Prompt Tuning in NLP, we incorporate these multimodal causal nodes as learnable prompts into the embeddings of CLIP. For the input text $\mathcal{T}$ and image $\mathcal{V}$, they are separately processed through the word embedding and patch embedding layer:

$$E_t = WordEmbedding(\mathcal{T}), \tag{3}$$

$$E_v = PatchEmbedding(\mathcal{V}), \tag{4}$$

where $E_t \in \mathbb{R}^{N\times d}$ and $E_v \in \mathbb{R}^{M\times d}$, in which N and M is the length of word embedding and patch embedding, respectively, and d denotes feature dimension. Specifically, $E_t = [E_t^1, E_t^2, \ldots, E_t^N]$, $E_v = [E_v^1, E_v^2, \ldots, E_v^M]$. For text prompts:

$$P_t = TextProjection(\mathcal{N}_t), \tag{5}$$

where $P_t \in \mathbb{R}^{n\times d}$, in which n denotes the number of prompts. To ensure the correlation between image prompts P_v and text prompts P_t, we apply a coupling function $\mathcal{F}(\cdot)$ to map P_t to P_v:

$$P_v = \mathcal{F}(P_t), \tag{6}$$

where $\mathcal{F}(\cdot)$ is a linear layer and $P_v \in \mathbb{R}^{n\times d}$. Next, P_t and P_v are inserted into E_t and E_v, respectively, to obtain $E_t^P = [P_t^1, P_t^2, \ldots, P_t^n, E_t^1, E_t^2, \ldots, E_t^N]$ and $E_v^P = [P_v^1, P_v^2, \ldots, P_v^n, E_v^1, E_v^2, \ldots, E_v^M]$ as replacements for the original two embeddings. Then, E_t^P and E_v^P are fed into text and image encoder, respectively, for encoding:

$$f_T = TextEncoder(E_t^P), \tag{7}$$

$$f_V = ImageEncoder(E_v^P), \tag{8}$$

where $f_T \in \mathbb{R}^{(n+N)\times d}$ represents the text feature and $f_V \in \mathbb{R}^{(n+M)\times d}$ denotes the image feature. Finally, we utilize the [CLS] embedding of f_T and f_V as global features for text and image, respectively, denoted as $T_{CLS} \in \mathbb{R}^d$ and $V_{CLS} \in \mathbb{R}^d$.

3.4 Image-Text Alignment

To align the global features of text (T_{CLS}) and image (V_{CLS}), we follow [13] adopt a triplet loss as our alignment objective:

$$L(T, V) = \sum_{\hat{V}}[\alpha - S(T, V) + S(T, \hat{V})]_+ + \sum_{\hat{T}}[\alpha - S(T, V) + S(\hat{T}, V)]_+, \tag{9}$$

where $[x]_+ = max(x, 0)$, S denotes the cosine similarity, T represents T_{CLS}, and V is V_{CLS}. For a positive pair (T, V), we aim for it to have a larger distance from all negative texts $\hat{T}$ and negative images $\hat{V}$ by a margin α. We optimize the triplet loss to enhance the model's capability to align text and image data.

4 Experiments

To verify the effectiveness and generalization of our proposed MMC-CLIP, in this section, we perform a series of experiments on the standard MSCOCO and Flickr30K datasets. Additionally, we also conducted ablation studies to demonstrate the superiority of our method.

Table 1. Results on MSCOCO. We report results by averaging over 5 folds of 1K test images and testing on the full 5K images. And the FT denotes fine-tuning.

Methods	MSCOCO-1K(5-fold)			MSCOCO-5K		
	R@1	R@5	R@10	R@1	R@5	R@10
VSE++ [4] (2017)	64.6	90.0	95.7	41.3	71.1	81.2
SCAN [13] (2018)	72.7	94.8	98.4	50.4	82.2	90.0
SGRAF [3] (2021)	79.6	96.2	98.5	57.8	-	91.6
Unicoder-VL [14] (2020)	84.3	97.3	99.3	70.0	90.7	95.6
ALADIN [24] (2022)	88.1	99.1	99.7	70.0	90.7	95.6
OSCAR [19] (2020)	-	-	-	70.0	91.1	96.3
ALBEF [17] (2021)	-	-	-	77.6	94.3	97.2
BLIP [16] (2022)	-	-	-	82.4	95.4	97.9
BLIP-2 [15] (2023)	-	-	-	85.4	97.0	98.5
CLIP(FT) [27] (2021)	84.3	95.3	98.7	71.1	88.0	93.3
Ours(MMC-CLIP)	**96.2**	**100.0**	**100.0**	**90.9**	**98.9**	**99.7**

4.1 Datasets and Settings

Datasets. We evaluate our model on two popular datasets: MSCOCO and Flickr30K. The MSCOCO dataset consists of 123,287 images, each accompanied by 5 annotated captions. We split the dataset into 113,287 training images, 5,000 validation images, and 5,000 test images. For a fair comparison, we report the results on both 5K and 1K test sets. In the case of 1K images, the results are computed by performing a 5-fold cross-validation and averaging the results. The Flickr30K dataset comprises 31,783 images, each with 5 corresponding captions. Following the split in a previous work [6], we use 1,000 images for validation, 1,000 images for testing, and the remaining images for training.

Metric. For evaluating image-to-text retrieval task, we adopt the Recall@K (R@K) metric, which measures the proportion of queries where the ground truth is ranked within the top K results. Specifically, we report R@1, R@5, and R@10 as our evaluation metrics to assess the performance of our method.

Settings. For training, we use SGD optimizer with a mini-batch size of 128. We conducted training for 10 epochs with a learning rate of 1e−3 on an NVIDIA 4090. All the results presented in our work are the average of 5 runs.

4.2 Results on MSCOCO

Table 1 lists the experimental results on MSCOCO 1K test set and 5K test set, along with a fair comparison with baselines. Our model, MMC-CLIP, significantly outperforms existing baseline models in the image-to-text retrieval task. Specifically, on the 1K test set, our model achieves R@1 of 96.2%, outperforming the baseline model by 8.1%. On the 5K test set, our model achieves an R@1 of 90.9%, outperforming the baseline model by 5.5%. Notably, our MMC-CLIP surpasses the original fine-tuned CLIP model by 7.9% and 19.8% in R@1, respectively. Compared to baseline models and original CLIP, our MMC-CLIP integrates the multimodal causal graph generated by MCD. By leveraging these causal relationships, effectively enhances the correlation between image and text data, thereby strengthening the model's ability to align multimodal data.

Table 2. Results on Flickr30K. The ZS and FT denote zero-shot and fine-tuning, respectively.

Methods	Flickr30K		
	R@1	R@5	R@10
VSE++ (2017)	52.9	80.5	87.2
SCAN (2018)	67.4	90.3	95.8
SGRAF (2021)	77.8	94.1	97.4
Unicoder-VL (2020)	86.2	96.3	99.0
ALIGN [9] (2021)	95.3	99.8	**100.0**
ALBEF (2021)	95.9	99.8	**100.0**
BLIP (2022)	97.3	99.9	**100.0**
BLIP-2(ZS) (2023)	97.6	**100.0**	**100.0**
CLIP (FT)	90.1	98.5	99.7
Ours (MMC-CLIP)	**99.2**	**100.0**	**100.0**

4.3 Results on Flickr30K

Table 2 compares the image-to-text retrieval results on the Flickr30K dataset with the latest algorithms. We can observe that MMC-CLIP surpasses all baselines by at least 1.6% in R@1 results. Furthermore, our model outperforms the original CLIP comprehensively, further demonstrating the advantages of causal semantics discovered by MCD for the base model. Moreover, our causal graph, derived from the MSCOCO dataset, exhibits effective generalization to the Flickr30K dataset. This highlights the robustness and generalization capabilities of the causal relationships obtained through the MCD approach.

4.4 Ablation Studies

In this section, we conduct a series of ablation studies to explore the impact of our proposed MCD. All the comparative experiments are performed on the MSCOCO 5K test set.

Table 3. Ablation study of the impact of MCD. Among them, CLIP represents the base model. CLIP+RP refers to the integration of randomly initialized prompts into the CLIP model. CLIP+CP denotes our proposed MMC-CLIP, where prompts with causal relationships obtained from MCD are integrated into CLIP.

Model	MSCOCO-5K		
	R@1	R@5	R@10
CLIP	71.1	88.0	93.3
CLIP+RP	80.8	93.5	96.7
CLIP+CP	**90.9**	**98.9**	**99.7**

Impact of MCD. To validate the impact of our proposed MCD method on enhancing the correlation between image and text data, we choose CLIP as our baseline model for comparison. As shown in Table 3, our proposed MMC-CLIP outperforms the other two experimental settings by a significant margin. Notably, even with randomly initialized prompts inserted into CLIP's image and text encoders, we observe a significant 9.7% improvement in R@1 compared to the base model. Moreover, we achieve an additional performance boost when replacing these random prompts with causal nodes extracted from the multimodal causal graph generated by the MCD method. This provides further evidence of the effectiveness of our MCD method in imparting causal semantics to the model, effectively enhancing the correlation between image-text multimodal data.

Impact of Prompt Length. Figure 3a illustrates the impact of prompt length of MMC-CLIP. When the prompt length is greater than 1, the models consistently outperform the base model CLIP, with the best performance achieved when the prompt length is 5. This finding suggests that appropriately increasing the prompt length can lead to improved model performance. However, when the prompt length exceeds 5, the model's performance starts to decline. This suggests that we need to maintain a balance between the original data and causal relationships in order to achieve optimal performance.

Few-Shot Learner. We experiment with various training ratios, including 1%, 5%, 10%, 20%, 50%, and 100% to investigate the impact of different amounts of training data on the performance gap. As shown in Fig. 3b, our MMC-CLIP outperforms the original CLIP in terms of performance across all training ratios. Furthermore, our MMC-CLIP achieves superior performance with only 5% of the training data compared to the fully trained CLIP, demonstrating the few-shot capability of our approach. Therefore, the multimodal causal graph obtained from our proposed MCD method not only provides additional causal relationships to the model but also enriches the model with substantial prior knowledge, leading to its few-shot capability.

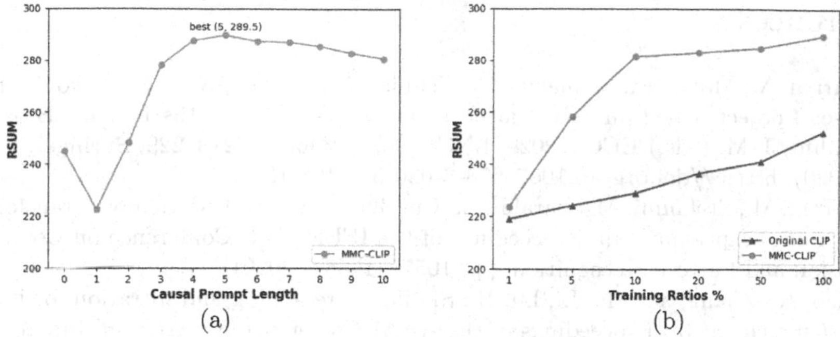

(a) (b)

Fig. 3. (a) Ablation on prompt length of MMCCLIP. We define RSUM as the sum of R@1, R@5, and R@10. Causal prompt length represents the number of causal graph nodes used as prompts. When the prompt length is 0, it represents the base model, which is the original CLIP model. (b) Impact of training data ratios. We compare the original CLIP with our MMC-CLIP. We select a set of training ratios, 1%, 5%, 10%, 20%, 50%, and 100%, to study how the performance gap changes along with different training data available.

5 Conclusion

In this paper, we propose the Multimodal Causal CLIP (MMC-CLIP) framework, which integrates causal semantic relationships from causal graph into CLIP for image-to-text retrieval task. Our approach begins with Multimodal Causal Discovery (MCD), where we mine the causal relationships within the causal variables of image and text data to construct a multimodal causal graph. Following that, we incorporate the causal nodes derived from the multimodal causal graph into the CLIP model as learnable prompts, resulting in the novel MMC-CLIP framework. By leveraging the causal semantics captured in the causal graph, MMC-CLIP effectively addresses the inherent weak correlation in multimodal data and enhances the alignment ability between different modalities. Through extensive experiments, we demonstrate the superiority and generalization of our method on image-to-text task. Furthermore, we observe that our causal graph endows the model with few-shot capability. In future work, we plan to explore larger and higher-quality multimodal datasets for causal discovery to obtain more generalizable causal graphs. We also intend to extend our method to various other multimodal downstream tasks.

Acknowledgement. This work is supported by the National Natural Science Foundation of China (No. 62076210, No. 81973752), the Natural Science Foundation of Xiamen city (No. 3502Z20227188) and the Open Project Program of The Key Laboratory of Cognitive Computing and Intelligent Information Processing of Fujian Education Institutions, Wuyi University (No. KLCCIIP2020203).

References

1. Carion, N., Massa, F., Synnaeve, G., Usunier, N., Kirillov, A., Zagoruyko, S.: End-to-end object detection with transformers. In: Vedaldi, A., Bischof, H., Brox, T., Frahm, J.-M. (eds.) ECCV 2020. LNCS, vol. 12346, pp. 213–229. Springer, Cham (2020). https://doi.org/10.1007/978-3-030-58452-8_13
2. Cornia, M., Stefanini, M., Baraldi, L., Cucchiara, R.: Meshed-memory transformer for image captioning. In: Proceedings of the IEEE/CVF Conference on Computer Vision and Pattern Recognition, pp. 10578–10587 (2020)
3. Diao, H., Zhang, Y., Ma, L., Lu, H.: Similarity reasoning and filtration for image-text matching. In: Proceedings of the AAAI Conference on Artificial Intelligence, vol. 35, pp. 1218–1226 (2021)
4. Faghri, F., Fleet, D.J., Kiros, J.R., Fidler, S.: VSE++: improving visual-semantic embeddings with hard negatives. arXiv preprint arXiv:1707.05612 (2017)
5. Feng, F., Zhang, J., He, X., Zhang, H., Chua, T.S.: Empowering language understanding with counterfactual reasoning. arXiv preprint arXiv:2106.03046 (2021)
6. Frome, A., et al.: DeViSE: a deep visual-semantic embedding model. In: Advances in Neural Information Processing Systems, vol. 26 (2013)
7. Girshick, R.: Fast R-CNN. In: Proceedings of the IEEE International Conference on Computer Vision, pp. 1440–1448 (2015)
8. Isinkaye, F.O., Folajimi, Y.O., Ojokoh, B.A.: Recommendation systems: principles, methods and evaluation. Egypt. Inf. J. **16**(3), 261–273 (2015)
9. Jia, C., et al.: Scaling up visual and vision-language representation learning with noisy text supervision. In: International Conference on Machine Learning, pp. 4904–4916. PMLR (2021)
10. Kim, W., Son, B., Kim, I.: ViLT: vision-and-language transformer without convolution or region supervision. In: International Conference on Machine Learning, pp. 5583–5594. PMLR (2021)
11. Kipf, T.N., Welling, M.: Semi-supervised classification with graph convolutional networks. arXiv preprint arXiv:1609.02907 (2016)
12. Landeiro, V., Culotta, A.: Robust text classification in the presence of confounding bias. In: Proceedings of the AAAI Conference on Artificial Intelligence, vol. 30 (2016)
13. Lee, K.-H., Chen, X., Hua, G., Hu, H., He, X.: Stacked cross attention for image-text matching. In: Ferrari, V., Hebert, M., Sminchisescu, C., Weiss, Y. (eds.) ECCV 2018. LNCS, vol. 11208, pp. 212–228. Springer, Cham (2018). https://doi.org/10.1007/978-3-030-01225-0_13
14. Li, G., Duan, N., Fang, Y., Gong, M., Jiang, D.: Unicoder-VL: a universal encoder for vision and language by cross-modal pre-training. In: Proceedings of the AAAI Conference on Artificial Intelligence, vol. 34, pp. 11336–11344 (2020)
15. Li, J., Li, D., Savarese, S., Hoi, S.: BLIP-2: bootstrapping language-image pre-training with frozen image encoders and large language models. arXiv preprint arXiv:2301.12597 (2023)
16. Li, J., Li, D., Xiong, C., Hoi, S.: BLIP: bootstrapping language-image pre-training for unified vision-language understanding and generation. In: International Conference on Machine Learning, pp. 12888–12900. PMLR (2022)
17. Li, J., Selvaraju, R., Gotmare, A., Joty, S., Xiong, C., Hoi, S.C.H.: Align before fuse: vision and language representation learning with momentum distillation. Adv. Neural. Inf. Process. Syst. **34**, 9694–9705 (2021)

18. Li, L., Gan, Z., Cheng, Y., Liu, J.: Relation-aware graph attention network for visual question answering. In: Proceedings of the IEEE/CVF International Conference on Computer Vision, pp. 10313–10322 (2019)
19. Li, X., et al.: Oscar: object-semantics aligned pre-training for vision-language tasks. In: Vedaldi, A., Bischof, H., Brox, T., Frahm, J.-M. (eds.) ECCV 2020. LNCS, vol. 12375, pp. 121–137. Springer, Cham (2020). https://doi.org/10.1007/978-3-030-58577-8_8
20. Lin, T.-Y., et al.: Microsoft COCO: common objects in context. In: Fleet, D., Pajdla, T., Schiele, B., Tuytelaars, T. (eds.) ECCV 2014. LNCS, vol. 8693, pp. 740–755. Springer, Cham (2014). https://doi.org/10.1007/978-3-319-10602-1_48
21. Liu, X., Yin, D., Feng, Y., Wu, Y., Zhao, D.: Everything has a cause: leveraging causal inference in legal text analysis. arXiv preprint arXiv:2104.09420 (2021)
22. Loper, E., Bird, S.: NLTK: the natural language toolkit. arXiv preprint cs/0205028 (2002)
23. Lu, J., Batra, D., Parikh, D., Lee, S.: ViLBERT: pretraining task-agnostic visiolinguistic representations for vision-and-language tasks. In: Advances in Neural Information Processing Systems, vol. 32 (2019)
24. Messina, N., et al.: ALADIN: distilling fine-grained alignment scores for efficient image-text matching and retrieval. In: Proceedings of the 19th International Conference on Content-Based Multimedia Indexing, pp. 64–70 (2022)
25. Nan, G., Zeng, J., Qiao, R., Guo, Z., Lu, W.: Uncovering main causalities for long-tailed information extraction. arXiv preprint arXiv:2109.05213 (2021)
26. Pennington, J., Socher, R., Manning, C.D.: GloVe: global vectors for word representation. In: Proceedings of the 2014 Conference on Empirical Methods in Natural Language Processing (EMNLP), pp. 1532–1543 (2014)
27. Radford, A., et al.: Learning transferable visual models from natural language supervision. In: International Conference on Machine Learning, pp. 8748–8763. PMLR (2021)
28. Ramsey, J., Glymour, M., Sanchez-Romero, R., Glymour, C.: A million variables and more: the fast greedy equivalence search algorithm for learning high-dimensional graphical causal models, with an application to functional magnetic resonance images. Int. J. Data Sci. Anal. **3**, 121–129 (2017)
29. Sharma, D., Shukla, R., Giri, A.K., Kumar, S.: A brief review on search engine optimization. In: 2019 9th International Conference on Cloud Computing, Data Science & Engineering (Confluence), pp. 687–692. IEEE (2019)
30. Tang, K., Niu, Y., Huang, J., Shi, J., Zhang, H.: Unbiased scene graph generation from biased training. In: Proceedings of the IEEE/CVF Conference on Computer Vision and Pattern Recognition, pp. 3716–3725 (2020)
31. Wang, Y., Liang, D., Charlin, L., Blei, D.M.: Causal inference for recommender systems. In: Proceedings of the 14th ACM Conference on Recommender Systems, pp. 426–431 (2020)
32. Yang, X., Zhang, H., Cai, J.: Deconfounded image captioning: a causal retrospect. IEEE Trans. Pattern Anal. Mach. Intell. (2021)
33. Young, P., Lai, A., Hodosh, M., Hockenmaier, J.: From image descriptions to visual denotations: new similarity metrics for semantic inference over event descriptions. Trans. Assoc. Comput. Linguist. **2**, 67–78 (2014)

Exploring Cross-Modal Inconsistency in Entities and Emotions for Multimodal Fake News Detection

Longzheng Wang[1,2], Chuang Zhang[1(✉)], Hongbo Xu[1], Yongxiu Xu[1], and Siqi Wang[1,2]

[1] Institute of Information Engineering, Chinese Academy of Sciences, Beijing, China
{wanglongzheng,zhangchuang,hbxu,xuyongxiu,wangsiqi2022}@iie.ac.cn
[2] School of Cyber Security, University of Chinese Academy of Sciences, Beijing, China

Abstract. The automatic detection of multimodal fake news has attracted significant attention recently. Numerous existing methods focus on the fusion of unimodal features to generate multimodal news representations. However, it is possible that these methods have not successfully acquired aligned modal information with sufficient accuracy and failed to effectively leverage the entity inconsistency present across modalities. Besides, there has been a lack of exploration regarding the emotional inconsistency across modalities. To address that, we propose CINEMA, a novel framework to explore cross-modal inconsistency in entities and emotions for multimodal fake news detection. We leverage the cross-modal contrastive learning objective to establish the alignment between the image and text modalities. An entity consistency learning module is developed to learn the cross-modality entity correlations. An emotional consistency learning module is implemented to effectively capture the emotional information within each modality. Finally, we evaluate the performance of CINEMA and conduct a comparative study using two extensively used datasets, Twitter and Weibo. The experimental results unequivocally demonstrate that our proposed CINEMA framework surpasses previous approaches by a substantial margin, establishing new state-of-the-art results on both datasets.

Keywords: Fake news detection · Multi-modal learning · Multi-modal fusion

1 Introduction

With the rapid expansion of Online Social Networks (OSNs) like Twitter and Weibo, individuals are empowered to openly disseminate daily information while articulating their perspectives and emotional states [3]. Nevertheless, the proliferation of OSNs and the dearth of effective oversight mechanisms to ascertain the veracity of online content have engendered the pervasive dissemination of sub-

© The Author(s), under exclusive license to Springer Nature Singapore Pte Ltd. 2024
Q. Liu et al. (Eds.): PRCV 2023, LNCS 14425, pp. 222–234, 2024.
https://doi.org/10.1007/978-981-99-8429-9_18

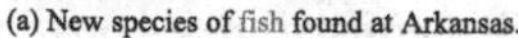

(a) New species of fish found at Arkansas.

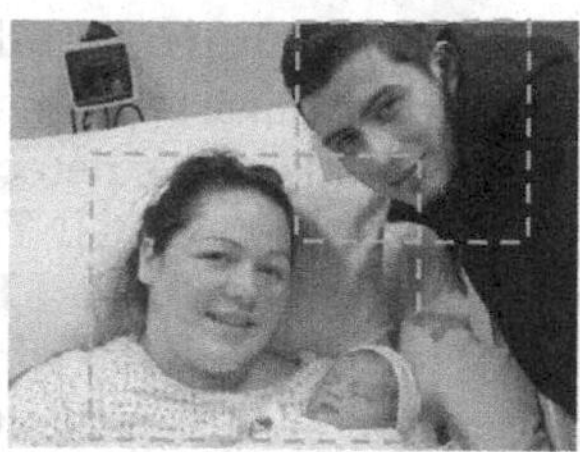

(b) Woman, 36, gives birth to 14 children from 14 different fathers.

(c) Little Syrian girl sells chewing gum on the street so she can feed herself.

Fig. 1. Some multimodal fake news from Twitter. Given the highlighted focus words in the text, the regions enclosed within the green box represent the entities exhibiting consistency. **(a) Entity Inconsistency**: The presence of the highlighted region within the red box indicates a strong likelihood of the news being categorized as fake. **(b & c) Emotion Inconsistency**: The visual-textual pairs in these two examples do not exhibit strong entity inconsistency. However, their images convey predominantly positive emotions, while the text expresses predominantly negative sentiment. This suggests that they may be indicative of fake news. (Color figure online)

stantial volumes of fabricated information [27]. Consequently, the detection of spurious news has garnered substantial interest and has emerged as a paramount concern in contemporary times.

Precisely identifying spurious news on social media platforms poses a highly arduous endeavor. Traditional methods [12,23] predominantly concentrate on scrutinizing the textual composition of news articles, encompassing aspects such as writing style and factual knowledge, with the aim of detecting fake news. In contradistinction to conventional textual news, the integration of visual elements such as images or videos in multimodal news offers the promise of providing readers with more captivating and enriched experiences [19]. Therefore, relying solely on unimodal features for detection purposes falls short of being adequate. Recent research endeavors aim to integrate textual and visual features to generate comprehensive multimodal representations of posts, thereby enhancing the efficacy of fake news detection methodologies [8,21]. Despite the notable progress achieved in the field, current approaches encounter two substantial challenges.

Firstly, numerous studies have proposed innovative fusion methods, allowing them to partially address the issue of entity inconsistency in multimodal fake news, as depicted in Fig. 1(a). However, the unaligned nature of visual tokens and word tokens poses a challenge for multimodal encoders to effectively capture image-text interactions [11]. Hence, we contend that the application of advanced multimodal representation learning paradigms is essential to obtain more refined aligned unimodal representations. Secondly, some multimodal fake news instances do not exhibit obvious entity inconsistency issues. However, there is a clear emotional inconsistency between their images and text, as demonstrated in Fig. 1(b & c). For instance, the textual contents in Fig. 1(c) describe a scenario resembling a war, while the corresponding image portrays a happy emotion. To the best of our knowledge, the investigation of cross-modal emotional inconsistency in the context of multimodal fake news detection remains limited. However, we posit that such inconsistency holds significant potential in enhancing the efficacy of multimodal fake news detection methods.

Taking the consideration above, we propose CINEMA, a novel framework to explore **C**ross-modal **I**nconsistency in e**N**tities and **E**motions for **M**ultimod**A**l Fake News Detection. We employ the cross-modal contrastive learning objective to establish the alignment between the image and text modalities. Subsequently, the aligned unimodal representations are fed into an entity consistency learning module to acquire the cross-modality entity correlations. In order to effectively capture the emotional information within each modality, we employ an emotional consistency learning module to assess the consistency of emotions across modalities. This enables us to acquire distinct emotional representations for each modality, thereby enhancing the detection of multimodal fake news. Through this approach, CINEMA achieves enhanced accuracy in aligning features across modalities, thereby facilitating improved learning of entity consistency between the modalities. Moreover, the integration of emotional consistency across modalities further enhances the overall performance of the detection process.

The main contributions of this paper are as follows:

- We present CINEMA, an innovative framework designed for the purpose of multimodal fake news detection by investigating the cross-modal inconsistency in entities and emotions.
- We explore the application of sophisticated multimodal representation learning methods for the purpose of detecting multimodal fake news, employing learned representations to capture the inconsistencies in entities across different modalities.
- Our study represents the pioneering attempt to investigate the influence of emotional inconsistency across modalities on the detection of multimodal fake news, which has not been previously explored to the best of our knowledge.
- We perform extensive experiments on two commonly employed datasets, namely Twitter and Weibo, to evaluate the performance of our proposed model. The experimental results showcase a significant improvement over existing systems, establishing a new state-of-the-art performance on both datasets.

2 Related Works

2.1 Single-Modality Fake News Detection

Existing unimodal fake news detection methods predominantly depend on analyzing either the textual content or the visual content within posts. In text content analysis, there are extant studies that specifically concentrate on uncovering the discernible emotional signals that differentiate between fake and real news. Wu and Rao [20] propose an adaptive fusion network for the purpose of detecting fake news, wherein they model emotion embeddings derived from both the content and the accompanying comments. Zhang et al. [25] propose and verify the distinctiveness of the dual emotion signal, encompassing both publisher emotion and social emotion, in discerning between fake and real news. However, to the best of our knowledge, the examination of emotional disparities between images and text within the realm of multimodal fake news detection remains unexplored.

For image content, Jin et al. [7] assert the existence of discernible discriminative features in the propagation patterns of image content that distinguish between real news and fake news. The proposed MVNN framework [14] synergistically harnesses visual features in the spatial domain along with image features in the frequency domain for forensic purposes. However, these methodologies neglect cross-modal attributes, including correlation and consistency, which have the potential to undermine their comprehensive performance in the context of multimodal news analysis.

2.2 Multimodal Fake News Detection

Recently, numerous approaches rooted in cross-modal discriminative patterns have emerged, aiming to achieve enhanced performance in multimodal fake news detection (MFND). To learn the cross-modal characteristics, EANN [19] combines textual and visual content features and then incorporates an auxiliary event discriminator to facilitate feature extraction. Khattar et al. [8] present MVAE, a multimodal variable autoencoder framework, designed to acquire knowledge of probabilistic latent variable models and subsequently reconstruct the original texts and low-level image features. MCAN [21] employs a stacked configuration of multiple co-attention layers to effectively integrate textual and visual features, thereby enhancing the capability of fake news detection. However, the integration of recently emerged paradigms in multimodal representation learning has been rarely explored in the context of multimodal fake news detection studies. Given the successful adoption of contrastive learning in vision-language representation learning [10,11,16], which has yielded a series of cutting-edge results, its potential application in multimodal fake news detection remains an underexplored avenue. In this study, we employ an image-text contrastive (ITC) learning objective to proficiently align visual and language representations within a straightforward dual-encoder framework, consequently yielding a cohesive latent embedding space.

3 Methodology

In this section, we present our proposed framework CINEMA, that investigates the entity consistency and emotional consistency between the image and text modalities. The overall model structure is illustrated in Fig. 2.

3.1 Feature Extraction

Let each input multimodal news $\mathbf{x} = [x^v, x^t] \in \mathcal{D}$, where x^v, x^t and $\mathcal{D}$ mean image, text and dataset, respectively. We commence the process by employing

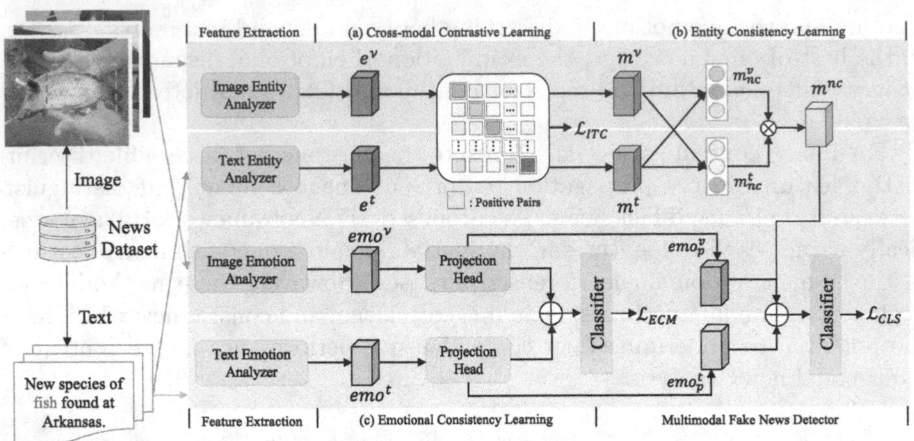

Fig. 2. Model Architecture Overview of CINEMA. The model consists of three main modules: **(a) Cross-modal Contrastive Learning**: The application of the cross-modal contrastive learning objective serves the purpose of aligning the visual representation and textual representation. **(b) Entity Consistency Learning**: We feed the aligned unimodal representations into the entity consistency learning module to capture the entity inconsistencies present between diverse modalities. **(c) Emotional Consistency Learning**: We utilize a binary classification task to enhance the capture of the primary emotional features in both the image and the text.

four analyzers dedicated to the examination of both visual and textual data, with the aim of extracting coarse-grained representations of entities and sentiments specific to each modality. We denote these representations as e^v, e^t, emo^v and emo^t. In the context of this study, we employ a pre-trained ResNet [5] model, which has been trained on the extensive ImageNet database, to extract regional features from the provided visual content. This approach enables us to effectively capture both the semantic information and contextualized representations inherent in the visual data. Additionally, for our textual language model, we leverage BERT [4] as the fundamental module. BERT's exceptional capabilities in capturing intricate language patterns and contextual dependencies make it an ideal choice for our research objectives.

3.2 Cross-Modal Contrastive Learning

Due to significant semantic gaps between features from different modalities, inspired by [18], we employ cross-modal contrastive learning to align the features across modalities by mapping the unimodal embeddings into a shared space. This alignment of image and text modalities greatly facilitates the subsequent learning process, particularly in terms of capturing entity consistency between them.

For a given batch of N image-text pairs, the modal-specific analyzers are employed to derive normalized embedded vectors $\{e_i^v, e_i^t\}_{i=1}^N$ with consistent dimensions. The primary objective of cross-modal contrastive learning is to predict the actual occurrences among the numerous image-text pairings, totaling $N \times N$, within a given batch. It is worth noting that a training batch comprises $N^2 - N$ negative image-text pairs. In the specific case of the i_{th} pair, the estimation of the predicted similarity between vision and text can be computed by evaluating the vision-to-text similarity $\boldsymbol{s}_i^{v \to t} = \{s_{ij}^{v \to t}\}_{j=1}^N$ and text-to-vision similarity $\boldsymbol{s}_i^{t \to v} = \{s_{ij}^{t \to v}\}_{j=1}^N$ using the following approach:

$$s_{ij}^{v \to t} = \frac{\exp(\text{sim}(e_i^v, e_j^t)/\tau)}{\sum_{j=1}^N \exp(\text{sim}(e_i^v, e_j^t)/\tau)} \tag{1}$$

Here, the calculation of the similarity scores involves a learnable temperature parameter, denoted as τ, which is initialized with a value of 0.07. The function sim$(\cdot)$ employs the dot product operation to quantify the similarity between the respective elements. The corresponding one-hot label vectors of the ground-truth $\boldsymbol{y}_i^{v \to t} = \{y_{ij}^{v \to t}\}_{j=1}^N$ and $\boldsymbol{y}_i^{t \to v} = \{y_{ij}^{t \to v}\}_{j=1}^N$, with positive pair denoted by 1 and negatives by 0, are used as the targets to calculate cross-entropy:

$$\mathcal{L}^{v \to t} = -\frac{1}{N} \sum_{i=1}^{N} \sum_{j=1}^{N} y_{ij}^{v \to t} \log s_{ij}^{v \to t} \tag{2}$$

Similarly, we can calculate $\mathcal{L}^{t \to v}$ and subsequently obtain:

$$\mathcal{L}_{ITC} = (\mathcal{L}^{v \to t} + \mathcal{L}^{t \to v})/2 \tag{3}$$

3.3 Entity Consistency Learning

In order to enhance the identification of entity relationships across modalities and contribute to the detection of fake news, we utilize an entity consistency learning module aimed at capturing the entity inconsistencies present between diverse modalities. Specifically, given the aligned unimodal representations m^v and m^t, the inter-modal attention weights are obtained by calculating the association between the unimodal representations.

$$\begin{aligned} Nc_{t \to v} &= \text{softmax}\left([m^v][m^t]^T / \sqrt{d}\right) \\ Nc_{v \to t} &= \text{softmax}\left([m^t][m^v]^T / \sqrt{d}\right) \end{aligned} \tag{4}$$

where d denotes the dimension size of the unimodal representation. Subsequently, the original unimodal embedding vectors are updated using the inter-modal attention weights, resulting in the generation of explicit correlation features.

$$\begin{aligned} m_{nc}^v &= Nc_{t \to v} \times m^v \\ m_{nc}^t &= Nc_{v \to t} \times m^t \end{aligned} \tag{5}$$

Finally, the entity consistency between the visual and textual modalities, denoted as m^{nc}, is defined by taking the outer product between m_{nc}^v and m_{nc}^t.

$$m^{nc} = m_{nc}^v \otimes m_{nc}^t \tag{6}$$

where $\otimes$ denotes outer product.

3.4 Emotional Consistency Learning

In addition to the previously described entity-level features, we also incorporate the consideration of emotional consistency across modalities. Typically, the sentiment score is a numerical value indicating the positive or negative polarity of the visual or textual content, reflecting the extent of positive or negative sentiment. As mentioned above, multimodal fake news may exhibit significant cross-modal emotional inconsistency. Therefore, we employ a binary classification task to learn whether there exists emotional inconsistency within an image-text pair, aiming to better capture the primary emotional features of both the image and the text. Specifically, given the coarse-grained image emotional features emo^v and text emotional features emo^t derived from dedicated modality-specific emotional analyzers, we utilize an MLP-based projection head to extract the primary emotional features and conduct classification:

$$emo_p^v = Proj(emo^v) \quad emo_p^t = Proj(emo^t) \tag{7}$$

Subsequently, the extracted features are fed into a fully-connected network to predict the emotional consistency.

$$\hat{ec} = \text{softmax}(MLP([emo_p^v \oplus emo_p^t]) \tag{8}$$

where $\oplus$ represents the concatenation operation. We use the cross-entropy loss function as:

$$\mathcal{L}_{ECM} = -(a \log(\hat{ec}) + (1 - a) \log(1 - \hat{ec})) \tag{9}$$

where a denotes the authenticity of the multimodal news.

3.5 Multimodal Fake News Detector

Given the entity consistency representations m^{nc}, the primary emotional features emo_p^v and emo_p^t, we feed them into a fully-connected network to predict the label:

$$\hat{y} = \text{softmax}(MLP([m^{nc} \oplus emo_p^v \oplus emo_p^t]) \tag{10}$$

We use the cross-entropy loss function as:

$$\mathcal{L}_{CLS} = -(y \log(\hat{y}) + (1 - y) \log(1 - \hat{y})) \tag{11}$$

where y denotes the ground-truth label. The final learning objective of the multimodal fake news detector is defined as:

$$\mathcal{L}_{FND} = \mathcal{L}_{CLS} + \lambda \mathcal{L}_{ECM} \tag{12}$$

where λ controls the the ratio of $\mathcal{L}_{ECM}$. We jointly train the emotion consistency learning module to uncover more accurate emotional inconsistencies, thereby effectively assisting in improving the performance of multimodal fake news detection.

The ultimate loss function for CINEMA is formulated as the amalgamation of the contrastive learning loss in Eq. 3 and the fake news detector loss in Eq. 12:

$$\mathcal{L} = \mathcal{L}_{ITC} + \mathcal{L}_{FND} \tag{13}$$

4 Experiments

4.1 Experimental Configurations

Datasets. Our model is evaluated on two real-world datasets: Twitter [1] and Weibo [6]. The Twitter dataset, which was made publicly available for the purpose of the Verifying Multimedia Use task at MediaEval, has gained recognition as a prominent multi-modal dataset widely utilized for fake news detection. In experiments, we keep the same data split scheme as the benchmark [1,2]. The training set contains 6, 840 real tweets and 5, 007 fake tweets, and the test set contains 1, 406 posts. The Weibo dataset collected by [6] contains 3749 fake news and 3783 real news for training, 1000 fake news and 996 real news for testing. During the experimental phase, we adhere to the identical procedures outlined in previous works [6,19] to eliminate duplicated and subpar images.

Implementation Details. The evaluation metrics include Accuracy, Precision, Recall, and F1-score. The model is trained using the Adam optimizer [9] with a batch size of 64, employing an initial learning rate of 0.001 for a duration of 50 epochs, with early stopping implemented to prevent overfitting. The λ in the fake news detector loss (Eq. 12) is set to 0.5 for both the Twitter and Weibo datasets.

4.2 Overall Performance

Table 1 provides a performance comparison between CINEMA and other existing methods on the Twitter and Weibo datasets. The results in the table demonstrate that our proposed model, CINEMA, achieves superior performance in terms of both *Acc* and *F1-score* compared to all the other methods across all datasets, which demonstrates the effectiveness of our proposed model. Specifically, CINEMA obtains a new state-of-the-art with an accuracy of **91.1%** on Twitter dataset, achieving significant improvements with **1.4%**. CINEMA also reaches an accuracy of **91.9%**, achieving a new state-of-the-art on Weibo dataset, which is **1.9%** higher than the previous best one.

Several methods for fake news detection, such as EANN [19] and MVAE [8], demonstrate inferior performance due to their separate learning of unimodal features, resulting in a lack of effective fusion and neglect of the semantic gap that exists across modalities, leading to distinct embedding spaces. State-of-the-art approaches, including LIIMR [17] and HMCAN [15], have directed their attention towards mitigating the semantic disparity between modalities by exploring their intermodal interactions. However, despite their progress, these methods still lack the desired level of simplicity and efficiency. Moreover, their primary emphasis lies on achieving entity consistency across modalities, thereby overlooking the crucial aspect of emotional consistency among modalities.

CINEMA demonstrates superior performance compared to these state-of-the-art methods primarily due to the following key factors: 1) The adoption of a cross-modal contrastive learning objective enables the attainment of optimal

Table 1. Performance comparison between CINEMA and other methods on Twitter and Weibo datasets.

	Method	Accuracy	Fake News			Real News		
			Precision	Recall	F1-score	Precision	Recall	F1-score
Twitter	EANN [19]	0.648	0.810	0.498	0.617	0.584	0.759	0.660
	MVAE [8]	0.745	0.801	0.719	0.758	0.689	0.777	0.730
	MKEMN [24]	0.715	0.814	0.756	0.708	0.634	0.774	0.660
	SAFE [26]	0.762	0.831	0.724	0.774	0.695	0.811	0.748
	MCNN [22]	0.784	0.778	0.781	0.779	0.790	0.787	0.788
	MCAN [21]	0.809	0.889	0.765	0.822	0.732	0.871	0.795
	CAFE [2]	0.806	0.807	0.799	0.803	0.805	0.813	0.809
	LIIMR [17]	0.831	0.836	0.832	0.830	0.825	0.830	0.827
	HMCAN [15]	0.897	**0.971**	0.801	0.878	0.853	**0.979**	0.912
	CINEMA	**0.911**	0.861	**0.961**	**0.908**	**0.964**	0.869	**0.914**
Weibo	EANN [19]	0.827	0.847	0.812	0.829	0.807	0.843	0.825
	MVAE [8]	0.824	0.854	0.769	0.809	0.802	0.875	0.837
	MKEMN [24]	0.814	0.823	0.799	0.812	0.723	0.819	0.798
	SAFE [26]	0.816	0.818	0.815	0.817	0.816	0.818	0.817
	MCNN [22]	0.823	0.858	0.801	0.828	0.787	0.848	0.816
	MCAN [21]	0.899	0.913	0.889	0.901	0.884	0.909	0.897
	CAFE [2]	0.840	0.855	0.830	0.842	0.825	0.851	0.837
	LIIMR [17]	0.900	0.882	0.823	0.847	0.908	**0.941**	**0.925**
	HMCAN [15]	0.885	0.920	0.845	0.881	0.856	0.926	0.890
	CINEMA	**0.919**	**0.923**	**0.919**	**0.921**	**0.914**	0.918	0.916

Table 2. Ablation study on the architecture design of CINEMA on two datasets.

	Method	Accuracy	F1 score	
			Fake News	Real News
Twitter	CINEMA	**0.911**	**0.908**	**0.914**
	- w/o ITC	0.861	0.847	0.873
	- w/o NCL	0.890	0.891	0.890
	- w/o ECM	0.901	0.893	0.908
Weibo	CINEMA	**0.919**	**0.921**	**0.916**
	- w/o ITC	0.910	0.915	0.905
	- w/o NCL	0.910	0.912	0.908
	- w/o ECM	0.914	0.916	0.912

alignment between visual and language representations, ensuring enhanced inter-modal coherence; 2) The incorporation of an entity consistency learning module enables accurate estimation of entity inconsistencies across different modalities, thus effectively capturing the variations in entity-related information; 3) The integration of an emotional consistency learning module facilitates the extraction of primary emotional features, which proves instrumental in assisting fake news detection, particularly in cases where the visual-textual pairs do not exhibit significant entity inconsistency.

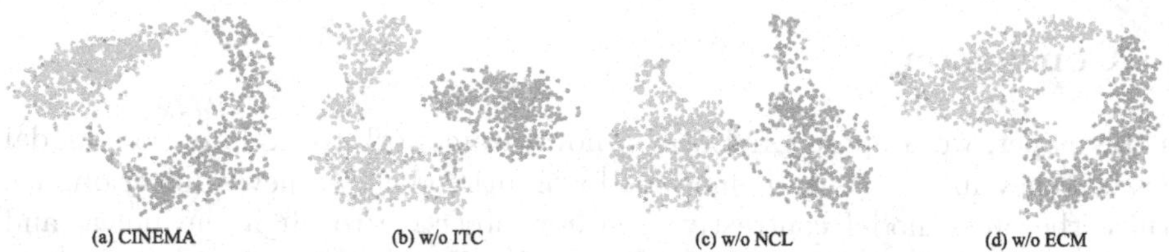

Fig. 3. T-SNE visualizations of the features before classifier that are learned by CINEMA and its three variants on the test dataset of Weibo.

4.3 Ablation Studies

Quantitative Analysis. In order to assess the individual contributions of each component within the proposed CINEMA model, we conduct a comprehensive evaluation by systematically removing each component and comparing the performance against the complete model. Specifically, the comparative variants of CINEMA are instantiated as follows: **1) w/o ITC:** we remove the cross-modal contrastive learning task and use a consistency learning task [2] to learn the aligned unimodal representations; **2) w/o NCL:** we remove the entity consistency learning module and replace it with simply concatenating m^v and m^t; **3) w/o ECM:** we remove the emotional consistency learning module.

Table 2 shows the results of ablation studies. It is evident that all the variant models exhibit inferior performance compared to the original CINEMA, thus highlighting the effectiveness of each component. By leveraging these distinct components, CINEMA surpasses existing methods by effectively addressing alignment challenges, accurately capturing entity inconsistencies, and leveraging emotional cues for improved fake news detection.

Qualitative Analysis. Moreover, we further analyze the proposed method using t-SNE [13] visualizations of the features before classifier in Fig. 3, which are learned by CINEMA and its three variants on the test dataset of Weibo.

Based on the observations depicted in Fig. 3, it is apparent that the boundaries separating different label dots in CINEMA are noticeably more distinct compared to its variant models. This observation suggests that the extracted features in CINEMA exhibit a higher level of discriminability, contributing to its superior performance. Note that, as shown in Fig. 3(b), numerous features acquired by CINEMA w/o ITC exhibit a notable susceptibility to misclassification. This observation highlights the importance of the cross-modal contrastive learning task in effectively capturing the distinctive characteristics inherent to multiple modalities, thereby enhancing the discrimination between fake news and real news. In addition, by comparing Fig. 3(a) and Fig. 3(d), we can see that the utilization of primary emotional features contributes to enhancing the representational capacity of the final features, as evidenced by the observed improvements in performance.

5 Conclusion

In this paper, we propose CINEMA, a novel framework to explore cross-modal inconsistency in entities and emotions for multimodal fake news detection. We utilize the cross-modal contrastive learning objective to align the image and text modalities effectively. Additionally, we develop an entity consistency learning module to capture the cross-modality entity correlations, and implement an emotional consistency learning module to accurately capture the emotional information within each modality. The experimental results on the Twitter and Weibo datasets substantiate the superior performance of CINEMA compared to previous approaches by a substantial margin, leading to the establishment of new state-of-the-art results on both datasets.

Acknowledgement. This work was supported by the National Key Research and Development of China (No. 2021YFB3100600), and Strategic Priority Research Program of Chinese Academy of Sciences (No. XDC02040400).

References

1. Boididou, C., Papadopoulos, S., Zampoglou, M., Apostolidis, L., Papadopoulou, O., Kompatsiaris, Y.: Detection and visualization of misleading content on twitter. Int. J. Multimed. Inf. Retr. **7**(1), 71–86 (2018)
2. Chen, Y., et al.: Cross-modal ambiguity learning for multimodal fake news detection. In: Proceedings of the ACM Web Conference 2022, pp. 2897–2905 (2022)
3. Del Vicario, M., et al.: The spreading of misinformation online. Proc. Natl. Acad. Sci. **113**(3), 554–559 (2016)
4. Devlin, J., Chang, M.W., Lee, K., Toutanova, K.: Bert: pre-training of deep bidirectional transformers for language understanding. arXiv preprint arXiv:1810.04805 (2018)
5. He, K., Zhang, X., Ren, S., Sun, J.: Deep residual learning for image recognition. In: Proceedings of the IEEE Conference on Computer Vision and Pattern Recognition, pp. 770–778 (2016)
6. Jin, Z., Cao, J., Guo, H., Zhang, Y., Luo, J.: Multimodal fusion with recurrent neural networks for rumor detection on microblogs. In: Proceedings of the 25th ACM International Conference on Multimedia, pp. 795–816 (2017)
7. Jin, Z., Cao, J., Zhang, Y., Zhou, J., Tian, Q.: Novel visual and statistical image features for microblogs news verification. IEEE Trans. Multimedia **19**(3), 598–608 (2016)

8. Khattar, D., Goud, J.S., Gupta, M., Varma, V.: MVAE: multimodal variational autoencoder for fake news detection. In: The World Wide Web Conference, pp. 2915–2921 (2019)
9. Kingma, D.P., Ba, J.: Adam: a method for stochastic optimization. arXiv preprint arXiv:1412.6980 (2014)
10. Li, J., Li, D., Xiong, C., Hoi, S.: Blip: bootstrapping language-image pre-training for unified vision-language understanding and generation. In: International Conference on Machine Learning, pp. 12888–12900. PMLR (2022)
11. Li, J., Selvaraju, R., Gotmare, A., Joty, S., Xiong, C., Hoi, S.C.H.: Align before fuse: vision and language representation learning with momentum distillation. Adv. Neural. Inf. Process. Syst. **34**, 9694–9705 (2021)
12. Liu, X., Nourbakhsh, A., Li, Q., Fang, R., Shah, S.: Real-time rumor debunking on twitter. In: Proceedings of the 24th ACM International on Conference on Information and Knowledge Management, pp. 1867–1870 (2015)
13. Van der Maaten, L., Hinton, G.: Visualizing data using t-SNE. J. Mach. Learn. Res. **9**(11) (2008)
14. Qi, P., Cao, J., Yang, T., Guo, J., Li, J.: Exploiting multi-domain visual information for fake news detection. In: 2019 IEEE International Conference on Data Mining (ICDM), pp. 518–527. IEEE (2019)
15. Qian, S., Wang, J., Hu, J., Fang, Q., Xu, C.: Hierarchical multi-modal contextual attention network for fake news detection. In: Proceedings of the 44th International ACM SIGIR Conference on Research and Development in Information Retrieval, pp. 153–162 (2021)
16. Radford, A., et al.: Learning transferable visual models from natural language supervision. In: International Conference on Machine Learning, pp. 8748–8763. PMLR (2021)
17. Singhal, S., Pandey, T., Mrig, S., Shah, R.R., Kumaraguru, P.: Leveraging intra and inter modality relationship for multimodal fake news detection. In: Companion Proceedings of the Web Conference 2022, pp. 726–734 (2022)
18. Wang, L., Zhang, C., Xu, H., Xu, Y., Xu, X., Wang, S.: Cross-modal contrastive learning for multimodal fake news detection. arXiv preprint arXiv:2302.14057 (2023)
19. Wang, Y., et al.: EANN: event adversarial neural networks for multi-modal fake news detection. In: Proceedings of the 24th ACM SIGKDD International Conference on Knowledge Discovery & Data Mining, pp. 849–857 (2018)
20. Wu, L., Rao, Y.: Adaptive interaction fusion networks for fake news detection. arXiv preprint arXiv:2004.10009 (2020)
21. Wu, Y., Zhan, P., Zhang, Y., Wang, L., Xu, Z.: Multimodal fusion with co-attention networks for fake news detection. In: Findings of the Association for Computational Linguistics: ACL-IJCNLP 2021, pp. 2560–2569 (2021)
22. Xue, J., Wang, Y., Tian, Y., Li, Y., Shi, L., Wei, L.: Detecting fake news by exploring the consistency of multimodal data. Inf. Process. Manag. **58**(5), 102610 (2021)
23. Yu, F., Liu, Q., Wu, S., Wang, L., Tan, T., et al.: A convolutional approach for misinformation identification. In: IJCAI, pp. 3901–3907 (2017)
24. Zhang, H., Fang, Q., Qian, S., Xu, C.: Multi-modal knowledge-aware event memory network for social media rumor detection. In: Proceedings of the 27th ACM International Conference on Multimedia, pp. 1942–1951 (2019)
25. Zhang, X., Cao, J., Li, X., Sheng, Q., Zhong, L., Shu, K.: Mining dual emotion for fake news detection. In: Proceedings of the Web Conference 2021, pp. 3465–3476 (2021)

26. Zhou, X., Wu, J., Zafarani, R.: SAFE: similarity-aware multi-modal fake news detection. In: Lauw, H.W., Wong, R.C.-W., Ntoulas, A., Lim, E.-P., Ng, S.-K., Pan, S.J. (eds.) PAKDD 2020. LNCS (LNAI), vol. 12085, pp. 354–367. Springer, Cham (2020). https://doi.org/10.1007/978-3-030-47436-2_27
27. Zubiaga, A., Aker, A., Bontcheva, K., Liakata, M., Procter, R.: Detection and resolution of rumours in social media: a survey. ACM Comput. Surv. (CSUR) **51**(2), 1–36 (2018)

Deep Consistency Preserving Network for Unsupervised Cross-Modal Hashing

Mengluan Li[1], Yanqing Guo[1,2], Haiyan Fu[1], Yi Li[2(✉)], and Hong Su[3]

[1] School of Information and Communication Engineering, Dalian University of Technology, Dalian, China
limengluan@mail.dlut.edu.cn

[2] School of Future Technology/School of Artificial Intelligence, Dalian University of Technology, Dalian, China
liyi@dlut.edu.cn

[3] Science and Technology on Communication Security Laboratory, Chengdu, China

Abstract. Given the proliferation of multimodal data in search engines and social networks, unsupervised cross-modal hashing has gained traction for its low storage consumption and fast retrieval speed. Despite the great success achieved, unsupervised cross-modal hashing still suffers from lacking reliable similarity supervision and struggles with reducing information loss caused by quantization. In this paper, we propose a novel deep consistency preserving network (DCPN) for unsupervised cross-modal hashing, which sufficiently utilizes the semantic information in different modalities. Specifically, we gain consistent features to fully exploit the co-occurrence information and alleviate the heterogeneity between different modalities. Then, a fusion similarity matrix construction method is proposed to capture the semantic relationship between instances. Finally, a fusion hash code reconstruction strategy is designed to fit the gap between different modalities and reduce the quantization error. Experimental results demonstrate the effectiveness of the proposed DCPN on unsupervised cross-modal retrieval tasks.

Keywords: Multimodal · Unsupervised deep hashing · Cross-modal retrieval

1 Introduction

The rapid development of information society has led to an explosive growth of multimedia data, which puts forward higher requirements for data storage and cross-modal semantic retrieval. Cross-modal hashing (CMH) has been a hot topic for its remarkable potential of compact representations and efficient similarity

This work is supported in part by the National Natural Science Foundation of China (No. 62106037, No. 62076052), in part by the Major Program of the National Social Science Foundation of China (No.19ZDA127), and in part by the Fundamental Research Funds for the Central Universities (No. DUT22YG205).

© The Author(s), under exclusive license to Springer Nature Singapore Pte Ltd. 2024
Q. Liu et al. (Eds.): PRCV 2023, LNCS 14425, pp. 235–246, 2024.
https://doi.org/10.1007/978-981-99-8429-9_19

calculations. In real-world scenarios, massive data is usually unlabeled, it is very time-consuming and difficult to manually label such large amounts of data. Thus, unsupervised cross-modal hashing (UCMH) has received extensive attention. Conventional UCMH [2,12,15,21] uses hand-crafted features to obtain projection functions and hash codes. As the excellent expressive and nonlinear processing ability of deep neural network, deep UCMH [3,4,6,19] can leverage features extracted from deep networks to obtain nonlinear hash functions. Recently, a vision-language pre-trained model CLIP [10] has been incorporated into deep UCMH [9,17,22] which can effectively bridge the gap between modalities.

Main challenge of UCMH is how to mine the semantic similarity between instances. Given the heterogeneity of features representation and data distribution between different modalities, many existing methods focus on constructing a unified semantic similarity matrix using features extracted by single-modal networks. However, the features fail to capture the co-occurrence semantic information of paired data, and the similarity matrix calculated by manually designed formula is not enough to represent the complex semantic relationship of cross-modal data. Although some approaches consider the distance distribution [8,13] or higher-order neighborhood information [18,20], they still lack sufficient interaction between different modalities, thus fail to fully grasp complex multimodal semantic information and bridge the heterogeneity in data characteristics.

Another challenge in UCMH is the information loss during binarization. In order to obtain high quality hash codes, many novel methods are proposed such as half-real and half-binary optimization strategy [18], min-max hashing layer [22] and bit-selection module [14], but they are not the end-to-end learning. Recently, modality-specific decoders are used to reconstruct the binary codes into original features [7,9] which could reduce the semantic loss of intra-modality. However, the reconstruction of single-modal hash codes fails to address the heterogeneity of different modalities and ignores the purpose of CMH that map various types of features into a common Hamming space.

To overcome the above two challenges, we propose a novel deep consistency preserving network (DCPN) for unsupervised cross-modal hashing, which sufficiently utilizes the semantic information in different modalities. An overview of the proposed method is illustrated in Fig. 1. Specifically, DCPN applies the pre-trained CLIP model to extract features from the original data. Then, the deep hash networks map the features to continuous hash codes. Furthermore, a feature fusion module is proposed to facilitate the learning of a consistent feature representation between different modalities, thereby bridging the heterogeneity between modalities and capturing latent semantic similarities. Simultaneously, a fusion hash code reconstruction strategy is designed to reduce the quantization loss. The main contributions of DCPN are as follows:

(1) A new unsupervised cross-modal hashing method DCPN is proposed to maintain semantic consistency of different modalities. We fuse the features extracted by CLIP model into a common latent space, which can fully exploit the co-occurrence information to guide the hash code learning.

(2) Different from previous methods that rely on manually designed formulas to calculate the similarity matrix, we leverage the consistent features to learn a fused similarity matrix, which effectively captures potential semantic information.
(3) We design a fusion hash code reconstruction strategy to align the common hash code with single-modal features as well as reduce quantization loss of the binarization process, thus further bridge the modal gap and generate high-quality hash code.

2 The Proposed Method

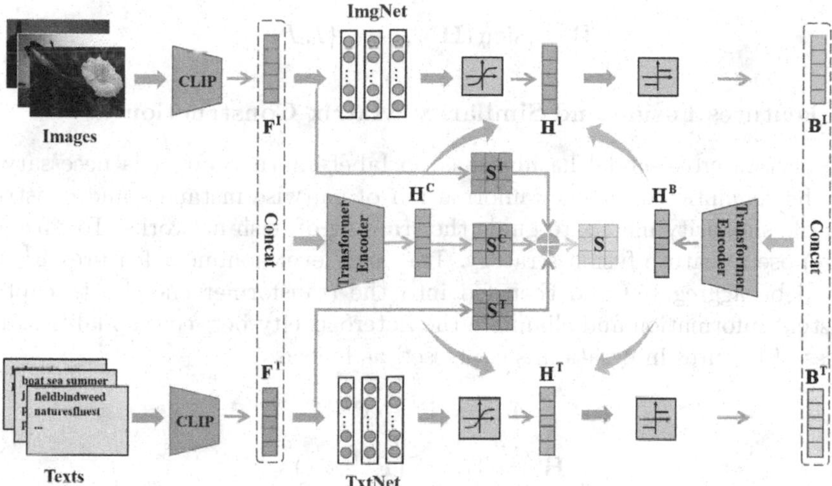

Fig. 1. The framework of DCPN. In this workflow, the green line and the blue line represent the processes of generating image and text modal hash codes, respectively. Two transformer networks are used as auxiliary modules to improve the performance of the hash networks. (Color figure online)

2.1 Problem Definition

To better illustrate, we apply DCPN to cross-modal retrieval tasks on two most commonly used modalities (image, text). We denote the training set with n image-text paired instances as $\mathbf{O} = \{\mathbf{o}_i\}_{i=1}^n$, where $\mathbf{o}_i = [I_i, T_i]$. DCPN aims to learn hash nets and generate unified binary codes $\mathbf{B}^* \in \mathbb{R}^{m \times K}$ for each modality. $* \in \{I, T\}$, m denotes the mini-batch size and K denotes the length of binary code. $\text{sign}(\cdot)$ is the element-wise sign function that generate the binary hash codes. $D_{\cos}(\mathbf{X}, \mathbf{Y})$ is pairwise cosine similarity matrix and each element is calculated as follows:

$$D_{\cos}(\mathbf{X}, \mathbf{Y})_{ij} = \frac{\mathbf{X}_i \mathbf{Y}_j^T}{\|\mathbf{X}_i\|_2 \|\mathbf{Y}_j\|_2} \in [-1, +1] \tag{1}$$

where $\|\cdot\|_2$ represents the L2-norm.

2.2 Deep Feature Extraction and Hashing Learning

Considering the powerful feature extraction ability of CLIP, we utilize pre-trained CLIP model to extract rich semantic information from raw data: $\mathbf{F}^* \in \mathbb{R}^{m\times d}$ where $* \in \{I,T\}$, $d = 1024$ denotes dimension of the extracted features. Then, we learn hash functions to generate relaxed real-valued hash features, effectively overcoming the gradient disappearance caused by binary codes:

$$\mathbf{H}^* = \mathcal{F}^*(\mathbf{F}^*;\theta^*),\ \ * \in \{I,T\} \tag{2}$$

where $\mathbf{H}^* \in \mathbb{R}^{m\times K}$, $\mathcal{F}^I(\cdot;\theta^I)$ and $\mathcal{F}^T(\cdot;\theta^T)$ are hash functions named ImgNet and TxtNet, θ^* are the parameters respectively for different models.

And then, the strict binary hash codes can be generate by sign function.

$$\mathbf{B}^* = \operatorname{sign}(\mathbf{H}^*),\ \ * \in \{I,T\} \tag{3}$$

2.3 Features Fusion and Similarity Matrix Construction

Unsupervised cross-modal hashing has no label information, it is necessary to mine the semantic consistency information of pairwise instances and construct a reliable similarity matrix to guide the training of hash networks. To this end, we propose a feature fusion strategy. The high-level nonlinear features $\mathbf{F^I}$ and $\mathbf{F^T}$ will be aggregated and then fed into the transformer encoder to capture consistent information and eliminate the heterogeneity between modalities. The process of features integration is expressed as follows:

$$\mathbf{F}^{IT} = \mathbf{F}^I \oplus \mathbf{F}^T \tag{4}$$

$$\mathbf{H}^C = \operatorname{Trans}^c\left(\mathbf{F}^{IT};\theta^C\right) \tag{5}$$

where $\oplus$ is the concatenation operation, θ^C denotes parameters of the transformer encoder.

Instead of calculating similarity matrix relying on the features extracted by single-modal network, we leverage the consistent features to construct a fusion similarity matrix, which maintains richer cross-modal semantic information and addresses the data heterogeneity. After normalize $\mathbf{F}^I$, $\mathbf{F}^T$ and $\mathbf{H}^C$, we calculate the cosine similarity between different instances to obtain the intra-modal similarity and the fused inter-modal similarity as follows:

$$\begin{aligned} \mathbf{S}^* &= D_{\cos}(\mathbf{F}^*,\mathbf{F}^*),\ \ * \in \{I,T\} \\ \mathbf{S}^C &= D_{\cos}\left(\mathbf{H}^C,\mathbf{H}^C\right) \end{aligned} \tag{6}$$

Subsequently, we obtain a unified similarity matrix $\mathbf{S} = \{s_{ij}\}_{i,j=1}^M$, which incorporates both inter- and intra-modal semantic information.

$$\begin{aligned} \mathbf{S} &= \alpha\mathbf{S}^I + \beta\mathbf{S}^T + \gamma\mathbf{S}^C \\ s.t.\ &\alpha,\beta,\gamma \geq 0, \alpha+\beta+\gamma = 1 \end{aligned} \tag{7}$$

Following [13], we use $2\mathbf{S} - 1$ as the final similarity matrix.

The consistent features and similarity matrix mutually reinforce each other, effectively directing the hash network to map diverse modal features into a common Hamming space. The fusion objective function is defined as follows:

$$\begin{aligned}\min_{\theta^C,\theta^I,\theta^T} L_c = &\left\|D_{\cos}\left(\mathbf{H}^C,\mathbf{H}^C\right)-\mathbf{S}\right\|_F+\left\|\mathrm{Tr}\left(D_{\cos}\left(\mathbf{H}^C,\mathbf{H}^C\right)-\eta\mathbf{I}\right)\right\|_2\\ &+\sum_{x,y}\left\|D_{\cos}\left(\mathbf{H}^C,\mathbf{H}^C\right)-D_{\cos}\left(\mathbf{H}^x,\mathbf{H}^y\right)\right\|_F\end{aligned} \tag{8}$$

where $\mathbf{I}$ is an identity matrix, $x, y \in \{I, T\}$, η is a hyperparameter that balances modal differences.

The first term in Eq. (8) serves to constrain the consistent features to retain semantic similarity while simultaneously optimizing the unified similarity $\mathbf{S}^C$. The second item ensures the similarity between pairs of instances. Due to modality gap, setting the diagonal similarity to 1 may not achieve optimal results, we introduce a balancing factor η. The third term maximizes the semantic similarity between the consistent features and cross-modal hash features to maintain the co-occurrence information between pairwise instances.

2.4 Hash Code Fusion and Reconstruction

The purpose of cross-modal hashing is to map high-dimensional multi-modal features into low-dimensional Hamming space, which will inevitably result in information loss. To eliminate the gap between real values and binary codes, DCPN proposes a more suitable method named fusion hash code reconstruction. First, we fuse the binary hash codes $\mathbf{B}^I$, $\mathbf{B}^T \in \mathbb{R}^{m\times K}$ of different modes by concatenate them into a joint hash code $\mathbf{B}^C \in \mathbb{R}^{m\times 2K}$, which contain multi-modal information and satisfy the demands of getting the same hash code for paired instances. The process of hash code fusion can be expressed as follows:

$$\mathbf{B}^C = \mathbf{B}^I \oplus \mathbf{B}^T \tag{9}$$

Second, a transformer encoder is applied to reconstruct the fused binary hash code to real-valued hash features:

$$\mathbf{H}^B = Trans^B\left(\mathbf{B}^C;\theta^B\right) \tag{10}$$

where $\mathbf{H}^B \in \mathbb{R}^{m\times K}$, $Trans^B\left(\cdot;\theta^B\right)$ denotes the transformer encoder and θ^B is the set of learnable parameters.

Similar to Eq. (8), we propose a hash code reconstruction loss:

$$\begin{aligned}\min_{\theta^B,\theta^I,\theta^T} L_b = &\left\|D_{\cos}\left(\mathbf{H}^B,\mathbf{H}^B\right)-\mathbf{S}\right\|_F+\left\|\mathrm{Tr}\left(D_{\cos}\left(\mathbf{H}^B,\mathbf{H}^B\right)-\eta\mathbf{I}\right)\right\|_2\\ &+\sum_{x,y}\left\|D_{\cos}\left(\mathbf{H}^B,\mathbf{H}^B\right)-D_{\cos}\left(\mathbf{H}^x,\mathbf{H}^y\right)\right\|_F\end{aligned} \tag{11}$$

2.5 Objective Function

To maintain the semantic similarity of hash features and maximizes the semantic similarity of paired instances, we formulate a semantic preservation objective function as follows:

$$\min_{\theta^I,\theta^T} L_s = \sum_{x,y} \|D_{\cos}(\mathbf{H}^x,\mathbf{H}^y) - \mathbf{S}\|_F + \left\|\mathrm{Tr}\left(D_{\cos}\left(\mathbf{H}^I,\mathbf{H}^T\right) - \eta\mathbf{I}\right)\right\|_2 \tag{12}$$

An intra- and inter-modal similarity preservation objective function is defined to generate unified hash codes for different modalities:

$$\min_{\theta^I,\theta^T} L_{it} = \sum_{x,y} \left\| D_{\cos}(\mathbf{H}^x,\mathbf{H}^y) - D_{\cos}\left(\mathbf{H}^{x'},\mathbf{H}^{y'}\right)\right\|_F \tag{13}$$

where $x, y, x', y' \in \{I, T\}$.

To achieve a common representation for different modal data, we propose a consistency features loss that aligns the hash codes of different modalities with fusion and reconstructed features:

$$\min_{\theta^I,\theta^T,\theta^C,\theta^B} L_f = \|\mathbf{H}^p - \mathbf{H}^q\|_2 \tag{14}$$

where $p, q \in \{C, B, I, T\}$.

Combining Eq. (8, 11–14) together, the final objective function is calculated as:

$$\min_{\theta^I,\theta^T,\theta^C,\theta^B} L = \lambda_1 L_c + \lambda_2 L_b + \lambda_3 L_s + \lambda_4 L_{it} + \lambda_5 L_f \tag{15}$$

where λ_1, λ_2, λ_3, λ_4 and λ_5 are hyper parameters to balance the total loss.

3 Experiments

3.1 Datasets and Baselines

We employ two benchmark datasets to evaluate our approach and follow the data setting strategy of [18].

MIRFlickr-25K [5] contains 25,000 image-text pairs belong to 24 distinct categories. We selected tags that appeared in at least 20 images, and then, split the remaining 20015 image-text pairs by randomly selecting 2,000 pairs as query set and 2,000 pairs as validation query set, the rest as retrieval set which contains 5,000 training pairs.

NUS-WIDE [1] is a multi-label dataset consisting of 26,9648 image-text pairs which belong to 81 categories. We selected 186577 image-text pairs from 10 most commonly used categories and randomly split them into query set with 2,000 pairs and validation query set with 2000 pairs, the rest as retrieval set with 5,000 training pairs included.

To evaluate the effectiveness of the proposed method, we conducted two cross-modal retrieval tasks: use an image query to retrieval corresponding text (I→T) and use a text query to retrieval corresponding image (T→I). We selected six unsupervised cross-modal hashing methods as baseline approaches to compared with DCPN: DJSRH [13], JDSH [8], DSAH [16], DGCPN [18], HNH [20], DAEH [11]. We adopt two common retrieval metrics to evaluate the performance: Mean Average Precision (MAP) and Precision@top-K curve. The MAP score is calculated by taking the mean of the average precision scores for all queries.

3.2 Implementation Details

In our experiment, we utilized a pre-trained CLIP (RN50) model to extract features from raw data. The ImgNet and TxtNet are three-layer Multi-layer Perception (MLP) ($d \rightarrow 1024 \rightarrow K$). They utilize the ReLU activation function for all layers except the final layer, which adopts the Tanh activation function. Notably, we fixed the parameters of CLIP during training and solely updated the parameters of two hash networks and two transformers. The training epoch is 100 and the batch size is 64. A mini-batch SGD optimizer with a 0.9 momentum and 0.0005 weight decay. The initial learning rate is 0.005. We determined the parameters by validation set, and all parameters are determined with 64 bit hash code. Through cross validation, the hyper-parameters finally set as: $\alpha = 0.6$, $\beta = 0.1$, $\gamma = 0.3$ and $\eta = 1.5$ for MIRFlickr-25K and $\alpha = 0.5$, $\beta = 0.2$, $\gamma = 0.3$ and $\eta = 1.4$ for NUS-WIDE, $\lambda_1 = \lambda_2 = \lambda_3 = \lambda_4 = \lambda_5 = 1$.

3.3 Results and Analysis

MAP Results. We evaluate the performance of the proposed DCPN against all baselines across a range of code lengths from 16 bit to 128 bit in both I→T and T→I retrieval tasks. The MAP results are reported in Table 1. It is evident that DCPN outperforms all baselines.

Specially, compared to DGCPN [18], the best deep baseline, our method achieves boosts of 4.43% and 4.35% in I→T and T→I task on MIRFlickr-25K and achieves boosts of 6.25% and 5.88% in two retrieval tasks on NUS-WIDE. For a fair comparison, we apply the pre-trained CLIP model on DGCPN which denoted as DGCPN∗. It can be seen that CLIP greatly promoted the retrieval accuracy and our method further improves the model performance.

Typically, most methods achieved better performance with longer code lengths compared to shorter hash codes. It is noteworthy that our proposed method consistently demonstrates superior results even with shorter hash code lengths. As can be seen from the MAP scores, when the hash code length is set to 32 bit, the performance of our method can still be comparable to the baselines on 128 bit hash code. This highlights the efficacy of our proposed method in preserving semantic information within the Hamming space.

Top-K Precision Curves. We also evaluate DCPN use top-K precision curves on two datasets, which are plotted in Fig. 2. Clearly, our method has achieved superior performance compared to all the baselines. The top-K precision curve

Table 1. The MAP results of two public datasets at various code lengths.

Task	Method	MIRFlickr-25K				NUS-WIDE			
		16bit	32bit	64bit	128bit	16bit	32bit	64bit	128bit
I→T	DJSRH	0.654	0.665	0.671	0.682	0.508	0.521	0.529	0.554
	JDSH	0.660	0.678	0.685	0.691	0.536	0.552	0.576	0.580
	DSAH	0.688	0.698	0.703	0.708	0.564	0.586	0.596	0.612
	DGCPN	0.709	0.721	0.730	0.746	0.627	0.635	0.651	0.654
	HNH	0.692	0.701	0.695	0.689	0.591	0.597	0.592	0.536
	DAEH	0.699	0.708	0.714	0.720	0.594	0.612	0.629	0.642
	DGCPN*	0.725	0.748	0.759	0.768	0.650	0.674	0.690	0.698
	DCPN	**0.759**	**0.773**	**0.775**	**0.776**	**0.685**	**0.706**	**0.712**	**0.714**
T→I	DJSRH	0.653	0.664	0.674	0.674	0.537	0.519	0.553	0.565
	JDSH	0.671	0.686	0.697	0.698	0.547	0.563	0.570	0.571
	DSAH	0.689	0.700	0.710	0.716	0.581	0.598	0.621	0.621
	DGCPN	0.711	0.715	0.725	0.746	0.631	0.647	0.662	0.665
	HNH	0.694	0.696	0.704	0.707	0.596	0.618	0.616	0.626
	DAEH	0.685	0.696	0.708	0.712	0.602	0.619	0.633	0.646
	DGCPN*	0.725	0.743	0.756	0.767	0.660	0.682	0.695	0.703
	DCPN	**0.754**	**0.770**	**0.772**	**0.775**	**0.690**	**0.712**	**0.717**	**0.721**

shows the precision at different numbers of retrieved instances (represented by varying values of K). Remarkably, our method demonstrates exceptional precision even when a larger number of points are retrieved.

Ablation Studies. To thoroughly analyze the influence of all the designed modules in our framework, we conducted extensive ablation studies on two datasets. Specifically, we introduced three variants of DCPN to examine the impact of the independently training strategy. In DCPN-1, the final similarity matrix is set as $S = \alpha S^I + (1 - \alpha) S^T$, aiming to explore the importance of constructing a fused similarity matrix. Additionally, we compare our method with six different similarity matrix construction methods to evaluate their performance. DCPN-2 is the variant that does not use fusion features to maintain semantic information and features consistency, which means that the fusion objective function L_C and the parts containing fused features in L_f are removed. DCPN-3 is the DCPN that removes the fusion hash code reconstruction module.

The MAP results of three variant are shown in Table 2. It is evident that the complete DCPN outperforms all other variants, while removing any component will lead to a relative decline in performance. The comparison between DCPN-1 and DCPN reflects the effectiveness of the fusion similarity matrix construction, which captures more accurate semantic similarity information. Figure 3 shows the top-K precision curves of six similarity matrix construction methods, the effectiveness of fusion similarity matrix construction method have been farther validated. The results of DCPN-2 show the important role of fusion features in bridging the gap between modalities. The results of DCPN-3 proves that the

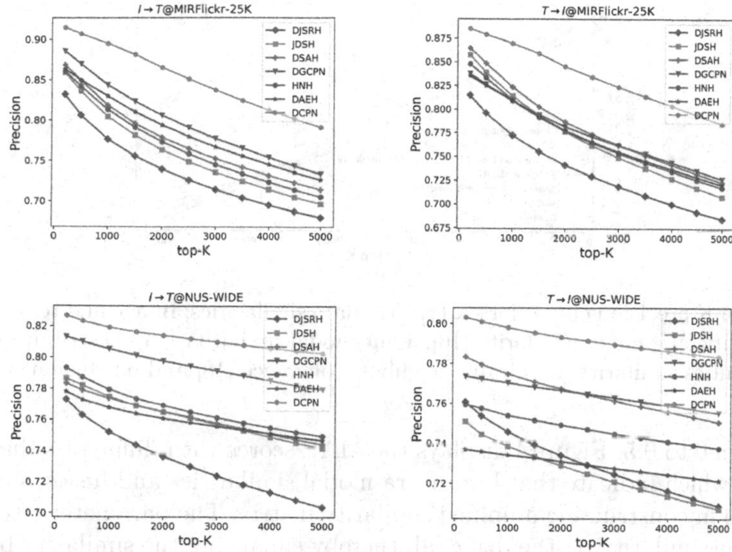

Fig. 2. Top-K precision curves on two datasets with 64 bit code length.

proposed method performs well, particularly when the number of hash bits is small. This demonstrates the role of our proposed fusion-reconstruction hashing strategy in reducing quantization errors.

Parameter Sensitivity Analysis. Four types of parameters namely, α, β, γ, η, may affect the performance of the proposed DCPN. Here, we evaluate the influence of parameters on validation set with 64 bit hash code length. Given the constraints: $\alpha, \beta, \gamma \geq 0$ and $\alpha + \beta + \gamma = 1$, the value of γ can be determined by α and β. Therefore, we only need to analyze the influence of α and β. Since images contain more semantic information than text, we choose α from 0.8 to 0.4

Table 2. The MAP for ablation experiments of different variants.

Task	Method	MIRFlickr-25K				NUS-WIDE			
		16bit	32bit	64bit	128bit	16bit	32bit	64bit	128bit
I→T	DCPN-1	0.719	0.729	0.748	0.753	0.648	0.658	0.674	0.683
	DCPN-2	0.716	0.738	0.751	0.754	0.642	0.656	0.675	0.690
	DCPN-3	0.707	0.765	0.773	0.773	0.593	0.703	0.705	0.716
	DCPN	0.759	0.773	0.775	0.776	0.685	0.706	0.712	0.714
T→I	DCPN-1	0.727	0.737	0.747	0.752	0.650	0.662	0.677	0.688
	DCPN-2	0.716	0.734	0.750	0.758	0.646	0.659	0.677	0.692
	DCPN-3	0.709	0.761	0.770	0.771	0.603	0.706	0.710	0.715
	DCPN	0.754	0.770	0.772	0.775	0.690	0.712	0.717	0.721

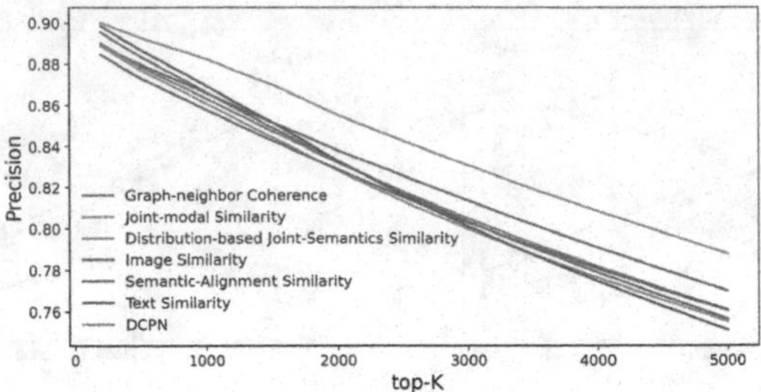

Fig. 3. Top-K precision curves for six types of data similarities: Image and text similarities, Semantic-alignment similarity [16], Joint-modal similarity [13], Distribution-based joint-semantics similarity [8], Graph-neighbor Coherence [18] and our fusion similarity

and β from 0 to 0.5. Figure 4 displays the MAP scores with different values of α, β, and γ, which indicate that both intra-modal similarities and fusion similarity are crucial in constructing a unified similarity matrix. The parameter η serves to restrict the similarity on the diagonal, thereby enhancing the similarity between pairs of instances. As demonstrated in Fig. 5, selecting an appropriate value for η can yield superior performance.

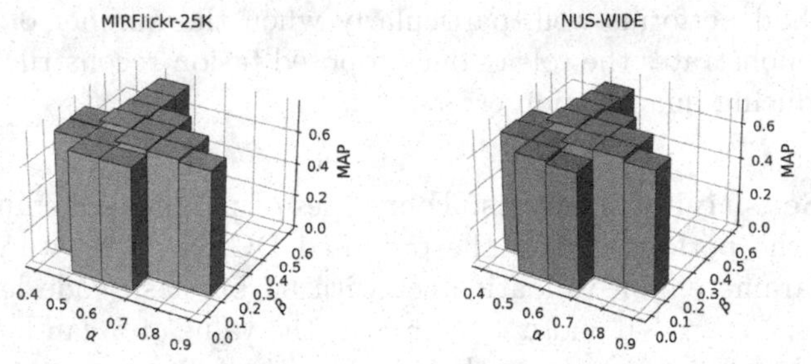

Fig. 4. The MAP of DCPN with different α, β and γ values.

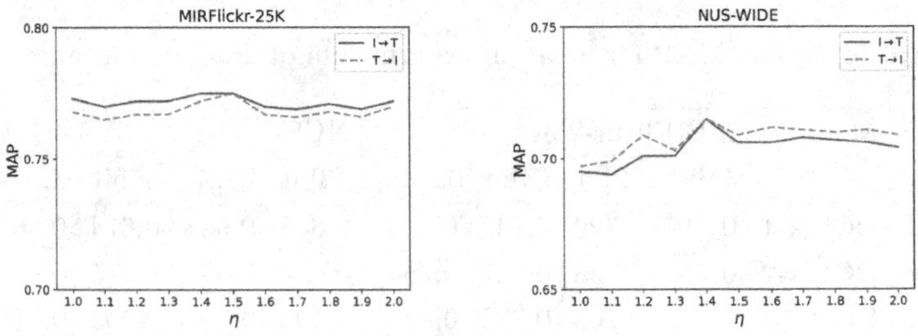

Fig. 5. The MAP of DCPN with different η values.

4 Conclusion

In this paper, we propose a novel deep consistency preserving network (DCPN) for unsupervised cross-modal hashing. DCPN uses the vision-language pre-trained model CLIP to extract rich semantic information from the original data. The feature fusion module is introduced to bridge the modality gap, and based on which, a more accurate similarity matrix is obtained. In addition, we design a fusion hash code reconstruction method to reduce the quantization error caused by the binarization process, while maintaining the consistency of hash codes for pairwise instances. Extensive experiments on two widely used datasets have demonstrated that DCPN can achieve excellent results compared with recent unsupervised deep cross-modal hashing methods.

References

1. Chua, T.S., Tang, J., Hong, R., Li, H., Luo, Z., Zheng, Y.: NUS-WIDE: a real-world web image database from national university of Singapore. In: Proceedings of the ACM International Conference on Image and Video Retrieval, pp. 1–9 (2009). https://doi.org/10.1145/1646396.1646452
2. Ding, G., Guo, Y., Zhou, J.: Collective matrix factorization hashing for multimodal data. In: Proceedings of the IEEE Conference on Computer Vision and Pattern Recognition, pp. 2075–2082 (2014). https://doi.org/10.1109/CVPR.2014.267
3. Hu, H., Xie, L., Hong, R., Tian, Q.: Creating something from nothing: unsupervised knowledge distillation for cross-modal hashing. In: Proceedings of the IEEE/CVF Conference on Computer Vision and Pattern Recognition, pp. 3123–3132 (2020). https://doi.org/10.1109/cvpr42600.2020.00319
4. Hu, P., Zhu, H., Lin, J., Peng, D., Zhao, Y.P., Peng, X.: Unsupervised contrastive cross-modal hashing. IEEE Trans. Pattern Anal. Mach. Intell. **45**(3), 3877–3889 (2023). https://doi.org/10.1109/TPAMI.2022.3177356
5. Huiskes, M.J., Lew, M.S.: The MIR flickr retrieval evaluation. In: Proceedings of the 1st ACM International Conference on Multimedia Information Retrieval, pp. 39–43 (2008). https://doi.org/10.1145/1460096.1460104
6. Li, C., Deng, C., Wang, L., Xie, D., Liu, X.: Coupled cyclegan: unsupervised hashing network for cross-modal retrieval. In: Proceedings of the AAAI Conference on Artificial Intelligence, vol. 33, pp. 176–183 (2019). https://doi.org/10.1609/aaai.v33i01.3301176
7. Li, T., Yang, X., Wang, B., Xi, C., Zheng, H., Zhou, X.: Bi-CMR: bidirectional reinforcement guided hashing for effective cross-modal retrieval. In: Proceedings of the AAAI Conference on Artificial Intelligence, vol. 36, pp. 10275–10282 (2022). https://doi.org/10.1609/aaai.v36i9.21268
8. Liu, S., Qian, S., Guan, Y., Zhan, J., Ying, L.: Joint-modal distribution-based similarity hashing for large-scale unsupervised deep cross-modal retrieval. In: Proceedings of the 43rd International ACM SIGIR Conference on Research and Development in Information Retrieval, pp. 1379–1388 (2020). https://doi.org/10.1145/3397271.3401086
9. Mingyong, L., Yewen, L., Mingyuan, G., Longfei, M.: Clip-based fusion-modal reconstructing hashing for large-scale unsupervised cross-modal retrieval. Int. J. Multimed. Inf. Retr. **12**(1), 2 (2023)

10. Radford, A., et al.: Learning transferable visual models from natural language supervision. In: International Conference on Machine Learning, pp. 8748–8763. PMLR (2021)
11. Shi, Y., et al.: Deep adaptively-enhanced hashing with discriminative similarity guidance for unsupervised cross-modal retrieval. IEEE Trans. Circuits Syst. Video Technol. **32**(10), 7255–7268 (2022)
12. Song, J., Yang, Y., Yang, Y., Huang, Z., Shen, H.T.: Inter-media hashing for large-scale retrieval from heterogeneous data sources. In: Proceedings of the 2013 ACM SIGMOD International Conference on Management of Data, pp. 785–796 (2013). https://doi.org/10.1145/2463676.2465274
13. Su, S., Zhong, Z., Zhang, C.: Deep joint-semantics reconstructing hashing for large-scale unsupervised cross-modal retrieval. In: Proceedings of the IEEE/CVF International Conference on Computer Vision, pp. 3027–3035 (2019). https://doi.org/10.1109/ICCV.2019.00312
14. Tu, R.C., Jiang, J., Lin, Q., Cai, C., Tian, S., Wang, H., Liu, W.: Unsupervised cross-modal hashing with modality-interaction. IEEE Trans. Circuits Syst. Video Technol. 1 (2023). https://doi.org/10.1109/TCSVT.2023.3251395
15. Wang, D., Wang, Q., Gao, X.: Robust and flexible discrete hashing for cross-modal similarity search. IEEE Trans. Circuits Syst. Video Technol. **28**(10), 2703–2715 (2017)
16. Yang, D., Wu, D., Zhang, W., Zhang, H., Li, B., Wang, W.: Deep semantic-alignment hashing for unsupervised cross-modal retrieval. In: Proceedings of the 2020 International Conference on Multimedia Retrieval, pp. 44–52 (2020). https://doi.org/10.1145/3372278.3390673
17. Yu, H., Ding, S., Li, L., Wu, J.: Self-attentive clip hashing for unsupervised cross-modal retrieval. In: Proceedings of the 4th ACM International Conference on Multimedia in Asia, pp. 1–7 (2022). https://doi.org/10.1145/3551626.3564945
18. Yu, J., Zhou, H., Zhan, Y., Tao, D.: Deep graph-neighbor coherence preserving network for unsupervised cross-modal hashing. In: Proceedings of the AAAI Conference on Artificial Intelligence, vol. 35, pp. 4626–4634 (2021). https://doi.org/10.1609/aaai.v35i5.16592
19. Zhang, J., Peng, Y., Yuan, M.: Unsupervised generative adversarial cross-modal hashing. In: Proceedings of the AAAI Conference on Artificial Intelligence, vol. 32 (2018). https://doi.org/10.1609/aaai.v32i1.11263
20. Zhang, P.F., Luo, Y., Huang, Z., Xu, X.S., Song, J.: High-order nonlocal hashing for unsupervised cross-modal retrieval. World Wide Web **24**, 563–583 (2021)
21. Zhou, J., Ding, G., Guo, Y.: Latent semantic sparse hashing for cross-modal similarity search. In: Proceedings of the 37th International ACM SIGIR Conference on Research & Development in Information Retrieval, pp. 415–424 (2014). https://doi.org/10.1145/2600428.2609610
22. Zhuo, Y., Li, Y., Hsiao, J., Ho, C., Li, B.: Clip4hashing: unsupervised deep hashing for cross-modal video-text retrieval. In: Proceedings of the 2022 International Conference on Multimedia Retrieval, pp. 158–166 (2022). https://doi.org/10.1145/3512527.3531381

Learning Adapters for Text-Guided Portrait Stylization with Pretrained Diffusion Models

Mintu Yang[1], Xianxu Hou[4], Hao Li[1], Linlin Shen[1,2,3](✉), and Lixin Fan[5]

[1] College of Computer Science and Software Engineering, Shenzhen University, Shenzhen, China
yangmintu2022@email.szu.edu.cn, llshen@szu.edu.cn

[2] WeBank Institute of Financial Technology, Shenzhen University, Shenzhen, China

[3] National Engineering Laboratory for Big Data System Computing Technology, Shenzhen University, Shenzhen, China

[4] School of AI and Advanced Computing, Xi'an Jiaotong-Liverpool University, Suzhou, China

[5] WeBank Co., Ltd., Shenzhen, China

Abstract. This paper presents a framework for text-guided face portrait stylization using a pre-trained large-scale diffusion model. To balance style transformation and content preservation, we introduce an adapter that modifies specific components of the diffusion model. By training the adapter to only modify these components, we reduce the tuning parameter space, resulting in an efficient solution for face portrait stylization. Our approach captures the target style and at the same time, preserves the source portrait content, making it an effective method for personalized image editing. Experimental results show its superiority over state-of-the-art techniques in various stylization tasks.

Keywords: Portrait stylization · Diffusion model

1 Introduction

Recently, large-scale text-to-image models, such as DALL.E2 [23], Imagen [26] and Stable Diffusion [24], have demonstrated impressive abilities in generating high-quality and diverse images with natural language guidance. This has led to an increased interest in exploring their potential for personalized image editing. One promising research field is generating face portraits that accurately reflect a given text prompt's style while preserving the content of the source image.

The existing methods [13,15] train DDPM-based models for human face domain transfer. However, the generated images suffer from poor quality with numerous artifacts and noise. Additionally, the edited images tend to have fewer stylized features and resemble the original image more closely. Recently, several methods such as [5] and [25] embed specific concepts into the text with Stable Diffusion. However, these methods struggle to retain the content of the face portrait.

© The Author(s), under exclusive license to Springer Nature Singapore Pte Ltd. 2024
Q. Liu et al. (Eds.): PRCV 2023, LNCS 14425, pp. 247–258, 2024.
https://doi.org/10.1007/978-981-99-8429-9_20

Fig. 1. Our method produces high-quality portrait stylization results while preserving the source content. The source prompt is ‘a photo of person’. The edited style prompt is ‘a photo of person with the style of <style>’. We can substitute <style> with different words to describe the style of the portrait.

To perform text-guided editing of a real image using diffusion models, it is necessary to invert the source image to find the initial noise input that, when combined with the prompt and fed into the diffusion process, reconstructs the input image. Denoising Diffusion Implicit Models (DDIM) [27] is a widely adopted method for achieving this inversion in unconditional diffusion models. However, when directly inputting the noised latent code obtained from a DDIM inversion and the target prompt to Stable Diffusion, there is a risk of changing the original image content. As demonstrated in [18], the classifier-free guidance method used in this approach can result in the loss of crucial face information, leading to an unnatural face generation. Therefore, it is crucial to explore alternative methods that can retain the original content and structural information while incorporating the desired style information.

With the above limitations in mind, we propose to learn an adapter to achieve face portrait stylization (Fig. 1) based on pretrained diffusion models. The main idea is to employ an adapter to modify specific components of diffusion models in order to simultaneously accomplish both style transformation and content preservation. Concretely, the main contributions of this work are summarized as follows:

- We propose a framework for stylizing face portraits guided by text prompts, using a pre-trained diffusion model. Our method effectively captures the target style while preserving the main source portrait content.
- We propose to train an adapter to only modify specific components of diffusion model for balanced style transformation and content preservation. This significantly reduces the space of parameter tuning, resulting in an efficient solution for face portrait stylization.

- We validate the effectiveness of our approach on various stylization tasks, demonstrating the superiority of our method over other state-of-the-art techniques.

2 Related Work

2.1 Text-to-Image Diffusion Models

In recent years, diffusion models have emerged as a powerful tool for text-to-image generating tasks, delivering state-of-the-art results. These models typically rely on pretrained language models like CLIP (Contrastive Language-Image Pre-Training) [22] to encode text inputs into latent vectors. For example, Glide [20] is a text-guided diffusion model that supports both image generation and editing. Disco Diffusion is a CLIP-guided implementation of [4] that processes text prompts, while Stable Diffusion is a large-scale implementation of latent diffusion [24] that achieves text-to-image generation. Imagen [23] is a text-to-image structure that directly diffuses pixels using a pyramid structure, without using latent images. Our method aims to leverage such pre-trained text-to-image diffusion models to enable text-guided stylization of face portraits.

2.2 Control of Pretrained Diffusion Model

Effective control of the pre-trained diffusion model offers a powerful tool for generating images that meet specific requirements. Recent advances, such as Textual Inversion [5] and DreamBooth [25], allow for image personalization using a small set of images with similar topics or objects. Custom Diffusion [14] and Cones [17] can generate images with multiple concepts, enabling users to create diverse and complex images. Additionally, ControlNet [29] and T2I-Adapter [19] incorporate additional conditions to control the generation process, allowing diverse outputs. Furthermore, other works have attempted to adapt text-guided diffusion models for image editing purposes. One such technique is Prompt-to-Prompt [8], which enables intuitive manipulation of local or global details by modifying only the text prompt. DiffEdit [3] employs text-conditioned diffusion models for the task of semantic image editing with mask guidance. To extend these techniques [18,27,28] to real images, several inversion methods have been proposed to invert an given image with a meaningful text prompt into the domain of the pretrained model. In this work, we propose training an adapter to modify specific components of the diffusion model for balanced style transformation and content preservation.

2.3 Text-Guided Portrait Stylizing

Many existing methods for text-guided portrait stylizing rely on pre-trained generative models to generate stylized portraits from textual descriptions. For example, StyleCLIP [21] modifies the latent code of a pre-trained StyleGAN [11] using CLIP [22], based on a text prompt. However, its image modification is limited to the pre-existing styles learned by the generator and does not extend to new stylization. StyleGAN-NADA [6] proposes an out-of-domain image manipulation method, which involves fine-tuning the StyleGAN model to shift it to new domains. DiffusionCLIP [13] fine-tunes a pre-trained human face diffusion model using identity and style losses. However, these methods require to fine-tune the entire pre-trained model, which can be computationally expensive and time-consuming. Another method, Asyrp [15], explores the potential semantic space of a frozen diffusion model to edit real pictures by training the potential space. However, this method has limitations in changing the overall style. In contrast, our method trains simple adapters for different styles and keeps the pre-trained model frozen, providing greater flexibility for generating diverse stylizations.

3 Method

3.1 Background and Preliminaries

Stable Diffusion. Our method is based on Stable Diffusion, a large-scale text-to-image model that operates in the latent space of an autoencoder instead of the image space. The autoencoder consists of an encoder $\mathcal{E}$, which maps a given image $x \in \mathcal{X}$ into a spatial latent code $z = \mathcal{E}(x)$, and a decoder $\mathcal{D}$, which decodes the latent code back into the image space. In the latent space, Stable Diffusion trains an improved UNet to implement the denoising process according to DDPM [9]. Stable Diffusion also uses the pre-trained CLIP [22] text encoder to embed text inputs as a condition during the denoising process.

DDIM Inversion. Inversion is the process of finding a noise map that can be fed into a diffusion process to produce a real image, essential for image editing with diffusion models. In our work, we adopt the deterministic DDIM [27] forward processes for inversion. Specifically, the inversion process is represented as:

$$z_{t+1} = \sqrt{\frac{\bar{\alpha}_{t+1}}{\bar{\alpha}_t}} z_t + \sqrt{\bar{\alpha}_{t+1}} \left(\sqrt{\frac{1}{\bar{\alpha}_{t+1}} - 1} - \sqrt{\frac{1}{\bar{\alpha}_t} - 1} \right) \cdot \varepsilon_\theta \left(z_t, t, c\left(y\right) \right) \quad (1)$$

$\epsilon_\theta \left(z_t, t, c\left(y\right) \right)$ is a UNet-based denoiser that predicts the added noise from latent code z_t at timestep t. $c(y)$ refers to the conditional embedding of a given text input y, which is generated using the CLIP text encoder c. $\bar{\alpha}_t$ is a noise scaling factor for timestep t. We can see that the diffusion process is performed in the reverse direction for T steps, that is $z_0 \rightarrow z_T$, where z_0 is the latent code of the given real image. Furthermore, we can perform DDIM denoising process to reconstruct the initial noise vector, z_0, from the final noise vector as z_T as follows:

$$z_{t-1} = \sqrt{\frac{\bar{\alpha}_{t-1}}{\bar{\alpha}_t}} z_t + \sqrt{\bar{\alpha}_{t-1}} \left(\sqrt{\frac{1}{\bar{\alpha}_{t-1}} - 1} - \sqrt{\frac{1}{\bar{\alpha}_t} - 1} \right) \cdot \varepsilon_\theta \left(z_t, t, c\left(y\right) \right) \quad (2)$$

3.2 Overview of Our Method

We propose a framework for stylizing face portraits using text prompts as guidance, as illustrated in Fig. 2. In Sect. 3.3, we introduce a face stylization method that leverages text prompts to guide the generation process. We then describe an adapter in Sect. 3.4 that optimizes a subset of weights in the UNet model to preserve the content information of the original portrait. In Sect. 3.5, we provide a detailed description of the training objective for the adapter. Our framework enables the creation of high-quality stylized face portraits that capture the desired style while preserving the crucial content information of the source portrait.

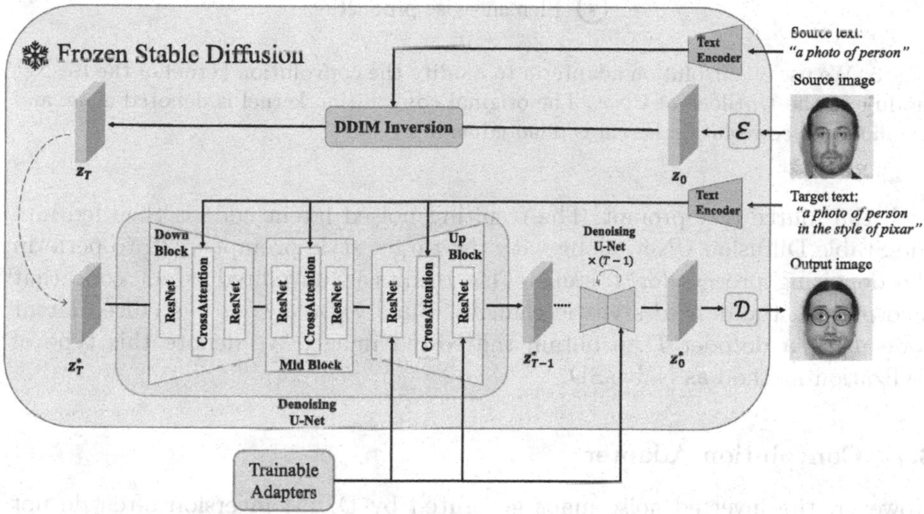

Fig. 2. Overview of our method. We first input the source text and a source image into Stable Diffusion to obtain a latent code z_T through DDIM inversion. We then modify the convolution kernels of the ResNet block in UpBlock using an adapter during DDIM sampling with the target text. This adapter balances style transformation and content preservation, generating a new latent code z_t^* for each step. We set $z_T^* = z_T$ at the beginning of DDIM sampling.

3.3 Portrait Stylization with Text Prompt

To stylize a portrait with a text prompt, our approach follows a sequence of steps that leverage Stable Diffusion. Specifically, we use two types of prompts: source text prompts y^{src} to describe the source image, and target style prompts y^{edit} to dictate the desired stylistic changes for the output image.

First, we use the deterministic DDIM inversion process Eq. 1 to gradually obtain a noised latent code after T steps. This process starts with an initial latent code z_0 obtained from the source image x^{src} using an encoder $\mathcal{E}$, along

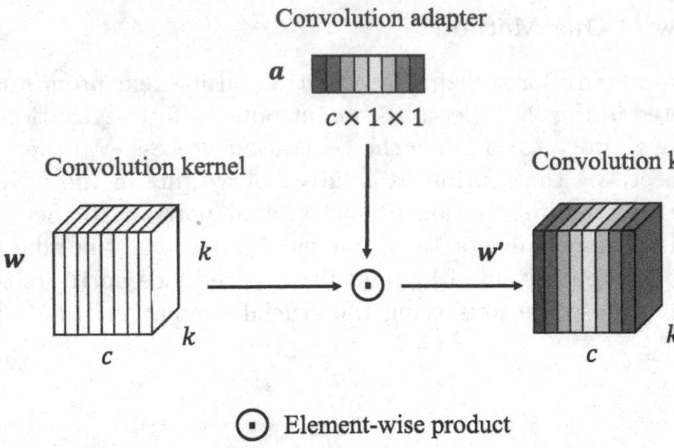

Fig. 3. We use a convolution adapter a to modify the convolution kernel of the ResNet module in the UpBlock of UNet. The original convolution kernel is denoted as w, and the modified convolution kernel is denoted as w'.

with the source text prompt. The resulting noised latent code is then fed into the Stable Diffusion UNet, along with the target style prompt y^{edit}, to perform the denoising process for T steps. This results in an edited latent code that incorporates the desired stylistic changes. Finally, we decode the edited latent code using a decoder $\mathcal{D}$ to obtain the edited image. We denote this type of stylization method as Naive-SD.

3.4 Convolution Adapter

However, the inverted noise maps generated by DDIM inversion often do not follow the statistical properties of uncorrelated, Gaussian white noise, resulting in poor editability. In such cases, directly using an edited text prompt can alter the content of the source image, leading to unnatural and distorted results. To address this issue, we propose an approach that optimizes a part of the weights of the UNet in the diffusion model. Specifically, we focus on adjusting only the convolution weights, which have been shown to be effective for stylizing [1,10,12,16]. To accomplish this, we propose a convolution adapter (Fig. 3) that modifies the convolution kernels of the ResNet module in the UpBlock of UNet. The formula for this process is as follows:

$$w_i' = a_i \odot w_i \tag{3}$$

where $a_i \in \mathbb{R}^{c\times 1\times 1}$ represents the convolution adapter, $w_i \in \mathbb{R}^{c\times k\times k}$ represents the weights of the convolution kernel, i denotes the ith ResNet convolution layer, k is the size of the convolution kernel, and c is the number of the channels.

As can be seen, our method introduces only a small number of additional parameters per edit and does not require any modification of the pre-trained weights. This reduces the storage requirements for the edited model, as it only needs to store the adapter.

3.5 Adapter Optimization

To optimize our convolution adapter, we use a combination of two loss functions: a perceptual loss and a CLIP direction loss.

Perceptual Loss. LPIPS (Learned Perceptual Image Patch Similarity) [30] is a technique for computing the perceptual similarity between two images. It uses a pretrained deep neural network to extract features from patches of images, and then computes the distance between the feature representations of two images to measure their perceptual similarity. We use LPIPS to measure the perceptual similarity between the source image x_t^{src} and the output image x_t^{edit} at each diffusion step t to evaluate how well the edited image preserves the perceptual characteristics of the source image during the diffusion process. However, since LPIPS is designed for clean images, we use the denoised $\hat{x}_0$ instead of x_t for each timestep t. The $\hat{x}_0$ is computed by z_t^* as follows:

$$\hat{z}_0 = \frac{z_t^* - \sqrt{1-\bar{\alpha}_t}\epsilon_\theta\left(z_t^*, t, c\left(y\right)\right)}{\sqrt{\bar{\alpha}_t}} \tag{4}$$

$$\hat{x}_0 = \mathcal{D}\left(\hat{z}_0\right) \tag{5}$$

For convenience, we represent perceptual loss as $\mathcal{L}_{\text{LPIPS}}\left(\hat{x}_0^{src}, \hat{x}_0^{edit}\right)$.

CLIP Direction Loss. CLIP learns multi-modal embeddings using an image encoder E_I and a text encoder E_T. The similarity between the embeddings reflects the semantic similarity between images and texts. To generate an image that matches a given text prompt, our method aligns image and text embeddings in the CLIP space using the CLIP direction loss. This loss function achieves homogeneous editing without mode collapse, commonly occurring when minimizing the cosine distance between the output image and the target text description. We use the CLIP directional loss to preserve the style information of the edited image for step t. Since the off-the-shelf CLIP models are trained on clean images, we also use the estimated clean image $\hat{x}_0^{src}$ and $\hat{x}_0^{edit}$ to help extract the image features.

$$\mathcal{L}_{\text{CLIP}}\left(\hat{x}_0^{\text{edit}}, y^{\text{edit}}; \hat{x}_0^{\text{src}}, y^{\text{src}}\right) = 1 - \frac{\Delta I \cdot \Delta T}{\|\Delta I\|\|\Delta T\|} \tag{6}$$

where $\Delta T = E_T\left(y^{\text{edit}}\right) - E_T\left(y^{\text{src}}\right)$ and $\Delta I = E_I\left(\hat{x}_0^{\text{edit}}\right) - E_I\left(\hat{x}_0^{\text{src}}\right)$

Total Loss. The overall optimization can be formulated by combining the loss components as

$$a^* = \arg\min_a \lambda_{\text{LPIPS}}\mathcal{L}_{\text{LPIPS}} + \lambda_{\text{CLIP}}\mathcal{L}_{\text{CLIP}} \tag{7}$$

where a is the parameters of the adapter, λ_{LPIPS} and λ_{CLIP} are the hyperparameters.

4 Experiments

4.1 Implementation Settings

We utilize the publicly available Stable Diffusion version 1.4 as the backbone text-to-image model. For the DDIM inversion process, we set the number of steps to $T = 50$ and the text guide scale to $w = 1$. For the DDIM sampling process, we find that setting the number of steps to $T = 50$ and the text guide scale to $w = 2$ achieves the best stylization performance. In addition, we set the hyperparameters for the training loss to $\lambda_{\text{LPIPS}} = 1$ and $\lambda_{\text{CLIP}} = 1$. During the inversion process, we use the prompt 'a photo of a person'. During the sampling process, we use the edited prompt 'a photo of a person in the style of < style>'. Our adapters are initialized with a value of 1, and we use the Adam solver is used for optimization with a constant learning rate of $1e^{-5}$. Our optimization process typically requires 7–9 iterations on a single NVIDIA A100 GPU.

4.2 Comparisons

Qualitative Comparison. We compare our portrait stylization method with recently proposed CLIP-based methods, including Diffusion-CLIP [13], StyleGAN-NADA [6], IPL [7], InstructPix2Pix [2] and Naive-SD. The comparison results are presented in Fig. 4. Diffusion-CLIP produces images with poor quality and distorted textures. Moreover, it is unable to produce the desired target styles, such as Anime and Cubism style. In the case of StyleGAN-NADA, some stylized images contain unexpected background information (in the case of Cartoon style) and an overall blurry and indistinct appearance, lacking in sharpness and detail (in the case of Pixar style). Additionally, some of the images generated by IPL could not be successfully transformed into the desired styles, such as Anime and Cartoon. Instructpix2pix which is based on diffusion model could produces unnatural images. The stylization process using Naive-SD results in images whose contents are substantially different from that of the original face. In contrast, our method generates high-quality images with clear, fine details, while retaining the original content information and successfully transformed styles. Notably, our method is the only one capable of producing truly Cubism-style images, with three-dimensional fragment composition characteristics that are not reflected in the results produced by other methods. Overall, our method represents a significant improvement over existing methods and offers a promising approach to portrait stylization.

User Study. To further evaluate the effectiveness of our portrait stylization method, we conduct a user study comparing it with other methods. Users were asked to rate the stylized images for image quality, style consistency, and content preservation. In Table 1, our approach obtains the highest scores in terms of image quality and style consistency. Although Diffusion-CLIP shows the best content score, it shows the worst score in style consistency due to minimal changes in the output from the source image. These results provide further evidence of the superior performance of our method in portrait stylization.

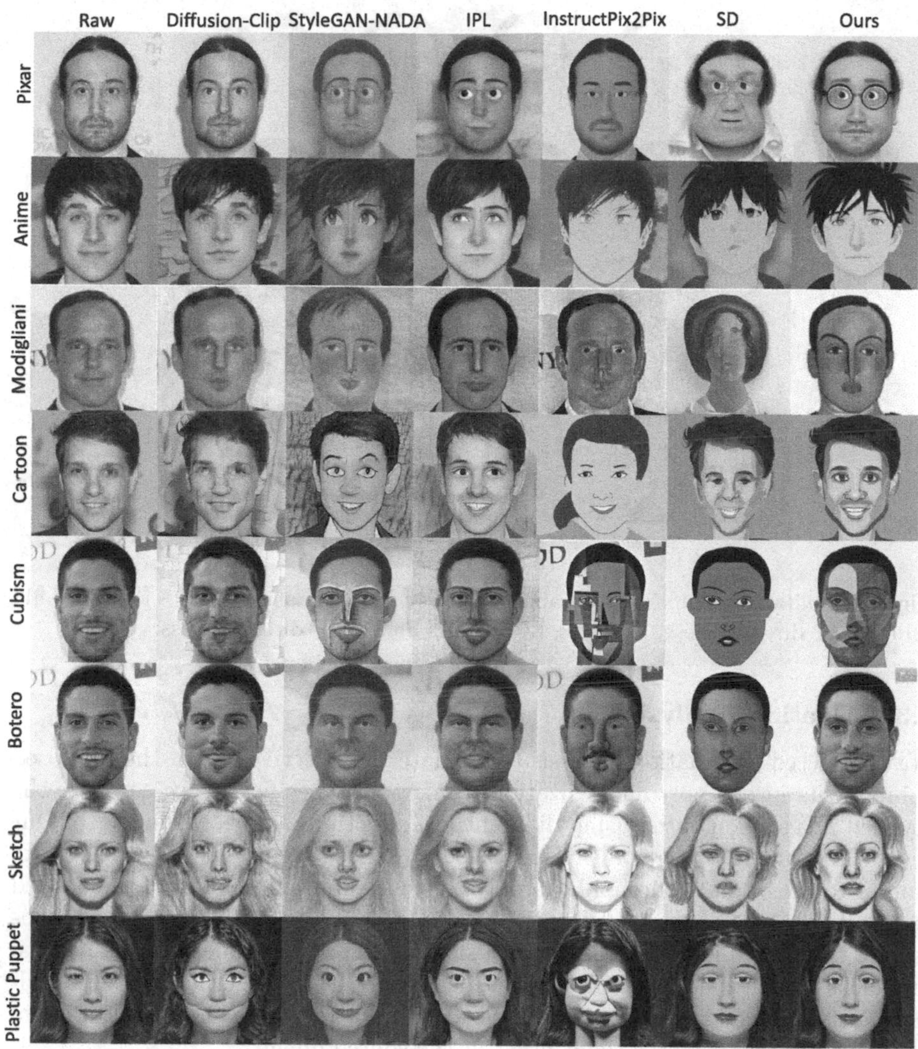

Fig. 4. Qualitative Comparison with Diffusion-CLIP [13], StyleGAN-NADA [6], IPL [7], InstructPix2Pix [2], and Naive-SD.

Table 1. User study results. We conduct evaluations based on image quality, style consistency, and content preservation.

Method	Quality↑	Style↑	Content↑
Diffusion-CLIP	2.86	2.22	**3.70**
StyleGAN-NADA	2.73	3.15	2.75
IPL	3.58	3.36	3.48
Naive-SD	1.90	2.51	1.72
Ours	**3.92**	**3.95**	3.62

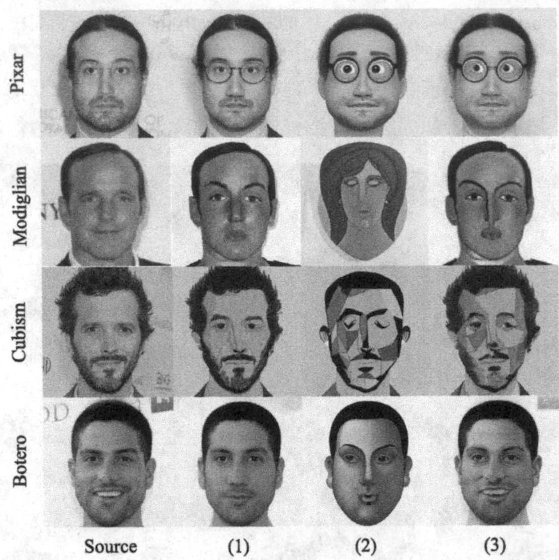

Fig. 5. Qualitative comparison on ablation study. (1) only using LPIPS loss, (2) only using CLIP direction loss, (3) using LPIPS loss and CLIP direction loss.

4.3 Ablation Analysis

We conducted an ablation analysis to validate the effectiveness of the proposed components in our portrait stylization framework. Specifically, we compare the stylized images generated using different loss functions, as shown in Fig. 5. Our results demonstrate that only using the $\mathcal{L}_{\text{LPIPS}}$ loss results in images that are similar to the original ones and can only capture a limited amount of the desired target style. On the other hand, only using the $\mathcal{L}_{\text{CLIP}}$ loss fails to capture the content of the source image, resulting in a stylized image that doesn't look natural and lacks overall coherence. In contrast, using both losses together produces stylized portraits that can capture the stylization information of the target domain, at the same time preserve the content of the source image.

5 Conclusion

In this paper, we propose a novel approach for face portrait stylization guided by text using a pre-trained diffusion model and a customized adapter. The adapter modifies specific components of the diffusion model to achieve both style transformation and content preservation. Experimental results demonstrate that our method outperforms existing techniques and has potential applications in various fields. Future work could explore its application beyond the domain of faces and design more complicated adapters for improved performance.

Acknowledgement. This work was supported by the National Natural Science Foundation of China under Grant 62206180, 82261138629; Guangdong Basic and Applied

Basic Research Foundation under Grant 2023A1515010688, 2020A1515111199 and 2022-A1515011018; and Shenzhen Municipal Science and Technology Innovation Council under Grant JCYJ20220531101412030, and Swift Fund Fintech Funding.

References

1. Alanov, A., Titov, V., Vetrov, D.: Hyperdomainnet: universal domain adaptation for generative adversarial networks. arXiv (2022). https://doi.org/10.48550/arxiv.2210.08884
2. Brooks, T., Holynski, A., Efros, A.A.: Instructpix2pix: learning to follow image editing instructions. In: Proceedings of the IEEE/CVF Conference on Computer Vision and Pattern Recognition, pp. 18392–18402 (2023)
3. Couairon, G., Verbeek, J., Schwenk, H., Cord, M.: Diffedit: diffusion-based semantic image editing with mask guidance. arXiv preprint arXiv:2210.11427 (2022)
4. Dhariwal, P., Nichol, A.: Diffusion models beat GANs on image synthesis. Adv. Neural. Inf. Process. Syst. **34**, 8780–8794 (2021)
5. Gal, R., et al.: An image is worth one word: personalizing text-to-image generation using textual inversion. arXiv preprint arXiv:2208.01618 (2022)
6. Gal, R., Patashnik, O., Maron, H., Chechik, G., Cohen-Or, D.: Stylegan-nada: clip-guided domain adaptation of image generators. arXiv (2021)
7. Guo, J., et al.: Zero-shot generative model adaptation via image-specific prompt learning. arXiv.org (2023). https://doi.org/10.48550/arxiv.2304.03119
8. Hertz, A., Mokady, R., Tenenbaum, J., Aberman, K., Pritch, Y., Cohen-Or, D.: Prompt-to-prompt image editing with cross attention control. arXiv preprint arXiv:2208.01626 (2022)
9. Ho, J., Jain, A., Abbeel, P.: Denoising diffusion probabilistic models. Adv. Neural. Inf. Process. Syst. **33**, 6840–6851 (2020)
10. Karras, T., Laine, S., Aila, T.: A style-based generator architecture for generative adversarial networks. In: 2019 IEEE/CVF Conference on Computer Vision and Pattern Recognition (CVPR). IEEE (2019). https://doi.org/10.1109/cvpr.2019.00453
11. Karras, T., Laine, S., Aittala, M., Hellsten, J., Lehtinen, J., Aila, T.: Analyzing and improving the image quality of stylegan. In: 2020 IEEE/CVF Conference on Computer Vision and Pattern Recognition (CVPR). IEEE (2020). https://doi.org/10.1109/cvpr42600.2020.00813
12. Karras, T., Laine, S., Aittala, M., Hellsten, J., Lehtinen, J., Aila, T.: Analyzing and improving the image quality of stylegan. In: Proceedings of the IEEE/CVF Conference on Computer Vision and Pattern Recognition, pp. 8110–8119 (2020)
13. Kim, G., Ye, J.C.: Diffusionclip: text-guided image manipulation using diffusion models (2021)
14. Kumari, N., Zhang, B., Zhang, R., Shechtman, E., Zhu, J.Y.: Multi-concept customization of text-to-image diffusion (2023)
15. Kwon, M., Jeong, J., Uh, Y.: Diffusion models already have a semantic latent space. arXiv preprint arXiv:2210.10960 (2022)
16. Laria, H., Wang, Y., van de Weijer, J., Raducanu, B.: Transferring unconditional to conditional GANs with hyper-modulation. In: Proceedings of the IEEE/CVF Conference on Computer Vision and Pattern Recognition, pp. 3840–3849 (2022)
17. Liu, Z., et al.: Cones: concept neurons in diffusion models for customized generation. arXiv preprint arXiv:2303.05125 (2023)

18. Mokady, R., Hertz, A., Aberman, K., Pritch, Y., Cohen-Or, D.: Null-text inversion for editing real images using guided diffusion models. In: Proceedings of the IEEE/CVF Conference on Computer Vision and Pattern Recognition, pp. 6038–6047 (2023)
19. Mou, C., et al.: T2i-adapter: learning adapters to dig out more controllable ability for text-to-image diffusion models. arXiv preprint arXiv:2302.08453 (2023)
20. Nichol, A., et al.: Glide: towards photorealistic image generation and editing with text-guided diffusion models. arXiv preprint arXiv:2112.10741 (2021)
21. Patashnik, O., Wu, Z., Shechtman, E., Cohen-Or, D., Lischinski, D.: Styleclip: text-driven manipulation of stylegan imagery. In: 2021 IEEE/CVF International Conference on Computer Vision (ICCV). IEEE (2021). https://doi.org/10.1109/iccv48922.2021.00209
22. Radford, A., et al.: Learning transferable visual models from natural language supervision. In: International Conference on Machine Learning, pp. 8748–8763. PMLR (2021)
23. Ramesh, A., Dhariwal, P., Nichol, A., Chu, C., Chen, M.: Hierarchical text-conditional image generation with clip latents. arXiv preprint arXiv:2204.06125 (2022)
24. Rombach, R., Blattmann, A., Lorenz, D., Esser, P., Ommer, B.: High-resolution image synthesis with latent diffusion models. In: Proceedings of the IEEE/CVF Conference on Computer Vision and Pattern Recognition, pp. 10684–10695 (2022)
25. Ruiz, N., Li, Y., Jampani, V., Pritch, Y., Rubinstein, M., Aberman, K.: Dreambooth: fine tuning text-to-image diffusion models for subject-driven generation. arXiv preprint arXiv:2208.12242 (2022)
26. Saharia, C., et al.: Photorealistic text-to-image diffusion models with deep language understanding. Adv. Neural. Inf. Process. Syst. **35**, 36479–36494 (2022)
27. Song, J., Meng, C., Ermon, S.: Denoising diffusion implicit models. In: International Conference on Learning Representations (2020)
28. Wallace, B., Gokul, A., Naik, N.: Edict: exact diffusion inversion via coupled transformations. In: Proceedings of the IEEE/CVF Conference on Computer Vision and Pattern Recognition, pp. 22532–22541 (2023)
29. Zhang, L., Agrawala, M.: Adding conditional control to text-to-image diffusion models. arXiv preprint arXiv:2302.05543 (2023)
30. Zhang, R., Isola, P., Efros, A.A., Shechtman, E., Wang, O.: The unreasonable effectiveness of deep features as a perceptual metric. In: Proceedings of the IEEE Conference on Computer Vision and Pattern Recognition, pp. 586–595 (2018)

EdgeFusion: Infrared and Visible Image Fusion Algorithm in Low Light

Zikun Song, Pinle Qin(✉), Jianchao Zeng, Shuangjiao Zhai, Rui Chai, and Junyi Yan

North University of China, Taiyuan, China
qpl@nuc.edu.cn

Abstract. Under low-light conditions, visible light imaging technology exhibits poor imaging performance, whereas infrared thermal imaging technology can effectively detect and identify targets. To solve the target imaging problem in low-light environments, multimodal image fusion technology can combine the advantages of both aforementioned methods. Existing fusion methods focus excessively on information from infrared images, obscuring the original texture details of the targets and resulting in low-quality images. Therefore, in this study, we propose a multiscale edge-fusion network for infrared and visible images called EdgeFusion, which can produce an edge-fusion image. Specifically, the network utilises infrared multiscale gradient information to enhance the edges of the thermal targets, thereby improving the ability to identify them. By designing a balanced loss, EdgeFusion suppresses the global information from infrared images that obscured the fine texture details of the original images. In addition, a residual gradient method is introduced to enhance the textural details of the generated images. After extensive experimentation on the public datasets LLVIP and TNO, the results indicate that EdgeFusion outperforms existing state-of-the-art methods in preserving fine-grained infrared edges and enhanced image texture details.

Keywords: Image fusion · Convolution network · Multiscale gradient retention module

1 Introduction

Image fusion techniques [1,2] integrate multimodal images into a single image, combining the advantages of different modalities. Under low-light conditions, visible images struggle to differentiate between background and targets, whereas infrared images lack texture information. Combining the two can generate images

Supported by the research project Study on the effectiveness of RF data and recognition models in wireless sensing' (No. 202203021222049) and Shanxi Province Major Scientific and Technological Project 'Revealing the List and Appointing the Leader' (No. 202101010101018).

© The Author(s), under exclusive license to Springer Nature Singapore Pte Ltd. 2024
Q. Liu et al. (Eds.): PRCV 2023, LNCS 14425, pp. 259–270, 2024.
https://doi.org/10.1007/978-981-99-8429-9_21

that highlight targets and preserve texture. These advantages enable a wide range of applications in military security [3], target detection [4], semantic segmentation [5], and other fields.

Over decades of development, various methods for infrared and visible image fusion have emerged. Among them, traditional methods [6–8] involve three steps: feature extraction, fusion, and inverse transformation. However, owing to coarse fusion rules such as the maximum and mean value strategies, these methods face challenges in adapting to feature specificity, leading to information distortion. In recent years, fusion methods based on deep learning have shown better results; they include autoencoders [4,9–12], CNN (convolutional neural networks) [13–15], and GAN (generative adversarial networks) [4,16–19]. Although current deep learning-based image fusion algorithms can generate images with complementary information, they face significant challenges in preserving edges and enhancing fine texture details. As shown in Fig. 1, Fig. 1(a) and Fig. 1(b) respectively represent the infrared image and visible image captured in low-light conditions, Fig. 1(c) represents the fusion results using existing methods, which focus on the overall infrared information of foreground thermal targets, leading to complete highlighting of the targets and obscuring certain color and texture details [20]. Figure 1(d) depicts the effect of edge-fusion image, which effectively distinguishes the targets from the background and enhances the person's color and texture details, and facilitates subsequent tasks.

(a) Infrared (b) Visible (c) Fusion image (d) Edge-fusion image

Fig. 1. Fusion results of infrared and visible images under low-light conditions.

To get the edge-fusion images, we propose a multiscale edge-fusion network EdgeFusion, for infrared and visible images. This method extracts the infrared thermal radiation gradient information and visible texture information for fusion. Specifically, a gradient operating module (GOM) has been introduced to enable feature reuse and enhance fine-grained details through residual gradient flow. This enables the extraction of rich texture information from visible images. Moreover, for the infrared image, a multiscale gradient retention module (MGRM) is designed to merge features from different scales of convolution and gradient operations, which allows the extraction of the gradient information from the infrared images. Additionally, to prevent the dominant infrared information from overshadowing the target texture, a balance loss is proposed to achieve convergence between the intensity and texture losses, allowing for the coexistence of intensity and texture information. This significantly preserved the rich information in the original image. In summary, this paper provides the following key contributions:

- EdgeFusion is designed for infrared and visible images. This network effectively combines fine texture information from visible images with edge gradient information from infrared images, producing fusion images that are more visually appealing in low-light environments.
- An MGRM is proposed to extract fine-grained infrared edge information and accurately distinguish targets from the background. The design of the balance loss helps preserve the internal texture details of the visible image while reducing the interference from infrared information and increasing the amount of preserved information.
- Extensive experiments were conducted on the LLVIP and TNO datasets, and the outcomes were qualitatively and quantitatively compared with that of the state-of-the-art methods. The experimental results verify the effectiveness of EdgeFusion, and the experimental results for target detection demonstrate the benefits of this method for downstream tasks.

2 Related Work

2.1 AE-Based Image Fusion Methods

Methods based on autoencoder networks have strong interpretability, and their feature fusion processes rely mainly on manually designed fusion rules. Li et al. proposed DenseFuse [10], which is composed of three layers: encoding, fusion, and decoding. Owing to the limited feature extraction capability of autoencoders, Li et al. proposed NestFuse [4], which introduces a nested connection network and a spatial attention model. They also proposed spatial and channel attention as fusion strategies. RFN-Nest [11] employs a novel loss function for detail preservation and a loss function for feature enhancement, integrating more information into the network. DRF [12], proposed by Xu et al., employs various fusion strategies to decompose the original image into representations associated with scene and sensor modalities.

2.2 CNN-Based Image Fusion Methods

Methods based on CNNs use specially designed network architectures and loss functions for feature extraction, fusion, and image reconstruction. DeepFuse [13] exhibits robustness to varying input conditions and can handle extreme exposure without artefacts, thereby addressing the limitations of multiexposure image fusion. Zhang et al. proposed the IFCNN [14], which is a generic image fusion framework that utilises a linear element fusion rule to fuse the convolutional features of multiple input images and designs a universal loss function for various fusion tasks. Xu et al. have designed a unified multitask fusion framework [21] that can perform different types of fusion tasks. STDFusionNet [15] leverages the salient characteristics of thermal targets in infrared images to guide fusion using saliency masks.

2.3 GAN-Based Image Fusion Methods

GANs are unsupervised learning frameworks highly suitable for image fusion. FusionGAN, proposed by Ma et al. [4], was the first model to introduce GANs into the field of image fusion. However, owing to the issue of mode collapse, Ma et al. further proposed DDcGAN [17], which utilises a dual-discriminator GAN to improve the robustness and ensure that the fused image retains the characteristics of both source images simultaneously. Additionally, Ma et al. introduced a multiclassification-constrained GAN [18]. Li et al. designed a GAN that incorporates a multiscale attention mechanism, enabling the network to focus more on the key information of both infrared and visible images [19].

However, current existing fusion methods of various kinds emphasise the usefulness of global infrared information while neglecting situations in which the texture information of the fused image is obscured. Therefore, it is necessary to design a method that preserves the internal details and highlights the edge information of thermal targets. This study innovatively extracts edge information from an infrared image and fuses it with texture information from a visible image, resulting in fused images that contain the infrared edges of thermal targets while preserving the details and textures of the image.

3 Design of EdgeFusion

3.1 Problem Formulation

In low-light environments, distinguishing between the targets and backgrounds in visible images is extremely challenging. To address this issue, an MGRM is specifically designed for infrared images to extract multiscale edges. Consider a pair of visible and infrared images denoted as I_{vi} and I_{ir}, respectively. The visible image is converted into the YCrCb colour space and only the Y channel is extracted and denoted as I_{vi}^{Y}. The GOM [22] feature extraction module, denoted as E_G, was used to extract texture features from the visible image. The MGRM is denoted as E_V. Thus, it can be expressed as:

$$F_{vi} = E_G\left(I_{vi}^{Y}\right), F_{ir} = E_V\left(I_{ir}\right) \tag{1}$$

where F_{vi} represents the extracted visible image feature information from the Y channel of the visible image and F_{ir} represents the extracted infrared image feature information from the infrared image.

The MGRM was used to extract edge features from the infrared image. With I_{ir} as the input infrared image, F_m^n as the single-layer output feature, and F_{ir} as the module output feature, the MGRM can be represented as:

$$F_m^n = \nabla\left(Conv_{n*n}\left(I_{ir}\right)\right), (n = 1, 3, 5) \tag{2}$$

$$F_{ir} = C\left(F_m^1, F_m^3, F_m^5\right) \tag{3}$$

The symbol ∇ represents the Sobel gradient operator, $Conv_{n*n}$ denotes the convolution operation, where n $\times$ n denotes the size of the convolutional kernel.

Function C(·) represents the concatenation along the channel dimension. The MGRM obtains rich edge texture information through multiscale convolution and gradient computations.

The input feature of the GOM was obtained by passing the visible Y-channel image through a convolutional layer and can be represented as:

$$F_{in\_vi} = Conv\left(I_{vi}^{Y}\right) \tag{4}$$

Taking F_{in_vi} as the input feature of GOM and F_{vi} as the output feature, the formula of GOM can be expressed as:

$$F_{vi} = Conv^{n}\left(F_{in\_vi}\right) \oplus Conv\left(\nabla F_{in\_vi}\right) \tag{5}$$

where $Conv(\cdot)$ represents a convolutional layer, $Conv^{n}(\cdot)$ represents n cascaded convolutional layers, and $\oplus$ denotes element-wise addition in features. Finally, image reconstruction was used to reconstruct the fused image, denoted as I_f, where F_f represents the fusion features. R(·) represents image reconstruction. This can be represented as:

$$I_f = R\left(F_f\right) \tag{6}$$

3.2 Network Architecture

The proposed network architecture for the EdgeFusion is shown in Fig. 2. In this architecture, two GOM modules are used to extract features from the visible image, and an MGRM is used to extract edge features from the infrared image. The visible image is first converted from the RGB colour space to the YCrCb colour space. The Y channel is then extracted and passed through a 3×3 convolutional layer using the activation function of LReLU. Subsequently, two GOM modules are employed to extract the texture and detailed features from the visible image.

The MGRM was used to extract edge features from an infrared image, and its detailed structure is shown in Fig. 3(a). This module first performs convolution operations using kernels of different scales to extract features from an infrared image. Specifically, 1×1, 3×3, and 5×5 convolutional kernels were applied to the infrared image, and the LReLU activation function was used. After multiscale convolution, Sobel gradient operations were performed on the infrared features to extract edge features at different scales. The obtained multiscale features were then concatenated along the channel dimensions. Subsequently, two 3×3 convolutional layers were applied to extract deep-level features. Finally, a 1×1 convolutional layer was used to eliminate the difference in channel dimensions.

The GOM was used to extract texture details from the visible image, as shown in the Fig. 3(b). It comprises two 3×3 LReLU convolutional layers and a 1×1 shared convolutional layer, utilising dense connections to fully leverage the features extracted by each convolutional layer. The residual flow undergoes gradient computation, followed by a 1×1 convolution and addition to the main-stream flow, achieving the fusion of deep and fine-grained features.

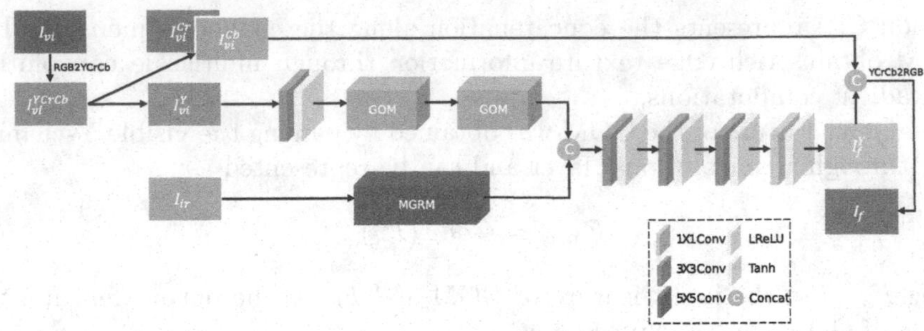

Fig. 2. EdgeFusion network structure diagram.

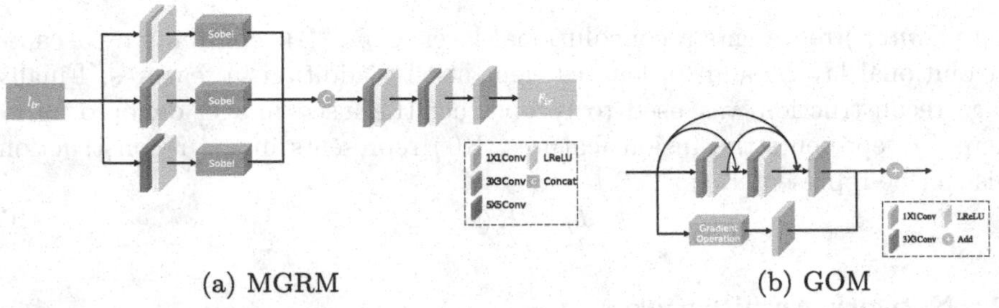

(a) MGRM (b) GOM

Fig. 3. Specific structure of MGRM and GOM modules.

Channel-wise concatenation was performed to combine the visible and infrared features, and the resulting feature maps were fed into the image reconstructor for image reconstruction. The decoder consists of three 3×3 convolutional layers and one 1×1 convolutional layer. The 3×3 convolutional layer used LReLU as the activation function, whereas the 1×1 convolutional layer used tanh as the activation function. No downsampling was introduced into the fusion network, and the feature maps during the fusion process were maintained consistent with the size of the source images, thus avoiding information loss.

3.3 Loss Function

To incorporate more essential and useful information into the fused image, detail loss was introduced in this study. The detail loss comprises three components: intensity loss $\mathcal{L}_{int}$ [22], texture loss $\mathcal{L}_{texture}$ [22], and balance loss $\mathcal{L}_{balance}$. The detailed loss is defined as follows:

$$\mathcal{L}_{detail} = \mathcal{L}_{int} + \alpha\mathcal{L}_{texture} + \beta\mathcal{L}_{balance} \tag{7}$$

where $\mathcal{L}_{int}$ is used to constrain the overall intensity performance of the fused image, $\mathcal{L}_{texture}$ is used to encourage the fused image to retain more intricate texture details, and $\mathcal{L}_{balance}$ is used to enforce a balanced proportion between the intensity and texture losses. The parameters α and β are used to adjust the weights of the texture loss and the balance loss, respectively. The intensity loss is defined as follows:

$$\mathcal{L}_{int} = \frac{1}{HW} \left\| I_f - max\left(I_{vi}, I_{ir}\right) \right\|_1 \tag{8}$$

In the formula, H and W represent the height and width of the image, $\|\cdot\|_1$ denotes the l_1 norm, and $max(\cdot)$ represents the selection of the maximum value among the elements. Here, it represents the fusion of salient pixels from the visible and infrared images. The texture loss is defined as follows:

$$\mathcal{L}_{texture} = \frac{1}{HW} \left\| |\nabla I_f| - max\left(|\nabla I_{vi}|, |\nabla I_{ir}|\right) \right\|_1 \tag{9}$$

The symbol ∇ represents the Sobel gradient operation, and $|\cdot|$ denotes the absolute value operation. This formulation indicates that the texture of the fused image tends to be the maximum union of textures from the visible and infrared images. The balance loss is defined as follows:

$$\mathcal{L}_{balance} = \left(1 - \left(\mathcal{L}_{int}/\gamma\mathcal{L}_{texture}\right)\right)^2 \tag{10}$$

The balance loss is defined as follows: The parameter γ is used to constrain the proportion between the intensity loss and texture loss. The intensity loss significantly affects the fusion result, and the proportion between it and the texture loss affects the visual performance of the fusion result. The balance loss effectively constrains the overall intensity of the fused image, allowing the coexistence of texture and edge information.

4 Experimental Validation

4.1 Experimental Configurations

This paper details the extensive experiments conducted on the LLVIP [23] and TNO [24] datasets to comprehensively evaluate the proposed method. A large number of images in the LLVIP dataset were taken on night roads and included 12,025 training pairs and 3,463 test pairs. The proposed method is trained using the LLVIP dataset. The TNO dataset contains grayscale versions of multispectral nighttime images of various military-related scenes. EdgeFusion was evaluated in comparison with four state-of-the-art deep learning methods, namely FusionGAN, GANMcC, U2Fusion [21], and SeAFusion. The implementations of the four compared methods were configured according to publicly available codes and the original paper parameters.

The training parameters were set as follows: batch size of 8 for the Adam optimiser, initial learning rate of 10^{-3}, learning rate updated by multiplying the initial learning rate by $power^{(iter-1)}$, where the initial power was set to 0.75, and iter represents the current training iteration, with a total of 10 iterations. All the experiments were conducted on an NVIDIA Station server with an Intel Xeon E5 2620 v4 processor, 128 GB of memory, and four Tesla V100 GPUs, each with 32 GB of VRAM.

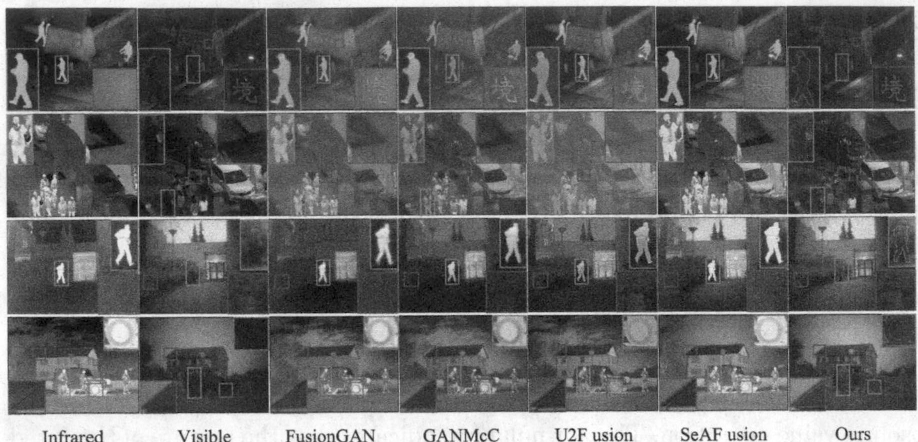

Fig. 4. Qualitative comparison of the EdgeFusion method in this paper on the LLVIP dataset (first and second rows) versus the TNO dataset (third and fourth rows).

4.2 Fusion Metrics

The selected metrics include peak signal-to-noise ratio (PSNR), average gradient (AG), $\mathbf{Q^{AB/F}}$, standard deviation (SD), and mutual information (MI). PSNR reflects the distortion during the fusion process; AG quantifies the gradient information of the fused image; $\mathbf{Q^{AB/F}}$ measures the amount of edge information from source to fused image; SD reflects the distribution and contrast of the fused image; MI measures the amount of information transmitted from source to fused image. Fusion algorithms with higher PSNR, AG, $\mathbf{Q^{AB/F}}$, SD, and MI values tend to exhibit better fusion performance.

4.3 Comparative Experiment

Qualitative Results. Two pairs of images from the LLVIP and TNO datasets were selected for fusion. Images #010011 and #190071 were chosen for the LLVIP dataset, whereas images Kapetein-1123 and Soldiers_with_jeep were selected for the TNO dataset. The results are shown in Fig. 4. The information provided by visible and infrared images in low light level is limited and the texture is weakened. However, except for our proposed method, other methods introduce global information from infrared images, which inevitably leads to coverage of the target and background details in the fused images. Examining the details within the colored box, it becomes apparent that the image generated by the edge fusion method exhibits favorable visual results. Targets and backgrounds can be distinctly differentiated, and the image overall retains a wealth of texture details, with all edges appearing sharp and clear.

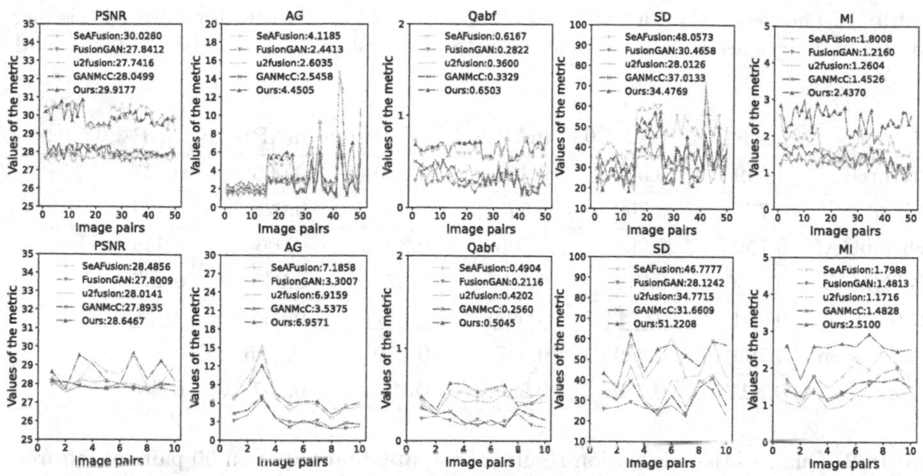

Fig. 5. A quantitative comparison of five metrics, namely PSNR, AG, Qabf, SD, and MI, was conducted on 50 images from the LLVIP dataset (row one) and 10 images from the TNO dataset (row two). The horizontal axis represents the image pairs, while the vertical axis represents the values of each fusion metric.

Quantitative Results. In this study, we conducted a quantitative comparison of five evaluation metrics on 50 image pairs from four different scenarios in the LLVIP dataset. The first row of Fig. 5 shows the results of all the methods. The method proposed in this study demonstrates significant advantages in three metrics: AG, Qabf, and MI. AG indicates that the fusion results of this method contain rich details and textures. Qabf suggests that the fusion images generated using this method contain considerable edge information. MI indicates that the method preserves most information from the source images. This is attributed to the strong edge preservation capability of the multiscale gradient preservation module. Because this method primarily selects the edge information from infrared images for fusion, it avoids the issue of completely highlighting the objects in the fused image. Therefore, it is understandable that the SD metric ranks third. In this study, we also conducted a quantitative evaluation of five metrics on a subset of 10 image pairs from the TNO dataset. The results are shown in the second row of Fig. 5. The various experimental results on different datasets indicate that EdgeFusion has a significant advantage in preserving infrared edges and enhancing texture details.

4.4 Detection Performance

This paper uses the LLVIP dataset for training with Yolov5, which has its own pedestrian detection label. The detection results for the LLVIP dataset are listed in Table 1. The average precision (mAP) metric was used to assess the detection performance, where a higher mAP value closer to 1 indicated a better pedestrian detection performance. The mAP_{iou} scores at different IoU thresholds. Additionally, mAP@[0.5:0.95] represents the average mAP across thresholds ranging from

Table 1. The pedestrian detection results on the LLVIP dataset for infrared, visible, and fused images are as follows (bold indicates the best, underline indicates the second best).

	mAP@0.5	mAP@0.6	mAP@0.7	mAP@0.8	mAP@0.9	mAP@[0.5:0.95]
Infrared	0.649	0.578	0.453	0.279	0.063	0.367
Visible	0.775	0.700	0.545	0.308	0.028	0.429
FusionGAN	0.756	0.698	0.606	0.321	<u>0.064</u>	0.449
GANMcC	<u>0.786</u>	0.717	**0.617**	0.370	0.046	0.467
U2Fusion	0.784	<u>0.743</u>	<u>0.609</u>	<u>0.433</u>	0.046	<u>0.485</u>
SeAFusion	0.769	0.729	0.597	0.379	**0.095**	0.475
Ours	**0.817**	**0.746**	0.586	**0.455**	0.057	**0.487**

Table 2. Quantitative evaluation results of the ablation study on 50 pairs of infrared and visible images from the LLVIP dataset (bold indicates the best, underline indicates the second best).

	without $\mathcal{L}_{balance}$	without MGRM	Ours
PSNR	**28.6715**	28.3948	<u>28.6467</u>
AG	<u>6.7383</u>	6.6774	**6.9571**
Qabf	**0.5167**	0.4443	<u>0.5045</u>
SD	44.9629	<u>49.9748</u>	**51.2208**
MI	1.8325	<u>1.9944</u>	**2.5100**

0.5 to 0.95 with a step size of 0.05. From the comprehensive point of view in the table, the method proposed in this paper has certain advantages in mAP, indicating that the method has a promoting effect on downstream tasks.

4.5 Ablation Studies

Important contributions in this paper are MGRM and balance loss. We conducted a series of ablation experiments on LLVIP and TNO datasets, and the

(a) Ours (b) Without $\mathcal{L}_{balance}$ (c) Without MGRM

Fig. 6. Visualization results of ablation experiments.

results are shown in Table 2 and Fig. 6. As can be seen from Fig. 6(c), balance loss significantly inhibits the intensity of infrared information. Without MGRM, the overall image becomes darker and edge information is missing. In Fig. 6(b), the absence of balance loss makes the fused image indistishable from the infrared image. In contrast, only EdgeFusion's fused images have both the bright edges of the thermal target and the enhanced texture details.

5 Conclusion

In this study, EdgeFusion was proposed for infrared and visible images. The network innovatively utilises infrared gradient information to annotate thermal targets, effectively avoiding the problem of infrared targets becoming completely highlighted. This was achieved through the gradient preservation capability of the infrared MGRM. Additionally, a balanced loss function was designed to introduce infrared image information while avoiding excessive interference from infrared information. This allowed the fused image to retain additional textural details. Experimental results on a public dataset demonstrate that this approach is beneficial for enhancing the visibility of textures in visible images, and the fused image retains clear and rich texture details. Quantitative experiments comparing four state-of-the-art methods using five evaluation metrics further confirm that the proposed approach preserves more gradients and textures, resulting in fused images with the highest information content. The results of the object detection experiments further validated the superior performance of EdgeFusion in advanced visual tasks, such as pedestrian detection.

References

1. Dogra, A., Goyal, B., Agrawal, S.: From multi-scale decomposition to non-multi-scale decomposition methods: a comprehensive survey of image fusion techniques and its applications. IEEE Access **5**, 16040–16067 (2017)
2. Ma, Y., Chen, J., Chen, C., Fan, F., Ma, J.: Infrared and visible image fusion using total variation model. Neurocomputing **202**, 12–19 (2016)
3. Das, S., Zhang, Y.: Color night vision for navigation and surveillance. Transp. Res. Rec. **1708**(1), 40–46 (2000)
4. Li, H., Wu, X.J., Durrani, T.: NestFuse: an infrared and visible image fusion architecture based on nest connection and spatial/channel attention models. IEEE Trans. Instrum. Meas. **69**(12), 9645–9656 (2020)
5. Ha, Q., Watanabe, K., Karasawa, T., Ushiku, Y., Harada, T.: MFNet: towards real-time semantic segmentation for autonomous vehicles with multi-spectral scenes. In: 2017 IEEE/RSJ International Conference on Intelligent Robots and Systems (IROS), pp. 5108–5115. IEEE (2017)
6. Zhou, Z., Wang, B., Li, S., Dong, M.: Perceptual fusion of infrared and visible images through a hybrid multi-scale decomposition with gaussian and bilateral filters. Inf. Fusion **30**, 15–26 (2016)
7. Li, H., Qi, X., Xie, W.: Fast infrared and visible image fusion with structural decomposition. Knowl.-Based Syst. **204**, 106182 (2020)

8. Ma, J., Zhou, Y.: Infrared and visible image fusion via gradientlet filter. Comput. Vis. Image Underst. **197**, 103016 (2020)
9. Long, Y., Jia, H., Zhong, Y., Jiang, Y., Jia, Y.: RXDNFuse: a aggregated residual dense network for infrared and visible image fusion. Inf. Fusion **69**, 128–141 (2021)
10. Li, H., Wu, X.j., Durrani, T.S.: Infrared and visible image fusion with resnet and zero-phase component analysis. Infrared Phys. Technol. **102**, 103039 (2019)
11. Li, H., Wu, X.J., Kittler, J.: RFN-Nest: an end-to-end residual fusion network for infrared and visible images. Inf. Fusion **73**, 72–86 (2021)
12. Xu, H., Wang, X., Ma, J.: DRF: disentangled representation for visible and infrared image fusion. IEEE Trans. Instrum. Meas. **70**, 1–13 (2021)
13. Ram Prabhakar, K., Sai Srikar, V., Venkatesh Babu, R.: DeepFuse: a deep unsupervised approach for exposure fusion with extreme exposure image pairs. In: Proceedings of the IEEE International Conference on Computer Vision, pp. 4714–4722 (2017)
14. Zhang, Y., Liu, Y., Sun, P., Yan, H., Zhao, X., Zhang, L.: IFCNN: a general image fusion framework based on convolutional neural network. Inf. Fusion **54**, 99–118 (2020)
15. Ma, J., Tang, L., Xu, M., Zhang, H., Xiao, G.: STDFusionNet: an infrared and visible image fusion network based on salient target detection. IEEE Trans. Instrum. Meas. **70**, 1–13 (2021)
16. Ma, J., Yu, W., Liang, P., Li, C., Jiang, J.: FusionGAN: a generative adversarial network for infrared and visible image fusion. Inf. Fusion **48**, 11–26 (2019)
17. Ma, J., Xu, H., Jiang, J., Mei, X., Zhang, X.P.: DDcGAN: a dual-discriminator conditional generative adversarial network for multi-resolution image fusion. IEEE Trans. Image Process. **29**, 4980–4995 (2020)
18. Ma, J., Zhang, H., Shao, Z., Liang, P., Xu, H.: GANMcC: a generative adversarial network with multiclassification constraints for infrared and visible image fusion. IEEE Trans. Instrum. Meas. **70**, 1–14 (2020)
19. Liu, D., Wen, B., Liu, X., Wang, Z., Huang, T.S.: When image denoising meets high-level vision tasks: a deep learning approach. arXiv preprint arXiv:1706.04284 (2017)
20. Ma, J., Ma, Y., Li, C.: Infrared and visible image fusion methods and applications: a survey. Inf. Fusion **45**, 153–178 (2019)
21. Xu, H., Ma, J., Jiang, J., Guo, X., Ling, H.: U2Fusion: a unified unsupervised image fusion network. IEEE Trans. Pattern Anal. Mach. Intell. **44**(1), 502–518 (2020)
22. Tang, L., Yuan, J., Ma, J.: Image fusion in the loop of high-level vision tasks: a semantic-aware real-time infrared and visible image fusion network. Inf. Fusion **82**, 28–42 (2022)
23. Jia, X., Zhu, C., Li, M., Tang, W., Zhou, W.: LLVIP: a visible-infrared paired dataset for low-light vision. In: Proceedings of the IEEE/CVF International Conference on Computer Vision, pp. 3496–3504 (2021)
24. Toet, A.: TNO image fusion dataset (2021). https://figshare.com/articles/dataset/TNO_Image_Fusion_Dataset/1008029

An Efficient Momentum Framework for Face-Voice Association Learning

Yuanyuan Qiu[1,2], Zhenning Yu[1], and Zhenguo Gao[1,2](✉)

[1] Department of Computer Science and Technology, Huaqiao University, Xiamen 361021, Fujian, China
{yyq,zny,zgh}@hqu.edu.cn

[2] Key Laboratory of Computer Vision and Machine Learning of Fujian Province University, Xiamen 361021, Fujian, China

Abstract. Cross-modal face-voice association is an active field that utilizes biometric features for cross-modal information retrieval. The primary approach for addressing this task involves utilizing contrastive learning to construct a modality-agnostic subspace. However, many existing contrastive learning methods in cross-modal research tend to neglect the significance of symmetrical information within heterogeneous data. This oversight leads to the generation of different negative examples for each identity in a random mini-batch. Furthermore, the length of negative examples in contrastive learning is coupled with the mini-batch size and is limited by the GPU memory size. To address these issues, this paper introduces an innovative Cross-Modal Momentum Contrast (CMMC) algorithm, which leverages queues to provide sufficient and symmetric information. Moreover, we propose an update strategy to maintain the consistency of negative example information throughout the training process. By combining the operations mentioned above, our proposed CMMC can effectively improve the correlation between face and voice data. Extensive experiments conducted on two datasets confirm the superiority of our framework and demonstrate its competitive performance compared to state-of-the-art methods.

Keywords: Cross-modal · Face-voice association · Momentum contrastive learning

1 Introduction

Many studies in the fields of cognitive science and artificial intelligence have illustrated a correlation between face and voice. Emerging evidence from the literature [1–3] suggests that humans are more sensitive to perceiving faces when presented with voice data, and vice versa. In recent years, face and voice have become the most prominent biometric features, greatly aiding in identity recognition, retrieval, and authentication. Simultaneously, the ability to automatically correlate face and voice presents exciting research opportunities in human-machine interaction systems. This area has given rise to numerous valuable applications such as virtual reality, criminal investigation, multimodal information

© The Author(s), under exclusive license to Springer Nature Singapore Pte Ltd. 2024
Q. Liu et al. (Eds.): PRCV 2023, LNCS 14425, pp. 271–283, 2024.
https://doi.org/10.1007/978-981-99-8429-9_22

retrieval, etc. [4,5]. Unlike traditional identity recognition tasks, cross-modal face and voice association learning aims to establish semantic correspondence and bridge the gap between different modalities, posing a significant challenge in learning modal-invariant features for data originating from various modalities.

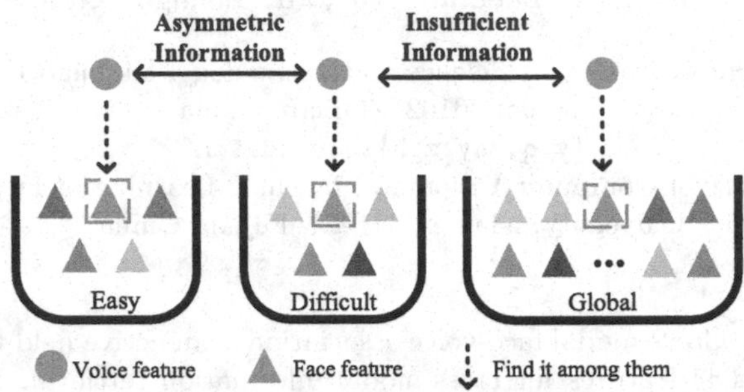

Fig. 1. Traditional cross-modal face and voice contrastive learning faces two challenges: (1) Asymmetric Information: In each identity involved in contrastive learning, randomly assigned negative samples exhibit varying levels of difficulty. (2) Insufficient information: In a small mini-batch, the number of negative examples is limited, which poses a challenge for cross-modal contrastive learning. The higher the similarity in colors, the more challenging the negative examples become.

Inspired by recent advances in multi-modal deep learning, a myriad of ingenious approaches have emerged within the realm of cross-modal face and voice association. For example, SVHF [6] defines this task as a binary classification problem. Subsequently, numerous extension methods [7–12] further combine metric learning and supervised learning to bridge the gap in heterogeneity and enhance modal-invariant features.

Although the existing methods have achieved promising results on the cross-modal face and voice matching tasks, they are still beset by the following limitations: (1) **Modality heterogeneity:** The features of face and voice are obtained by different feature extractors, creating a significant gap between the two modalities. (2) **Asymmetric information:** Many existing face-voice association approaches tend to overlook the importance of maintaining consistent information during cross-modal association learning. This results in unbalanced difficulty levels for negative examples within a mini-batch. (3) **Insufficient information:** Current methods often struggle due to limitations in GPU memory size, requiring the global information to be divided into random parts. While random strategies mitigate the absence of global information, each identity encounters deviations from their real-life characteristics when engaging in the process of contrastive learning.

Two challenges of traditional cross-modal face and voice in contrastive learning are illustrated in Fig. 1. Here is an interesting and concise example to illustrate the aforementioned challenges. When we encounter an unfamiliar mature male voice, it's relatively easy to find a matching face from a pool of face candidates, especially if the pool includes only one male and a group of women or children. This highlights the importance of gender and age as robust cues for cross-modal face and voice association. However, if we replace all the women and children in the candidate pool with male faces, the task becomes considerably more challenging. This illustrates how asymmetric information can introduce unfairness into the association process. Furthermore, when the number of face candidates in the pool increases, finding the matching face becomes akin to locating a needle in a haystack. While this complexity adds to the retrieval process, it allows an individual's identity to be positioned in a broader context by sensing other data. In this regard, an effective learning method should aim to eliminate asymmetric information and maximize the number of negative examples to ensure the successful reduction of modal heterogeneity.

Based on the above considerations, this paper introduces an efficient momentum framework for cross-modal face and voice association learning. Our method is directly inspired by traditional unimodal Momentum Contrast, which constructs a dynamic dictionary with a queue and a moving-averaged encoder to maintain consistent information. In comparison to understanding a single modality, comprehending multiple modalities presents a more complex scenario. Cross-modal contrastive learning is viewed as a bidirectional dictionary lookup process in this context. We gather negative examples into large and consistent queues to address the issues of asymmetric and insufficient information. Specifically, after each iteration, the current cross-modal data is enqueued, while the oldest data is dequeued. This approach enables each individual's identity to actively engage in contrastive learning with abundant information about negative examples. It helps the mature male face confidently stand out from candidate pool by minimizing the distance between data of the same identity across modalities, while maximizing the distance between the identity and the global negative examples. Meanwhile, the queues are separated from the random mini-batch, and the slowly momentum encoder is used to ensure the symmetrical information for each identity. Face and voice from the same identity are natural positive pair. Furthermore, Where there is a cross-modal positive pair, there is a smooth contrastive learning process. The effectiveness of the proposed CMMC framework is evaluated on various voice-face association tasks. Our method significantly outperforms the recent state-of-the-art methods. The main contributions of this paper are summarized as follows:

- We propose a novel CMMC framework, designed specifically for face-voice association learning, enabling the acquisition of effective cross-modal representations for various tasks.
- We propose the strategy of utilizing queues to store negative samples, addressing the challenges of information asymmetry and insufficiency in the cross-modal contrastive learning process.
- Experiments are conducted under various settings to demonstrate the effectiveness of the proposed framework.

2 Related Work

2.1 Face and Voice Association Learning

The primary objective of face-voice association learning is to acquire discriminative and modality-invariant feature representations for data across different modalities. The pioneering method SVHF [6] employs CNN architectures to learn representations of both face and voice, effectively transforming the modal association problem into a binary matching problem. However, it is observed that this framework may lack the flexibility needed to adapt to various cross-modal tasks. Benefiting from advancements in deep learning and contrastive learning, Pins [9] addresses this limitation by mining positive and negative examples from the video stream. Building upon these earlier methods, many researchers [8,12] utilize metric learning [13–15] to develop joint representations of face and voice. DIMNet [7], for instance, leverages supervised learning to map disjoint embeddings to shared modal covariates, including identity, gender, age, and nationality, thus enriching the semantic information. More recently, RIC [10] attempts to learn discriminative joint embeddings by employing bi-directional ranking constraints, identity constraints, and center constraints, thereby further exploring the relationships between samples within a mini-batch. Additionally, SSS [11] introduces a two-level modality alignment to compensate for the limitations of contrastive learning in a mini-batch.

2.2 Contrastive Learning

Contrastive learning represents a groundbreaking approach in unsupervised representation learning. Its exceptional performance has been observed across various fields, including computer vision and natural language processing. In many unsupervised learning works [16,17], contrastive learning finds use in diverse prediction scenarios, such as colorization, jigsaw solving, temporal coherence, and future prediction. Furthermore, it can also be employed to derive discriminative information from individual instances, effectively transforming them into an N-classification task, where N represents the number of instances. Contrastive losses, as defined in Hadsell et al.'s work [13], gauge the similarities among individual instances within a representation space. Different from classification tasks that assign instances to targets, contrastive learning operates in real-time. Momentum contrastive learning [18] serves as a general mechanism for implementing contrastive losses. It transcends the limitations of traditional end-to-end and dictionary-based methods in contrastive learning, establishing a large and consistent queue to facilitate the contrastive learning process.

Generally, in cross-modal face and voice association learning, many existing contrastive learning methods encounter challenges related to asymmetric and insufficient information. Additionally, the MoCo method, originally designed for traditional single-modal visual tasks, has limitations in bridging the heterogeneity gap between cross-modal data. Therefore, we introduce a new CMMC framework to address these issues.

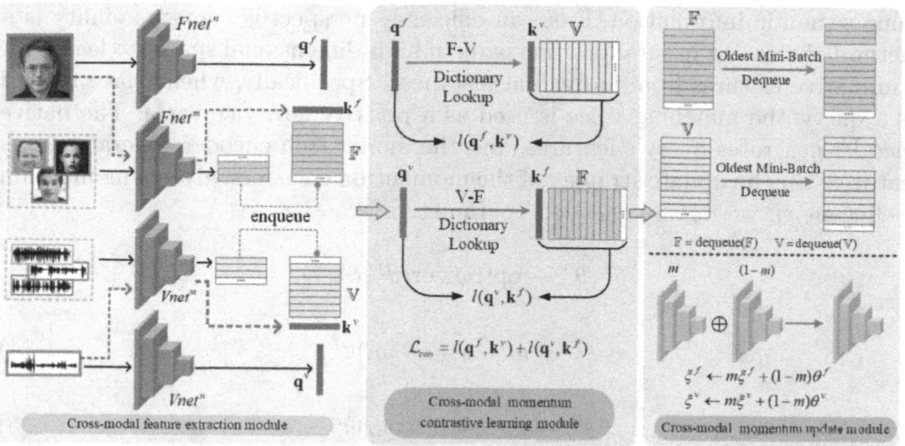

Fig. 2. The schematic architecture of the proposed CMMC framework.

3 Methodology

In this section, we will provide a detailed explanation of our proposed CMMC framework. The concept of the face-voice association framework we propose is depicted in Fig. 2.

3.1 General Definition

Without loss of generality, let X $=\{(f_n, v_n)\}_{n=1}^{N}$ represent the matched face and voice data, where N is the size of the training dataset. We have two native modal encoders $Fnet^n(\cdot)$ and $Vnet^n(\cdot)$, as well as two momentum encoders $Fnet^m(\cdot)$ and $Vnet^m(\cdot)$. The momentum encoders are copied from the native encoders. The parameters of $Fnet^n(\cdot)$ and $Vnet^n(\cdot)$ are denoted as θ^f and θ^v, respectively. The parameters of $Fnet^m(\cdot)$ and $Vnet^m(\cdot)$ are denoted as ξ^f and ξ^v, respectively. The MoCo framework proposes that contrastive learning can be seen as training an encoder for a dictionary look-up task. In the cross-modal setting, we treat $\mathbf{q}^f$, $\mathbf{q}^v$ as queries and $\mathbf{k}^f$, $\mathbf{k}^v$ as keys. The above procedure can be formulated as:

$$\mathbf{q}^f = Fnet^n(f), \mathbf{q}^v = Vnet^n(v), \tag{1}$$

$$\mathbf{k}^f = Fnet^m(f), \mathbf{k}^v = Vnet^m(v). \tag{2}$$

3.2 Cross-modal MoCo

Formally, traditional MoCo method employs momentum-updated dictionaries to provide a new mechanism for contrastive learning. It employs a data augmentation process to define positive pairs. In cross-modal face and voice, we define positive pairs in contrastive learning with two modalities which have the same semantic information. From an efficiency perspective, cross-modality is a method of creating natural positive pairs in high-dimensional spaces to learn discriminative features from individual instances. Specifically, when a face is used as a query, the matching voice is used as a positive key, vice versa. The native encoder generates query's features, and the momentum encoder generates key's features. For the update strategy of the momentum encoders, given a momentum coefficient m, we update encoders' parameters by:

$$\theta^f, \theta^v \leftarrow \text{optimizer}(\theta^f, \theta^v, r), \tag{3}$$

$$\xi^f \leftarrow m\xi^f + (1-m)\theta^f, \tag{4}$$

$$\xi^v \leftarrow m\xi^v + (1-m)\theta^v. \tag{5}$$

where r represents the learning rate of the optimizer and $m \in (0,1]$ is a momentum coefficient. θ^f and θ^v participate in back-propagation to update themselves.

Similar to MoCo, CMMC needs to maintain two queues $\mathbb{F} = \{\mathbf{k}_1^f, \mathbf{k}_2^f, ..., \mathbf{k}_K^f\}$ and $\mathbb{V} = \{\mathbf{k}_1^v, \mathbf{k}_2^v, ..., \mathbf{k}_K^v\}$, where K is the maximum length of the queue. The queues are filled by momentum encoders and all of them are treated as negative instances in the training moment. Although considering label information here may mitigate the impact of potential positive instances, we adopt the unsupervised setting to train the encoders using only real-time cross-modal instance without the exact identity labels. Moreover, the queues are updated with iteration. During training phase, when the queues reach the maximum value K, the current mini-batch enters the queue and the oldest mini-batch exits the queue. This enables decoupling it from the mini-batch size.

In short, to briefly describe the process of contrastive learning in cross-modal, we adopt a pair of matched face and voice as an example. We collect $\{\mathbf{q}^f, \mathbf{q}^v\}$ and $\{\mathbf{k}^f, \mathbf{k}^v\}$ from native decoders and momentum encoders at the same moment. Next, we form two positive pairs: $\{\mathbf{q}^f, \mathbf{k}^v\}$ and $\{\mathbf{q}^v, \mathbf{k}^f\}$. As for $\{\mathbf{q}^f, \mathbf{k}^v\}$, we complete the contrastive learning with voice queue $\mathbb{V}$. If we regard $\mathbf{k}^v$ and $\mathbb{V}$ as a large voice dictionary, the task of contrastive learning for face modality is transformed into a voice dictionary look-up. Finding the key $\mathbf{k}^v$ corresponding to $\mathbf{q}^f$ is a significant task for dictionary look-up. Similarly, as for $\{\mathbf{q}^v, \mathbf{k}^f\}$, we complete the contrastive learning with face queue $\mathbb{F}$.

Finally, in this paper, cross-modal face-voice association learning loss is defined as:

$$l(\mathbf{q}^f, \mathbf{k}^v) = -\log \frac{\exp(\mathbf{q}^f \cdot \mathbf{k}^v/\tau)}{\exp(\mathbf{q}^f \cdot \mathbf{k}^v/\tau) + \sum_{j=1}^{K} \exp(\mathbf{q}^f \cdot \mathbf{k}_j^v/\tau)}, \tag{6}$$

$$l(\mathbf{q}^v, \mathbf{k}^f) = -\log \frac{\exp(\mathbf{q}^v \cdot \mathbf{k}^f/\tau)}{\exp(\mathbf{q}^v \cdot \mathbf{k}^f/\tau) + \sum_{j=1}^{K} \exp(\mathbf{q}^v \cdot \mathbf{k}_j^f/\tau)}, \tag{7}$$

$$\mathcal{L}_{con} = l(\mathbf{q}^f, \mathbf{k}^v) + l(\mathbf{q}^v, \mathbf{k}^f). \tag{8}$$

4 Experiment

4.1 Dataset

VoxCeleb and VGGFace. As recommended in most baseline studies [6,7,10, 11], we have chosen to evaluate our method using the publicly available VoxCeleb [19] and VGGFace [20] datasets. The VoxCeleb dataset comprises short videos collected from 1251 celebrity interviews, while the VGGFace dataset contains identity information for 2622 individuals. In alignment with previous research [11], we have selected 1225 overlapping identities from these two datasets. The data has been split into separate train, validation, and test sets, ensuring there is no identity overlap among them. The statistical data information is presented in Table 1.

Table 1. Statistical numbers for splitting datasets.

Details	Train	Val	Test	Total
face images	106584	12260	20076	138920
audio clips	124740	14182	21850	160772
identity	924	112	189	1225
V-F test pairs	–	42546	65550	108096
F-V test pairs	–	36780	60228	97008

4.2 Implementation Details

Network Architecturer: The face extractor is constructed using the SE-ResNet-50 architecture. The cropped RGB face image is resized to a size of 112×112×3 and subjected to preprocessing operations similar to those outlined in the Pins [8]. The voice extractor is designed using the Thin-ResNet-34 architecture. Voice data is extracted from the original video using a voice activity detector, and silent periods are removed. Subsequently, 64-dimensional log mel-spectrograms are generated (with a window size of 25ms and a hop size of 10ms) and followed by mean and variance normalization. Pre-training weights strictly adhere to the SSS [11]method: the face network is pre-trained for the face recognition task on MS-1M [21], while the voice network is pre-trained for audio speaker recognition on VoxCeleb2 [22].

Training Strategy: Based on the pairwise input of data, our model is guided by a training strategy akin to the operations described in the SSS study, randomly selects several identities in an iteration, and then randomly sample one face image and one voice audio clip for each identity. It is important to note that throughout this process, we exclusively treat the randomly sampled paired data as positive pairs, with their labeling having no influence. Parameter settings in this article are as follows: momentum coefficient $m = 0.99$, temperature hyper-parameter $\tau = 0.07$, length of the queue $K = 4096$. We adopt the Adam optimizer, which learning rate is set 10^{-5}, and the batch size is 32.

Table 2. Comparison with baselines on 1:2 matching task, verification task and retrieval task on the VoxCeleb and VGGFace datasets. The best results of our models and competitors are highlighted in soft red and soft green, respectively. Δ denotes the improvement comparing the previous state-of-the-art results.

VoxCeleb + VGGFace											
Methods	Years	1:2 Matching(ACC)				Verification(AUC)				Retrieval(mAP)	
		V-F(U)	F-V(U)	V-F(G)	F-V(G)	V-F(U)	F-V(U)	V-F(G)	F-V(G)	V-F	F-V
SVHF	CVPR18	81.0	79.5	63.9	63.4	–	–	–	–	–	–
DIMNet	ICLR19	81.3	81.9	70.6	69.9	81.0	81.2	70.4	69.3	4.3	3.8
RIC	SIGIR20	83.4	84.2	71.7	71.1	82.6	82.9	70.3	70.1	4.4	3.4
SSS	CVPR21	87.2	86.5	77.7	75.3	87.2	87.0	77.5	76.1	5.5	5.8
Ours	–	87.7	88.2	78.2	77.5	88.2	88.4	77.7	77.3	6.1	6.8
Δ	–	**0.5**	**1.7**	**0.5**	**2.2**	**1.0**	**1.4**	**0.2**	**1.2**	**0.6**	**1.0**

4.3 Evaluation Protocols

To evaluate the effectiveness of the proposed CMMC framework, we conduct experiments on VoxCeleb and VGGFace with four different protocols: verification, 1:2 matching, 1:N matching and retrieval.

(1) **Verification:** Given a face and a voice data example, the purpose of cross-modal verification task is to determine whether they are collected from the same person or not. The verification performance is evaluated with AUC quantitative indicator (area under the ROC curve).
(2) **1:2 matching**: Given an anchor example from one modality and two candidates from the other modality, the objective of 1:2 matching task is to determine which candidate shares the same identity information as the anchor. Performance is evaluated using the accuracy metric(ACC).
(3) **1:N matching**: This task is an extension of the aforementioned 1:2 matching task, with the only difference being that there are now N candidates instead of just two. Similarly, accuracy is used to measure the matching performance (ACC).
(4) **Retrieval**: This task extends the matching work to a cross-modal retrieval scenario. In this scenario, when provided with a query data point from one modality, the goal is to index more candidates with the same identity as those matched in another modality. Performance is evaluated using the standard mean average precision (mAP).

To assess the face-voice association performance, we categorize all cross-modal association tasks into two cases: face matching voice (**F-V**) and voice matching face (**V-F**). Similar to most baselines, the symbol (**G**) indicates that all data in the task share the same gender, while the symbol (**U**) denotes an unstratified group. In the experiments, we reference classical methods such as SVHF [6], DIMNet [7], and recent methods like RIC [10] and SSS [11] as meaningful competitors to our framework.

4.4 Performance Analysis and Comparison

The results for 1:2 matching, verification, and retrieval on the VoxCeleb and VGGFace datasets are shown in the following Table 2, with the best results highlighted in bold. It's evident that our unsupervised learning method yields promising results when compared to previous supervised methods. According to the testing metrics proposed by SSS [11], our method has an average improvement of about 1.08% compared with the previous state-of-the-art results on matching and verification tasks. In particular, for the modes V-F and F-V, some experimental results show that our method narrows the performance gap. That is, the symmetric information makes the model insensitive to the modality. This provides a smoother cross-modal representation for various tasks.

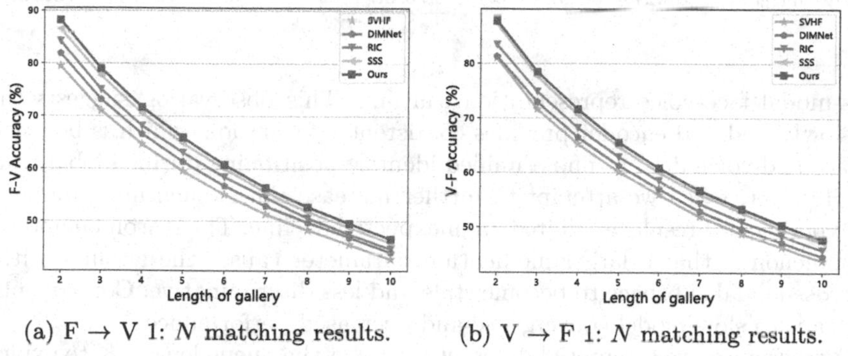

(a) F → V 1: N matching results. (b) V → F 1: N matching results.

Fig. 3. Comparison of 1: N matching performance.

Furthermore, the representative 1:N matching results are depicted in the Fig. 3, and it is evident that our method consistently produces the best results, even as the natural accuracy decreases with increasing N. This highlights that the ample information from the global negative examples provided by the modal queue enables our model to handle varying values of N more effectively.

4.5 Ablation Study

Cross-modal MoCo employs cross-modal queues to provide negative examples for each identity, effectively addressing the issue of asymmetric and insufficient information. To evaluate the impact of different parameters in cross-modal MoCo, we have conducted additional experiments with different values of parameters, m and K. Additionally, we assessed the effectiveness of the cross-modal momentum contrastive learning mechanism by designing an end-to-end contrastive learning process with the same batch size.

Table 3 shows the performance of various variant experiments on 1:2 matching and verification tasks. We assessed different values of the momentum parameter m and observed that a larger momentum parameter is advantageous for

Table 3. The effects of CMCC parameters m and K on 1:2 matching and verification. The best results and parameters for each section are highlighted in **blue** and red.

Parameters		1:2 Matching(ACC)				Verification(AUC)			
m	K	V-F(U)	F-V(U)	V-F(G)	F-V(G)	V-F(U)	F-V(U)	V-F(G)	F-V(G)
0.9999	4096	85.70	85.98	75.17	73.23	85.42	85.52	73.65	72.50
0.999	4096	87.67	87.56	78.09	76.37	87.75	87.34	**77.85**	76.50
0.99	4096	**87.76**	**88.23**	**78.29**	**77.59**	**88.25**	**88.44**	77.73	**77.35**
0.9	4096	87.58	87.47	78.23	76.81	88.02	88.08	77.64	77.12
0.99	16384	85.20	85.44	74.50	72.99	85.81	86.05	73.32	72.75
0.99	4096	**87.76**	**88.23**	**78.29**	**77.59**	**88.25**	**88.44**	**77.73**	**77.35**
0.99	1024	87.11	87.63	77.58	76.72	87.60	87.89	77.04	76.87
0.99	32	86.20	**86.41**	**75.76**	74.70	**86.55**	**86.56**	74.44	73.31
end-to-end	32	**86.27**	86.26	75.45	**75.08**	86.20	86.40	**74.73**	**73.57**

cross-modal face-voice representation learning. This observation suggests that the slowly updated encoder provides consistent information, which is beneficial for face and voice data to find a unified identity positioning during global training. However, when we attempt to further increase the momentum parameter, the experimental results exhibited an unexpected decline. The reason behind this phenomenon is that a larger momentum parameter causes the features within the cross-modal subspace to become stale and less discriminative. Consequently, this leads to slow model convergence and decreased performance.

Furthermore, we attempted different values of the queue length K to explore the association between face and voice. The effects appear to stabilize when $K = 16384$. We believe that the value of K is directly proportional to the dataset's order of magnitude and the number of categories. It is worth noting that the datasets of cross-modal face-voice are far less than a million level. Simultaneously, the number of potential positive examples also increases with the rise of K. The existence of numerous potential positive examples can misguide the objectives of contrastive learning, thus affecting the model's performance in association tasks. Therefore, configuring a large value for K is unnecessary.

To validate the effectiveness of our proposed CMMC framework, we also designed an end-to-end variant experiment using native encoders. The results showed that both mechanisms perform similarly with the same batch size. Nonetheless, the end-to-end learning mechanism is constrained by GPU memory and cannot support further increases in batch size. In contrast, within our framework, the length of the global queue can be extended to provide ample information, thereby enhancing the correlation between face and voice.

4.6 Visualization Study

As shown in Fig. 4, we use the t-SNE [23] algorithm to visualize the learned embedding vectors from ten randomly selected individuals. The visualization reveals that clusters of faces and voices of the same color are closely grouped

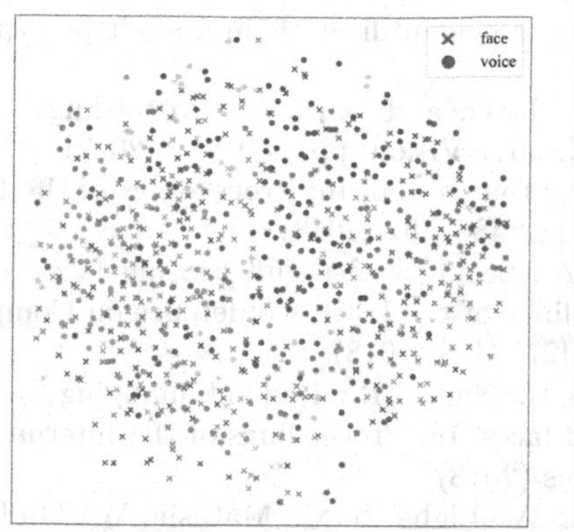

(a) Embeddings before pre-process.

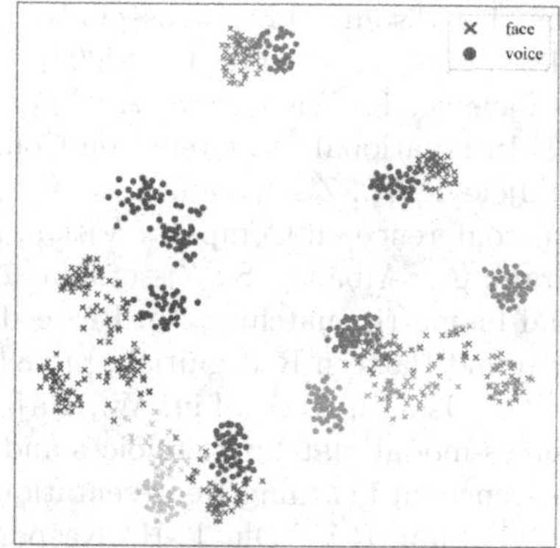

(b) Embeddings after pre-process.

Fig. 4. Visualization of the learned embeddings.

together, while clusters representing different identities are distinctly separated. This suggests that our method not only enhances the separation between features of different identities but also reduces the distance among the same identity modalities. Therefore, cross-modal momentum contrastive learning fosters a deeper understanding of global semantic information, which can be advantageous for various face-voice association tasks.

5 Conclusion

In this paper, we propose an efficient momentum cross-modal face-voice association learning framework. This framework employs cross-modal queues to store negative examples, providing sufficient and symmetric information for contrastive learning process. In addition, the momentum encoder is updated gradually to ensure information consistency. Consequently, this approach outperforms the state-of-the-art methods, and extensive experiments evaluated on the popular VoxCeleb and VGGFace datasets have shown its competitive performance.

Acknowledgements. This work was jointly supported by Natural Science Foundation of China under Grants 61972166 and 62372190, and Industry University Cooperation Project of Fujian Province under Grant 2021H603.

References

1. Joassin, F., Pesenti, M., Maurage, P., Verreckt, E., Bruyer, R., Campanella, S.: Cross-modal interactions between human faces and voices involved in person recognition. Cortex **47**(3), 367–376 (2011)
2. Kamachi, M., Hill, H., Lander, K., Vatikiotis-Bateson, E.: Putting the face to the voice': matching identity across modality. Curr. Biol. **13**(19), 1709–1714 (2003)

3. Lachs, L., Pisoni, D.B.: Crossmodal source identification in speech perception. Ecol. Psychol. **16**(3), 159–187 (2004)
4. Arandjelovic, R., Zisserman, A.: Look, listen and learn. In: Proceedings of the IEEE International Conference on Computer Vision, pp. 609–617 (2017)
5. Arandjelovic, R., Zisserman, A.: Objects that sound. In: Proceedings of the European Conference on Computer Vision, pp. 435–451 (2018)
6. Nagrani, A., Albanie, S., Zisserman, A.: Seeing voices and hearing faces: cross-modal biometric matching. In: Proceedings of the IEEE Conference on Computer Vision and Pattern Recognition, pp. 8427–8436 (2018)
7. Wen, Y., Ismail, M.A., Liu, W., Raj, B., Singh, R.: Disjoint mapping network for cross-modal matching of voices and faces. In: Proceedings of the International Conference on Learning Representations (2018)
8. Kim, C., Shin, H.V., Oh, T.-H., Kaspar, A., Elgharib, M., Matusik, W.: On learning associations of faces and voices. In: Proceedings of the Asian Conference on Computer Vision, pp. 276–292 (2019)
9. Nagrani, A., Albanie, S., Zisserman, A.: Learnable pins: cross-modal embeddings for person identity. In: Proceedings of the European Conference on Computer Vision, pp. 71–88 (2018)
10. Wang, R., Liu, X., Cheung, Y.-M., Cheng, K., Wang, N., Fan, W.: Learning discriminative joint embeddings for efficient face and voice association. In: Proceedings of the International ACM SIGIR Conference on Research and Development in Information Retrieval, pp. 1881–1884 (2020)
11. Wen, P., Xu, Q., Jiang, Y., Yang, Z., He, Y., Huang, Q.: Seeking the shape of sound: an adaptive framework for learning voice-face association. In: Proceedings of the IEEE Conference on Computer Vision and Pattern Recognition, pp. 16347–16356 (2021)
12. Horiguchi, S., Kanda, N., Nagamatsu, K.: Face-voice matching using cross-modal embeddings. In: Proceedings of the ACM International Conference on Multimedia, pp. 1011–1019 (2018)
13. Hadsell, R., Chopra, S., LeCun, Y.: Dimensionality reduction by learning an invariant mapping. In: Proceedings of the IEEE Computer Society Conference on Computer Vision and Pattern Recognition, pp. 1735–1742 (2006)
14. Weinberger, K.Q., Saul, L.K.: Distance metric learning for large margin nearest neighbor classification. J. Mach. Learn. Res. **10**(2), 207–244 (2009)
15. Sohn, K.: Improved deep metric learning with multi-class n-pair loss objective. Adv. Neural Inf. Process. Syst. **29** (2016)
16. Han, T., Xie, W., Zisserman, A.: Memory-augmented dense predictive coding for video representation learning. In: Proceedings of the European Conference on Computer Vision, pp. 312–329 (2020)
17. Pan, T., Song, Y., Yang, T., Jiang, W., Liu, W.: Videomoco: contrastive video representation learning with temporally adversarial examples. In: Proceedings of the IEEE Conference on Computer Vision and Pattern Recognition, pp. 205–214 (2021)
18. He, K., Fan, H., Wu, Y., Xie, S., Girshick, R.: Momentum contrast for unsupervised visual representation learning. In: Proceedings of the IEEE Conference on Computer Vision and Pattern Recognition, pp. 9729–9738 (2020)
19. Nagrani, A., Chung, J.S., Xie, W., Zisserman, A.: Voxceleb: large-scale speaker verification in the wild. Comput. Speech Lang. **60**, 101027 (2020)
20. Parkhi, O.M., Vedaldi, A., Zisserman, A.: Deep face recognition. In: Proceedings of the British Machine Vision Conference, pp. 41.1–41.12 (2015)

21. Guo, Y., Zhang, L., Hu, Y., He, X., Gao, J.: Ms-celeb-1m: a dataset and benchmark for large-scale face recognition. In: Proceedings of the European Conference on Computer Vision, pp. 87–102 (2016)
22. Chung, J.S., Nagrani, A., Zisserman, A.: Voxceleb2: deep speaker recognition. In: Interspeech, pp. 1086–1090 (2018)
23. van der Maaten, L., Hinton, G.: Visualizing data using t-sne. J. Mach. Learn. Res. **9**, 2579–2605 (2008)

Multi-modal Instance Refinement for Cross-Domain Action Recognition

Yuan Qing[1], Naixing Wu[2], Shaohua Wan[1], and Lixin Duan[1(✉)]

[1] Shenzhen Institute for Advanced Study, University of Electronic Science and Technology of China, Shenzhen, China
yuanqing.shawn@gmail.com, {shaohua.wan,lxduan}@uestc.edu.cn
[2] China Unicom Shenzhen Branch, Shenzhen, China
1320544942@qq.com

Abstract. Unsupervised cross-domain action recognition aims at adapting the model trained on an existing labeled source domain to a new unlabeled target domain. Most existing methods solve the task by directly aligning the feature distributions of source and target domains. However, this would cause negative transfer during domain adaptation due to some negative training samples in both domains. In the *source domain*, some training samples are of low-relevance to target domain due to the difference in viewpoints, action styles, etc. In the *target domain*, there are some ambiguous training samples that can be easily classified as another type of action under the case of source domain. The problem of negative transfer has been explored in cross-domain object detection, while it remains under-explored in cross-domain action recognition. Therefore, we propose a *Multi-modal Instance Refinement (MMIR)* method to alleviate the negative transfer based on reinforcement learning. Specifically, a reinforcement learning agent is trained in both domains for every modality to refine the training data by selecting out negative samples from each domain. Our method finally outperforms several other state-of-the-art baselines in cross-domain action recognition on the benchmark EPIC-Kitchens [4] dataset, which demonstrates the advantage of MMIR in reducing negative transfer.

Keywords: Negative Transfer · Cross-domain Action Recognition · Unsupervised Domain Adaptation · Reinforcement Learning

1 Introduction

Action recognition is one of the most important tasks in video understanding, which aims at recognizing and predicting human actions in videos [1,4,22,24]. The majority of the works in action recognition are carried out on the basis of

Supplementary Information The online version contains supplementary material available at https://doi.org/10.1007/978-981-99-8429-9_23.

© The Author(s), under exclusive license to Springer Nature Singapore Pte Ltd. 2024
Q. Liu et al. (Eds.): PRCV 2023, LNCS 14425, pp. 284–296, 2024.
https://doi.org/10.1007/978-981-99-8429-9_23

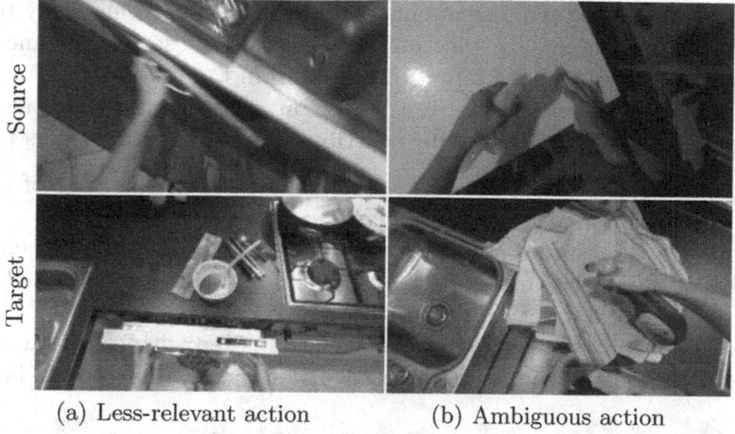

(a) Less-relevant action (b) Ambiguous action

Fig. 1. Negative samples that cause negative transfer from source domain **D2** (first row) and target domain **D3** (second row) defined in [17].

supervised learning, which involves using a large number of labeled video/action segments to train a model for a specified scene setting. However, obtaining a large number of annotated video data for a certain scene would be very costly and sometimes difficult due to the environment setup and video post-processing as well as labeling. To fully leverage existing labeled data and reduce the cost of acquiring new data, Unsupervised Domain Adaptation (UDA) [14,18,20] has been introduced to generalize a model trained on a source domain with adequate annotations to a target domain with no labels, where the two domains differentiate from each other but are partially related. For action recognition, though there are several UDA methods [10,17,26] proposed, most of them achieve this by directly aligning the feature distribution of source and target domains. However, this could lead to negative transfer during domain adaptation due to some negative training samples in both domains [3]. For instance, there might be some ambiguous actions in target domain that do not belong to the defined action categories in the source domain or are very similar to another kind of action in the source domain. Additionally, there might also be some less-relevant actions in source domain that have completely different viewpoints/action styles compared with samples in the target domain. To be specific, Fig. 1 shows these two types of negative samples in domain D2 (source) and D3 (target) defined in [17] from EPIC-Kitchens dataset [4]. In Fig. 1(a), action *open* in source domain is considered less-relevant to that of the target domain since the trajectory of motion and way of opening are dissimilar. In Fig. 1(b), a spraying action in target domain that does not belong to a predefined action type is likely to be mistakenly recognized as *wipe* due to the similarity in action style and appearance.

To alleviate the impact of negative transfer brought by these negative training samples, we propose Multi-modal Instance Refinement (MMIR) based on deep Q-learning (DQN) [15], under the framework of MM-SADA [17]. Our MMIR trains reinforcement learning agents in both domains in each modality to refine source and target samples by selecting out less-relevant source instances from source domain and ambiguous target instances from target domain. To the best of our knowledge, there's no previous work on reducing negative transfer in cross-domain action recognition. Our contributions are summarised as follows:

- As far as we know, we are the first to define and tackle the issue of negative transfer in cross-domain action recognition.
- We adopt a novel instance refinement strategy using deep reinforcement learning to select outlier instances in source domain and target domain within two modalities (RGB and optical flow).
- Our method achieves superior performance compared with other state-of-the-art methods in cross-domain action recognition on EPIC-Kitchens dataset.

2 Related Work

2.1 Action Recognition

In action recognition, early works use 2D/3D convolution [8,9] for feature extraction only in a single modality, i.e., RGB. Later, optical flow of video segments is used as auxiliary training data which carries more temporal and motion information compared with RGB [21]. Therefore, current popular CNN-based methods adopt a two-stream 3D convolutional neural network structure [1,19] for feature extraction which could utilize the information contained in multiple modalities and model the temporal information. Most recently, vision transformer [6] based approaches have excelled CNN-based methods on many benchmarks. MVD [22] builds a two-stage masked feature modeling framework, where the high-level features of pretrained models learned in the first stage will serve as masked prediction targets for student model in the second stage. Although these methods show promising performance in a supervised manner, we are going to focus on action recognition under the setting of UDA.

2.2 Unsupervised Domain Adaptation for Action Recognition

Though both RGB and optical flow have been studied for domain adaptation in action recognition, there are only a limited number of works attempted to conduct multi-modal domain adaptation [10,17,26]. Munro and Dame [17] propose MM-SADA, a multi-modal 3D convolutional neural network with a self-supervision classifier between modalities. It uses Gradient Reversal Layer (GRL) [7] to implement domain discriminator within different modalities. Kim *et al.* [10] apply contrastive learning to design a unified framework using transformer for multi-modal UDA in video understanding. Xu *et al.* [26] propose a source-free UDA model to learn temporal consistency in videos between source domain and target domain. Similar to [17], our work adopts a multi-modal 3D ConvNet for feature extraction and utilizes domain adversarial learning, but we focus on a different task by incorporating deep reinforcement learning into our action recognition framework to eliminate the effect of negative transfer.

2.3 Deep Reinforcement Learning

Deep reinforcement learning has been applied to various tasks in computer vision [23,27]. Wang *et al.* [23] design a reinforcement learning-based two-level framework for video captioning, in which a low-level module recognizes the original actions to fulfill the goals specified by the high-level module. ReinforceNet [27] incorporates region selection and bounding box refinement networks to form a reinforcement learning framework based on CNN to select optimal proposals and refine bounding box positions. Recently, several works apply reinforcement learning to action recognition [5,25]. Dong *et al.* [5] design a deep reinforcement learning framework to capture the most discriminative frames and delete confusing frames in action segments. Weng *et al.* [25] improve recognition accuracy by designing agents that learn to produce binary masks to select out interfering categories. All these methods adopt deep reinforcement learning to refine negative frames in the action segment within the same domain, while our method uses deep reinforcement learning to refine negative action segments across domains to handle negative transfer in cross-domain action recognition.

3 Proposed Method

In unsupervised cross-domain action recognition, two domains are given, namely source and target. A labeled source domain $\mathcal{D}^s$ is denoted as $\mathcal{D}^s = \{(x_i^s, y_i^s)|_{i=1}^{n_s}\}$, where x_i^s is the i-th action segment and y_i^s is its verbal label. Similarly, the unlabeled target domain $\mathcal{D}^t$ is represented by $\mathcal{D}^t = \{x_i^t|_{i=1}^{n_t}\}$, where x_i^t is the i-th action segment. For action segments, each segment is formed as a sequence of L frames. Therefore, we have $x_i^s = \{x_{i,1}^s, x_{i,2}^s, x_{i,3}^s, \ldots, x_{i,L}^s\}$ and $x_i^t = \{x_{i,1}^t, x_{i,2}^t, x_{i,3}^t, \ldots, x_{i,L}^t\}$, respectively.

To reduce negative transfer during domain adaptation, two reinforcement learning agents, S-agent and T-agent, are defined under a deep-Q learning network (DQN) to make selections in source and target domains. S-agent learns policies to select out less-relevant action segments from source action segments x^s, while T-agent is trained to select out ambiguous action segments from target action segments x^t. After refinement, we use the refined instances $\hat{x^s}$ and $\hat{x^t}$ to train our domain discriminator to learn domain invariant features.

The following sections give detailed explanations of our proposed method *Multi-model Instance Refinement (MMIR)*. The architecture of our MMIR is shown in Fig. 2(a), which is composed of a two-stream 3D convolutional feature extractor in two modalities: **RGB** and **Optical Flow** together with a domain discriminator and instance refinement module in each modality followed by an action classifier. The structure of domain discriminator and instance refinement module is depicted in Fig. 2.

3.1 Two-Stream Action Recognition

For multiple modalities, a feature fusion layer is applied after the action classifiers for summing the prediction scores from different modalities as shown in Fig. 2(b).

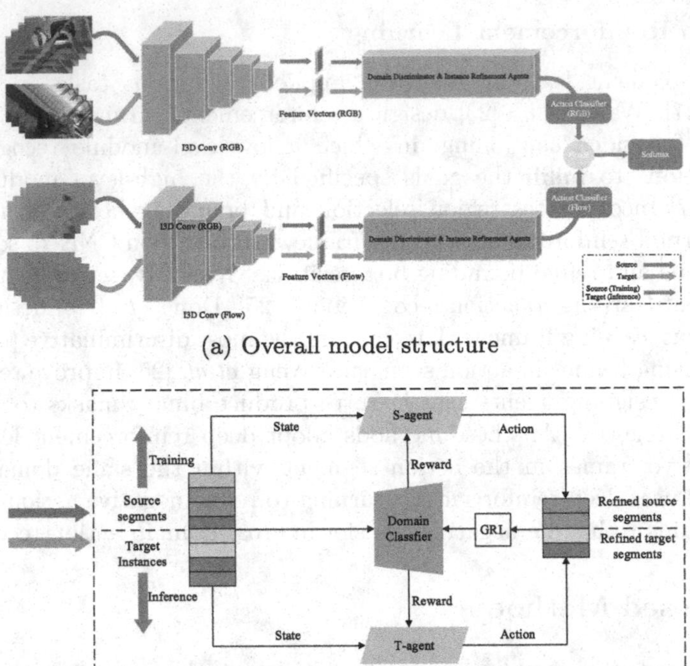

(a) Overall model structure

(b) Domain discriminator and instance refinement agents

Fig. 2. Structure of proposed method MMIR: (a) An I3D [1] feature extractor in each modality is shared by both domains. The output feature vectors of I3D network are fed to the instance refinement and domain adversarial learning modules as well as action classifiers. (b) S-agent and T-agent are built for source and target domain, respectively. They take input feature vectors as state and make selections in the training instances to select out noisy samples. A domain classifier with GRL is optimized with refined instances, which gives rewards to agents according to their selections.

For input $\mathcal{X}$ with multiple modalities, we have $\mathcal{X} = (\mathcal{X}^1, \mathcal{X}^2, \ldots, \mathcal{X}^K)$, where $\mathcal{X}^k$ represents the input from the k-th modality. Therefore, we can define the classification loss as follows:

$$\mathcal{L}_{cls} = \sum_{x \in S} -y \cdot \log \left(Softmax \left(\sum_{k=1}^{K} C^k \left(F^k \left(x^k \right) \right) \right) \right) \tag{1}$$

where y represents the class label, C^k is the action classifier of the k-th modality, F^k denotes the feature extractor of the k-th modality and x^k represents source action segments from the k-th modality which are labeled.

3.2 Instance Refinement

We visualize the overall workflow of instance refinement module in Fig. 3. In each modality, we select negative instances from the i-th batch of action seg-

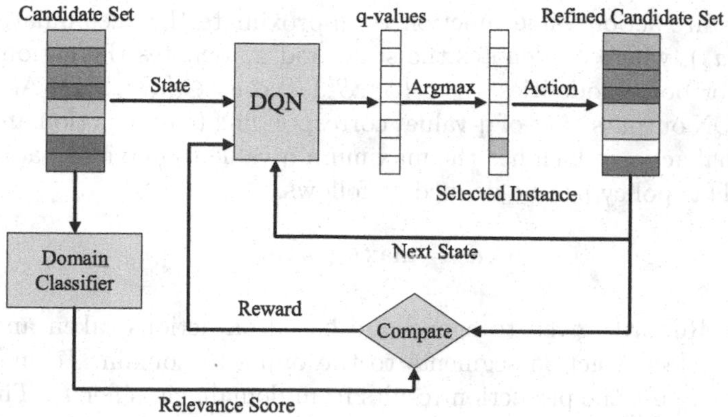

Fig. 3. Workflow of the instance refinement module.

ments $\mathcal{F}_i^s$ and $\mathcal{F}_i^t$ in source and target domain, respectively. We divide a batch into several sub batches, namely, candidate set $\mathcal{F}_C$ for iterating the agents over more episodes. Therefore, we can have a total number of $\mathcal{E}$ candidate sets in a batch. Each episode is responsible for selecting out E negative samples, thus the terminal time of an episode is defined as E. Take time e in the selection process as an example, S-agent takes an action $\mathcal{A}_e^s$ by observing current state $\mathcal{S}_e^s$. Then, the current state is updated as $\mathcal{S}_{e+1}^s$ since an action segment has been selected out. In the meantime, S-agent would receive a reward $\mathcal{R}_e^s$ for taking action $\mathcal{A}_e^s$. After arriving at terminal time E, S-agent has done selection for this episode and the candidate set $\mathcal{F}_C$ is optimized as $\hat{\mathcal{F}}_C$. Then, the batch $\mathcal{F}_i^s$ would become $\hat{\mathcal{F}}_i^s$ at the end of the last episode, and similarly, we can reach a $\hat{\mathcal{F}}_i^t$ for $\mathcal{F}_i^t$. We give detailed illustrations on *State*, *Action*, *Reward* and *DQN Loss* in the following part of this section.

- **State.** Agents make selections on the level of candidate set and take feature vectors of all the action segments inside the candidate set as state. In this case, the state $\mathcal{S}_k^s$ of S-agent in the k-th candidate set $\mathcal{C}_k^s$ could be defined as $\mathcal{S}_k^s = [f_{k,1}^s, f_{k,2}^s, f_{k,3}^s, \cdots, f_{k,n_c}^s] \in \mathbb{R}^{d \times n_c}$ where $f_{k,n}^s$ is the feature vector of $\mathcal{F}_{i,k.n}^s$ that has d dimensions and n_c is the number of action segments inside a candidate set. Once an action segment $f_{k,n}^s$ is selected out from $\mathcal{S}_k^s$, it will be replaced by a d-dimensional zero vector to keep the state shape unchanged. This is the same for T-agent where $\mathcal{S}_k^t = [f_{k,1}^t, f_{k,2}^t, f_{k,3}^t, \cdots, f_{k,n_c}^t] \in \mathbb{R}^{d \times n_c}$ is the state and $f_{k,n}^t$ is the feature vector of target action segment.
- **Action.** For a candidate set of size n_c, we can have n_c actions to perform in each episode. Therefore, we can define the set of actions that can be taken by S-agent as $\mathcal{A}^s = \{1, 2, \ldots, n_c\}$ and T-agent as $\mathcal{A}^t = \{1, 2, \ldots, n_c\}$, which represents the index of the action segment that is to be selected out. The aim of the DQN agents is to maximize the accumulated reward of the actions taken. We define the accumulated reward at time e as $\mathcal{R}_e = \sum_{t=e}^{E} \gamma^{t-e} r_e$, where γ is the discount factor and r_e represents the instant reward at time e. In DQN, we

define a state-action value function to approximate the accumulated reward as $Q(\mathcal{S}_e, a_e)$, where $\mathcal{S}_e$ denotes the state and a_e denotes the action taken at time e. For both modalities, $\mathcal{S}_e \in \{\mathcal{S}_e^s, \mathcal{S}_e^t\}$ and $a_e \in \{\mathcal{A}^s, \mathcal{A}^t\}$. As shown in Fig. 3, DQN outputs a set of q-values corresponding to each action and chooses the optimal action which has the maximum q-value to maximize accumulated reward. This policy can be defined as follows:

$$\hat{a_e} = \max_{a_e} Q(\mathcal{S}_e, a_e). \tag{2}$$

- **Reward.** Rewards given to agents are based on actions taken and the relevance of selected action segments to the opposite domain. To measure the relevance, we use the prediction results from domain classifier D. The domain logits of an action segment are processed by a sigmoid function and the relevance measure $\Delta(f)$ is defined as:

$$\Delta(f) = \begin{cases} Sigmoid(D(f)), & f \in \hat{\mathcal{F}}^s \\ 1 - Sigmoid(D(f)), & f \in \hat{\mathcal{F}}^t. \end{cases} \tag{3}$$

In Eq. 3, we unify the relevance measure in both source and target domains by defining the domain label of source to be 0 and target to be 1. Then, the predefined threshold τ and the relevance measure $\Delta(f)$ can be compared to give rewards to agents according to the criterion defined below:

$$r = \begin{cases} 1, & \Delta(f) < \tau, \\ -1, & otherwise. \end{cases} \tag{4}$$

This criterion is quite intuitive for an agent to recognize if it has made the right selection. Besides, we can set different thresholds for agents in different domains as τ^s and τ^t in source and target, respectively.
- **DQN Loss.** For a DQN, the target output is defined as:

$$y_e = r_e + \gamma \cdot \max_{a_{e+1}} Q(\mathcal{S}_{e+1}, a_{e+1} | \mathcal{S}_e, a_e) \tag{5}$$

where y_e represents the temporal difference target value of the Q function $Q(\mathcal{S}_e, a_e)$. Based on this, the loss of DQN can be defined as:

$$\mathcal{L}_q = \mathbb{E}_{\mathcal{S}_e, a_e}[(y_e - Q(\mathcal{S}_e, a_e))^2]. \tag{6}$$

Then, we can have the overall deep Q-learning loss defined as follows:

$$\mathcal{L}_{dqn} = \sum_{k=1}^{K} (\mathcal{L}_q^s + \mathcal{L}_q^t)_k \tag{7}$$

which is the sum of losses from S-agents and T-agents from all modalities.

3.3 Domain Adversarial Alignment

We realize feature alignment across domains in an adversarial way by connecting the domain discriminator with a GRL as shown in Fig. 2(b). We apply a domain discriminator in each modality rather than using a single discriminator for all modalities after late fusion since aligning domains in a combined way might lead the network to focus on a less robust modality and lose the ability to generalize to other modalities [17]. Then, we can define our domain adversarial loss as:

$$\mathcal{L}_{adv} = \sum_{x^k \in \{S,T\}} -d \cdot \log\left(D^k\left(F^k\left(x^k\right)\right)\right) - (1-d) \cdot \log\left(1 - D^k\left(F^k\left(x^k\right)\right)\right) \quad (8)$$

where d is the domain label, D^k is the domain discriminator for the k-th modality, F^k is the feature extractor of the k-th modality, and x^k denotes the action segments from source domain or target domain of the k-th modality.

3.4 Training

With the losses defined in previous sections, we can have an overall loss function:

$$\mathcal{L} = \mathcal{L}_{cls} + \mathcal{L}_{dqn} + \mathcal{L}_{adv}. \quad (9)$$

For DQN agents, we use experience replay [13] and ϵ-greedy strategy [16] during training. An experience replay pool to store actions, states, rewards, etc. is established for every agent, which ensures that data given to them is uncorrelated. The ϵ-greedy strategy introduces a probability threshold of random action ϵ to control whether an action is predicted by DQN agent or just randomly selected. This helps to balance the exploitation and exploration of an agent. The strategy is implemented as follows:

$$a_e = \begin{cases} \max_{a_e} Q(\mathcal{S}_e, a_e) & \text{if } \lambda \geq \epsilon, \\ a_e^* & \textit{otherwise}, \end{cases} \quad (10)$$

where λ is a random variable. If λ is larger than ϵ, the action would be predicted by the agents, or the action would be randomly chosen from the pool of actions.

4 Experiments

4.1 Implementation Details

Dataset. We use EPIC-Kitchens [4] to set up the cross domain environment for fine-grained action recognition as it includes action segments captured in 32 different scenes. Following the domain split in [17], we sample videos taken in 3 different kitchens to form 3 different domains, which are denoted as D1, D2 and D3. We have a total of 8 classes of action and according to [17], the distribution of training and testing samples from the 8 action classes are highly unbalanced. However, we use this unbalanced dataset to prove that our method could achieve competitive performance even on an unbalanced dataset since the unbalanced distribution of data makes domain adaptation more challenging.

Model Architecture. We set a two-stream I3D [1] feature extractor as our backbone. A trianing instance is composed of a temporal window of 16 frames sampled from an action segment. The domain discriminator D^k in each modality takes in the feature vectors and flattens them to pass through a GRL and two fully connected layers with the dimension of a hidden layer to be 128 and an intermediate LeakyReLU activation function. For data augmentation, we follow the setup in [2] where random cropping, scale jittering and horizontal flipping are applied to training data. For testing data, only center cropping is applied.

Hyperparameter and Training Setting. The overall dropout rate of F^k is set to 0.5 and a weight decay of 10^{-7} is applied for model parameters. We divide training process into two stages. In stage 1, our network is trained without domain discriminator and DQN agents. Then, the loss is optimized as follows:

$$\mathcal{L}_{stage1} = \mathcal{L}_{cls}. \tag{11}$$

The learning rate of this stage is set to 0.01 and the network is trained for 4000 steps. In the second stage, the domain discriminator and DQN agents are incorporated and the objective function for this stage is defined as:

$$\mathcal{L}_{stage2} = \mathcal{L}_{cls} + \mathcal{L}_{adv} + \mathcal{L}_{dqn}. \tag{12}$$

The learning rate in this stage is reduced to 0.001 and the model is further trained for 8000 steps. Note that for both stages, the action classifier is optimized only using labeled source data. For the hyperparameters of DQN, we set the discount factor $\gamma = 0.9$, ϵ-greedy factor $\epsilon = 0.5$, relevance threshold $\tau^s = \tau^t = 0.5$, terminal time $E = 1$ and candidate size $n_c = 5$. Besides, Adam [11] optimizer is used for both stages and the batch size is set to 96 in stage 1 and 80 in stage 2, which is equally divided for source and target domains. It takes 6 h to train our model using 4 NVIDIA RTX 3090 GPUs.

4.2 Results

For all the experimental results, we follow [17] to report the average top-1 accuracy on target domain over the last 9 epochs during training. In the meantime, the experimental results of our model trained with only source data are reported as a lower limit. Also, we report results of supervised learning on target domain as the upper bound.

We have a total of 6 domain adaptation settings based on D1, D2 and D3. We make a comparison between our method, several baseline methods and two state-of-the-art methods under every domain setting in Table 1. On average, our method outperforms all other state-of-the-art methods. MMIR has an overall performance improvement of 5.8% compared with Source-only, 4.5% compared with MMD [14], 4.1% compared with AdaBN, 4.0% compared with MCD [18], 1.0% compared with MM-SADA [17] and 0.3% compared with Kim *et al.* [10].

Table 1. Top-1 Accuracy of the experimental results of different baselines and our MMIR under different domain settings.

Method	D2 → D1	D3 → D1	D1→ D2	D3 → D2	D1 → D3	D2 → D3	Mean
Source-only	42.5	44.3	42.0	56.3	41.2	46.5	45.5
MMD [14]	43.1	48.3	46.6	55.2	39.2	48.5	46.8
AdaBN [12]	44.6	47.8	47.0	54.7	40.3	48.8	47.2
MCD [18]	42.1	47.9	46.5	52.7	43.5	51.0	47.3
MM-SADA [17]	48.2	50.9	49.5	56.1	44.1	52.7	50.3
Kim *et al.* [10]	**49.5**	51.5	**50.3**	56.3	**46.3**	52.0	51.0
MMIR	46.1	**53.5**	49.7	**61.5**	44.5	**52.6**	**51.3**
Supervised	71.7	74.0	62.8	74.0	62.8	71.7	69.5

4.3 Ablation Study

Effects of RL Agents. We evaluate the performance of reinforcement learning agents according to modality and domain. In particular, we investigate the effect of S-agent and T-agent only in the modality of RGB as this modality contributes more during the feature alignment process. The results are shown in Table 2 and we denote "without" as "w/o". We further elaborate on the results in the following part.

- *RGB vs Optical flow.* Our method without agents in RGB has a performance drop of 1.8% while the case without agents in Optical flow has only a drop of 0.7%. This indicates that agents in RGB play a major part in refining feature alignment compared with that of Optical flow since RGB frames contain more spatial information while flow frames contain more temporal information which contributes less to the feature alignment process.
- *S-agent vs T-agent in RGB.* In RGB, when S-agent is removed, we can observe a performance drop of 0.3%. While by removing the T-agent, the performance drop is 0.8%. This shows that in the modality of RGB, T-agent weighs more than S-agent in alleviating the issue of negative transfer.

Table 2. Ablation study of the effect of reinforcement learning agents on D2 → D1.

Method	D2 → D1
Source-only	42.5
MMIR (w/o) RGB agents	44.3
MMIR (w/o) Flow agents	45.4
MMIR (w/o) S-agent (RGB)	45.8
MMIR (w/o) T-agent (RGB)	45.3
MMIR	**46.1**

Overall Evaluation. In addition, we also evaluate the overall effect of our instance refinement strategy (**IR**) by comparing it with the case of **Adversarial-only** in Table 3. We give a detailed illustration of our results as follows.

- *Adversarial-only.* Compared with Source-only, an improvement of 4.4% in top-1 accuracy can be observed. This shows that our domain adversarial alignment is effective in improving model performance on target domain through directly training a domain discriminator in an adversarial way.
- *Adversarial + IR.* Compared with Adversarial-only, it has an overall performance boost of 1.4% and shows an improvement in every domain setting as depicted in Table 3. This is an effective demonstration of the successful implementation of our instance refinement strategy and its capability to help alleviate negative transfer during cross-domain action recognition.

Table 3. Overall evaluation of our instance refinement strategy.

Method	Adversarial	IR	D2 → D1	D3 → D1	D1→ D2	D3 → D2	D1 → D3	D2 → D3	Mean
Source-only			42.5	44.3	42.0	56.3	41.2	46.5	45.5
MMIR	✓		43	51.8	49.3	59.9	43.5	51.6	49.9
MMIR	✓	✓	**46.1**	**53.5**	**49.7**	**61.5**	**44.5**	**52.6**	**51.3**

5 Conclusion

We design a multi-modal instance refinement framework to help alleviate the problem of negative transfer during cross-domain action recognition. The reinforcement learning agents are trained to learn policies to select out negative training samples, thus resulting in a better-aligned feature distribution via domain adversarial learning. Experiments show that our method successfully addresses the negative transfer in multi-modal cross-domain action recognition and outperforms several competitive methods on a benchmark dataset. We believe in the future, it is worth conducting experiments on a spectrum of datasets to validate if our MMIR could be generalized to all use cases and even in different modalities such as text, speech, depth and so on.

Acknowledgements. This work is supported by the Major Project for New Generation of AI under Grant No. 2018AAA0100400, National Natural Science Foundation of China No. 82121003, and Shenzhen Research Program No. JSGG20210802153537009.

References

1. Carreira, J., Zisserman, A.: Quo vadis, action recognition? A new model and the kinetics dataset. In: Proceedings of the IEEE Conference on Computer Vision and Pattern Recognition, pp. 6299–6308 (2017)

2. Chen, C.F., et al.: Deep analysis of cnn-based spatio-temporal representations for action recognition. In: The IEEE Conference on Computer Vision and Pattern Recognition (CVPR) (2021)
3. Chen, J., Wu, X., Duan, L., Chen, L.: Sequential instance refinement for cross-domain object detection in images. IEEE Trans. Image Process. **30**, 3970–3984 (2021)
4. Damen, D., et al.: Scaling egocentric vision: the epic-kitchens dataset. In: Proceedings of the European Conference on Computer Vision (ECCV), pp. 720–736 (2018)
5. Dong, W., Zhang, Z., Tan, T.: Attention-aware sampling via deep reinforcement learning for action recognition. In: Proceedings of the AAAI Conference on Artificial Intelligence, vol. 33, pp. 8247–8254 (2019)
6. Dosovitskiy, A., et al.: An image is worth 16×16 words: transformers for image recognition at scale. arXiv preprint arXiv:2010.11929 (2020)
7. Ganin, Y., et al.: Domain-adversarial training of neural networks. J. Mach. Learn. Res. **17**(1), 2096–2030 (2016)
8. Ji, S., Xu, W., Yang, M., Yu, K.: 3d convolutional neural networks for human action recognition. IEEE Trans. Pattern Anal. Mach. Intell. **35**(1), 221–231 (2012)
9. Karpathy, A., Toderici, G., Shetty, S., Leung, T., Sukthankar, R., Fei-Fei, L.: Large-scale video classification with convolutional neural networks. In: Proceedings of the IEEE Conference on Computer Vision and Pattern Recognition, pp. 1725–1732 (2014)
10. Kim, D., et al.: Learning cross-modal contrastive features for video domain adaptation. In: Proceedings of the IEEE/CVF International Conference on Computer Vision, pp. 13618–13627 (2021)
11. Kingma, D.P., Ba, J.: Adam: a method for stochastic optimization. arXiv preprint arXiv:1412.6980 (2014)
12. Li, Y., Wang, N., Shi, J., Hou, X., Liu, J.: Adaptive batch normalization for practical domain adaptation. Pattern Recogn. **80**, 109–117 (2018)
13. Lin, L.J.: Self-improving reactive agents based on reinforcement learning, planning and teaching. Mach. Learn. **8**, 293–321 (1992)
14. Long, M., Cao, Y., Wang, J., Jordan, M.: Learning transferable features with deep adaptation networks. In: International Conference on Machine Learning, pp. 97–105. PMLR (2015)
15. Mnih, V., et al.: Playing atari with deep reinforcement learning. arXiv preprint arXiv:1312.5602 (2013)
16. Mnih, V., et al.: Human-level control through deep reinforcement learning. Nature **518**(7540), 529–533 (2015)
17. Munro, J., Damen, D.: Multi-modal domain adaptation for fine-grained action recognition. In: Proceedings of the IEEE/CVF Conference on Computer Vision and Pattern Recognition, pp. 122–132 (2020)
18. Saito, K., Watanabe, K., Ushiku, Y., Harada, T.: Maximum classifier discrepancy for unsupervised domain adaptation. In: Proceedings of the IEEE Conference on Computer Vision and Pattern Recognition, pp. 3723–3732 (2018)
19. Tran, D., Bourdev, L., Fergus, R., Torresani, L., Paluri, M.: Learning spatiotemporal features with 3d convolutional networks. In: Proceedings of the IEEE International Conference on Computer Vision, pp. 4489–4497 (2015)
20. Tzeng, E., Hoffman, J., Saenko, K., Darrell, T.: Adversarial discriminative domain adaptation. In: Proceedings of the IEEE Conference on Computer Vision and Pattern Recognition, pp. 7167–7176 (2017)

21. Wang, H., Schmid, C.: Action recognition with improved trajectories. In: Proceedings of the IEEE International Conference on Computer Vision, pp. 3551–3558 (2013)
22. Wang, R., et al.: Masked video distillation: rethinking masked feature modeling for self-supervised video representation learning. arXiv preprint arXiv:2212.04500 (2022)
23. Wang, X., Chen, W., Wu, J., Wang, Y.F., Wang, W.Y.: Video captioning via hierarchical reinforcement learning. In: Proceedings of the IEEE Conference on Computer Vision and Pattern Recognition, pp. 4213–4222 (2018)
24. Wang, Y., et al.: Internvideo: general video foundation models via generative and discriminative learning. arXiv preprint arXiv:2212.03191 (2022)
25. Weng, J., Jiang, X., Zheng, W.L., Yuan, J.: Early action recognition with category exclusion using policy-based reinforcement learning. IEEE Trans. Circ. Syst. Video Technol. **30**(12), 4626–4638 (2020)
26. Xu, Y., Yang, J., Cao, H., Wu, K., Wu, M., Chen, Z.: Source-free video domain adaptation by learning temporal consistency for action recognition. In: Computer Vision-ECCV 2022: 17th European Conference, Tel Aviv, Israel, 23–27 October 2022, Proceedings, Part XXXIV. pp. 147–164. Springer, Heidelberg (2022). https://doi.org/10.1007/978-3-031-19830-4_9
27. Zhou, M., et al.: Reinforcenet: a reinforcement learning embedded object detection framework with region selection network. Neurocomputing **443**, 369–379 (2021)

Modality Interference Decoupling and Representation Alignment for Caricature-Visual Face Recognition

Yang Xu[1,2,3], Junyi Wu[3,4], Yan Yan[1,2(✉)], Xinsheng Du[3], Huiji Zhang[3], Jianqiang Zhao[3], and Zhipeng Gao[3]

[1] Institute of Artificial Intelligence, Xiamen University, Xiamen, China
[2] State Key Laboratory of Integrated Services Networks (Xidian University), Xi'an, China
yanyan@xmu.edu.cn
[3] AI Research Center, Xiamen Meiya Pico Information Co., Ltd., Xiamen, China
[4] Xiamen Meiya Pico Information Security Research Institute Co., Ltd., Xiamen, China

Abstract. Cross-modality face recognition aims to match facial images across different modalities. This task becomes very challenging when one of the modalities is the facial caricature, which enhances instinctive facial features through extreme distortions and exaggerations with diverse styles by artists. In this paper, we develop a novel modality interference decoupling and representation alignment (MIR) method for visual-caricature face recognition. Our MIR method consists of a backbone network, an identity-interference orthogonal decoupling (IIOD) module, and a modality feature alignment (MFA) module. The IIOD module adopts a three-branch structure to decouple the deep semantic features extracted by the backbone network into identity features and modality features. In IIOD, we design an identity subspace alignment (ISA) module to align the identity features from different branches. Moreover, we design the MFA module to perform feature alignment between the modality feature from the IIOD module and that from the pre-trained modality interference information encoder (MIE) via adversarial learning, extracting the modality-specific information. Based on the above designs, we can effectively alleviate the interference of modality differences and style differences, improving the final performance. Extensive experimental results on multiple datasets show that our proposed method outperforms several state-of-the-art cross-modality face recognition methods.

Keywords: Cross-modality face recognition · Feature decomposition · Representation Alignment

1 Introduction

With the rapid growth of multi-modality data, cross-modality face recognition has received increasing attention. A great number of methods [3,6–8,19,21],

© The Author(s), under exclusive license to Springer Nature Singapore Pte Ltd. 2024
Q. Liu et al. (Eds.): PRCV 2023, LNCS 14425, pp. 297–308, 2024.
https://doi.org/10.1007/978-981-99-8429-9_24

Fig. 1. Examples of two facial images and their corresponding facial caricatures with diverse artistic styles. The images are from the CaVI dataset [4].

which aim to match cross-modality facial images (such as near-infrared images and visible-light images) collected from the real world, have been developed and achieved promising performance in many practical applications.

Despite much progress, cross-modality face recognition is still challenging, especially when one of the modalities is a facial caricature. Facial caricatures are often created by artists, which enhance certain instinctive facial features through extreme distortions and exaggerations. Moreover, facial caricatures also exhibit significant internal morphological differences due to the diversity of artistic styles. Some examples are given in Fig. 1. Therefore, both intra-class and inter-class differences in facial caricatures can be significant. In this paper, we study caricature-visual face recognition, which matches facial images between caricature and visual (i.e., visible-light) modalities.

Facial caricatures involve exaggerated deformations and the stylistic diversity of the caricatures (such as oil painting and sketching). Hence, it is difficult to completely eliminate the interference of modality information when extracting modality invariant features or projecting into a common subspace by existing cross-modality face recognition methods. Moreover, converting facial caricatures (or visual facial images) to visual facial images (or facial caricatures) is not a trivial task. Therefore, how to develop a method specifically designed for caricature-visual face recognition merits further study.

In this paper, we develop a novel modality interference decoupling and representation alignment (MIR) method, which consists of a backbone network, an identity-interference orthogonal decoupling (IIOD) module, and a modality feature alignment (MFA) module, for caricature-visual face recognition.

Specifically, a lightweight modality interference information encoder (MIE) is first pre-trained as a teacher model to encode the modality interference information. Then, an identity-interference orthogonal decoupling (IIOD) module is developed to decouple the modality interference information (contained in the caricature face recognition branch and the visual face recognition branch) and the identity information. In IIOD, we design an identity subspace alignment (ISA) module to align the identity features from different branches based on a consistency loss. Next, the modality features are fed into a modality feature alignment (MFA) module to perform alignment with the modality features obtained from the pre-trained model via adversarial learning. In this way, the modality-specific knowledge from different modalities can be effectively captured. Based on the above designs, we obtain identity features by removing the modality differences and style differences for cross-modality caricature-visual face recognition.

In summary, our contributions are summarized as follows:

- We develop a novel MIR method to significantly alleviate the interference caused by modality differences by taking advantage of a pre-trained MIE.
- We extract different identity features from subspace mapping and feature decoupling while minimizing their distances with a consistency loss. Such a way can effectively extract modality-irrelevant identity information.
- We extensively evaluate our method on several popular caricature-visual facial datasets and show its superiority against state-of-the-art methods.

2 Related Work

Cross-Modality Face Recognition. Existing methods can be divided into three main categories: modality-invariant feature-based, common subspace learning-based, and image generation-based methods.

Modality-invariant feature-based methods [6,16,18,19] extract identity features invariant across different modalities. Conventional methods mainly focus on designing hand-crafted features. Recently, deep learning-based methods have been developed by reducing the distances between cross-modality features with the triplet loss [16] or the center loss [18]. W-CNN [6] captures modality-invariant depth features by minimizing the Wasserstein distances between different modality features. Wu *et al.* [19] design a disentangled variational representation (DVR) framework on the top of W-CNN to separate the identity information and modality variations from heterogeneous faces.

The common subspace-based methods [5,7,20] project facial images from different modalities into a compact common subspace, making the feature distances between the same identities close and those between different identities far apart. He *et al.* [5] map high-level facial features onto two orthogonal subspaces to encode domain-invariant identity information and domain-dependent spectral information. Wu *et al.* [20] introduce coupled deep learning (CDL) based on nuclear norm constraints on the fully-connected layer to alleviate overfitting. Hu *et al.* [7] perform disentanglement with an orthogonal constraint and then learn residual-invariant representations by aligning high-level features of non-neutral and neutral faces.

Image generation-based methods [3,12,21] usually transform facial images from one modality to another so that the images can be matched in the same modality. Yang *et al.* [21] generate facial images using generative adversarial networks, which greatly enrich the diversity of attributes of synthetic images. HFIDR [12] explicitly interprets semantic information of facial representations to synthesize cross-modality facial images. DVG-face [3] generates heterogeneous facial images with the same identity.

The above methods are mainly developed based on cross-modality facial images captured from the real world. However, facial caricatures involve exaggerated and distorted transformations. This substantially increases the difficulty of learning suitable common spaces/modality-invariant features or generating new modality features. In this paper, we consider using the modality features obtained from a pre-trained model to guide the decoupling process of identity features, where we combine feature decoupling with the common subspace mapping to extract modality-irrelevant identity features.

Caricature-Visual Face Recognition. Existing caricature-visual face recognition methods mainly focus on extracting modality-irrelevant identity information. Huo *et al.* [9] perform feature extraction from different scales and angles according to facial landmarks for caricature and visual facial feature alignment. Li *et al.* [11] use multiple Siamese sub-networks to extract local and global features to construct a unified feature representation and similarity learning framework. Dai *et al.* [1] introduce gating to fuse local and global features and use a convolutional attention module to improve feature discrimination. Mishra *et al.* [14] propose a nonlinear transformation method to map features from different modalities onto a common subspace. Garg *et al.* [4] decouple shared and modality-specific representations with orthogonal constraints and classifies facial identities by using a combination of shared and modality-specific representations.

The above methods can partially address the problem of the large modality gap between caricature and visual facial images. But the facial caricature itself contains a variety of different styles, such as sketch, oil painting, watercolor, etc. These intra-class differences pose a great challenge for identity decoupling. In this paper, we not only perform feature decoupling guided by a pre-trained model, but also construct the consistency loss between the identity features obtained from the modality-specific branch and those obtained from the common subspace, further eliminating the influence of caricature style diversity on identity features.

3 Proposed Method

3.1 Overview

In this paper, we study caricature-visual face recognition. Formally, the whole training set consists of a facial caricature subset $\mathcal{D}_c = \{\mathbf{x}_{i,c}, y_{i,c}\}_{i=1}^{N_c}$ and a visual facial image subset $\mathcal{D}_v = \{\mathbf{x}_{j,v}, y_{j,v}\}_{j=1}^{N_v}$, where $\mathbf{x}_{i,c}$ and $y_{i,c}$ denote the i-th facial caricature and its corresponding identity label in $\mathcal{D}_c$, respectively; $\mathbf{x}_{j,v}$ and $y_{j,v}$ denote the j-th visual facial image and its corresponding identity label in $\mathcal{D}_v$, respectively; N_c and N_v represent the numbers of images in $\mathcal{D}_c$ and $\mathcal{D}_v$, respectively. The test set $\mathcal{D}_t$ also contains both facial caricatures and visual facial images. During each training batch, three images are randomly selected from two modality subsets to form a triplet set $\{\mathbf{x}_k, \mathbf{x}_{k,p}, \mathbf{x}_{k,n}\}_{k=1}^{N}$ as the input to the model, where we randomly select $\mathbf{x}_k$ as the anchor from $\mathcal{D}_v$, $\mathbf{x}_{k,p}$ with the same identity label as $\mathbf{x}_k$ from $\mathcal{D}_c$ as the positive image of the triplet, and $\mathbf{x}_{k,n}$ with a different identity label from $\mathcal{D}_v$ as the negative image of the triplet. N is the number of triplets constructed in each batch.

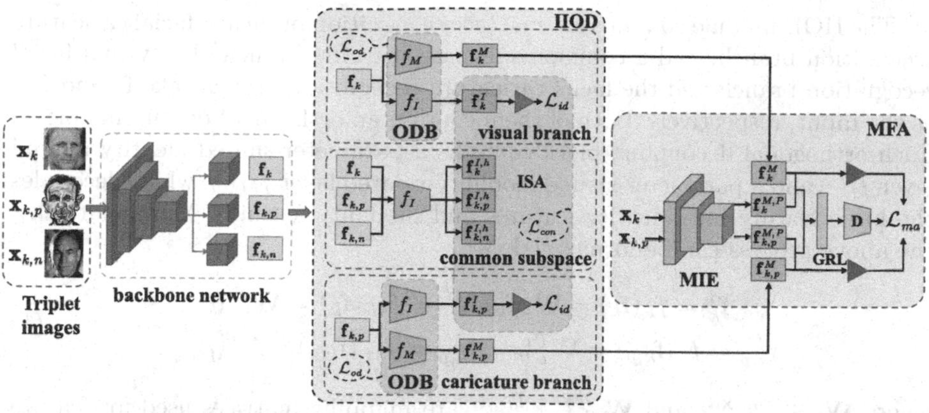

Fig. 2. Overview of our MIR method. It consists of a backbone network, an identity-interference orthogonal decoupling (IIOD) module, and a modality feature alignment (MFA) module for cross-modality caricature-visual face recognition. ISA denotes identity subspace alignment.

In this paper, we develop a new modality interference decoupling and representation alignment (MIR) method consisting of a backbone network, an identity-interference orthogonal decoupling (IIOD) module, and a modality feature alignment (MFA) module for caricature-visual face recognition. The overview of our method is shown in Fig. 2.

More specifically, a lightweight modality interference information encoder (MIE) is first pre-trained as a teacher model to encode the modality interference information. Given a triplet, we pass through it into the backbone network to extract the basic features of different modality images. In this paper, we adopt Inception-ResNet-v1 [17] as the backbone network. Then, the basic features of the triplet are fed into the IIOD module, which decouples the caricature facial and the visual facial features to separate the identity information and modality interference information. At the same time, the different modality facial images are fed into a common feature space in the shared branch. The consistency between the identity features in the common feature space and the decoupled identity features is constructed to align identity features by an identity subspace alignment (ISA) module. Next, an MIE is used in the MFA module to learn modality interference information. Based on it, the consistency of the distribution between the pre-trained modality features and the decoupled modality features is constructed to perform feature alignment via adversarial learning, improving the separation of modality interference information. Finally, we can obtain identity features by removing the modality differences and style differences for caricature-visual face recognition.

3.2 Identity-Interference Orthogonal Decoupling (IIOD) Module

The triplet images $\{\mathbf{x}_k, \mathbf{x}_{k,p}, \mathbf{x}_{k,n}\}$ are passed through the backbone network to obtain the basic features $\mathbf{f}_k$, $\mathbf{f}_{k,p}$, and $\mathbf{f}_{k,n}$, respectively.

The IIOD module contains a visual face recognition branch, a facial caricature recognition branch, and a common subspace learning branch. The visual facial recognition branch and the facial caricature recognition branch take $\mathbf{f}_k$ and $\mathbf{f}_{k,p}$ as the input, respectively. Both of them contain an orthogonal decoupling block. Each orthogonal decoupling block contains a parameter-shared identity-related layer $f_I(\cdot)$ and a parameter-shared modality-related layer $f_M(\cdot)$, which decouples the basic feature into identity features and modality features. Mathematically, the above process can be formulated as

$$\begin{aligned} \mathbf{f}_k^I &= f_I(\mathbf{f}_k) = \mathbf{W}_I^T \mathbf{f}_k,\ \mathbf{f}_k^M = f_M(\mathbf{f}_k) = \mathbf{W}_M^T \mathbf{f}_k, \\ \mathbf{f}_{k,p}^I &= f_I(\mathbf{f}_{k,p}) = \mathbf{W}_I^T \mathbf{f}_{k,p},\ \mathbf{f}_{k,p}^M = f_M(\mathbf{f}_{k,p}) = \mathbf{W}_M^T \mathbf{f}_{k,p}, \end{aligned} \tag{1}$$

where $\mathbf{W}_I \in \mathbb{R}^{d_0 \times d_1}$ and $\mathbf{W}_M \in \mathbb{R}^{d_0 \times d_1}$ are mapping matrices used for feature decomposition. $\mathbf{f}_k^I \in \mathbb{R}^{d_1}$ and $\mathbf{f}_{k,p}^I \in \mathbb{R}^{d_1}$ denote the identity features. $\mathbf{f}_k^M \in \mathbb{R}^{d_1}$ and $\mathbf{f}_{k,p}^M \in \mathbb{R}^{d_1}$ denote the decoupled visual facial modality features and caricature modality features, respectively. $\mathbf{f}_k \in \mathbb{R}^{d_0}$ and $\mathbf{f}_{k,p} \in \mathbb{R}^{d_0}$ denote the basic features from the backbone network.

To minimize the correlation between identity features and modality features, an orthogonality constraint is introduced, which is expressed as

$$\mathcal{L}_{od} = \cos\left(\theta_{ij}^{IM}\right) = \frac{{\mathbf{W}_I^i}^T \mathbf{W}_M^j}{\|\mathbf{W}_I^i\| \|\mathbf{W}_M^j\|} = 0\,(i, j = 1, 2, , d_1), \tag{2}$$

where θ_{ij}^{IM} indicates angle between the mapping vectors $\mathbf{W}_I^i$ (the i-th column of $\mathbf{W}_I$) and $\mathbf{W}_M^j$ (the j-th column of $\mathbf{W}_M$). The identity-related layer and modality-related layer aim to learn identity-specific representations and modality-specific representations, respectively. The orthogonal representation decorrelation is imposed on these two layers to enhance representations.

To effectively extract identity features, we leverage the commonly used cross-entropy loss, that is,

$$\mathcal{L}_{id} = -\frac{1}{N} \sum_{k=1}^{N} \sum_{c=1}^{C} \mathbb{I}_{[c=y_k]} \log\left(\mathcal{P}_{id}^v\left(\mathbf{f}_k^I\right)\right) + \mathbb{I}_{[c=y_k]} \log\left(\mathcal{P}_{id}^c\left(\mathbf{f}_{k,p}^I\right)\right), \tag{3}$$

where y_k denotes the corresponding label of identity features $\mathbf{f}_k^I$ and $\mathbf{f}_{k,p}^I$, respectively. C denote the total number of facial identities in the dataset $\mathcal{D}$. $\mathcal{P}_{id}^v(\cdot)$ and $\mathcal{P}_{id}^c(\cdot)$ are the prediction functions (an FC layer) for identity classification. $\mathbb{I}_{[c=y_k]}$ equals to 1 when $c = y_k$, and 0 otherwise.

The common subspace learning branch takes the basic features of the triplet images as the input and maps all features into a common subspace through an identity-related layer, obtaining the identity features $\mathbf{f}_k^{I,h}$, $\mathbf{f}_{k,p}^{I,h}$, and $\mathbf{f}_{k,n}^{I,h}$. In the common subspace, the distances between identity features from the same identity are reduced while those from different identities are enlarged by the triplet loss [16]. In addition, we expect that after separating the modality interference information, the remaining identity features also exist in the common identity subspace. Therefore, we propose to use the identity subspace alignment

(ISA), which constrains the consistency between the identity features between the visual face recognition branch (the facial caricature recognition branch) and the common subspace in the common subspace learning branch. The final identity consistency loss form is given as

$$\begin{aligned}\mathcal{L}_{con} &= \frac{1}{N}\sum_{k=1}^{N}\left[\left\|\mathbf{f}_k^{I,h} - \mathbf{f}_{k,p}^{I,h}\right\|_2^2 - \left\|\mathbf{f}_k^{I,h} - \mathbf{f}_{k,n}^{I,h}\right\|_2^2 + \alpha\right]_+ \\ &+ \frac{1}{N}\sum_{k=1}^{N}\left(\mathbf{f}_k^{I,h} - \mathbf{f}_k^{I}\right)^2 + \left(\mathbf{f}_{k,p}^{I,h} - \mathbf{f}_{k,p}^{I}\right)^2,\end{aligned} \tag{4}$$

where α denotes the margin parameter.

3.3 Modality Feature Alignment (MFA) Module

Based on the pre-trained MIE, the modality feature is fed into the MFA module for feature alignment.

MIE is pre-trained by using all the images in the training set $\mathcal{D}_{train} = \{\mathbf{x}_i, m_i\}_{i=1}^{N_p}$, where m_i denotes the modality label corresponding to $\mathbf{x}_i$, indicating that the image is a visual facial image or a facial caricature image. More specifically, MIE uses ResNet-18 as the feature extraction network $\mathcal{E}$, and converts the extracted features to modality features using a modality correlation layer $f_M(\cdot) : \mathbf{f}_i^M = f_M(\mathcal{E}(\mathbf{x}_i))$. The modality features are optimized using the cross-entropy loss, i.e.,

$$\mathcal{L}_{modal} = -\frac{1}{N}\sum_{i=1}^{N} m_i \log\left(\mathcal{P}_m\left(\mathbf{f}_i^M\right)\right) + (1-m_i)\log\left(1 - \mathcal{P}_m\left(\mathbf{f}_i^M\right)\right), \tag{5}$$

where m_i indicates that the feature belongs to the caricature modality ($m_i = 1$) or the visual modality ($m_n = 0$). $\mathcal{P}_m(\cdot)$ is the prediction function (an FC layer) for modality classification. After pre-training, the parameters in MIE are fixed.

Based on the pre-trained MIE, it extracts the modality features $\mathbf{f}_k^{M,P} = f_M(\mathcal{E}(\mathbf{x}_k))$ and $\mathbf{f}_{k,p}^{M,P} = f_M(\mathcal{E}(\mathbf{x}_{k,p}))$ of the anchor image and the positive image in the triplet, respectively. Then, the modality features $\mathbf{f}_k^{M,P}$, $\mathbf{f}_{k,p}^{M,P}$ obtained by MIE and the decoupled modality features $\mathbf{f}_k^M$, $\mathbf{f}_{k,p}^M$ obtained by IIOD are fed into MFA. MFA contains a discriminator $\mathcal{D}$ with a gradient reversal layer, and a modality classifier $\mathcal{S}$. We leverage adversarial learning to ensure that the distribution of the decoupled modality features is consistent with the modality features obtained from MIE. A negative log-likelihood loss $\mathcal{L}_d(\cdot,\cdot)$ is used at the end of the discriminator $\mathcal{D}$, and a cross-entropy loss $\mathcal{L}_s(\cdot,\cdot)$ is used at the end of the modality classifier $\mathcal{S}$. Finally, the modality distribution alignment loss is expressed as

$$\begin{aligned}\mathcal{L}_{ma} &= \frac{1}{N}\sum_{k=1}^{N}\mathcal{L}_s\left(\mathbf{f}_k^M, m_k\right) + \mathcal{L}_s\left(\mathbf{f}_{k,p}^M, m_{k,p}\right) \\ &- \lambda_p\frac{1}{N}\sum_{k=1}^{N}\mathcal{L}_d\left(\mathbf{f}_k^M, s\right) + \mathcal{L}_d\left(\mathbf{f}_k^{M,P}, s^P\right) + \mathcal{L}_d\left(\mathbf{f}_{k,p}^M, s\right) + \mathcal{L}_d\left(\mathbf{f}_{k,p}^{M,P}, s^P\right),\end{aligned} \tag{6}$$

where m_k and $m_{k,p}$ denote the modality labels corresponding to $\mathbf{f}_k^M$ and $\mathbf{f}_{k,p}^M$, respectively; s indicates that the feature is derived from IIOD, and s^P indicates that the feature is derived from MIE; $\lambda_p = \frac{2}{1+\exp(-\gamma \cdot p)} - 1$ is the dynamic weight. Here, p is the ratio of the number of current iterations to the total number of iterations, and γ is a constant (we set it to 10 by default).

3.4 Joint Loss Function

Finally, the joint loss function is given as

$$\mathcal{L}_{vcr} = \mathcal{L}_{id} + \mathcal{L}_{od} + \mathcal{L}_{con} + \mathcal{L}_{ma}. \tag{7}$$

4 Experiments

4.1 Datasets

The WebCaricature dataset [10] is a popular caricature-visual facial dataset consisting of 6,042 facial caricatures and 5,974 visual facial images from 252 identities collected from the web. **The CaVI dataset** [4] contains images of 205 identities. There are 5,091 facial caricatures ranging from 10 to 15 images per identity and 6,427 visual facial images ranging from 10 to 15 images per identity. **The IIIT-CFW dataset** [15] includes a total of 8,928 cartoon characters (including caricatures, sketches, and paintings) from 100 identities. Each identity also provides 10 visual facial images.

4.2 Implementation Details

In our experiments, we use RetinaFace [2] to automatically detect all facial caricatures and visual facial images in the dataset. The RetinaFace model is pre-trained on the WebCaricature dataset to ensure the detection of exaggerated facial caricatures. Each image is then cropped, aligned, and finally resized to the size of 160 × 160. The backbone network is modified based on Inception-Resnet-v1, using three identically structured "block8" modules to capture deep semantic features of the images specific to different branches. Inception-ResNet-v1 is pre-trained on the CASIA-WebFace dataset [22] to enable the model to extract the identity feature effectively. The deep semantic features extracted by the backbone network are fed into the pooling and FC layers to obtain the basic features of the size of $256 \times 3 \times 3$, and the 128-Dim identity feature and modality feature are obtained through an identity-related layer and a modality-related layer, respectively. MIE directly converts the features extracted by ResNet-18 into a 128-Dim modality features through the pooling and FC layers.

All the experiments are implemented by Pytorch and run on a single NVIDIA GTX3090. The model is trained for 200 epochs. We use the Adam optimizer with $\beta_1 = 0.9$ and $\beta_2 = 0.999$. The initial learning rate is set to $1 \times e^{-3}$, and the learning rate is decayed every 10 epochs until it reaches $1 \times e^{-5}$. In each

Table 1. The details of different variants of our method.

Methods	TL	MT	ODB	ISA	MFA
TL (Baseline)	✓				
TL+MT	✓	✓			
TL+MT+ODB	✓	✓	✓		
TL+MT+ODB+ISA	✓	✓	✓	✓	
TL+MT+ODB+MFA	✓	✓	✓		✓
TL+MT+ODB+ISA+MFA	✓	✓	✓	✓	✓

TL denotes the triplet loss. MT indicates the multi-branch structure.

Table 2. The average accuracy (%) obtained by different variants of our method on the WebCaricature, CaVI and IIIT-CFW datasets.

Methods	WebCaricature		CaVI		IIIT-CFW	
	C-to-V	V-to-C	C-to-V	V-to-C	C-to-V	V-to-C
TL (Baseline)	78.62	66.18	77.13	64.85	79.76	67.57
TL+MT	82.63	72.89	81.27	71.64	83.13	73.74
TL+MT+ODB	83.21	78.62	82.14	75.83	83.82	77.12
TL+MT+ODB+ISA	84.34	79.52	83.53	77.53	84.10	80.84
TL+MT+ODB+MFA	85.97	81.62	84.49	80.71	85.66	82.72
TL+MT+ODB+ISA+MFA	**86.69**	**82.12**	**85.77**	**82.05**	**86.64**	**83.42**

batch, we randomly select 64 anchor images with different identities and use them to construct 64 sets of triplet images for training. The test is validated using a 10-fold cross-validation. The final Rank-1 accuracy is obtained by the average of the ten-round tests. The cosine similarity is used to match the identity features in the gallery with the identity features extracted from the test images. We evaluate our method with two modes: C-to-V (Caricature-to-Visual) and V-to-C (Visual-to-Caricature), where C-to-V represents that facial caricatures and visual facial images are used as the probe set and the gallery set, respectively, and the other way around for V-to-C.

4.3 Ablation Studies

We evaluate the performance of several variants of our method on three datasets, where the baseline method is the FaceNet [16] fine-tuned on the caricature dataset. Table 1 summarizes the details of these variants. Table 2 gives the comparison results on these datasets.

Table 3. Performance comparison in terms of recognition accuracy (%) on WebCaricature, CaVI, IIIT-CFW datasets. The best results are marked in bold.

Methods	WebCaricature		CaVI		IIIT-CFW	
	C-to-V	V-to-C	C-to-V	V-to-C	C-to-V	V-to-C
DHFML [14]	75.48	72.62	75.52	73.08	77.26	73.95
GFDF [1]	60.39	65.41	60.12	66.04	62.61	66.40
CaVINet [4]	85.12	80.54	83.72	79.18	85.81	82.40
DDML [13]	85.85	81.94	84.26	80.38	86.13	82.59
MIR (Ours)	**86.69**	**82.12**	**85.77**	**82.05**	**86.64**	**83.42**

Influence of the Multi-Branch Structure. From Table 2, TL+MT outperforms the Baseline method, validating the effectiveness of the multi-branch structure. This is because the multi-branch structure extracts identity features from the common subspace branch and two modality-specific branches. This encourages the backbone to focus on extracting identity-related information.

Influence of the Orthogonal Decoupling Block (ODB). Compared with the TL+MT method, the variants that adopt the ODB structure achieve significant performance improvements in terms of the average recognition accuracy on three datasets. Note that the facial images contain identity-related information and identity-independent interference information. For facial images with different modalities, the modality interference information caused by modality differences further increases the difficulty of extracting identity-related information. ODB constructs the orthogonal loss between the identity feature mapping matrix and the modality feature mapping matrix to remove the redundant modality features from the identity features effectively.

Influence of the Identity Subspace Alignment (ISA) Module. From Table 2, compared with TL+MT+ODB, TL+MT+ODB+ISA achieves 1.02%, 1.55%, and 2.00% improvements in terms of the average accuracy on the Caricature, CaVI, and IIIT-CFW datasets, respectively. TL+MT+ODB+ISA+MFA also outperforms TL+MT+ODB+MFA. This shows the necessity of ISA. The identity features obtained by orthogonal decoupling are aligned with the identity features in the common subspace. On the one hand, ISA reduces the distances between the identity features obtained by decoupling in different modes. On the other hand, ISA encourages feature mapping in the common subspace to focus more on identity-related information.

Influence of the Modality Feature Alignment (MFA) Module. From Table 2, we can see that TL+MT+ODB+MFA outperforms TL+MT+ODB by a large margin and TL+MT+ODB+ISA+MFA achieves better performance than TL+MT+ODB+ISA. This shows the importance of the MFA module, which can guide the decoupling process of ODB through pre-trained MIE. The decoupling of modality features can thus effectively focus on the interference information due to modality differences.

4.4 Comparison with State-of-the-Art Methods

We compare our method with several state-of-the-art caricature-visual face recognition methods. Table 3 shows the results obtained by all competing methods on the WebCaricature, CaVI, and IIIC-CFW datasets.

Our proposed MIR method outperforms current caricature-visual face recognition methods on all three datasets. DHFML [14] directly maps different modality images into a common subspace. GFDF [1] combines local and global features to learn identity features. Both methods do not explicitly deal with the interference information caused by modality differences, leading to performance degradation. CaVINet [4] and DDML [13] try to extract modality-specific identity features in the common subspace, which still inevitably suffers from the interference of identity-irrelevant information. Our proposed method uses the MFA module and the IIOD module to separate the identity features and identity-irrelevant modality features, which effectively alleviates the influence of modality interference information on cross-modality caricature-visual face recognition.

5 Conclusion

In this paper, we propose a novel MIR method for caricature-visual face recognition. In our MIR method, an IIOD module decouples the identity features and modality features, where the identity features are aligned in a common subspace based on the ISA module. The MFA module transfers the learned modality interference information to the modality interference decoupled feature representation with the help of MIE. In this way, we are able to obtain modality-irrelevant identity features. Extensive experiments show the effectiveness of our method.

Acknowledgements. This work was supported by the National Natural Science Foundation of China under Grants 62372388, 62071404 and by the Natural Science Foundation of Fujian Province under Grant 2020J01001.

References

1. Dai, L., et al.: Gated fusion of discriminant features for caricature recognition. In: Proceedings of the International Conference on Intelligent Science and Big Data Engineering, pp. 563–573 (2019)
2. Deng, J., Guo, J., Ververas, E., Kotsia, I., Zafeiriou, S.: RetinaFace: single-shot multi-level face localisation in the wild. In: Proceedings of the IEEE Conference on Computer Vision and Pattern Recognition, pp. 5203–5212 (2020)
3. Fu, C., Wu, X., Hu, Y., Huang, H., He, R.: DVG-Face: dual variational generation for heterogeneous face recognition. IEEE Trans. Pattern Anal. Mach. Intell. **44**(6), 2938–2952 (2021)
4. Garg, J., Peri, S.V., Tolani, H., Krishnan, N.C.: Deep cross modal learning for caricature verification and identification (CaVINet). In: Proceedings of the ACM International Conference on Multimedia, pp. 1101–1109 (2018)

5. He, R., Wu, X., Sun, Z., Tan, T.: Learning invariant deep representation for NIR-VIS face recognition. In: Proceedings of the AAAI Conference on Artificial Intelligence, pp. 2000–2006 (2017)
6. He, R., Wu, X., Sun, Z., Tan, T.: Wasserstein CNN: learning invariant features for NIR-VIS face recognition. IEEE Trans. Pattern Anal. Mach. Intell. **41**(7), 1761–1773 (2018)
7. Hu, W., Hu, H.: Orthogonal modality disentanglement and representation alignment network for NIR-VIS face recognition. IEEE Trans. Circ. Syst. Video Technol. **32**(6), 3630–3643 (2021)
8. Hu, W., Hu, H.: Domain-private factor detachment network for NIR-VIS face recognition. IEEE Trans. Inf. Forensics Secur. **17**, 1435–1449 (2022)
9. Huo, J., Gao, Y., Shi, Y., Yin, H.: Variation robust cross-modal metric learning for caricature recognition. In: Proceedings of the on Thematic Workshops of ACM Multimedia, pp. 340–348 (2017)
10. Huo, J., Li, W., Shi, Y., Gao, Y., Yin, H.: WebCaricature: a benchmark for caricature recognition. arXiv preprint arXiv:1703.03230 (2017)
11. Li, W., Huo, J., Shi, Y., Gao, Y., Wang, L., Luo, J.: A joint local and global deep metric learning method for caricature recognition. In: Proceedings of the Asian Conference on Computer Vision, pp. 240–256 (2019)
12. Liu, D., Gao, X., Peng, C., Wang, N., Li, J.: Heterogeneous face interpretable disentangled representation for joint face recognition and synthesis. IEEE Trans. Neural Netw. Learn. Syst. **33**(10), 5611–5625 (2021)
13. Ming, Z., Burie, J.C., Muzzamil Luqman, M.: Dynamic deep multi-task learning for caricature-visual face cecognition. In: Proceedings of the International Conference on Document Analysis and Recognition Workshops, pp. 92–97 (2019)
14. Mishra, A.: DHFML: deep heterogeneous feature metric learning for matching photograph and caricature pairs. Int. J. Multimedia Inf. Retr. **8**(3), 135–142 (2019)
15. Mishra, A., Rai, S.N., Mishra, A., Jawahar, C.: IIIT-CFW: a benchmark database of cartoon faces in the wild. In: Proceedings of the European Conference on Computer Vision, pp. 35–47 (2016)
16. Schroff, F., Kalenichenko, D., Philbin, J.: FaceNet: a unified embedding for face recognition and clustering. In: Proceedings of the IEEE Conference on Computer Vision and Pattern Recognition, pp. 815–823 (2015)
17. Szegedy, C., Ioffe, S., Vanhoucke, V., Alemi, A.: Inception-v4, Inception-ResNet and the impact of residual connections on learning. In: Proceedings of the AAAI Conference on Artificial Intelligence, pp. 4278–4284 (2017)
18. Wen, Y., Zhang, K., Li, Z., Qiao, Y.: A discriminative feature learning approach for deep face recognition. In: Proceedings of the European Conference on Computer Vision, pp. 499–515 (2016)
19. Wu, X., Huang, H., Patel, V.M., He, R., Sun, Z.: Disentangled variational representation for heterogeneous face recognition. In: Proceedings of the AAAI Conference on Artificial Intelligence, pp. 9005–9012 (2019)
20. Wu, X., Song, L., He, R., Tan, T.: Coupled deep learning for heterogeneous face recognition. In: Proceedings of the AAAI Conference on Artificial Intelligence, pp. 1679–1686 (2018)
21. Yang, Z., Liang, J., Fu, C., Luo, M., Zhang, X.Y.: Heterogeneous face recognition via face synthesis with identity-attribute disentanglement. IEEE Trans. Inf. Forensics Secur. **17**, 1344–1358 (2022)
22. Yi, D., Lei, Z., Liao, S., Li, S.Z.: Learning face representation from scratch. arXiv preprint arXiv:1411.7923 (2014)

Plugging Stylized Controls in Open-Stylized Image Captioning

Jie Wang, Yixiao Zheng, Ruoyi Du, Yiming Zhang, Kongming Liang(✉), and Zhanyu Ma

Beijing University of Posts and Telecommunications, Beijing 100876, China
liangkongming@bupt.edu.cn

Abstract. Image captioning is a classical multi-modal task for vision-language understanding. In recent years, researchers have begun to focus on generating captions with personalized styles, but the range of available styles is often fixed. The existing methods for Stylized Image Captioning Generation are mainly done by reinforcement learning or contrastive learning. Even with the assistance of large models such as CLIP and GPT, previous methods still require fine-tuning to generate targeted style captions and these methods necessitate a certain amount of computational resources and training costs. In this paper, we design a Plug-in Stylized Controls Module (PSCM), which can be directly inserted into the text-generation procedure of a well-trained model to generate open-stylized captions. Specifically, PSCM uses the style factor and fluency factor to guide the text-generation decoder. The style factor helps to generate text in a specified style, while the fluency factor helps to improve the fluency of generated texts. PSCM is a straightforward yet effective plug-and-play module that can readily produce open-stylized captions for each image without fine-tuning of backbones and text-image paired training data in target style. We add PSCM to two existing Stylized Image Captioning models and conduct experiments on four datasets to demonstrate the effectiveness of this module.

Keywords: open-stylized image captioning · language modelling

1 Introduction

Image captioning (IC) is a foundational task that combines natural language processing and computer vision. The IC task aims to generate texts to describe images (e.g. visual attributes [9,10] or relationships [11]), and there is growing interest in describing images with texts in different styles (called stylized IC). This research has practical applications in areas like children's education and personalized conversations. For instance, when generating poetry or advertising slogans, it's necessary not only to accurately describe the content of the image but also to reflect a specific style in word choice and grammatical structure.

In earlier studies [15,21], Convolutional Neural Networks (CNNs) or object detectors are employed as encoder for extracting image features, which are then

© The Author(s), under exclusive license to Springer Nature Singapore Pte Ltd. 2024
Q. Liu et al. (Eds.): PRCV 2023, LNCS 14425, pp. 309–320, 2024.
https://doi.org/10.1007/978-981-99-8429-9_25

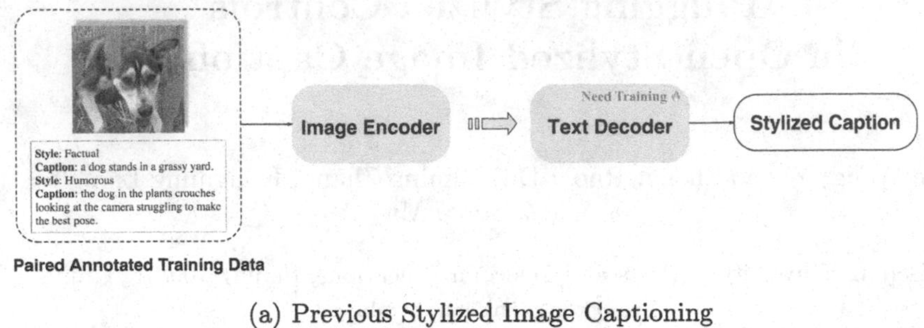

(a) Previous Stylized Image Captioning

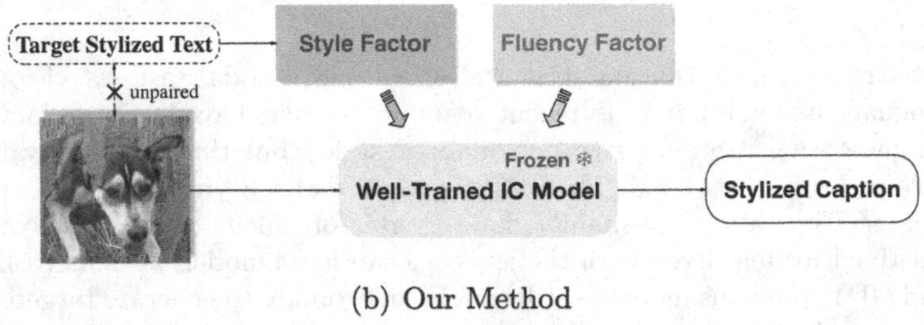

(b) Our Method

Fig. 1. Our method compared with previous stylized IC model. We do not need paired datasets or additional training of text generation models.

used to generate captions through language models built on long short-term memory networks (LSTM [7]) or Transformer [20]. However, with the emergence of pre-trained models in the multimodal field of text and image, such as CLIP [16] and ALIGN [8], coupled with the ongoing advancement of large language models (LLM) like BERT [3] and GPT [17], a number of researchers [13,18,19] have begun integrating these pre-trained models to fulfill image captioning tasks. CLIP is used to extract image features, which are then fed into the LLM along with style control to generate image captions in the desired style. In terms of style control, Memcap [23] employs reinforcement learning and SACO [24] uses contrastive learning to generate stylized captions.

Training the model to comprehend style information typically requires multiple iterations which is computationally intensive and time-consuming. Additionally, previous studies rely on using factual and stylized corpus pairs or image-text pairs as training data for the model. In reality, styled caption datasets offer only a limited range of styles, and it is challenging to build corpus or image-text pairs with desired additional styles. As a result, most Stylized IC approaches can only generate captions with styles limited by existing datasets, such as positive or negative emotions provided in [22–24] dataset, and humor or romance provided in [2,5,14]dataset. Nevertheless, in practical scenarios, it is crucial to generate captions in various styles rather than being confined to fixed style categories found in existing Stylized IC datasets. In the text-generation phase, most existing approaches utilize maximization-based decoding methods like beam search to aid LLM in generating captions [13,14]. However, these decoding methods often result in text outputs with poor fluency (such as duplicate words) which negatively impact the quality of IC sentences.

To address these limitations, in this paper, we introduce a plug-and-play module named Plug-in Stylized Controls Module (PSCM). The PSCM comprises two essential components: a style factor and a fluency factor. With the help of an additional text style classifier, the style factor guides the backbone model to form texts in target style. In the meanwhile, the fluency factor is computed based on the similarity between predicted tokens and previous tokens. The combination of these two factors empowers the backbone model to generate stylized results. As shown in Fig. 1, PSCM offers the advantage of not requiring fine-tuning of text generation models to imbue LLMs with target style knowledge. Moreover, our method is not resource-intensive. PSCM can be seamlessly integrated with any pre-trained IC model, facilitating straightforward inference during the text decoding stage. Importantly, the incorporation of this module enhances the IC model's capability to generate diverse stylized captions. By leveraging just a few target stylized texts instead of text-image pairs in target style, we can swiftly generate more fluent stylized captions.

The main contributions of this paper include: (1) We propose a plug-and-play module, PSCM, that can be used on encoder-decoder-shaped IC models to generate open-stylized caption simply and efficiently. (2) Unlike traditional methods, our module does not rely on image-text pairs or factual-stylized corpus pairs for training. Working with style factor and fluency factor, it only requires target style text to generate captions in the desired style. (3) We conduct experiments on four datasets, SentiCap [12], FlickrStyle10k [5], News Articles Corpus [4] and Shakespeare plays [14], to demonstrate the effectiveness of PSCM.

2 Related Work

2.1 Image Captioning

Earlier work [5,6,23] used CNNs or object detection model as encoder to extract image features and then used text generation models (such as RNN/LSTM) as decoder. Most of them require a large number of text-image pairs to train the model. The CLIP model [16] was proposed to break the modality difference between text and image and recent works [13,18,23] used the CLIP model to extract image features instead of CNNs and object detectors. Subsequently, the emergence of LLMs [3,17] with superb text comprehension ability made researchers realize that they can generate captions directly using pre-trained LLMs, which is more efficient and simpler. Therefore, how to smoothly transform the feature embedding extracted by CLIP into the input of LLMs to generate image-related text becomes a key part of the task.

The ClipCap model [13] proposed by Mokady et al. combines CLIP and GPT2 for the first time. MAGIC [18] and CapDec [14], also use CLIP to fine-tune language model (LM) with only text data. While these methods reduce the dependence on image-text pairs, they still require significant training. Tewel et al. proposed ZeroCap [19], a zero-shot model that updates the context cache using a language model to compute similarity between predicted text and image. These methods directly use LLM as decoder to generate results, most of them use maximization-based decoding methods in the process of generating text. These decoding methods might produce duplicate tokens, which would produce wrong sentences.

In this paper, we study a different setting for IC that generates image captions with specific target styles. The captions produced by traditional IC models typically constitute "average" sentences lacking style information [1]. To address this issue, we propose the Plug-in Stylized Controls Module (PSCM), and insert the PSCM into the text decoding process of the well-trained IC model to govern the selection of predicted tokens and get the open-stylized image captions.

2.2 Stylized Image Captioning

With the development of IC, some researchers are focusing on the topic of creating stylistic captions. The generated text in stylized IC must not only be relevant to the image but also adhere to a particular style. How to stylize the text has become another key issue. Memcap [23] used a Style Memory Module to store style information and optimize the model with a self-critical training strategy. SACO [24] used contrastive learning with pre-trained models to generate texts. Both methods require supervised training with text-image pairs or normal-stylized text pairs as input. CapDec [14] reduced model's dependence on data by transforming captions into different styles, but requires extra training to adapt to new styles. Conzic [22] avoided extensive training by using a BERT encoder to incorporate text-image similarity, sentence fluency, and style into the decoding process, but requires fixed sentence length and a complicated, inference procedure.

In this paper, we design style factor and fluency factor to control LM to generate more diverse stylized captions. Compared with the previous works, our method dose not rely on stylized captions paired with images. Instead, the proposed PSCM can help LM generate captions in desired styles with only target-stylized texts. Moreover, our approach eliminates the need to fine-tune the LM. We simply train a target-style text classifier, which is used to guide the LM to generate captions with the desired style. This method reduces device dependency and cuts training costs, given that we can even train the text classifier on a CPU.

3 Method

Aiming for Stylized IC task, we create a plug-and-play module called Plug-in Stylized Controls Module (PSTM), which consists of two parts, the style factor and the fluency factor. As shown in Fig. 2, PSCM is plugged directly into the text decoding part of the IC model and chooses the next predicted word by combining the scores of the style factor and the fluency factor. In the text decoding process of the IC model, x_t denotes the selection of the output token at time step t with two factors. v denotes the candidate which belongs to the set of top-k

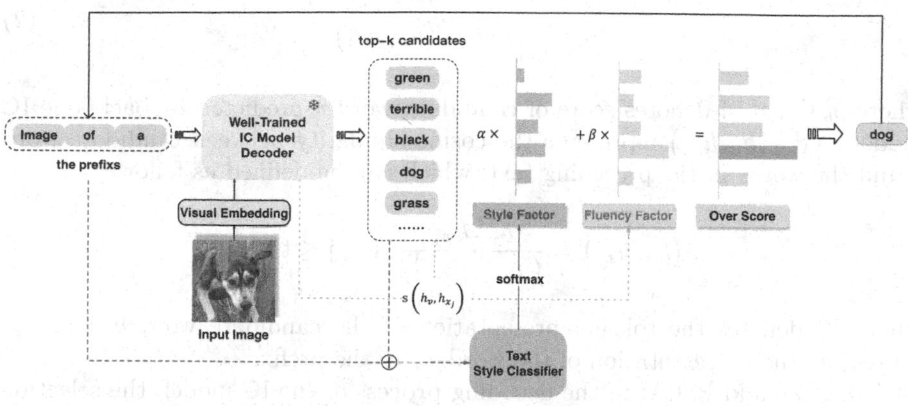

Fig. 2. An overview of our method PSCM, which is used in the well-trained IC model's decoding process. ⊕ denotes concating the prefixs and the candidate word. And $\mathrm{s}\left(h_v, h_{x_j}\right)$ is defined by Eq. 3.

predictions $V^{(k)}$ from the language model's confidence. The hyperparameters α and β control the respective degrees of action for the two factors. The style factor and fluency factor will each be introduced in turn in the sections that follow.

3.1 Style Factor

Given the image i, the prefix $x_{<t}$ and the target style L, the style factor $S_{x_{<t},v}^{i,L}$ is defined as the score that measures how closely adding candidate word v to the prefix match the style L. We use a text classifiers with Softmax function to evaluate the correlation between the candidate captions and the target style L. The style factor is then defined as follow:

$$S_{x_{<t},v}^{i,L} = \mathrm{Softmax}\left(\mathrm{Classifier}^{L}\left[x_{<t} : v\right]\right), \tag{1}$$

where $[x_{<t} : v]$ denotes the concatenation of prefix $x_{<t}$ and candidate word v. For the generation of captions in different target styles, Our method is simpler and more effective since we only need to train the text style classifier specifically for the target type of text, rather than fine-tuning the model using image-text pairs as in the previous approach.

3.2 Fluency Factor

The fluency factor aims to calculate the similarity between the candidate word v and the preceding text. This factor incorporates a penalty term, the similarity, to minimize the occurrence of duplicate words by selecting tokens that exhibit low similarity to the preceding text. Given time step t, candidate word v, and the prefix $x_{<t}$, the fluency factor $F_{x_{<t},v}^{i}$ is defined as

$$F_{x_{<t},v}^{i} = (1-\beta) \times p_{\theta}(v \mid x_{<t}) - \beta \times \max_{1 \leq j \leq t-1} \mathrm{s}(h_v, h_{x_j}), \tag{2}$$

where $p_{\theta}(v \mid x_{<t})$ denotes score of candidate word v predicted by backbone IC model, and $\mathrm{s}(h_v, h_{x_j})$ represents the cosine similarity between candidate word v and the words in the preceding text, which can be defined as follow:

$$\mathrm{s}(h_v, h_{x_j}) = \frac{h_v^{\top}, h_{x_j}}{(|h_v| \cdot |h_{x_j}|)}, 1 \leq \mathrm{j} \leq \mathrm{t}-1, \tag{3}$$

where h_v denotes the token representation of the candidate word v, and h_{x_j} represents the representation of the word x_j in the prefix.

When we add PSCM in the decoding process of the IC model, the selection of the output token x_t at time step t with style controls parameter α and fluency controls parameter β can be formulated as follow:

$$x_t = \operatorname{argmax}\left\{\alpha \times S_{x_{<t},v}^{i,L} + F_{x_{<t},v}^{i}\right\}. \tag{4}$$

The two factors work together to affect the selection of stylized words made by the captions generating model, which ultimately produces image captions in the desired style.

4 Experimental Results and Discussion

4.1 Datasets

This paper uses two stylized caption datasets: SentiCap [12] and FlickrStyle10k [5] and two stylized texts: News Articles Corpus [4] and Shakespeare plays [14]. SentiCap is a sentiment dataset with 2360 images, which captions are split into positive part and negative part. Each image is associated with three positive or/and three negative captions. FlickrStyle10k dataset comprises 10,000 stylized captioned images, of which only 7,000 are publicly available. Each of the 7,000 images is associated with both a humorous and a romantic caption. And 1000 images were randomly selected for testing. News Articles Corpus is a dataset of more than 800k TXT files, we randomly selected 6781 texts to train the news style text classifier. The Shakespeare Plays dataset contains all the plays written by Shakespeare, and we randomly take 4500 texts to train a text classifier for Shakespearean style.

4.2 Metric

Follow [23], we measure the relevance between images and captions by Bleu-n, METEOR, and CIDEr, and the performance of styles by classifier accuracy and style accuracy. Bleu-n counts the frequency of shared words between generated and reference captions, while METEOR considers synonyms. CIDEr combines

Table 1. Comparison results on SentiCap dataset. Bn, M, C, cls, and sty denote Bleu-n, METEOR, CIDEr, and classifier accuracy and style accuracy respectively. The styles "pos", "neg", "roman", and "humor" are the abbreviations for positive, negative, romantic, and humorous.

method	style	B1	B3	M	C	cls	sty
StyleNet	pos	45.3	12.1	12.1	36.3	45.2	-
	neg	43.7	10.6	10.9	36.6	56.6	-
MSCap	pos	46.9	16.2	16.8	55.3	92.5	-
	neg	45.5	15.4	16.2	51.6	93.4	-
MemCap	pos	51.1	17	16.6	52.8	96.1	-
	neg	48.7	18.1	15.7	59.4	98.9	-
CapDec	pos	39.3	10.1	10.3	25.3	97.8	68.5
	neg	39	10	10.2	25.9	98.6	68.9
CapDec with PSTCm	pos	40.4	10.4	11.1	27.1	98.4	76.1
	neg	39.6	10.2	10.4	27	99.7	74.1
conzic	pos	-	1.89	5.39	-	97.2	76.4
	neg	-	1.78	5.54	-	99.1	75.3
conzic with PSTCm	pos	-	2.3	8.2	-	98.9	99.8
	neg	-	2.2	8.5	-	99.4	99.3

Bleu-n with a vector space model to evaluate consistency between image, reference captions, and generated captions. Classifier accuracy and style accuracy are obtained from the text classifier. For different categories, we use different classifiers. The text classifier is obtained by fine-tuning language model and getting more than 98% accuracies in four styles. The proportion of sentences that accurately match the target style is used to calculate classifier accuracy. To provide a clearer measurement of the text style degree, we propose a novel style metric called style accuracy, which is measured by the score of text classifier. The higher the score, the greater the stylization of the caption, which contains more stylistic information.

4.3 Implementation Details

We choose the unsupervised text-only training Capdec [14] method and zero-shot ConZIC [22] method as backbones. We add PSCM to the text decoding phase of each of the two baselines to conduct the predictive inference of the captions generation phase. In baseline CapDec, we directly use the framework of CapDec: CLIP image encoding in RN-50x4 backbone, language model is GPT-2 (large), without any fine-tuning and parameter modification. In baseline ConZic, we use the experimental setup of ConZic: the whole network is based on CLIP-ViT-B/32, using the BERT-Base language model. For the discriminators used to implement different control tasks, we did not modify them.

Table 2. Results on FlickrStyle10K dataset.

method	style	B1	B3	M	C	cls	sty
StyleNet	roman	13.3	1.5	4.5	7.2	37.8	-
	humor	13.4	0.9	4.3	11.3	41.9	-
MSCap	roman	17	2	5.4	10.1	88.7	-
	humor	16.3	1.9	5.3	15.2	91.3	-
MemCap	roman	21.2	4.8	8.4	22.4	91.7	-
	humor	19.9	4.3	7.4	19.4	97.1	-
CapDec	roman	23	4.6	9.1	27.4	95.4	59.4
	humor	22.7	4.3	9.7	29	92.3	61.5
CapDec with PSTCm	roman	27.8	7.9	11.2	37.1	98.8	76.7
	humor	28.2	8.5	11.6	38.9	99.4	80.5
conzic	roman	-	1.2	6.1	-	96.3	63.2
	humor	-	1.2	6.1	-	91.4	60.6
conzic with PSTCm	roman	-	2.9	6.4	-	97.6	68.8
	humor	-	2.8	6.4	-	95.7	64.8

Table 3. Results on News Articles Corpus and Shakespeare plays datasets.

method	style	B1	B3	M	C	cls	sty
CapDec	news	16.2	3.4	8.7	26.4	87.9	58.6
	Shakespeare	14.3	2.7	6.4	24.8	85.5	56.2
CapDec with PSTCm	news	20.1	6.4	10.1	35.4	90.3	65.7
	Shakespeare	18.5	4.7	8.6	33.7	89.4	68.3
conzic	news	-	1.4	4.8	-	92.3	68.4
	Shakespeare	-	1.1	4.1	-	91.8	67.1
conzic with PSTCm	news	-	2.3	5.7	-	96.5	72.3
	Shakespeare	-	1.9	5.3	-	97.3	74.8

The style factor is obtained from the pre-trained text classifier. The text classifier is fine-tuned from GPT2, and for different styles, we provide different text classifiers. In the decoding process of IC model, the starts of the LM is a special start-of-sequence token, and we set k equals 45, α equals 1, and β equals 0.1 on the test set for CapDec and ConZic.

4.4 Quantitative Analysis

Table 1, 2, 3 present the comparison between our method and the baselines, highlighting the differences in Quantitative metrics. In the four datasets, the insertion of PSCM shows higher accuracy compared to the baseline. When utilizing the CapDec framework on the SentiCap dataset, our method performs similarly

Table 4. Analysis of Style Factor and Fluency factor based on CapDec.

Factor	style	B1	B3	M	C	sty
PSTCm	roman	27.8	7.9	11.2	37.1	76.7
	humor	28.2	8.5	11.6	38.9	80.5
PSTCm w/o Fluency factor	roman	26.6	7.4	10.9	35.8	80.2
	humor	26.9	7.9	11.1	36.4	83.7
PSTCm w/o Style factor	roman	28.1	8.1	11.4	39	70.2
	humor	28.3	8.6	11.8	39.9	70.4
backbone	roman	23	4.6	9.1	27.4	59.4
	humor	22.7	4.3	9.7	29	61.5

to the supervised method previously discussed. However, on the FlickrStyle10k dataset, our findings surpass the performance of the supervised method.

The enhancement in Bleu-n, METEOR, and CIDEr metrics indicates that the sentences generated using our method exhibit greater similarity and semantic compatibility with the reference captions. This suggests that the captions produced under PSCM effectively incorporate image information and demonstrate improved fluency, thereby mitigating the issue of duplicate token generation. This validates the effectiveness of the fluency factor.

Furthermore, the increase in classifier accuracy and style accuracy shows that our method generates captions with stronger style information, highlighting the significance of the style factor. The combined action of both factors allows our method to generate fluent captions that are contextually relevant to the image and possess a specific style.

4.5 Qualitative Analysis

As shown in Fig. 3, we generate four different styles of captions by inserting the PSCM into the decoding process of CapDec model. Without the help of PSCM, the captions produced by CapDec encompass redundant and repetitive words, resulting in compromised sentence fluency and readability. Furthermore, these captions only weakly express stylistic features, thereby inhibiting the effective generation of target-style captions. Compared to the baseline, the captions generated by PSCM-assisted CapDec minimize the occurrence of repetitive words and maintain sentence fluency. Additionally, our results not only precisely depict the content of the image, but they also exhibit a greater adherence to the target style characteristics.

4.6 Ablation Study

We conduct ablation experiments on the Flickr-Style10k dataset using the CapDec-based structure with PSCM to evaluate the effectiveness of the style factor and fluency factor separately.

image	style	CapDec with Our PSCM	CapDec
	Humorous	a brown dog rushes hastily through a field of yellow flowers, just to find a bone.	a black and white dog is running through a grassy field to search for bones no bones found no bones found.
	Romantic	a dog leaps through the grass to catch a toy to impress his love.	a brown dog is running through the grass to meet his lover she is waiting for him to return to her loving master.
	News	A dog in a grassy field near trees. by. daily mail reporter. published:. 11:29.	By. daily mail reporter. published:. 11:35 est, 5 November 2013. |. updated:. 11:35 est, 5 november 2013. A dog appears on the lawn.
	Shakespeare	The dog walked through the brier : i'll be sworn i love them all.	i am a dog: i am a dog: i am a dog: i am a dog: i am a dog: i am
	Humorous	a girl in a red dress is being held up like a baton.	a girl in a red shirt is climbing like a lizard to reach outer space. she is about to fall down like a lizard. she is about to crash like a lizard.
	Romantic	a group of girls successfully lifted a girl and they completed a difficult challenge together.	a girl in a pink dress jumping in the air to win the competition proud of her accomplishment and proud of her talent as a performer herself is competing in a competition.
	News	A group of girls are performing acrobatics by. liz davies. published:. 04:32 est.	By. daily mail reporter. published:. 03:33 est, 13 december 2013. |. updated:. 03:33 est, 13 december 2013. A local acrobatic show is being held.
	Shakespeare	They are cheering for victory.	enter lady capulet and others with drum and trumpets and the drum a little before them and the drum a little before them.
	Humorous	a boy in a blue shirt is climbing up a pole like a lizard.	a boy in a shirt is climbing a rock wall like a lizard to reach outer space. he is about to crash into a rock wall like a lizard.
	Romantic	the boy is bravely climbing a rope ladder.	a boy in a shirt is climbing a large rock to conquer the high challenging challenge of life as he is called upon to do so by his lover on a mountaintop to conquer the high challenging challenge of life.
	News	(cnn) -- A boy is using a rope ladder to climb up.	By. daily mail reporter. published:. 09:32 est, 5 July 2013. |. updated:. 09:32 est, 5 July 2013. A child is crawling into the air.
	Shakespeare	A boy climbed up in thethe pendent sway of the tempestuous wind.	Little warrior is crawling along the rope to the sky wearing a helmet and crawling along the rope to the sky wearing a helmet and crawling along the rope.
	Humorous	a skateboarder jumps the ramp and prepares to fly freely.	a skateboarder doing a jump on his skateboard. he fell off of his skateboard and landed on his face like a frog in a swimming pool. he fell off of his skateboard and landed on his.
	Romantic	a skateboarder jumps over a bar proud of his skills and accomplishments.	a skateboarder doing a trick on a rail to show his courage to his friends on the ground that he is brave enough to do it again later on in life proud to show his courage to his friends on the ground.
	News	There was a local skateboarding competition. by. associated press. published:. 10:29 est.	By. daily mail reporter. published:. 09:33 est, 13 August 2013. |. updated:. 09:33 est, 13 August 2013. A man participates in a skateboarding competition.
	Shakespeare	All brought drums and colours to watch the skateboarding competition.	exeunt all came to the skateboard competition but antipholus of ephesus and biondello of ephesus and sicinius of ephesus and sicinius of ephesus.

Fig. 3. Examples of our image captioning in four styles with PSCM compared against baseline CapDec.

When we only use the style factor and omit the fluency factor, all semantically related metrics decline, as depicted in Table 4. But style-related indicators increases significantly, showing the validity of the style factor. The addition of the style factor leads the caption generation model to prefer tokens with a specific style. Since text style and text-image similarity are distinct characteristics, sentences with a strong style often contain more words unrelated to the images, which in turn affects these semantical metrics. Nevertheless, all metrics have been improved compared to backnone.

On the contrary, when we use only the fluency factor, Bleu-n, METEOR, and CIDEr metrics go up and the style accuracy goes down, which proves the effectiveness of the fluency factor. By reducing the generation of duplicate tokens, the quality of the output text improves, enhancing the coherence between result captions, reference captions, and images. As the reference captions already possessed the intended style, the style accuracy of the result captions improves compared to backnone.

To further customize the control of generated text and achieve varying degrees of desired style captions, we can also adjust the α and β accordingly. A higher α signifies a more pronounced stylization of the generated captions, while a larger β implies enhanced sentence fluency and reduced instances of repetitive words. However, the excess of style characteristics can potentially render captions less descriptive of the image. Similarly, an overly high degree of fluency may result in overly brief sentences that lack variety. Therefore, we recommend maintaining the parameters α within a range of 0 to 3 and β within a range of 0.1 to 0.3.

5 Conclusion

In this study, we introduce a plug-and-play module named Plug-in Stylized Controls Module for generating Open-Stylized Image Captioning. PSCM is a straightforward and efficient module that can be seamlessly incorporated into any well-trained caption models, enabling the generation of captions in the desired style. To demonstrate the effectiveness of PSCM, we integrate this module into two baseline models and evaluate their performance on four datasets. Looking ahead, our future work aims to integrate PSCM with more advanced IC models that have access to larger training data. This integration will enable the production of stylistic captions of even higher quality.

Acknowledgements. This work was supported in part by National Natural Science Foundation of China (NSFC) No. 62106022, 62225601, U19B2036, U22B2038, in part by Beijing Natural Science Foundation Project No. Z200002, in part by Youth Innovative Research Team of BUPT No. 2023QNTD02, in part by scholarships from China Scholarship Council (CSC) under Grant CSC No. 202206470055, and in part by BUPT Excellent Ph.D. Students Foundation No. CX2022152.

References

1. Chen, Q., Deng, C., Wu, Q.: Learning distinct and representative modes for image captioning. arXiv preprint arXiv:2209.08231 (2022)
2. Cho, J., Yoon, S., Kale, A., Dernoncourt, F., Bui, T., Bansal, M.: Fine-grained image captioning with clip reward. arXiv preprint arXiv:2205.13115 (2022)
3. Devlin, J., Chang, M.W., Lee, K., Toutanova, K.: Bert: pre-training of deep bidirectional transformers for language understanding. arXiv preprint arXiv:1810.04805 (2018)
4. Fabbri, A.R., Li, I., She, T., Li, S., Radev, D.R.: Multi-news: a large-scale multi-document summarization dataset and abstractive hierarchical model. arXiv preprint arXiv:1906.01749 (2019)
5. Gan, C., Gan, Z., He, X., Gao, J., Deng, L.: Stylenet: generating attractive visual captions with styles. In: Proceedings of the IEEE Conference on Computer Vision and Pattern Recognition, pp. 3137–3146 (2017)
6. Guo, L., Liu, J., Yao, P., Li, J., Lu, H.: Mscap: multi-style image captioning with unpaired stylized text. In: Proceedings of the IEEE/CVF Conference on Computer Vision and Pattern Recognition, pp. 4204–4213 (2019)

7. Hochreiter, S., Schmidhuber, J.: Long short-term memory. Neural Comput. **9**(8), 1735–1780 (1997)
8. Jia, C., et al.: Scaling up visual and vision-language representation learning with noisy text supervision. In: International Conference on Machine Learning, pp. 4904–4916. PMLR (2021)
9. Liang, K., Chang, H., Ma, B., Shan, S., Chen, X.: Unifying visual attribute learning with object recognition in a multiplicative framework. IEEE Trans. Pattern Anal. Mach. Intell. **41**(7), 1747–1760 (2018)
10. Liang, K., Chang, H., Shan, S., Chen, X.: A unified multiplicative framework for attribute learning. In: Proceedings of the IEEE International Conference on Computer Vision, pp. 2506–2514 (2015)
11. Liang, K., Guo, Y., Chang, H., Chen, X.: Visual relationship detection with deep structural ranking. In: Proceedings of the AAAI Conference on Artificial Intelligence, vol. 32 (2018)
12. Mathews, A., Xie, L., He, X.: Senticap: generating image descriptions with sentiments. In: Proceedings of the AAAI Conference on Artificial Intelligence, vol. 30 (2016)
13. Mokady, R., Hertz, A., Bermano, A.H.: Clipcap: clip prefix for image captioning. arXiv preprint arXiv:2111.09734 (2021)
14. Nukrai, D., Mokady, R., Globerson, A.: Text-only training for image captioning using noise-injected clip. arXiv preprint arXiv:2211.00575 (2022)
15. Pedersoli, M., Lucas, T., Schmid, C., Verbeek, J.: Areas of attention for image captioning. In: Proceedings of the IEEE International Conference on Computer Vision, pp. 1242–1250 (2017)
16. Radford, A., et al.: Learning transferable visual models from natural language supervision. In: International Conference on Machine Learning, pp. 8748–8763. PMLR (2021)
17. Radford, A., Wu, J., Child, R., Luan, D., Amodei, D., Sutskever, I., et al.: Language models are unsupervised multitask learners. OpenAI Blog **1**(8), 9 (2019)
18. Su, Y., et al.: Language models can see: plugging visual controls in text generation. arXiv preprint arXiv:2205.02655 (2022)
19. Tewel, Y., Shalev, Y., Schwartz, I., Wolf, L.: Zerocap: zero-shot image-to-text generation for visual-semantic arithmetic. In: Proceedings of the IEEE/CVF Conference on Computer Vision and Pattern Recognition, pp. 17918–17928 (2022)
20. Vaswani, A., et al.: Attention is all you need. In: Advances in Neural Information Processing Systems, vol. 30 (2017)
21. Yu, J., Li, J., Yu, Z., Huang, Q.: Multimodal transformer with multi-view visual representation for image captioning. IEEE Trans. Circuits Syst. Video Technol. **30**(12), 4467–4480 (2019)
22. Zeng, Z., Zhang, H., Lu, R., Wang, D., Chen, B., Wang, Z.: Conzic: controllable zero-shot image captioning by sampling-based polishing. In: Proceedings of the IEEE/CVF Conference on Computer Vision and Pattern Recognition, pp. 23465–23476 (2023)
23. Zhao, W., Wu, X., Zhang, X.: Memcap: memorizing style knowledge for image captioning. In: Proceedings of the AAAI Conference on Artificial Intelligence, vol. 34, pp. 12984–12992 (2020)
24. Zhou, Y., Long, G.: Style-aware contrastive learning for multi-style image captioning. arXiv preprint arXiv:2301.11367 (2023)

MGT: Modality-Guided Transformer for Infrared and Visible Image Fusion

Taoying Zhang[1], Hesong Li[1], Qiankun Liu[1], Xiaoyong Wang[1(✉)], and Ying Fu[1,2]

[1] MIIT Key Laboratory of Complex-Field Intelligent Sensing, Beijing Institute of Technology, Beijing 100081, China
wangxiaoyong@bit.edu.cn

[2] Yangtze Delta Region Academy of Beijing Institute of Technology, Jiaxing 314019, China

Abstract. Infrared and visible image fusion aims to generate high-quality fused images containing thermal radiation information from infrared images and texture information from visible images. Most deep learning-based methods are simple stacks of Transformer or convolution blocks and fail to further integrate the feature information of source images that may be missed in the fusion stage after generating the fused features. In this work, we develop a cross-attention-based macro framework, named Modality-Guided Transformer (MGT), that reintroduces detailed information from the two input images across multiple feature extraction layers into the initially obtained fused image. For efficiency, our MGT also introduces shared attention and multi-scale windows to reduce the computational costs of attention. Experimental results show that the proposed MGT outperforms state-of-the-art methods, especially in preserving salient targets and infrared texture details. Our code is publicly available at https://github.com/TaoYing-Zhang/MGT.

Keywords: Infrared and visible image fusion · Cross-attention · Modality-guided · Transformer

1 Introduction

Limited by the theoretical and technical development of hardware devices, the information of a single modality image cannot characterize the imaging scene comprehensively. Therefore, image fusion technology was born to fuse the scene information obtained from multi-modal sensors or under different shooting setting into a single image. Infrared and visible image fusion (IVF) technology is one of the most widely used image fusion technologies, and has abundant applications in many fields such as object detection [1,2,7], tracking [10,16,26], recognition [9,32], semantic segmentation [4,40], and remote sensing [6,11].

In recent decades, IVF technology has made great progress, and a large number of advanced methods [13,21,27] have been proposed. These methods can

© The Author(s), under exclusive license to Springer Nature Singapore Pte Ltd. 2024
Q. Liu et al. (Eds.): PRCV 2023, LNCS 14425, pp. 321–332, 2024.
https://doi.org/10.1007/978-981-99-8429-9_26

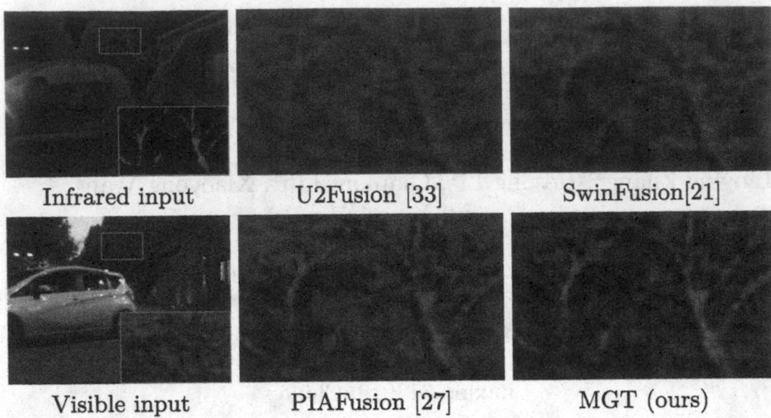

Fig. 1. An example of preserving infrared texture details. For a clearer comparison, we mark the infrared texture details while increasing the brightness.

be roughly divided into two categories, *i.e.*, traditional methods [13,20] and deep learning-based methods [12,33,38]. Traditional methods extract features of images from different modalities use the same extraction method without considering the differences between multi-modal images. Meanwhile, manually designed rules to measure activity levels and fusion process reduces the generalization ability of the methods [20]. The early deep learning-based methods can be roughly divided into three categories, *i.e.*, Auto-Encoder (AE)-based methods [12,14], Convolutional Neural Network (CNN)-based methods [27,38], and Generative Adversarial Networks (GAN)-based methods [19,22]. While existing data-driven IVF methods can produce excellent fusion images, their performance is limited by convolutional layers and can only mine the interaction relationships within single modality images. Therefore, some image fusion methods [24,29,30] introduce the Transformer into their frameworks to achieve promising results via Transformer's long-range dependency modeling capabilities.

Although existing deep learning-based methods can obtain great fusion images, they have two main problems. Firstly, most existing methods are simple stacks of Transformers and convolutional blocks. They do not integrate the missing feature information in the source images again after generating the fused features. Secondly, some work [15,37,39] prove the strong potential of Transformers in computer vision tasks. However, the existing Transformer-based IVF methods have not fully exploited the ability of the Transformer and increased the amount of calculation while introducing Transformer.

To address these issues, we design an end-to-end fusion network named Modality-Guided Transformer (MGT). Our design is primarily developed from the following two aspects. Firstly, we design a modality-guided fusion module to reduce information loss during the fusion process. It first uses features from infrared and visible images to generate fused features via several convolutional layers. Then, it integrates the extracted features from both shallow and deep layers into fused features directly via Modality-Guided Cross-Attention Blocks (MGCAB) to enhance the fused features while avoiding the information loss caused by the continuous deepening process of feature extraction. Therefore, the fused image output by our MGT can fully maintain the visible background and infrared key objects while preserving the information of infrared texture as shown in Fig. 1. Secondly, to reduce the computation resource cost and achieve better performance, our MGT introduces several mechanisms, including shared attention, shift convolution, and multi-scale windows. To sum up, the major contributions of this work can be summarized as follows:

- We propose a modality-guided fusion module based on cross-attention, which can integrate the information of the source image into the fused image directly while avoiding the loss of detailed information.
- We design an end-to-end fusion framework, which can fully use the Transformer's long-range dependency modeling capabilities and reduce computation resource costs.
- Experimental results show that the proposed MGT outperforms state-of-the-art methods, especially in preserving infrared texture details.

2 Related Work

Traditional IVF. Artificially designed feature extraction algorithms and fusion rules are important components of traditional image fusion methods. Feature extraction methods can be simply classified into three types, *i.e.*, multi-scale transform-based methods, sparse representation-based methods, and subspace-based methods [20]. Multi-scale transform-based methods [28] decompose images into components of different scales, which conform to human visual characteristics. Sparse representations-based methods [17,31] can effectively characterize the human visual system. Subspace-based methods [5,23] project high-dimensional input images into low-dimensional spaces or subspaces, which help to remove redundant information existing in natural images and capture the intrinsic structure of the original images. Certain fusion rules are applied to the extracted features or representations for fusion, such as choose-max and weighted average, according to the needs of different levels of features. Traditional methods can achieve good fusion images. However, artificially designed fusion rules and using the same feature extraction method for images of different modalities limit the generalization ability of traditional methods.

Early Deep Learning-Based IVF. Early deep learning-based IVF includes three categories, *i.e.*, Auto-Encoder (AE)-based IVF, Convolution Neural Network (CNN)-based IVF, and Generative Adversarial Network (GAN)-based IVF. The AE-based IVF methods use the powerful nonlinear fitting ability of the neural network to learn the features of multi-modal data adaptively. In DenseFuse [12], Li *et al.* build an encoder with convolutional layers and dense blocks to extract image features and reconstruct images using a fusion strategy and a decoding network that includes four CNN layers. The CNN-based IVF methods

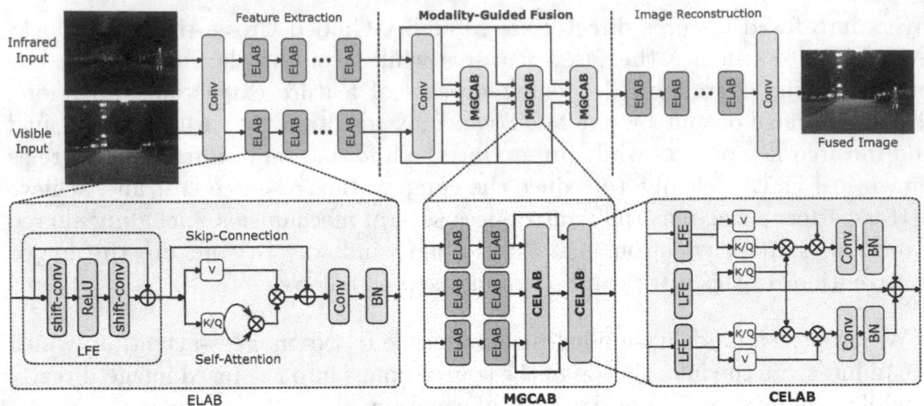

Fig. 2. The overall framework of Modality-guided Transformer (MGT) and structures of Modality-Guided Cross-Attention Block (MGCAB), Efficient Long-range Attention Block (ELAB), and Cross Efficient Long-range Attention Block (CELAB). To reduce information loss, MGCAB integrates the extracted features from the infrared and visible images into fused features directly.

construct an end-to-end image fusion framework without manually designing fusion rules. Xu *et al.* introduced a densely connected network to generate fused images [34]. Meanwhile, a weight block is applied in the fusion process to obtain two data-driven weights as the retention degree of feature in different source images. The GAN can train the network through the adversarial loss of the discriminator and the generator. Ma *et al.* innovatively introduced GAN into IVF in FusionGAN [22]. They model the generation of fused images as an adversarial game between keeping the thermal radiation information from infrared images and maintaining the texture structure information from visible images.

Transformer-Based IVF. In recent years, many researchers explore the Transformer-based image fusion methods [21,24,29], and achieve better performance than the framework composed of convolutional structures. VS *et al.* propose an Image Fusion Transformer (IFT) with a novel Spatio-Transformer (ST) fusion strategy, where a spatial CNN branch and a Transformer branch are employed to use both local and global features to fuse the given images better [29]. Ma *et al.* propose a joint CNN-Transformer framework called SwinFusion [21]. They design a cross-attention-based fusion module to integrate the interactive relationship between infrared and visible input image features. However, these Transformer-based methods cannot fully use the potential of the Transformer and reduce the network efficiency because of the need to calculate attention.

The above deep learning-based fusion methods fail to integrate the feature information in the source images into the fused features again after generating the preliminary fused features. It leads to inevitable information loss in the fusion stage. In this work, we propose an end-to-end framework with Modality-Guided Cross-Attention Blocks (MGCAB) to directly and effectively integrate the features of shallow to deep levels into the fused features.

3 Methodology

In this section, we comprehensively introduce our MGT from four aspects. First, we provide our motivation to design the framework. Then, we describe the overall pipeline of the proposed MGT. Next, the details of the Modality-Guided Cross-Attention Block (MGCAB) are presented. Finally, we introduce the loss function.

3.1 Motivation

Previous IVF methods usually first concatenate infrared features with visible features in the channel dimension and input them into convolutional layers to generate fused features. Then, the fused features are enhanced by a series of convolutional structures or Transformer blocks. After the fused features are generated, these fusion frameworks do not re-mine the connection between the fused features and the extracted source image features. As a result, the information that is not integrated into the fused features during the fusion stage is lost. We propose a Modality-Guided Cross-Attention Block (MGCAB), which extracts long-range dependencies between different modality image features by computing the cross-attention to guide the infrared and visible image features to be further integrated into fused features. It can reduce information loss while enhancing the fused features to improve the fusion image quality.

3.2 Overall Pipeline of MGT

Given an infrared image $\boldsymbol{I}_{\mathrm{ir}} \in \mathbb{R}^{\mathrm{H}\times\mathrm{W}\times 1}$ and a visible image $\boldsymbol{I}_{\mathrm{vi}} \in \mathbb{R}^{\mathrm{H}\times\mathrm{W}\times 3}$ that are both registered, the fused image $\boldsymbol{I}_{\mathrm{f}} \in \mathbb{R}^{\mathrm{H}\times\mathrm{W}\times 3}$ is generated via feature extraction, fusion, and image reconstruction. To reduce the information loss during the fusion stage and enhance the fused feature, we design a modality-guided fusion module based on cross-attention. The overall framework of our modality-guided Transformer is presented in Fig. 2.

First, we obtain the shallow features and deep features ($\boldsymbol{F}_{\mathrm{ir}}^1$, $\boldsymbol{F}_{\mathrm{ir}}^2$, $\boldsymbol{F}_{\mathrm{ir}}^3$, $\boldsymbol{F}_{\mathrm{vi}}^1$, $\boldsymbol{F}_{\mathrm{vi}}^2$, and $\boldsymbol{F}_{\mathrm{vi}}^3$) from infrared and visible images through the feature extraction module. The larger numbers represent blocks on deeper layers.

Then, these features are fed into a modality-guided fusion module. We first use deepest features $\boldsymbol{F}_{\mathrm{ir}}^3$ and $\boldsymbol{F}_{\mathrm{vi}}^3$ as inputs to generate a preliminary fused feature $\boldsymbol{F}_{\mathrm{f}}^0$ through two convolutional layers $\mathrm{Conv}_{\mathrm{F}}$, which can be defined as

$$\boldsymbol{F}_{\mathrm{f}}^0 = \mathrm{Conv}_{\mathrm{F}}\left(\mathrm{Cat}\left(\boldsymbol{F}_{\mathrm{ir}}^3, \boldsymbol{F}_{\mathrm{vi}}^3\right)\right), \tag{1}$$

where $\mathrm{Cat}(\cdot)$ denotes to concatenation in the channel dimension.

To prevent loss of information caused by continuous deepening of feature extraction, we employ Modality-Guided Cross-Attention Blocks (MGCAB) to directly integrate the extracted features from both shallow and deep layers into fused features. For the i-th ($i = 1, 2, 3$) MGCAB, given input features $\boldsymbol{F}_{\mathrm{ir}}^i$, $\boldsymbol{F}_{\mathrm{vi}}^i$, and $\boldsymbol{F}_{\mathrm{f}}^{i-1}$, we can define the output $\boldsymbol{F}_{\mathrm{f}}^i$ as

$$\boldsymbol{F}_{\mathrm{f}}^i = \mathrm{MGCAB}^i\left(\boldsymbol{F}_{\mathrm{ir}}^i, \boldsymbol{F}_{\mathrm{vi}}^i, \boldsymbol{F}_{\mathrm{f}}^{i-1}\right). \tag{2}$$

The output of the last MGCAB $\boldsymbol{F}_{\mathrm{f}}^{3}$ is used as the final fused feature $\boldsymbol{F}_{\mathrm{f}}$, which is input to the next module. The final fused image $\boldsymbol{I}_{\mathrm{f}}$ is reconstructed from the fused features using an image reconstruction module.

3.3 Modality-Guided Cross-Attention Block

MGCAB uses long-range dependencies across different modality features to guide the fusion stage, which can effectively reduce information loss during the fusion process and enhance the quality of the fused features. Specifically, the three input features are first enhanced with global features by two Efficient Long-range Attention Blocks (ELAB) [37] and then entered into Cross Efficient Long-range Attention Block (CELAB) to further integrate feature information.

First, the infrared, visible, and fused features are respectively enhanced with global features by two ELABs with shared attention. Second, these three enhanced features are fed into CELAB to enhance fused features. The CELAB first performs local feature extraction LFE through two shift-convs. Then, it divides feature $\boldsymbol{F}$ with size H × W × C into Z groups equally in the C dimensions and uses different window sizes M_z for window division in each group. In this work, Z is set to 3 and the window sizes used by the three groups are 4, 8, and 16, respectively. Next, it calculates the cross-attention of fused features and infrared features while computing the cross-attention of fused features and visible features, to obtain the output of MGCAB. We set $\boldsymbol{Q} = \boldsymbol{K}$ in this work, which means that we use two learnable weight matrices $\boldsymbol{W}^{Q/K}$, $\boldsymbol{W}^{V}$ to project a divided window into query $\boldsymbol{Q}$, key $\boldsymbol{K}$, and value $\boldsymbol{V}$. The first CELAB shares attention with the second CELAB.The whole process can be defined as

$$\left\{\boldsymbol{F}_{\mathrm{ir}}^{i*'}, \boldsymbol{F}_{\mathrm{vi}}^{i*'}, \boldsymbol{F}_{\mathrm{f}}^{i-1*'}\right\} = \left\{\mathrm{LFE}\left(\boldsymbol{F}_{\mathrm{ir}}^{i*}\right), \mathrm{LFE}\left(\boldsymbol{F}_{\mathrm{vi}}^{i*}\right), \mathrm{LFE}\left(\boldsymbol{F}_{\mathrm{f}}^{i-1*}\right)\right\}, \quad (3)$$

$$\left\{\boldsymbol{Q}_{\mathrm{ir}}/\boldsymbol{K}_{\mathrm{ir}}, \boldsymbol{V}_{\mathrm{ir}}\right\} = \left\{\boldsymbol{X}_{\mathrm{ir}}\boldsymbol{W}_{1}^{Q/K}, \boldsymbol{X}_{\mathrm{ir}}\boldsymbol{W}_{1}^{V}\right\}, \quad (4)$$

$$\left\{\boldsymbol{Q}_{\mathrm{vi}}/\boldsymbol{K}_{\mathrm{vi}}, \boldsymbol{V}_{\mathrm{vi}}\right\} = \left\{\boldsymbol{X}_{\mathrm{vi}}\boldsymbol{W}_{2}^{Q/K}, \boldsymbol{X}_{\mathrm{vi}}\boldsymbol{W}_{2}^{V}\right\}, \quad (5)$$

$$\left\{\boldsymbol{Q}_{\mathrm{f}}/\boldsymbol{K}_{\mathrm{f}}, \boldsymbol{V}_{\mathrm{f}}\right\} = \left\{\boldsymbol{X}_{\mathrm{f}}\boldsymbol{W}_{3}^{Q/K}, \boldsymbol{X}_{\mathrm{f}}\boldsymbol{W}_{3}^{V}\right\}, \quad (6)$$

$$\boldsymbol{F}_{\mathrm{f}}^{i} = \mathrm{BN}\left(\mathrm{CA}\left(\boldsymbol{Q}_{\mathrm{f}}, \boldsymbol{K}_{\mathrm{ir}}, \boldsymbol{V}_{\mathrm{ir}}\right)\right) + \mathrm{BN}\left(\mathrm{CA}\left(\boldsymbol{Q}_{\mathrm{f}}, \boldsymbol{K}_{\mathrm{vi}}, \boldsymbol{V}_{\mathrm{vi}}\right)\right) + \boldsymbol{F}_{\mathrm{f}}^{i-1*'}, \quad (7)$$

where BN denotes batch normalization, and CA denotes cross-attention. $\boldsymbol{X}_{\mathrm{ir}}$, $\boldsymbol{X}_{\mathrm{vi}}$, and $\boldsymbol{X}_{\mathrm{f}}$ are divided windows from $\boldsymbol{F}_{\mathrm{ir}}^{i*'}$, $\boldsymbol{F}_{\mathrm{vi}}^{i*'}$, and $\boldsymbol{F}_{\mathrm{f}}^{i-1*'}$.

3.4 Loss Function

Inspired by SwinFusion [21], we use a loss function consisting of three parts, *i.e.*, SSIM loss $\mathcal{L}_{\mathrm{ssim}}$, texture loss $\mathcal{L}_{\mathrm{text}}$, and intensity loss $\mathcal{L}_{\mathrm{int}}$, to promote that our MGT adaptively integrates meaningful information. They are defined as

$$\mathcal{L}_{\mathrm{ssim}} = w_{\mathrm{ir}}\left(1 - \mathrm{ssim}\left(\boldsymbol{I}_{\mathrm{ir}}, \boldsymbol{I}_{\mathrm{f}}\right)\right) + w_{\mathrm{vi}}\left(1 - \mathrm{ssim}\left(\boldsymbol{I}_{\mathrm{vi}}, \boldsymbol{I}_{\mathrm{f}}\right)\right), \quad (8)$$

$$\mathcal{L}_{\text{text}} = \frac{1}{HW} \| \mid \nabla \boldsymbol{I}_{\text{f}} \mid - \max\left(\mid \nabla \boldsymbol{I}_{\text{ir}} \mid, \mid \nabla \boldsymbol{I}_{\text{vi}} \mid\right) \|_1, \tag{9}$$

$$\mathcal{L}_{\text{int}} = \frac{1}{HW} \left\| \boldsymbol{I}_{\text{f}} - \max\left(\boldsymbol{I}_{\text{ir}}, \boldsymbol{I}_{\text{vi}}\right) \right\|_1, \tag{10}$$

where ssim$(\cdot)$ denotes the structural similarity operation. The w_{ir} and w_{vi} are the weights of the two source images, respectively. In this work, we set $w_{\text{ir}} = w_{\text{vi}} = 0.5$. ∇ denotes the Sobel gradient operator, which measures the texture information of images. $\left\|\cdot\right\|_1$ denotes the ℓ_1-norm, $\left|\cdot\right|$ denotes the absolute operation, and $\max(\cdot)$ denotes the element-wise maximum selection.

Finally, the full loss function for our MGT is a weighted sum of all sub-loss terms from Eq. (8) to Eq. (10), which can be defined as

$$\mathcal{L}_{\text{total}} = \lambda_1 \mathcal{L}_{\text{ssim}} + \lambda_2 \mathcal{L}_{\text{text}} + \lambda_3 \mathcal{L}_{\text{int}}, \tag{11}$$

where λ_1, λ_2, and λ_3 are the hyper-parameters that weigh the impact of different loss items on the fusion results. In this work, we set $\lambda_1 = 10$, $\lambda_2 = 50$, and $\lambda_3 = 50$ to get fused images of higher quality.

4 Experiments

In this section, we first present the experimental configuration. Then, we perform comparative experiments with nine other state-of-the-art methods and test the operating efficiency. Finally, we perform ablation experiments to verify the effectiveness of the modality-guided fusion module.

4.1 Experimental Configurations

Comparison Methods. We compare our method with a total of nine state-of-the-art methods in experiments, which include two conventional methods (GTF [18] and MDLatLRR [13]), three AE-based methods (DenseFuse [12], RFN-Nest [14], and CSF [35]), three CNN-based methods (IFCNN [38], PIAFusion [27], and U2Fusion [33]), and one Transformer-based method (SwinFusion [21]). Importantly, all of these nine methods are made publicly available, and we follow the parameter settings as described in the corresponding original papers.

Data Set and Evaluation. In this work, four distinct evaluation metrics are chosen to evaluate the effectiveness of image fusion methods on the MSRS dataset. The training and testing sets of the MSRS dataset contain 1083 and 361 pairs of registered infrared and visible images, respectively. To speed up the training process and increase sample diversity, we randomly crop a 128×128 window on the image each time as input in the training stage. The four metrics include Average gradient (AG) [3], standard deviation (SD) [25], visual information fidelity (VIF) [8], and $Q^{AB/F}$ [36]. It should be noted that the larger values of AG, SD, VIF, and $Q^{AB/F}$ indicate superior fusion performance.

Implementation Details. Following the PIAFusion [27], we transform visible images from RGB color space into the YCrCb color space to reduce the modal

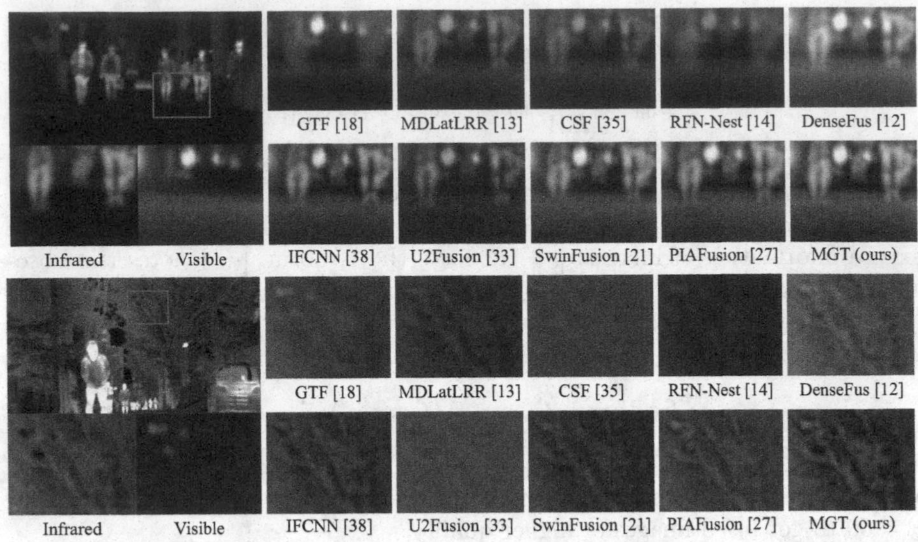

Fig. 3. Visualized results of the qualitative comparison.

discrepancy between infrared and visible images. Then, the Y channel of the visible images is fused with infrared images through MGT. The final color fusion image is obtained by converting the fused Y channel image and Cr, Cb channels of visible images back to RGB color space. During the training procedure, we specify a batch size of $b = 4$ and a total of $N = 100$ epochs. For parameter updates, we employ the Adam optimizer while setting the learning rate as 0.0002. Our method is implemented on the PyTorch framework. All training processes are conducted on the NVIDIA T4 GPU and Intel Cascade Lake CPU provided by Amazon SageMaker Studio Lab, while testing is conducted on local NVIDIA GeForce RTX 2060 GPU and 2.60GHz Intel(R) Core(TM) i7-9750H CPU.

4.2 Comparative Experiment

To comprehensively evaluate our method, we quantitatively and qualitatively compare MGT with nine other methods on the MSRS dataset.

Qualitative Results. As shown in Fig. 3, GTF, MDLatLRR, CSF, and RFN-Nest exhibit poor performance in preserving details of the input images. DenseFuse, IFCNN, U2Fusion, and SwinFusion maintain significant infrared targets but compromise their sharp edges. GTF, MDLatLRR, CSF, RFN-Nest, and U2Fusion cannot maintain the detailed texture of branches. Although DenseFuse, IFCNN, SwinFusion, and PIAFion can integrate salient targets and detailed textures, they fail to maintain the sharp edges of salient targets and texture details. In contrast, our method preserves the reasonable intensity and sharp edges of infrared targets, while retaining the visible background details.

Table 1. Quantitative comparisons of the four metrics, *i.e.*, AG, SD, VIF, $Q^{AB/F}$ and average running times of ten methods, on the MSRS dataset. The best and second-best values are bolded and underlined, respectively.

Methods	Params (M)↓	FLOPs (G)↓	Time (s)↓	AG [3] ↑	SD [25] ↑	VIF [8] ↑	$Q^{AB/F}$ [36] ↑
GTF [18]	**0**	N/A	4.1828	2.5444	6.3578	0.5479	0.3886
MDLatLRR [13]	**0**	N/A	28.063	2.4510	7.4766	0.7367	0.5280
CSF [35]	0.1853	<u>52.4999</u>	3.8651	1.8811	6.8074	0.5196	0.2549
RFN-Nest [14]	7.7431	544.4238	0.4210	2.1477	7.7931	0.7348	0.3902
DenseFuse [12]	<u>0.0742</u>	54.1512	0.5674	3.9065	<u>8.4144</u>	0.8075	0.5787
IFCNN [38]	0.0836	**39.9325**	**0.0659**	3.9845	7.9829	0.8515	0.6078
U2Fusion [33]	0.6592	405.1745	0.4137	2.5161	7.0138	0.6446	0.4150
SwinFusion [21]	0.9737	292.5376	3.2519	3.5745	8.3935	1.0049	<u>0.6428</u>
PIAFusion [27]	1.1761	360.9919	<u>0.2548</u>	<u>3.9677</u>	8.4057	<u>1.0422</u>	0.6416
MGT (ours)	1.4672	474.5272	1.5955	**4.3337**	**8.5875**	**1.0837**	**0.6569**

Quantitative Results. In quantitative experiments, our MGT outperforms nine state-of-the-art methods on all four metrics, as shown in Table 1. The highest SD metric represents our fusion results with high contrast, which means better visual quality. The best AG represents that the proposed method can generate images containing rich detailed information. Achieving the best performance on these two metrics proves that the fusion images generated by our method are of high quality. The highest VIF metric represents that our method retains as much ideal information as possible from the source image, which makes the fusion result more consistent with the human visual system. The highest Qabf metric means the proposed MGT preserves more edge information from both infrared and visible images, benefiting from the designed modality-guided fusion module. In addition, our MGT introduces Transformer, which means that it takes more time to calculate the window attention. For efficiency, we introduced some mechanisms to reduce the computation, so that our running time is half of the Transformer-based method SwinFusion.

4.3 Ablation Study

In this work, we design a Modality-Guided Cross-Attention Block (MGCAB) to reduce the information loss in the fusion stage. We conduct an ablation experiment to further verify the effectiveness of our designed MGCAB. The visualized experimental results are shown in Fig. 4. Visualization results show that after removing MGCAB, the fusion result weakens the infrared thermal radiation information, especially a part of the infrared texture information is lost. In addition, as shown in Fig. 4, the fusion image generated by the model with MGCAB has infrared texture details more similar to the source images because MGCAB integrates more feature information from source images into the fusion image. In contrast, due to MGCAB further integrating infrared and visible features, our MGT better retains infrared salient target and background information, while retaining clearer infrared detail textures which are closer to the part in the infrared source image than the result without MGCAB.

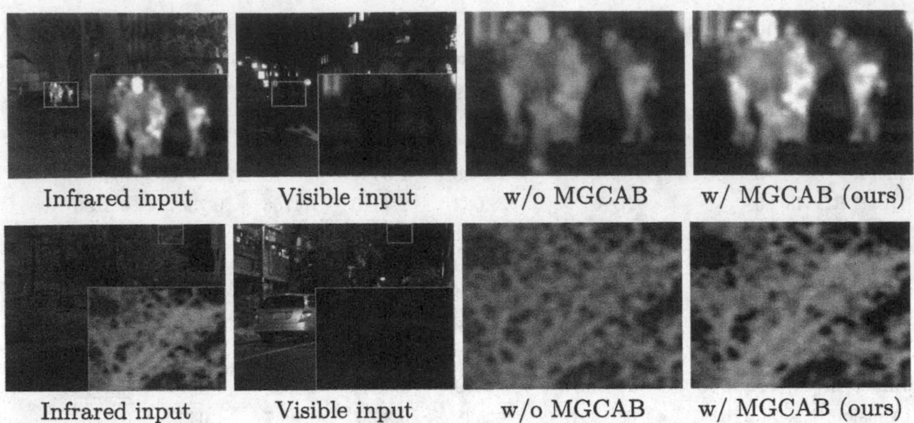

Fig. 4. Ablation study of Modality-Guided Cross-Attention Block on MSRS dataset. We increase the brightness of the region in the red box. (Color figure online)

Table 2. Ablation study of Modality-Guided Cross-Attention Block (MGCAB) on MSRS dataset. We remove the MGCAB from MGT while adjusting the depth of other modules to make the models have similar computational costs.

Methods	Params (M)↓	FLOPs (G)↓	Time (s)↓	AG [3] ↑	SD [25] ↑	VIF [8] ↑	$Q^{AB/F}$ [36] ↑
w/o MGCAB	**1.3797**	**426.7515**	1.8061	3.6743	8.3716	0.9994	**0.6663**
w/ MGCAB (ours)	1.4672	474.5272	**1.5955**	**4.3337**	**8.5875**	**1.0837**	0.6569

The quantitative results of the ablation experiments are shown in Table 2. Except for $Q^{AB/F}$, the other three metrics AG, SD, and VIF all declined. The lower AG, SD, and VIF represent the degradation of fused image quality.

5 Conclusion

In this work, we propose a modality-guided Transformer for IVF. Firstly, we design a Modality-Guided Cross-Attention Block, which computes the cross-attention between infrared, visible, and fused features at the same time, to integrate the feature information from the source images again into the fused features. This design significantly enhances the information integration ability of the network and reduces information loss during the fusion stage. Secondly, we introduce various mechanisms to improve efficiency, which can reduce the computational costs of attention. Experimental results show that our MGT outperforms state-of-the-art methods, especially in terms of preserving salient targets and background information while maintaining infrared texture details.

Acknowledgement. This work was supported by the National Natural Science Foundation of China (62331006, 62171038, and 62088101), and the Fundamental Research Funds for the Central Universities.

References

1. Han, J., Bhanu, B.: Fusion of color and infrared video for moving human detection. Pattern Recognit. **40**(6), 1771–1784 (2007)
2. Cao, Y., Guan, D., Huang, W., Yang, J., Cao, Y., Qiao, Y.: Pedestrian detection with unsupervised multispectral feature learning using deep neural networks. Inf. Fusion **46**, 206–217 (2019)
3. Cui, G., Feng, H., Xu, Z., Li, Q., Chen, Y.: Detail preserved fusion of visible and infrared images using regional saliency extraction and multi-scale image decomposition. Optics Commun. **341**, 199–209 (2015)
4. Fu, Y., Liang, S., Chen, D., Chen, Z.: Translation of aerial image into digital map via discriminative segmentation and creative generation. IEEE Trans. Geosci. Remote Sens. **60**, 1–15 (2021)
5. Fu, Z., Wang, X., Xu, J., Zhou, N., Zhao, Y.: Infrared and visible images fusion based on RPCA and NSCT. Infrared Phys. Technol. **77**, 114–123 (2016)
6. Meng-Yin, F., Cheng, Z.: Fusion of infrared and visible images based on the second generation curvelet transform. J. Infrared Millimeter Waves **28**(4), 254–258 (2009)
7. Gao, S., Cheng, Y., Zhao, Y.: Method of visual and infrared fusion for moving object detection. Opt. Lett. **38**(11), 1981–1983 (2013)
8. Han, Y., Cai, Y., Cao, Y., Xu, X.: A new image fusion performance metric based on visual information fidelity. Inf. Fusion **14**(2), 127–135 (2013)
9. Heo, J., Kong, S.G., Abidi, B.R., Abidi, M.A.: Fusion of visual and thermal signatures with eyeglass removal for robust face recognition. In: ICIP, pp. 122–122 (2004)
10. Li, C., Zhu, C., Huang, Y., Tang, J., Wang, L.: Cross-modal ranking with soft consistency and noisy labels for robust RGB-T tracking. In: ECCV (2018)
11. Li, H., Ding, W., Cao, X., Liu, C.: Image registration and fusion of visible and infrared integrated camera for medium-altitude unmanned aerial vehicle remote sensing. Remote Sens. **9**(5), 441 (2017)
12. Li, H., Wu, X.J.: Densefuse: a fusion approach to infrared and visible images. IEEE Trans. Image Process. **28**(5), 2614–2623 (2019)
13. Li, H., Wu, X.J., Kittler, J.: MDLatLRR: a novel decomposition method for infrared and visible image fusion. IEEE Trans. Image Process. **29**, 4733–4746 (2020)
14. Li, H., Wu, X.J., Kittler, J.: RFN-Nest: an end-to-end residual fusion network for infrared and visible images. Inf. Fusion **73**, 72–86 (2021)
15. Liang, J., Cao, J., Sun, G., Zhang, K., Van Gool, L., Timofte, R.: SwinIR: image restoration using swin transformer. In: CVPR, pp. 1833–1844 (2021)
16. Liu, H., Sun, F.: Fusion tracking in color and infrared images using joint sparse representation. SCIENCE CHINA Inf. Sci. **55**, 590–599 (2012)
17. Liu, Y., Chen, X., Ward, R.K., Jane Wang, Z.: Image fusion with convolutional sparse representation. IEEE Signal Process. Lett. **23**(12), 1882–1886 (2016)
18. Ma, J., Chen, C., Li, C., Huang, J.: Infrared and visible image fusion via gradient transfer and total variation minimization. Inf. Fusion **31**, 100–109 (2016)
19. Ma, J., et al.: Infrared and visible image fusion via detail preserving adversarial learning. Inf. Fusion **54**, 85–98 (2020)
20. Ma, J., Ma, Y., Li, C.: Infrared and visible image fusion methods and applications: a survey. Inf. Fusion **45**, 153–178 (2019)
21. Ma, J., Tang, L., Fan, F., Huang, J., Mei, X., Ma, Y.: Swinfusion: cross-domain long-range learning for general image fusion via swin transformer. IEEE/CAA J. Automatica Sinica **9**(7), 1200–1217 (2022)

22. Ma, J., Yu, W., Liang, P., Li, C., Jiang, J.: Fusiongan: a generative adversarial network for infrared and visible image fusion. Inf. Fusion **48**, 11–26 (2019)
23. Mou, J., Gao, W., Song, Z.: Image fusion based on non-negative matrix factorization and infrared feature extraction. In: CISP, vol. 2, pp. 1046–1050 (2013)
24. Rao, D., Xu, T., Wu, X.J.: TGFuse: an infrared and visible image fusion approach based on transformer and generative adversarial network. IEEE Trans. Image Process. (2023)
25. Rao, Y.J.: In-fibre bragg grating sensors. Meas. Sci. Technol. **8**(4), 355 (1997)
26. Smith, D., Singh, S.: Approaches to multisensor data fusion in target tracking: a survey. IEEE Trans. Knowl. Data Eng. **18**(12), 1696–1710 (2006)
27. Tang, L., Yuan, J., Zhang, H., Jiang, X., Ma, J.: Piafusion: a progressive infrared and visible image fusion network based on illumination aware. Inf. Fusion **83–84**, 79–92 (2022)
28. Vanmali, A.V., Gadre, V.M.: Visible and NIR image fusion using weight-map-guided laplacian-gaussian pyramid for improving scene visibility. Sādhanā **42**, 1063–1082 (2017)
29. Vs, V., Valanarasu, J.M.J., Oza, P., Patel, V.M.: Image fusion transformer. In: ICIP, pp. 3566–3570 (2022)
30. Wang, Z., Chen, Y., Shao, W., Li, H., Zhang, L.: Swinfuse: a residual swin transformer fusion network for infrared and visible images. IEEE Trans. Instrum. Meas. **71**, 1–12 (2022)
31. Wu, M., Ma, Y., Fan, F., Mei, X., Huang, J.: Infrared and visible image fusion via joint convolutional sparse representation. J. Opt. Soc. Am. A **37**(7), 1105–1115 (2020)
32. Xiang, Y., Fu, Y., Huang, H.: Global relative position space based pooling for fine-grained vehicle recognition. Neurocomputing **367**, 287–298 (2019)
33. Xu, H., Ma, J., Jiang, J., Guo, X., Ling, H.: U2Fusion: a unified unsupervised image fusion network. IEEE Trans. Pattern Anal. Mach. Intell. **44**(1), 502–518 (2022)
34. Xu, H., Ma, J., Le, Z., Jiang, J., Guo, X.: Fusiondn: a unified densely connected network for image fusion. In: AAAI, vol. 34, pp. 12484–12491 (2020)
35. Xu, H., Zhang, H., Ma, J.: Classification saliency-based rule for visible and infrared image fusion. IEEE Trans. Comput. Imaging **7**, 824–836 (2021)
36. Xydeas, C.S., Petrovic, V., et al.: Objective image fusion performance measure. Electron. Lett. **36**(4), 308–309 (2000)
37. Zhang, X., Zeng, H., Guo, S., Zhang, L.: Efficient long-range attention network for image super-resolution. In: Avidan, S., Brostow, G., Cissé, M., Farinella, G.M., Hassner, T. (eds.) ECCV 2022. LNCS, vol. 13677, pp. 649–667. Springer, Cham (2022). https://doi.org/10.1007/978-3-031-19790-1_39
38. Zhang, Y., Liu, Y., Sun, P., Yan, H., Zhao, X., Zhang, L.: IFCNN: a general image fusion framework based on convolutional neural network. Inf. Fusion **54**, 99–118 (2020)
39. Zheng, S., et al.: Rethinking semantic segmentation from a sequence-to-sequence perspective with transformers. In: CVPR, pp. 6881–6890 (2021)
40. Zhou, W., Liu, J., Lei, J., Yu, L., Hwang, J.N.: GMNet: graded-feature multilabel-learning network for RGB-thermal urban scene semantic segmentation. IEEE Trans. Image Process. **30**, 7790–7802 (2021)

Multimodal Rumor Detection by Using Additive Angular Margin with Class-Aware Attention for Hard Samples

Chenyu Zhou[1], Xiuhong Li[1](✉), Zhe Li[2], Fan Chen[1], Xiaofan Wang[1], Dan Yang[1], Bin Chen[3], and Songlin Li[1]

[1] School of Information Science and Engineering, Xinjiang University, Xinjiang, China
Zhou_cy@stu.xju.edu.cn, xjulxh@xju.edu.cn

[2] Department of Electrical and Electronic Engineering, The Hong Kong Polytechnic University, Hong Kong, Hong Kong SAR, China
lizhe.li@connect.polyu.hk

[3] School of Mechanical Engineering, University of Shanghai for Science and Technology, Shanghai, China

Abstract. Currently, several factors limit the practicality of multimodal rumor detection (MRD). These include incomplete feature fusion in multimodal data, the weak discriminative power in the softmax-based loss, and the detrimental impact of hard negative samples on the learning process. To address these issues, we propose a MRD framework that combines a supervised contrastive loss with an additive angular margin and incorporates class-aware attention. We propose a multi-layer fusion (MLF) module to enhance the multimodal feature fusion to align and fuse token-level features from text and image modalities. And also, by adding an angular margin to the loss function, we bolster the discriminative power of the contrastive loss. Additionally, the class-aware attention module effectively mitigates the impact of hard negative samples on the supervised contrastive loss. Extensive experiments on three real-world multimodal datasets demonstrate that our proposed learning objective leads to an embedding space that effectively distinguishes between rumors and truths. Furthermore, our work has significantly improved the efficacy of rumor detection, enabling us to promptly identify and curtail rumors' propagation.

Keywords: Rumor Detection · Multimodal Fusion · Contrastive Learning · Additive Angular Margin · Class-aware attention

Thanks to the open project of key laboratory, Xinjiang Uygur Autonomous Region (No. 2023D04079).

© The Author(s), under exclusive license to Springer Nature Singapore Pte Ltd. 2024
Q. Liu et al. (Eds.): PRCV 2023, LNCS 14425, pp. 333–344, 2024.
https://doi.org/10.1007/978-981-99-8429-9_27

1 Introduction

Multimodal rumor detection (MRD) has gained significant attention in recent years [20]. However, there remain three key challenges in this field. An essential challenge is multimodal fusion [23]. Utilizing a rudimentary concatenation approach for integrating multimodal data may inadvertently lead to diminished access to crucial information, consequently undermining the efficacy of the classification process. Moreover, the decision boundary in many researches tends to be narrow, meaning that even slight perturbations around this boundary may lead to incorrect decisions and a subsequent drop in accuracy. Additionally, hard samples pose another challenge in MRD. Hard samples refer to instances in which deep learning models encounter difficulty in accurately predicting the correct label. Most contrastive approaches focus on determining the positive and negative pairs and improving the learning framework [25]. However, the optimization objective often overlooks the inclusion of hard samples, which remains an understudied aspect of current research.

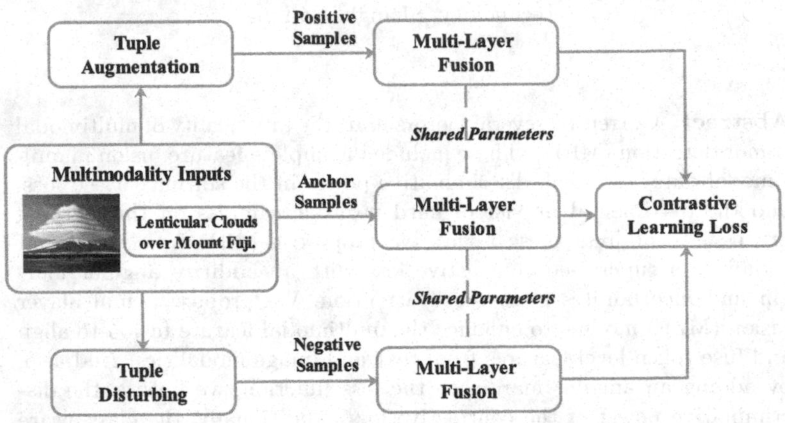

Fig. 1. Illustration of our basic idea.

Distinct modalities possess unique representations, and several approaches have been developed to learn their joint embeddings [9]. However, these approaches typically rely on simple concatenation strategies with limited potential for improvement. Drawing inspiration from the success of Transformer architecture in domains like natural language processing and computer vision [13,16,17], we propose a Multi-Layer Fusion (MLF) module that harnesses the power of the Transformer architecture. By effectively aligning and fusing text and image features, the proposed MLF module overcomes the challenge of multimodal fusion, resulting in enhanced performance in MRD.

MRD involves learning the representation of events, wherein tweets are encoded as fixed-sized embedding vectors. The objective of MRD is to ensure that truth tweets have closely located embedding vectors, whereas fake tweets have distant ones. For MRD tasks, contrastive representation learning, such as InstDisc [31], SimCLR [1], and MoCo [10], are natural fits to this requirement. The underlying principle of these methods revolves around discovering representations by maximizing the concordance among different augmentations or perspectives of the same instance in the latent space. To further bolster the discriminative capacity in contrastive learning, we propose the incorporation of an angular margin, especially for samples that are more susceptible to incorrect classification near the decision boundary. This addition strengthens the separability of embeddings, allowing for better differentiation between rumor and truth tweets.

Hard samples are data instances that pose significant challenges for machine learning models regarding classification. The apprehension regarding hard samples may be outliers and could increase the loss during training, which can impede the overall training process. Consequently, the impact of difficult samples has been largely neglected in previous research on contrastive learning for rumor detection [8]. However, ignoring these samples may lead to suboptimal results. To mitigate this, we incorporate class-aware attention specifically designed for handling challenging instances. This attention mechanism enables the model to allocate more focus and resources to these hard samples, thus improving the model's ability to address and classify hard samples effectively.

To alleviate the above challenges, we propose the **MarCAMLF**, which stands for a contrastive learning architecture by using additive angular **Mar**gin with **C**lass-aware **A**ttention for hard samples, and **M**ulti-**L**ayer **F**usion (**MLF**) module that fuses text and image features to obtain deep features in multimodal data. Specifically, an additive angular margin used in the supervised contrastive loss enhances the discrimination ability of rumors by maximizing the margin in the angular space. Our proposed methodology has proven effective in other areas of our original work [18,19,27]. Figure 1 provides a high-level overview of our approach.

The contributions of this paper are three folds:

- To improve the discrimination capacity of our rumor detection system, we introduce an additive angular margin into the supervised contrastive loss, maximizing the decision margin in the angular space.
- To address the issue of very hard negatives, we propose a class-aware attention mechanism that assigns a lower attention score to such samples. A lower attention score is assigned to hard samples to alleviate a sub-optimal local minimum in the trained model.
- We adopt a Transformer-based MLF module with multi-headed self-attention to align and fuse text and image features.

2 Related Work

2.1 Multimodal Rumor Detection

In the earlier research conducted before 2018, most rumor detection efforts focused on unimodal approaches [21,22]. However, with the widespread dissemination of multimodal tweets on social networks, multimodal rumor detection is gradually becoming the mainstream of research nowadays. Wu *et al.* [30] used multimodal co-attention networks to better fuse textual and visual features. Ying *et al.* [34] utilized bootstrapping multi-view representations for rumor detection. Hua *et al.* [11] used contrastive learning method for classifying tweets as truth or rumor. A common limitation of the above methods is that they do not adequately consider the presence of hard samples, so in this paper, we use a class aware attention (CA) module to solve this challenge.

2.2 Contrastive Learning

Since SimCLR [1] and MOCO [10] have successfully utilized contrastive learning, self-supervised contrastive learning has attracted many researchers due to its impressive performance in representation learning over the past several years. Contrastive learning aims to learn an embedding space where positive pairs are close while negative pairs are far apart. It is also closely related to metric learning, which aims to learn a distance function in an embedding space. Our work draws inspiration from self-supervised contrastive learning [15]. However, the key differentiating factor lies in utilizing the expanded decision space perspective, which aims to improve classification accuracy.

2.3 Multimodal Fusion

Recent research on the fusion of multimodal features has made significant progress. Jin *et al.* [12] used the recurrent neural network with an attention mechanism to fuse textual, visual, and social context features and to pool out distinct temporal-linguistic features with a particular focus. Khattar *et al.* [14] employed autoencoders in conjunction with a detector to acquire a shared representation of both the textual content and the accompanying image. Wu *et al.* [30] used BERT and VGG-19 to extract features from text and images and fused them through a co-attention mechanism. Peng *et al.* [24] used contrastive pairwise learning to capture the inter-modal relationship across multimodal. Nonetheless, ample scope remains for refinement in the abstraction and fusion of features within the aforementioned models proposed by other researchers. On the other hand, our proposed model leverages the depth of the MLF module and harnesses the efficacy of the multi-headed attention mechanism. This enables us to enhance the abstraction capabilities of our model while effectively aligning the graphic features.

3 Methodology

Our objective is to distinguish the truths and rumors in tweets. We train a text-image-based feature embedding network using labeled samples to achieve this goal. We use the Vision Transformer (ViT) and the BERT models to process images and text. As shown in Fig. 2, the encoder network produces normalized

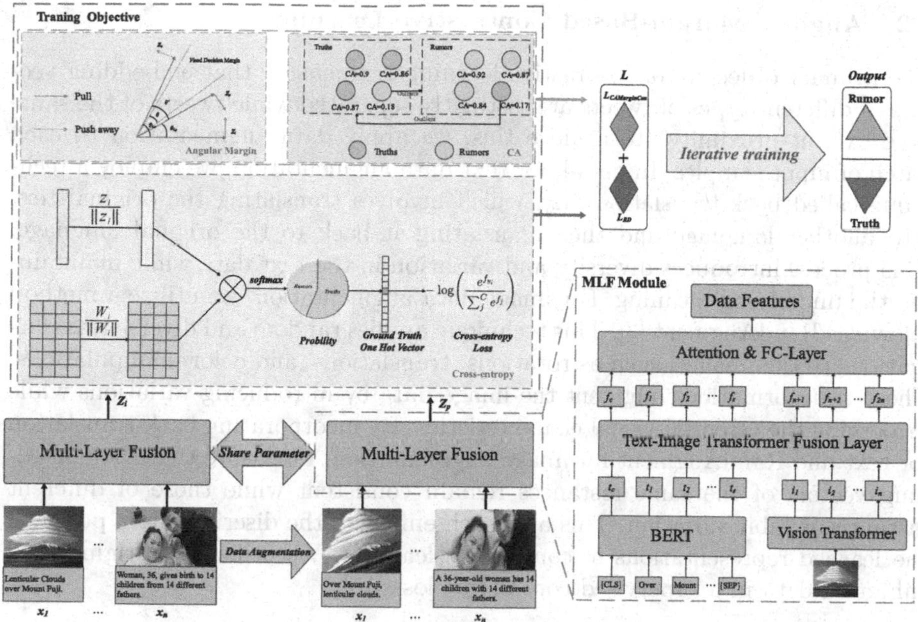

Fig. 2. Our model leverages Vision Transformer and BERT for supervised contrastive learning. An embedding space is learned in which the same-sample pairs stay close to each other while different-sample pairs remain far apart.

embeddings based on tweet text and image inputs. We employ an attention-based contrast loss with an additive angular margin at the network's output to maximize the decision margin in the angular space.

3.1 Multi-layer Fusion Module

To align and combine the textual and visual characteristics, we employ the Multi-Layer Fusion module, as illustrated in Fig. 2. First, we use BERT [6] and Vision transformer (ViT) [7] to encode the text and image for the hidden representation of the text $T = \{\boldsymbol{t}_c, \boldsymbol{t}_1, \boldsymbol{t}_2 ..., \boldsymbol{t}_s\}$ and the hidden representation of the image $I = \{\boldsymbol{i}_1, \boldsymbol{i}_2, \ldots, \boldsymbol{i}_n\}$. Then, to align and fuse the text and image feature features, we concatenate the text feature and the image sequence feature by the multi-layer Transformer encoder. Now that we have the text and image fusion sequence features, it is clear that these features cannot be used for classification. Thus, we employ a straightforward attention layer to get the multimodal representation $\boldsymbol{R}$. At the network's output, we compute an attention-based class-aware contrastive loss with an additive angular margin. This loss function allows us to enhance the discriminative power of the model and further improve the separation between different classes.

3.2 Angular Margin-Based Contrastive Learning

The primary objective of contrastive learning is to ensure that embedding vectors of different types of tweets are distinctly separated while tweets of the same type exhibit proximity. To achieve this, we apply data augmentation to each batch of input samples. In detail, for text data augmentation, we employ a technique called back-translation [32], which involves translating the original text into another language and then translating it back to the original language. This process introduces diversity and variation in the text data while maintaining the underlying meaning. For image data augmentation, we utilize a method known as RandAugment [3]. This technique applies random and diverse transformations to the images, such as rotations, translations, and color manipulations. These transformations augment the image data by introducing variations while preserving the essential visual characteristics. By incorporating back-translation for text and RandAugment for image augmentation, we ensure that the embedding vectors of the same instances remain consistent while those of different instances exhibit variation. This approach enhances the discriminative power of the learned representations in contrastive learning. We utilize the original and enhanced data in a supervised contrastive loss [15]:

$$L_{Con} = \sum_{i=1}^{N} \frac{-1}{|\mathcal{P}(i)|} \sum_{p \in \mathcal{P}(i)} \log \frac{\exp\left(\operatorname{sim}\left(\boldsymbol{z}_i, \boldsymbol{z}_p\right)/\tau\right)}{\sum_{a \in \mathcal{A}(i)} \exp\left(\operatorname{sim}\left(\boldsymbol{z}_i, \boldsymbol{z}_a\right)/\tau\right)}. \tag{1}$$

In Eq. 1, $\boldsymbol{z}_i$ is an anchor, $\boldsymbol{z}_a$ is a negative sample, $\mathcal{A}(i)$ is an index set of negative samples concerning $\boldsymbol{z}_i$, $\boldsymbol{z}_p$ is a positive sample concerning $\boldsymbol{z}_i$, $\mathcal{P}(i)$ contains the indices of positive samples in the augmented batch (original + augmentation), and $sim(\boldsymbol{z}_i, \boldsymbol{z}_p)$ is the cosine similarity. Although the training target attempts to distinguish positive and negative samples, these representations may not be discriminative or resilient enough to noise. Set the cosine similarity as:

$$\cos\theta_{i,p} = \frac{\boldsymbol{z}_i^{\top} \boldsymbol{z}_p}{\|\boldsymbol{z}_i\| \, \|\boldsymbol{z}_p\|}. \tag{2}$$

where $\theta_{i,p}$ is the angle between the embedding $\boldsymbol{z}_i$ and $\boldsymbol{z}_p$. The decision boundary of $\boldsymbol{z}_i$ for specific p and a is $\theta_{i,p} = \theta_{i,a}$, where p and a are indexes to positive and negative samples, respectively. If there is no decision margin, a minor modification of the embedding vectors near the decision boundary may result in an incorrect conclusion. To address this issue, we propose a novel training objective for rumor detection learning that includes an additive angular margin m. We name it **Margin Con**trastive (MarginCon), which can be expressed as follows:

$$L_{\text{MarginCon}} = \sum_{i=1}^{N} \frac{-1}{|P(i)|} \sum_{p \in P(i)} \log \frac{\exp\left(\cos\left(\theta_{i,p} + m\right)/\tau\right)}{\sum_{a \in A(i)} \exp\left(\cos\left(\theta_{i,a}\right)/\tau\right)}. \tag{3}$$

In this loss function, the decision boundary for a specific pair of rumors and truths, represented by z_i, is defined as $\theta_{i,p+m} = \theta_{i,a}$. This pushes the value of z_i

towards the area where the value of $\theta_{i,p}$ decreases and the value of $\theta_{i,a}$ increases. As a result, adding a margin can increase the disparity between representations of the rumors and truths. Furthermore, the decision boundary also includes an additional margin m from the boundary $\theta_{i,p+m} = \theta_{i,a}$, which makes it more tolerant to noise and more resilient during inference.

3.3 Class-Aware Attention Module

As discussed in prior research [4,26], hard samples can significantly impact sample mining, potentially leading to suboptimal local minima for the trained model. While focusing on more difficult negatives is common, some may be outliers. To evaluate the relationship between tweet embeddings and their class labels, we compute the compatibility between an embedding vector and its corresponding class vector. The dot product of the two vectors can be used as a measure of compatibility.

We use a classification branch after the tweet embeddings to learn the class vectors. The class vectors can be trained and updated using gradient descent. Make $\{c_k\}_{k=1}^{C}$ for C. In our study, C is the number of tweets in a mini-batch. We compute the CA score:

$$\alpha_{i,j} = \frac{\exp\left(\boldsymbol{z}_i^\top \boldsymbol{c}_{y_j}\right)}{\sum_{k=1}^{C} \exp\left(\boldsymbol{z}_i^\top \boldsymbol{c}_k\right)}. \tag{4}$$

where $\boldsymbol{y}_j$ is the class label of $\boldsymbol{z}_j$. Through gradient descent, $\{c_k\}_{k=1}^{C}$ will be able to represent the C classes. We incorporated this module into the MarginCon loss to derive a new loss function, which we name it **Class-aware Attention Margin Contrastive** (CAMarginCon). It can be expressed as follows:

$$L_{CAMarginCon} = \sum_{i=1}^{n} \frac{-1}{|P(i)|} \sum_{p \in P(i)} \log \frac{\exp\left(\cos\left(\theta_{i,p} + m\right) \cdot \alpha_{i,p}/\tau\right)}{\sum_{a \in A(i)} \exp\left(\cos\left(\theta_{i,a}\right) \cdot \alpha_{i,a}/\tau\right)}. \tag{5}$$

where $\alpha_{i,p}$ is the CA score of pair $(\boldsymbol{z}_i, \boldsymbol{z}_p)$ and $\alpha_{i,a}$ is the CA score of pair $(\boldsymbol{z}_i, \boldsymbol{z}_a)$. According to Eq. 4, a very hard negative will make $\alpha_{i,a}$ close to 0, causing a smaller denominator for that i in the supervised contrastive loss in Eq. 5. As a result, the contribution of that negative will be suppressed.

3.4 Weighted Loss with Cross-Entropy

For the rumor detection shown in Fig. 2, we feed the multimodal representation $\boldsymbol{R}$ into the fully connected layer and utilize the softmax algorithm to detect rumors. We use the cross-entropy loss as the classification loss, and it is as follows:

$$L_{RD} = -\frac{1}{N} \sum_{i=1}^{N} \log\left(\frac{\exp(R \cdot \boldsymbol{w}_{RD} + \boldsymbol{b}_{RD})}{\sum_{i=1}^{n} \exp(R \cdot \boldsymbol{w}_{RD} + \boldsymbol{b}_{RD})}\right). \tag{6}$$

The total loss function can be combined like follows, λ represents weights:

$$L = \lambda_1 \cdot L_{RD} + \lambda_2 \cdot L_{CAMarginCon}. \tag{7}$$

Table 1. Statistic of three databases.

Statistics		Weibo	Twitter
Training Set	Rumors	3749	7272
	Truths	3783	5926
Testing Set	Rumors	1000	927
	Truths	996	755
Total		9528	14480

4 Experiments and Results

4.1 Datasets

To validate our approach and statistically assess the outcomes, We employ Weibo [12] and Twitter [5]. All of these datasets emanate from the paramount social media platforms in China and the United States. By leveraging these datasets, we aim to establish the remarkable efficacy of our proposed model across different languages and demonstrate its effectiveness in diverse linguistic contexts. Table 1 shows the data information for all datasets. In the division of the datasets, we adhere to the delineation expounded in the original articles. For more details, we partitioned the Weibo and Twitter datasets into distinct subsets of training and testing data, employing ratios of approximately 8:2 and 9:1, respectively.

4.2 Implementation Details

We implement our method and baselines for our experiments using the Hugging-Face Transformers and Pytorch on RTX3090. We use the BERT-base and ViT as the text and image encoders in the Multi-Layer Fusion module. Each augmentation provides a unique data perspective and comprises some of the original sample's information. For texts, we use a random mask. For images, we perform cropping, rotating, adjusting contrast, inverting, flipping, solarizing, posterizing, brightness, and sharpness.

4.3 Result

Table 2 presents the performance of our proposed MarCAMLF approach on two datasets, evaluated using accuracy, precision, recall, and F1 score. The results show that our method outperforms other state-of-the-art methods on all datasets, achieving an average accuracy of 92.1% on Weibo and 82.3% on Twitter. This can be attributed to the effective handling of challenging samples and our approach's expansion of the decision boundary.

On the Weibo dataset, MarCAMLF performs better than other existing methods across various evaluation metrics. However, compared to the CMC method, our approach exhibits slightly lower in both the Precision metric for rumors and the Recall metric for truths. This difference can be attributed to

Table 2. The result of the experiment. For each metric, the best is highlighted in bold.

Dataset	Methods	Accuracy	Rumors			Truths		
			Precision	Recall	F1-score	Precision	Recall	F1-score
Weibo	att-RNN [12]	0.772	0.854	0.656	0.742	0.720	0.889	0.795
	EANN [28]	0.827	0.847	0.812	0.829	0.807	0.843	0.825
	MCNN [33]	0.846	0.809	0.857	0.832	0.879	0.837	0.858
	MCAN [30]	0.899	0.913	0.889	0.901	0.884	0.909	0.897
	CAFE [2]	0.840	0.855	0.830	0.842	0.825	0.851	0.837
	CMC [29]	0.893	**0.940**	0.869	0.899	0.876	**0.945**	0.907
	MarCAMLF(Ours)	**0.921**	0.920	**0.919**	**0.919**	**0.921**	0.922	**0.922**
Twitter	att-RNN [12]	0.664	0.749	0.615	0.676	0.589	0.728	0.651
	EANN [28]	0.648	0.810	0.498	0.617	0.584	0.759	0.660
	MVAE [14]	0.745	0.801	0.719	0.758	0.689	0.777	0.730
	MCAN [30]	0.809	0.889	0.765	0.822	0.732	0.871	0.795
	MarCAMLF(Ours)	**0.856**	**0.895**	**0.783**	**0.835**	**0.766**	**0.889**	**0.815**

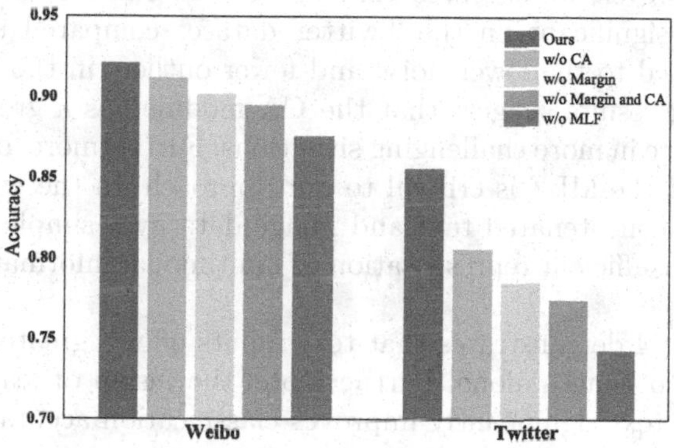

Fig. 3. Ablation Analysis in Accuracy.

the effectiveness of the cross-modal distillation objective function proposed by CMC, which notably improves their performance in these aspects. Nevertheless, it is important to emphasize that our proposed model outperforms the approach presented in CMC due to incorporating discriminative margins and reduced attention towards hard negative samples.

Our proposed framework on the Twitter dataset comprehensively surpasses the current baseline models, particularly with a notable 4.7% lead in overall accuracy. This substantial improvement further highlights the effectiveness of our proposed approach.

4.4 Ablation Study

To assess the effectiveness of our approach, we conducted ablation studies to evaluate the impact of our proposed CA module, margin, and MLF on model performance. We performed experiments on two datasets, and Fig. 3 presents the results. Our findings indicate that all three factors improve the model's

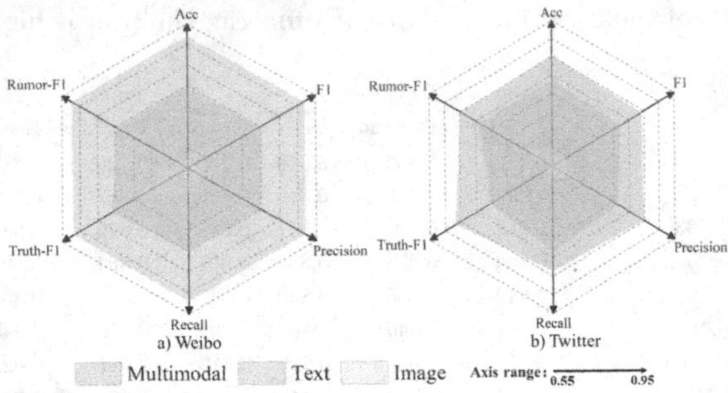

Fig. 4. Ablation Analysis in different modalities.

performance. However, we observed that the CA module's contribution to performance is less significant on the Twitter dataset compared to Weibo. This could be attributed to the lower noise and fewer outliers in the Weibo dataset. Nevertheless, our results suggest that the CA module has a greater impact on model performance in more challenging situations. Furthermore, our experiments demonstrate that the MLF is critical to our approach. In the w/o MLF experiment, where we concatenated text and image data by a simple concatenation, we observed an insufficient representation of multimodal information and a drop in accuracy.

Moreover, Fig. 4 demonstrates that text inputs offer a greater wealth of features compared to images alone. Furthermore, the fusion of features extracted from images and text significantly improves classification accuracy.

5 Conclusion

This paper proposes an improved multimodal rumor detection model based on contrastive learning and multi-layer fusion. To reduce the probability of misclassification, we incorporate an angular margin and a class-aware attention mechanism into the supervised contrastive loss. Experimental results on publicly available datasets demonstrate that our proposed model is competitive with strong baseline models. In addition, our work is crucial as we have significantly improved the efficacy of rumor detection, allowing us to promptly detect and curb the spread of rumors, thereby mitigating potential adverse effects.

References

1. Chen, T., Kornblith, S., Norouzi, M., Hinton, G.: A simple framework for contrastive learning of visual representations. In: International Conference on Machine Learning, pp. 1597–1607. PMLR (2020)
2. Chen, Y., et al.: Cross-modal ambiguity learning for multimodal fake news detection. In: Proceedings of the ACM Web Conference 2022, pp. 2897–2905 (2022)

3. Cubuk, E.D., Zoph, B., Shlens, J., Le, Q.V.: RandAugment: practical automated data augmentation with a reduced search space. In: Proceedings of the IEEE/CVF Conference on Computer Vision and Pattern Recognition Workshops, pp. 702–703 (2020)
4. Cui, Y., Zhou, F., Lin, Y., Belongie, S.: Fine-grained categorization and dataset bootstrapping using deep metric learning with humans in the loop. In: Proceedings of the IEEE Conference on Computer Vision and Pattern Recognition, pp. 1153–1162 (2016)
5. Boididou, C., Papadopoulos, S., Zampoglou, M., Apostolidis, L., Papadopoulou, O., Kompatsiaris, Y.: Detection, visualization of misleading content on Twitter. Int. J. Multimedia Inf. Retrieval **7**(1), 71–86 (2018)
6. Devlin, J., Chang, M.W., Lee, K., Toutanova, K.: BERT: pre-training of deep bidirectional transformers for language understanding. ArXiv abs/1810.04805 (2019)
7. Dosovitskiy, A., et al.: An image is worth 16×16 words: transformers for image recognition at scale. ArXiv abs/2010.11929 (2020)
8. Gao, Y., Wang, X., He, X., Feng, H., Zhang, Y.: Rumor detection with self-supervised learning on texts and social graph. Front. Comp. Sci. **17**(4), 174611 (2023)
9. Han, W., Chen, H., Poria, S.: Improving multimodal fusion with hierarchical mutual information maximization for multimodal sentiment analysis. arXiv preprint arXiv:2109.00412 (2021)
10. He, K., Fan, H., Wu, Y., Xie, S., Girshick, R.: Momentum contrast for unsupervised visual representation learning. In: Proceedings of the IEEE/CVF Conference on Computer Vision and Pattern Recognition, pp. 9729–9738 (2020)
11. Hua, J., Cui, X., Li, X., Tang, K., Zhu, P.: Multimodal fake news detection through data augmentation-based contrastive learning. Appl. Soft Comput. **136**, 110125 (2023)
12. Jin, Z., Cao, J., Guo, H., Zhang, Y., Luo, J.: Multimodal fusion with recurrent neural networks for rumor detection on microblogs. In: Proceedings of the 25th ACM International Conference on Multimedia, pp. 795–816 (2017)
13. Ke, Z., Sheng, J., Li, Z., Silamu, W., Guo, Q.: Knowledge-guided sentiment analysis via learning from natural language explanations. IEEE Access **9**, 3570–3578 (2021)
14. Khattar, D., Goud, J.S., Gupta, M., Varma, V.: MVAE: multimodal variational autoencoder for fake news detection. In: The World Wide Web Conference, pp. 2915–2921 (2019)
15. Khosla, P., et al.: Supervised contrastive learning. Adv. Neural. Inf. Process. Syst. **33**, 18661–18673 (2020)
16. Li, X., Li, Z., Sheng, J., Slamu, W.: Low-resource text classification via cross-lingual language model fine-tuning. In: Sun, M., Li, S., Zhang, Y., Liu, Y., He, S., Rao, G. (eds.) CCL 2020. LNCS (LNAI), vol. 12522, pp. 231–246. Springer, Cham (2020). https://doi.org/10.1007/978-3-030-63031-7_17
17. Li, Z., Li, X., Sheng, J., Slamu, W.: AgglutiFiT: efficient low-resource agglutinative language model fine-tuning. IEEE Access **8**, 148489–148499 (2020)
18. Li, Z., Mak, M.W.: Speaker representation learning via contrastive loss with maximal speaker separability. In: 2022 Asia-Pacific Signal and Information Processing Association Annual Summit and Conference (APSIPA ASC), pp. 962–967. IEEE (2022)
19. Li, Z., Mak, M.W., Meng, H.M.L.: Discriminative speaker representation via contrastive learning with class-aware attention in angular space. In: ICASSP 2023–2023 IEEE International Conference on Acoustics, Speech and Signal Processing (ICASSP), pp. 1–5. IEEE (2023)

20. Lu, M., Huang, Z., Li, B., Zhao, Y., Qin, Z., Li, D.: SIFTER: a framework for robust rumor detection. IEEE/ACM Trans. Audio Speech Lang. Process. **30**, 429–442 (2022)
21. Ma, J., Gao, W., Wong, K.F.: Detect rumors in microblog posts using propagation structure via kernel learning. Association for Computational Linguistics (2017)
22. Ma, J., Gao, W., Wong, K.F.: Rumor detection on twitter with tree-structured recursive neural networks. Association for Computational Linguistics (2018)
23. Ngiam, J., Khosla, A., Kim, M., Nam, J., Lee, H., Ng, A.Y.: Multimodal deep learning. In: Proceedings of the 28th International Conference on Machine Learning (ICML-2011), pp. 689–696 (2011)
24. Peng, L., Jian, S., Li, D., Shen, S.: MRML: multimodal rumor detection by deep metric learning. In: ICASSP 2023–2023 IEEE International Conference on Acoustics, Speech and Signal Processing (ICASSP), pp. 1–5. IEEE (2023)
25. Sang, M., Li, H., Liu, F., Arnold, A.O., Wan, L.: Self-supervised speaker verification with simple Siamese network and self-supervised regularization. In: ICASSP 2022–2022 IEEE International Conference on Acoustics, Speech and Signal Processing (ICASSP), pp. 6127–6131. IEEE (2022)
26. Schroff, F., Kalenichenko, D., Philbin, J.: FaceNet: a unified embedding for face recognition and clustering. In: Proceedings of the IEEE Conference on Computer Vision and Pattern Recognition, pp. 815–823 (2015)
27. Sheng, J., et al.: Multi-view contrastive learning with additive margin for adaptive nasopharyngeal carcinoma radiotherapy prediction. In: Proceedings of the 2023 ACM International Conference on Multimedia Retrieval, pp. 555–559 (2023)
28. Wang, Y., et al.: EANN: event adversarial neural networks for multi-modal fake news detection. In: Proceedings of the 24th ACM SIGKDD International Conference on Knowledge Discovery & Data Mining, pp. 849–857 (2018)
29. Wei, Z., Pan, H., Qiao, L., Niu, X., Dong, P., Li, D.: Cross-modal knowledge distillation in multi-modal fake news detection. In: ICASSP 2022–2022 IEEE International Conference on Acoustics, Speech and Signal Processing (ICASSP), pp. 4733–4737. IEEE (2022)
30. Wu, Y., Zhan, P., Zhang, Y., Wang, L., Xu, Z.: Multimodal fusion with co-attention networks for fake news detection. In: Findings of the Association for Computational Linguistics: ACL-IJCNLP 2021, pp. 2560–2569 (2021)
31. Wu, Z., Xiong, Y., Yu, S.X., Lin, D.: Unsupervised feature learning via non-parametric instance discrimination. In: 2018 IEEE/CVF Conference on Computer Vision and Pattern Recognition, pp. 3733–3742 (2018)
32. Xie, Q., Dai, Z., Hovy, E., Luong, T., Le, Q.: Unsupervised data augmentation for consistency training. Adv. Neural. Inf. Process. Syst. **33**, 6256–6268 (2020)
33. Xue, J., Wang, Y., Tian, Y., Li, Y., Shi, L., Wei, L.: Detecting fake news by exploring the consistency of multimodal data. Inf. Process. Manag. **58**(5), 102610 (2021)
34. Ying, Q., Hu, X., Zhou, Y., Qian, Z., Zeng, D., Ge, S.: Bootstrapping multi-view representations for fake news detection. In: Proceedings of the AAAI Conference on Artificial Intelligence (2023)

An Effective Dynamic Reweighting Method for Unbiased Scene Graph Generation

Lingfeng Hu[1], Si Liu[2], and Hanzi Wang[1(✉)]

[1] Fujian Key Laboratory of Sensing and Computing for Smart City, School of Informatics, Xiamen University, 361005 Xiamen, People's Republic of China
hanzi.wang@xmu.edu.cn

[2] Institute of Artificial Intelligence, Beihang University, Beijing, China

Abstract. Despite the remarkable advancements in Scene Graph Generation (SGG) in recent years, the precise capture and modeling of long-tail object relationships remain persistent challenges in the field. Conventional methods generally employ resampling and reweighting techniques to achieve unbiased predictions. Existing reweighting methods in SGG calculate weights based on the class distribution of the dataset. And they focus on the reweighting of the related samples while overlooking the reweighting of the samples whose objects are unrelated. However, the sample distribution during the training process is inconsistent with the class distribution of the dataset, and the reweighting of samples whose objects are unrelated should not be overlooked. In this paper, we propose a novel method named Dynamic Reweighting based on the Sample Distribution (DRSD). The DRSD method calculates the weights of classes based on the sample distribution during the training process and incorporates reweighting for the samples whose objects are unrelated. Specifically, we utilize a sample queue mechanism to record and update the sample distribution and introduce a transition mechanism to ensure training stability. The experiments conducted on the Visual Genome dataset demonstrate the effectiveness of our method. Our method exhibits model-agnostic characteristics and yields significant performance improvements on three benchmark models (Motif, VCTree, and Transformer). Specifically, it achieves an increase of 23.4%, 25.1%, and 27.6% on the mR@100 metric for the Predicate Classification task, achieving 40.9%, 41.2%, and 43.4%, respectively. Moreover, our method outperforms the state-of-the-art reweighting method in SGG, *i.e.* FGPL, by 3%.

Keywords: Scene graph generation · Reweighting method · Long-tail

1 Introduction

Scene graph generation (SGG) [24] is fundamental towards achieving a comprehensive understanding of a scene, playing a crucial role in many high-level vision tasks such as image captioning [1,2,26,27], image/video retrieval [6,11,15,17,32],

© The Author(s), under exclusive license to Springer Nature Singapore Pte Ltd. 2024
Q. Liu et al. (Eds.): PRCV 2023, LNCS 14425, pp. 345–356, 2024.
https://doi.org/10.1007/978-981-99-8429-9_28

and visual question answering [8,16,21,31]. A typical SGG process can be formulated as follows: Given an input image (a scene), the objective is to generate a graph network consisting of nodes and edges. The nodes represent various objects in the scene, and the edges represent the relations between the objects. In SGG, the output is represented as triplets: (o_i, r_{ij}, o_j), for example, "person wearing shirt". We categorize the r_{ij} classes into two groups. We collectively refer to the different annotated relationship categories as the relation classes. For object pairs without a relationship, we assign them the label of the non-relation class.

The unbiased prediction is a challenging problem in SGG [19]. The long-tailed distribution of the Visual Genome dataset [9] makes the model prone to fitting the head classes but ignoring the tail classes. Class rebalancing strategies, such as resampling and reweighting, are effective techniques for addressing the long-tailed problem. Studies [7,12,13,28] have employed these strategies in scene graph generation. However, their sampling and weighting strategies are based on the class distribution of the dataset. We argue that the sample distribution during the training process is inconsistent with the statistical class distribution of the dataset. Additionally, existing reweighting methods [13,28] ignore the reweighting of the non-relation class, even though the number of samples in the non-relation class is nearly 100 times than that in the relation classes.

Based on the aforementioned observations, we propose a novel method named Dynamic Reweighting based on the Sample Distribution (DRSD), which contains two components aimed at addressing the aforementioned issues. The first component is the Dynamic Reweighting Mechanism (DRM), designed to address the issue of inconsistency between the distribution derived from dataset statistics and the actual distribution observed during training. This mechanism performs reweighting based on the sample distribution during the training process. Specifically, we incorporate a sample queue mechanism to record the recent sample information during the training process. The weights calculated based on the sample queue mechanism are referred to as *queue weights*. Considering the instability of *queue weights* in the early stages of training, we calculate *initial weights* based on the class distribution of the dataset and introduce a transition mechanism to gradually shift towards *queue weights* as the training progresses. The second component is the Non-relation Class Reweighting (NCR) strategy, which involves applying reweighting to the non-relation class. This strategy aims to enhance the model's emphasis on the relation classes.

The contributions of our work are summarized as follows: 1) We review prior works and reveal two shortcomings of existing reweighting methods that hinder unbiased prediction in SGG. 2) We propose a model-agnostic reweighting method to address the aforementioned issues. 3) Extensive experiments conducted on Visual Genome dataset demonstrate the effectiveness of our method.

2 Related Work

Scene Graph Generation: In the field of scene graph generation [9], researchers initially focus on improving the Recall@K metric, which aims to enhance the overall predicate recall rate. However, Motifs [29] demonstrates that even when the model predicts based solely on the categories of the subject and object, it can still achieve superior Recall@K performance compared to state-of-the-art methods. This observation can be attributed to the long-tailed distribution of the Visual Genome dataset, the model primarily focuses on the head classes. To address the long-tailed problem, VCTree [20] proposes the concept of unbiased scene graph generation, advocating for equal treatment of every class by the model. Studies have been devoted to achieving this challenging objective. Zhang et al. [30] aim to refine the dataset by annotating potentially missed relations and relabeling potentially mislabeled relations into more fine-grained ones. Goel et al. formulate the SGG task as a multi-label task and explore hidden labels to mitigate the bias impact. Additionally, Deng et al. [4] adopt an iterative approach, progressively transferring from coarse-grained prediction to fine-grained prediction. These methods often involve making modifications to annotations, datasets, or network architecture, whereas the class rebalancing methods are typically implemented in a plug-and-play manner.

Class Rebalancing Methods: Class rebalancing methods in SGG can be categorized into resampling methods and reweighting methods. For resampling methods, Deng et al. [12] propose a dual-level resampling approach that includes image-level upsampling and instance-level downsampling. Dong et al. [5] divide all classes into five groups and get a balance sample space inside each group. Regarding reweighting methods, Wang et al. [23] utilize the inverse of category frequencies to balance rare and common categories. Lyu et al. [13] construct a predicate lattice to present the correlation between the predicates and employ reweighting based on this lattice. However, the aforementioned methods compute the weights or sampling rates based on the class distribution of the dataset and disregard the reweighting of the non-relation class. We propose the DRSD (Dynamic Reweighting based on the Sample Distribution) method to address the limitations.

3 Method

3.1 Basic Reweighting Formula

We adopt Inverse Class Frequency (ICF) [10] as the prototype for the basic reweighting formula of our method. Specifically, we calculate the weight of the i-th class as follows:

$$w_i = \frac{C * \hat{w}_i}{\sum_{k=1}^{C} \hat{w}_k}, \qquad \text{with } \hat{w}_i = \left(\frac{\sum_{k=1}^{C} n_k}{n_i} \right)^{\theta} \tag{1}$$

where C indicates the number of predicate categories, and n_i presents the sample number of the i-th class. The parameter θ is a hyperparameter. We compute w_i, which represents the loss weight assigned to the i-th class in the reweighting method, and finally calculate the loss using the following equation:

$$L_{rwt}(\vec{z}) = -\sum_{i=1}^{C} w_i y_i \log(\sigma_i), \quad \text{with } \sigma_i = \frac{e^{z_i}}{\sum_{j=1}^{C} e^{z_j}} \tag{2}$$

where $\vec{z} = [z_1, z_2, \ldots, z_c]$ is the predicted logits. And y_i is the i-th value of the one-hot ground truth $\vec{y}$.

Despite the basic reweighting formula being straightforward, our reweighting method achieves impressive results when combined with the two proposed components: the Dynamic Reweighting Mechanism (DRM) and the Non-relation Class Reweighting (NCR) strategy.

3.2 Dynamic Reweighting Mechanism

The existing reweighting methods in the SGG field calculate weights based on the class distribution of the dataset [13,23,28]. These methods actually implicitly assume that the sample distribution during the training process is always consistent with the class distribution of the dataset. However, due to the long-tailed property of the Visual Genome dataset, the distribution of the tail classes undergoes significant changes during the training process, resulting in the weights computed based on the static data being unable to accommodate dynamic training scenarios. Therefore, we propose the Dynamic Weighting Mechanism (DRM) to calculate the weights based on the actual training conditions. Specifically, DRM consists of two components: A Sample Queue Mechanism that records information about the recent training samples and a Weight Transition Mechanism that ensures the stability of the weights during the training process.

Sample Queue Mechanism: To capture the sample distribution during the training process, we adopt the data structure "queue", which is known for its "first-in, first-out" property. This property allows for continuous updates and meets our requirement for recording the recent sample distribution. We create an empty queue, named $\mathcal{Q}$, at the beginning of training. As the training progresses, more and more samples are fed into $\mathcal{Q}$ until it reaches its maximum size. Once $\mathcal{Q}$ reaches the maximum size, the oldest sample is popped out as a new sample arrives. We calculate the weights using Eq. (1) with the specific data derived from the information recorded in $\mathcal{Q}$.

Weight Transition Mechanism: Although the sample queue mechanism can reflect the sample distribution during the training process, there are still some issues with it. At the beginning of training, only a few samples are recorded in $\mathcal{Q}$, which can lead to unstable weights. Additionally, at this stage, the model has not yet learned the basic patterns, so utilizing these unstable weights may hinder the training process. To mitigate these issues, we proposed to use *initial weights* as a starting point. As the training progresses, the weights gradually transfer to *queue weights*. We update *dynamic weights* as follows:

$$w_i^d \leftarrow \alpha w_i^i + (1-\alpha) w_i^q, \quad \text{with } \alpha = \min(0.999^{\gamma}, 0.9) \tag{3}$$

where w_i^i represents the *initial weight* of i-th class, which is calculated based on the class distribution of the dataset. And w_i^q represents the *queue weight* of i-th class, which is calculated based on the information recorded in $\mathcal{Q}$. We combine them to obtain w_i^d, which is the *dynamic weight* that we utilize to reweight the loss. The parameter γ represents the number of samples recorded in $\mathcal{Q}$. This number increases as the training progresses until it reaches the maximum size of $\mathcal{Q}$.

The transition factor $\alpha \in (0, 0.9]$ indicates the extent to which *dynamic weights* have transferred towards *queue weights*. Initially, when γ is relatively small, the value of α is set to 0.9, indicating that 90% of *dynamic weights* are determined by the *initial weights*. As the training progresses, γ increases, and *dynamic weights* gradually transfer to *queue weights*. The pipeline of the Dynamic Reweighting Mechanism (DRM) is illustrated in Fig. 1.

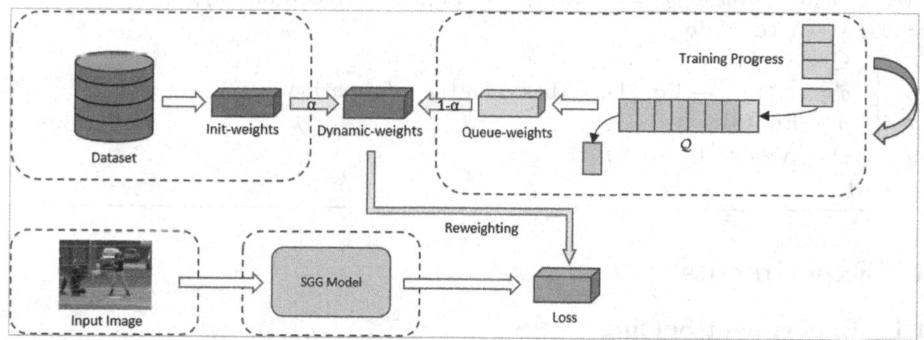

Fig. 1. Pipeline of Dynamic Reweighting Mechanism (DRM). *Dynamic weights* are combined by two weights, i.e. *initial weights* and *queue weights*. *Initial weights* are calculated based on the class distribution of the dataset. *Queue weights* are calculated based on the sample distribution recorded by $\mathcal{Q}$. *Dynamic weights* will be utilized to reweight the loss during the training process.

3.3 Non-relation Class Reweighting

Existing reweighting methods [13,28] in SGG mainly focus on the reweighting of the relation classes while overlooking the reweighting of the non-relation class. However, through the analysis of the sample distribution during the middle stage of training, we found that the number of the non-relation samples is approximately 100 times that of relation samples, which highlights the importance of not overlooking the reweighting of the non-relation class. In fact, applying a small weight to the non-relation class allows the model to focus more on the relation classes. Moreover, this does not hinder the learning of the non-relation class as the number of its samples is significantly large.

Therefore, we propose the Non-relation Class Reweighting (NCR) strategy to apply reweighting to the non-relation class. After the reweighting process, the model focuses more on distinguishing the relation class samples.

We summarize the workflow of the proposed DRSD method in Algorithm 1. For output, w_i^d represents the dynamic weight of the i-th class. Here, $i = 0$

corresponds to the non-relation class. We calculate weights from 0 to C, indicating the utilization of the Non-relation Class Reweighting (NCR) strategy. The pipeline of the Dynamic Reweighting Mechanism (DRM) is illustrated in lines 1 to 9.

Algorithm 1 . Dynamic Reweighting based on the Sample Distribution (DRSD).

Input: Training samples $\mathcal{S}_{all} = \{s_k\}_{k=1}^{N}$; Sample numbers of each class in the dataset $\{n_i\}_{i=0}^{C}$

Output: Weights for each class $\vec{w} = \{w_i^d\}_{i=0}^{C}$

1 $\mathcal{Q} \leftarrow \{\}$;

2 Initialize *queue weights* $\vec{q} = \{w_i^q\}_{i=0}^{C}$ as $[1, 1, \ldots, 1]$

3 Calculate *initial weights* $\vec{i} = \{w_i^i\}_{i=0}^{C} \leftarrow$ Eq. (1) (based on $\{n_i\}_{i=0}^{C}$)

4 **for** $k \leftarrow 1$ **to** N **do**

5 $\quad \mathcal{Q} \leftarrow \mathcal{Q} \cup \{s_k\}$;

6 $\quad \vec{q} = \{w_i^q\}_{i=0}^{C} \leftarrow$ Eq. (1); (based on the information in $\mathcal{Q}$)

7 $\quad \alpha \leftarrow$ Eq. (3);

8 $\quad \vec{w} \leftarrow \alpha * \vec{i} + (1 - \alpha) * \vec{q}$;

9 **end**

---

4 Experiments

4.1 Experiment Setting

Dataset: We conduct the evaluation on the Visual Genome (VG) dataset[1]. Following previous work [3,19,29], we split the VG dataset to get the widely-used subset called VG150, which only keeps the most frequent 150 object categories and 50 predicate categories.

Model Configuration: To validate the model-agnostic property of our method, following the recent work [13], we integrate our method with three baseline models, which are VCTree [20], Motif [29], and Transformer [22], in the SGG benchmark [18].

Tasks: We evaluate our method on three sub-tasks of SGG: (1) Predicate Classification (**PredCls**); (2) Scene graph classification (**SGCls**); (3) Scene graph generation (**SGGen**).

Evaluation Metrics: Following the previous work [13,19,20], we use the mean Recall@K (**mR@K**) to evaluate our method. This metric was proposed by Tang et al. [20], which calculates the average recall of all predicates. Due to the extremely uneven distribution of the VG dataset, the Recall@K (**R@K**) metric drives the model only to predict the high-frequency predicates and ignore the low-frequency predicates. However, mean Recall@K (**mR@K**) can make high-frequency predicates and low-frequency predicates equally important to achieve comprehensive learning, for which it is widely used as an unbiased evaluation metric.

[1] https://homes.cs.washington.edu/~ranjay/visualgenome/index.html.

4.2 Implementation Details

We utilize the frozen pre-trained Faster R-CNN by [19] for object detection and utilize Glove [14] to obtain the semantic embedding. The baseline model [18] uses the vanilla cross-entropy loss, and we replace it with our proposed reweighting method DRSD to evaluate the performance.

The parameter "VAL_PERIOD" was modified to 3607 because calculations revealed that iterating 3607 times precisely completes one pass through the training set. Meanwhile, the size of Q is set as 3607 for the same reason. We have also explored various alternative configurations of Q size. Keeping the parameter around 3607 ensures relatively consistent performance. Performance tends to decline when the size is either too large or too small. A smaller size restricts the sample distribution's representation, while a larger size slows down queue updates, hampering the capture of real-time training dynamics.

The batch size is 16 and θ in Eq. (1) is set to 0.9. Regarding the learning schedule, we begin with a learning rate of 0.002 and apply the WarmupMultiStepLR strategy as in [18]. All experiments are conducted on two RTX 2080 Ti GPUs. Apart from the parameters mentioned above, we keep the rest of the parameter settings for the baseline model unchanged, as we incorporate the DRSD method as an additional optional loss function attached to the baseline model.

4.3 Comparisons with State-of-the-Art Methods

Following [13], we incorporate our proposed reweighting method DRSD with three benchmark models, namely VCTree [20], Motif [29], and Transformer [22]. We compare the results of the DRSD method with other state-of-the-art reweighting methods on three benchmark models. Additionally, we also compare the DRSD method with non-reweighting methods [12,19,25,30]. Both aspects of the comparison demonstrate the superior performance of our proposed DRSD method. The Quantitative results are shown in Table 1.

In comparison with the other reweighting methods, our method DRSD achieves state-of-the-art (SOTA) performance on all three tasks and across three benchmark models [20,22,29], which confirms that our method is model-agnostic. The proposed DRSD method outperforms the previous state-of-the-art (SOTA) [13] by a large margin of 3% on mR@100 for the PredCls task on the Transformer benchmark model. Besides, when compared to other non-reweighting methods, our method maintains a performance advantage.

The superior performance demonstrates the effectiveness of our method, validating the significance of the dynamic reweighting and the reweighting of the non-relation class.

Table 1. Comparison between existing methods and the proposed DRSD method. Results for Seesaw-related methods are obtained from [13]. The best result is highlighted in bold.

Method	PredCls		SGCls		SGDet	
	mR@50	mR@100	mR@50	mR@100	mR@50	mR@100
TDE-VCTree [19]	25.4	28.7	12.2	14.0	9.3	11.1
PCPL [25]	35.2	37.8	18.6	19.6	9.5	11.7
BGNN [12]	30.4	32.9	14.3	16.5	10.7	12.6
IETrans-Rwt [30]	35.0	38.0	20.8	22.3	15.0	18.1
Motif [29]	16.0	17.5	9.6	10.2	7.3	8.8
CogTree-Motif [28]	26.4	29.0	14.9	16.1	10.4	11.8
Seesaw-Motif [23]	28.1	33.7	15.6	18.3	10.5	13.2
FGPL-Motif [13]	33.0	37.5	21.3	22.5	15.4	18.2
DRSD-Motif(Ours)	**38.8**	**40.9**	**23.7**	**24.6**	**18.1**	**20.9**
VCTree [20]	14.9	16.1	7.5	7.9	5.7	6.9
CogTree-VCTree [28]	27.6	29.7	18.8	19.9	10.4	12.1
Seesaw-VCTree [23]	29.6	35.3	19.9	23.5	10.5	13.1
FGPL-VCTree [13]	37.5	40.2	**26.2**	**27.6**	16.2	19.1
DRSD-VCTree(Ours)	**38.9**	**41.2**	25.2	26.0	**17.8**	**20.8**
Transformer [22]	14.6	15.8	8.0	8.5	5.5	6.8
CogTree-Transformer [28]	28.4	31.0	15.7	16.7	11.1	12.7
Seesaw-Transformer [23]	28.6	34.4	17.2	20.7	11.5	14.9
FGPL-Transformer [13]	36.4	40.3	22.6	24.0	17.4	20.3
DRSD-Transformer(Ours)	**40.3**	**43.4**	**24.8**	**25.7**	**18.0**	**20.6**

4.4 Ablation Study

As previously mentioned, we address the limitations of existing reweighting methods in the SGG domain by proposing two components. Firstly, we introduce the Dynamic Weighting Mechanism (DRM) to address the inconsistency between the class distribution calculated on the dataset and the sample distribution during the training process. Secondly, we point out that previous reweighting methods have overlooked the reweighting of the non-relation class. So we assign weights to the non-relation class, thus enhancing the model's focus on the relation classes. We name it the Non-relation Class Reweighting (NCR) strategy.

To validate the effectiveness of the above components, we conducted ablation experiments on VG150. The experiments are based on the transformer benchmark model [22] and conducted on the PredCls task. The settings are as follows: 1) To ensure consistency and minimize the influence of other factors, we use the same weight calculation formula as DRSD, i.e. Eq. (1). However, in this set, the calculation is based on the class distribution of the dataset and no weight is assigned to the non-relation class (without DRM and NCR). 2) Transformer with NCR (without DRM). 3) Transformer with DRM (without NCR). 4) Transformer with DRSD (with both DRM and NCR).

Results are presented in Table 2. The experimental results show that the integration of NCR and DRM individually contributes to performance improvements. Moreover, combining them brings greater enhancement. This ablation experiment has demonstrated that reweighting the non-relation class can make the model perform better on the relation classes. And calculating weights based on the sample distribution during the training process can obtain better performance than on the class distribution of the dataset.

Table 2. Ablation study for the two components of the proposed DRSD method. The benchmark model is Transformer [22]. The best result is highlighted in bold.

DRSD		PredCls		
NCR	DRM	mr@20	mr@50	mr@100
x	x	30.37	37.04	39.33
√	x	31.81	37.82	40.22
x	√	32.81	39.41	41.58
√	√	**34.04**	**40.29**	**43.41**

Table 3. Comparison between calculating weights based on the class distribution of the dataset or the sample distribution during the training process. The best result is highlighted in bold.

Reweight	PredCls		
	head	body	tail
Static Reweighting	**47.98**	43.50	32.58
Dynamic Reweighting(Static Reweighting + DRM)	42.22	**44.43**	**42.82**

We conduct another experiment to demonstrate the effectiveness of DRM in achieving unbiased predictions. Following [12], we divide the predicate classes into three groups: Head, body, and tail, according to the instance numbers in the training split. We compare the mean recall of these groups between Transformer [22] with static reweighting and Transformer [22] with dynamic reweighting. Results are shown in Table 3. It can be observed that dynamic reweighting achieves a large margin of improvement by 10.24% on mR@100 for the tail class, demonstrating that calculating based on the sample distribution during the training process indeed improves the performance on the tail classes. While there is a decrease in mean recall for the head classes, it is important to consider that the head classes only constitute 7 out of the total 50 classes, whereas the tail classes account for 22 classes. Furthermore, the predicates in the tail classes contain richer information than those in the head classes. Therefore, sacrificing a portion of the performance in the head classes in exchange for a greater improvement in the tail classes is a worthwhile trade-off.

4.5 Visualization Results

To visually illustrate the effect of the proposed DRSD method on unbiased scene graph generation, we present the visualized comparison between Transformer-DRSD and Transformer. Results are shown in Fig. 2. It can be observed that the vanilla Transformer method gains high recall rates for the predicates in the head classes, but it achieves low or even zero recall rates for semantically related predicates in the body and tail classes. For example, the recall rate for the head predicate "on" is 80.0%, whereas the recall rates for semantically related predicates in the body and tail classes such as "standing on", "parked on", and "laying on" are 4.3%, 2.5%, and 3.2%, respectively. The results reveal the limitations of the vanilla Transformer method. When encountering semantically similar predicates, the model fails to distinguish between them and simply predicts the class with the most samples, indicating that the model lacks the ability to discriminate semantically related predicates.

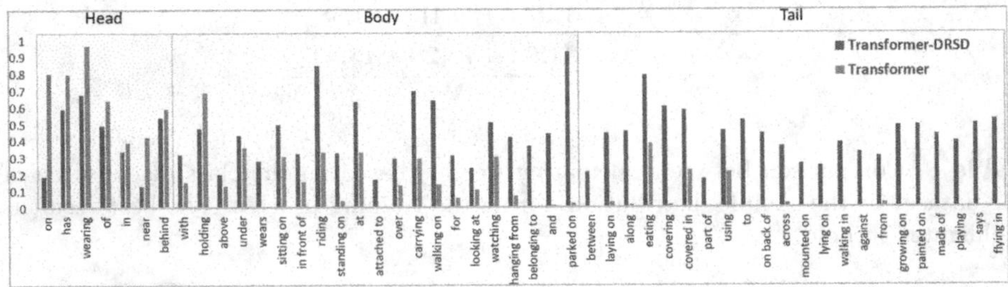

Fig. 2. Comparison between Transformer-DRSD and Tansformer.

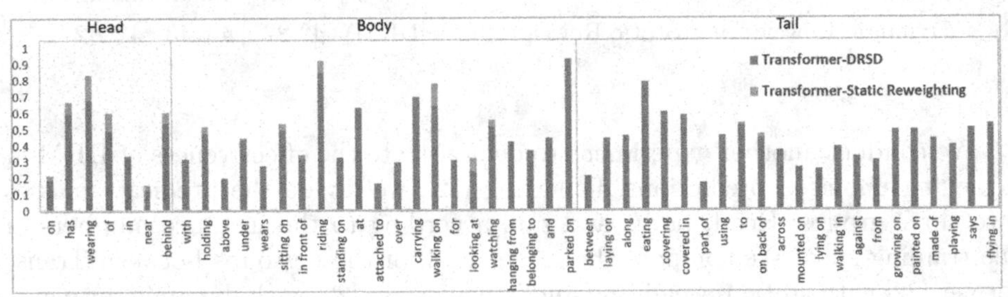

Fig. 3. Comparison between Transformer-DRSD and Transformer-Static Reweighting.

In contrast, our method successfully distinguishes semantically related predicates. Although the recall rate for "on" is 20.0%, the recall rates for the fine-grained predicates related to "on" in the body and tail classes, such as "standing on", "parked on", and "laying on", are 32.5%, 93.0%, and 44.3%, respectively. These results demonstrate that our method effectively discriminates semantically related predicates and achieves unbiased predictions.

We also compare the proposed Transformer-DRSD with Transformer-Static Reweighting. Specifically, both methods utilize the same reweighting formula.

The latter method uses static weights calculated from the class distribution of the dataset and overlooks the reweighting of the non-relation class. In contrast, the former method incorporates the two proposed components in this paper. It calculates weights based on the sample distribution during the training process and employs reweighting on the non-relation class. Results are shown in Fig. 3. It can be observed that the two proposed components (DRM and NCR) effectively improve the performance of the body and tail classes.

5 Conclusion

In this work, we address two issues with the existing reweighting methods in scene graph generation. Firstly, these methods calculate weights solely based on the class distribution of the dataset, which does not align with the sample distribution during the training process. Secondly, these methods ignore the reweighting for the non-relation class. To overcome these limitations, we propose a new reweighting method named Dynamic Reweighting based on the Sample Distribution (DRSD). The proposed DRSD method calculates weights based on the sample distribution during the training process and applies reweighting to the non-relation classes. Comprehensive experiments have demonstrated that our method can significantly enhance the performance in both the body and tail classes, thus achieving unbiased scene graph generation.

Acknowledgments. This work was supported by National Key R&D Program of China under Grant 2022ZD0115502, by the National Natural Science Foundation of China under Grant U21A20514 and 62122010, and by the FuXiaQuan National Independent Innovation Demonstration Zone Collaborative Innovation Platform Project under Grant 3502ZCQXT2022008.

References

1. Abedi, A., Karshenas, H., Adibi, P.: Multi-modal reward for visual relationships-based image captioning. arXiv preprint arXiv:2303.10766 (2023)
2. Chen, S., Jin, Q., Wang, P., Wu, Q.: Say as you wish: fine-grained control of image caption generation with abstract scene graphs. In: CVPR (2020)
3. Chen, T., Yu, W., Chen, R., Lin, L.: Knowledge-embedded routing network for scene graph generation. In: CVPR (2019)
4. Deng, Y., et al.: Hierarchical memory learning for fine-grained scene graph generation. In: Avidan, S., Brostow, G., Cissé, M., Farinella, G.M., Hassner, T. (eds.) ECCV. LNCS, vol. 13687, pp. 266–283. Springer, Cham (2022). https://doi.org/10.1007/978-3-031-19812-0_16
5. Dong, X., Gan, T., Song, X., Wu, J., Cheng, Y., Nie, L.: Stacked hybrid-attention and group collaborative learning for unbiased scene graph generation. In: CVPR (2022)
6. Guo, Y., Chen, J., Zhang, H., Jiang, Y.G.: Visual relations augmented cross-modal retrieval. In: ICMR (2020)
7. Guo, Y., et al.: From general to specific: informative scene graph generation via balance adjustment. In: ICCV (2021)

8. Hildebrandt, M., Li, H., Koner, R., Tresp, V., Günnemann, S.: Scene graph reasoning for visual question answering. arXiv preprint arXiv:2007.01072 (2020)
9. Krishna, R., et al.: Visual genome: connecting language and vision using crowdsourced dense image annotations. IJCV (2017)
10. Lertnattee, V., Theeramunkong, T.: Analysis of inverse class frequency in centroid-based text classification. In: ISCIT (2004)
11. Li, M., Qi, Y.: XPNet: cross-domain prototypical network for zero-shot sketch-based image retrieval. In: Yu, S., et al. (eds.) PRCV. LNCS, vol. 13534, pp. 394–410. Springer, Cham (2022). https://doi.org/10.1007/978-3-031-18907-4_31
12. Li, R., Zhang, S., Wan, B., He, X.: Bipartite graph network with adaptive message passing for unbiased scene graph generation. In: CVPR (2021)
13. Lyu, X., et al.: Fine-grained predicates learning for scene graph generation. In: CVPR (2022)
14. Pennington, J., Socher, R., Manning, C.D.: GloVe: global vectors for word representation. In: EMNLP (2014)
15. Schroeder, B., Tripathi, S.: Structured query-based image retrieval using scene graphs. In: CVPRW (2020)
16. Song, J., Zeng, P., Gao, L., Shen, H.T.: From pixels to objects: cubic visual attention for visual question answering. arXiv preprint arXiv:2206.01923 (2022)
17. Song, X., Chen, J., Wu, Z., Jiang, Y.G.: Spatial-temporal graphs for cross-modal Text2Video retrieval. IEEE T-MM (2021)
18. Tang, K.: A scene graph generation codebase in Pytorch (2020). https://github.com/KaihuaTang/Scene-Graph-Benchmark.pytorch
19. Tang, K., Niu, Y., Huang, J., Shi, J., Zhang, H.: Unbiased scene graph generation from biased training. In: CVPR (2020)
20. Tang, K., Zhang, H., Wu, B., Luo, W., Liu, W.: Learning to compose dynamic tree structures for visual contexts. In: CVPR (2019)
21. Teney, D., Liu, L., van Den Hengel, A.: Graph-structured representations for visual question answering. In: CVPR (2017)
22. Vaswani, A., et al.: Attention is all you need. In: NeurIPS (2017)
23. Wang, J., et al.: Seesaw loss for long-tailed instance segmentation. In: CVPR (2021)
24. Xu, P., Chang, X., Guo, L., Huang, P.Y., Chen, X., Hauptmann, A.G.: A survey of scene graph: generation and application. TNNLS (2020)
25. Yan, S., et al.: PCPL: predicate-correlation perception learning for unbiased scene graph generation. In: ACM MM (2020)
26. Yang, X., et al.: Transforming visual scene graphs to image captions. arXiv preprint arXiv:2305.02177 (2023)
27. Yang, X., Tang, K., Zhang, H., Cai, J.: Auto-encoding scene graphs for image captioning. In: CVPR (2019)
28. Yu, J., Chai, Y., Wang, Y., Hu, Y., Wu, Q.: CogTree: cognition tree loss for unbiased scene graph generation. arXiv preprint arXiv:2009.07526 (2020)
29. Zellers, R., Yatskar, M., Thomson, S., Choi, Y.: Neural motifs: scene graph parsing with global context. In: CVPR (2018)
30. Zhang, A., et al.: Fine-grained scene graph generation with data transfer. In: Avidan, S., Brostow, G., Cissé, M., Farinella, G.M., Hassner, T. (eds) ECCV. LNCS, vol. 13687, pp. 409–424. Springer, Cham (2022). https://doi.org/10.1007/978-3-031-19812-0_24
31. Zhang, C., Chao, W.L., Xuan, D.: An empirical study on leveraging scene graphs for visual question answering. arXiv preprint arXiv:1907.12133 (2019)
32. Zhou, X., Li, S., Chen, H., Zhu, A.: Disentangled OCR: a more granular information for "text"-to-image retrieval. In: PRCV (2022)

Multi-modal Graph and Sequence Fusion Learning for Recommendation

Zejun Wang, Xinglong Wu, Hongwei Yang, Hui He, Yu Tai, and Weizhe Zhang(✉)

Harbin Institute of Technology, Harbin, China
{zejunwang,xlwu}@stu.hit.edu.cn,
{yanghongwei,hehui,taiyu,wzzhang}@hit.edu.cn

Abstract. Multi-modal recommendation aims to leverage multi-modal information for mining users' latent preferences. Existing multi-modal recommendation approaches primarily exploit graph structures and multi-modal information to explore the graph information derived from user-item interactions, overlooking the underlying sequence information. Furthermore, by treating items solely as coarse-grained entities, the latent relationships of items within each modality are disregarded, impeding the effective extraction of latent user preferences. To address the limitations, we propose a novel approach called **M**ulti-**m**odal **G**raph and Sequence **F**usion Learning Architecture for Recommendation (**MMGCF**). In MMGCF, we first construct dynamic item-item graphs to enhance item features and capture relationships within each modality. Subsequently, according to the influence between modalities, we design a self attention network to fuse multi-modal features. Finally, in addition to regular graph convolution, we also devise a sequence-aware learning layer to preserve and capture sequence information for model to learn user preferences from a sequential perspective. Extensive experiments conducted on three real-world datasets demonstrate the superiority of our method over various state-of-the-art baselines.

Keywords: Multi-modal Recommendation · Graph Neural Network · Recurrent Neural Network · Self Attention Network

1 Introduction

The growing abundance of network information and the evolving user demands prompt platforms to heavily rely on accurate recommendation systems for delivering personalized services. However, performance of existing models reaches a bottleneck due to the restricted information available solely from user interactions with items. To address such limitation, multi-modal recommendation

This work was supported in part by the Joint Funds of the National Natural Science Foundation of China (Grant No. U22A2036) and the Key-Area Research and Development Program of Guangdong Province (2020B0101360001).

© The Author(s), under exclusive license to Springer Nature Singapore Pte Ltd. 2024
Q. Liu et al. (Eds.): PRCV 2023, LNCS 14425, pp. 357–369, 2024.
https://doi.org/10.1007/978-981-99-8429-9_29

emerges as a viable solution, integrating visual, textual and other multi-modal features of items to effectively learn user preferences.

We generally divide multi-modal recommender systems into two categories: (1) Collaborative Filtering (CF) based models and (2) Graph Neural Networks (GNN) based models. CF-based methods [5,9,10] incorporate multi-modal features as auxiliary information to enhance user-item interaction learning. However, CF-based models disregard modality differences and their specific influence on user preferences, leading to suboptimal results. In contrast, GNN-based models [1,4,7] have gained popularity in multi-modal recommendation by leveraging the graph structure of user-item interactions to capture high-level connections. GNN facilitates feature aggregation and propagation, aligning with the concept of multi-modal recommendation.

Multi-modal recommendation primarily integrates diverse multi-modal item features to enhance the representation of entities. However, most existing multi-modal recommendation models [15,20–22,24] only focus on the coarse-grained fusion in units of entity, we call it *entity-oriented feature fusion.* Since entity-oriented feature fusion method can only capture the intra-entity fusion information, let alone significant inter-entity context information. Therefore, we argue that such fusion introduces significant information loss and we denote them as *modality-oriented information loss* and *sequence-aware information loss.* Specifically, most methods focus on directly exploring the user-item click information without considering the fine-grained intra-modal connection between items within each modality, which causes *modality-oriented information loss.* Furthermore, the main intention of users can be more entirely understood by the model with sequence information, which is commonly ignored, leading to *sequence-aware information loss.* These drawbacks are also explained in the following example.

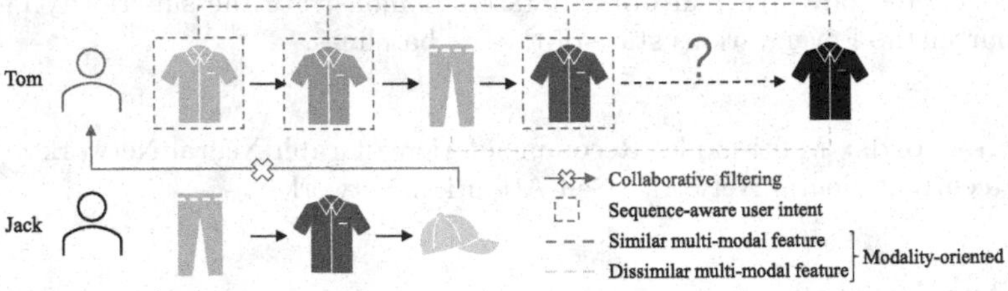

Fig. 1. A toy example: Tom's main intention is to pick shirts and will click on the black shirt ( ) in the next time. Under the influence of collaborative relationships, conventional models will incorrectly recommend caps ( ) to Tom based on the interactions of Jack. The sequence-aware information among shirts and modality-oriented similarity between clicked items can enhance the model to make more accurate recommendations.

A Motivating Example. As Fig. 1 shows, Tom is searching for shirts and clicks on various colors (, ,), accidentally clicking on trousers ( ). In this case, if we only consider the collaborative filtering of graph information, items such as caps ( ) are likely to be recommended to Tom, because many users like Jack click on caps after clicking on trousers ( ) and shirts ( ). In contrast, by taking *sequence-aware information* into consideration, we can discern Tom's main intention and accurately determine his preference for shirts. However, even if it is clear that Tom prefers shirts, existing models may only recommend Tom the shirts (, ,) that Tom clicked on, while black shirts ( ) with similar multi-modal features will not be recommended. This flaw arises from the models' failure to establish connections between shirts with similar multi-modal features, neglecting potential *modality-oriented information* within each modality. Hence, it is essential to design a novel multi-modal recommendation model that can capture and learn both graph and sequence information in user-item interactions while conducting fine-grained processing of multi-modal item features to uncover potential connections between items within each modality.

With the significant power to model graph information, we opt for GNN to accomplish the recommendation task. However, as the above example shows, sequence information is ignored in the GNN method, leading to inaccurate recommendations. Therefore, the first challenge **CH1** is: 'how to effectively extract and learn both graph and sequence information?'.

On the other hand, we need to capture fine-grained information within each modality to improve the model's prediction of user preferences. Therefore, there is another challenge **CH2**: 'How to explore potential relationships and enhance the representation of items within each modality?'.

Our Approach and Contributions: In this paper, we propose a Multi-modal Graph and Sequence Fusion Learning Architecture for Recommendation (MMGCF). Our work contributes in the following ways:

(1) For **CH1**, we introduce a sequence-aware learning layer that incorporates sequence information of user-item interactions into the user and item representations allowing MMGCF to leverage both sequence and graph information and enhance recommendation performance for all users.
(2) For **CH2**, we propose a multi-modal-aware fusion layer constructing the top-K similarity adjacency graph to enhance item features and explore potential item relationships within each modality. Additionally, the multi-modal-aware fusion layer employs self attention network [14] to integrate different item information across various modalities, which assists MMGCF to make more accurate recommendation according to the different effects of various modalities on user preferences.
(3) Extensive experiments conducted on TikTok, Amazon-Baby, Amazon-Sports datasets demonstrate the superiority of MMGCF.

2 Related Work

In this section, we review some classic approaches to GNN-based recommendation and multi-modal recommendations.

GNN-Based Recommendation: GNN has demonstrated their effectiveness in various recommendation domains. In classical collaborative filtering models, GNN learn user preferences and represent users and items using a user-item interaction graph. An example is lightgcn [6], which achieves remarkable results with lower computational costs by employing a simple graph convolution operation. In session recommendation [18,25], GNN-based models investigate the potential transformation patterns of items by constructing an item transformation graph. Such as GCE-GNN constructing both local session graphs and global session graphs enables the learning of user interests from diverse perspectives.

Multi-modal Recommendation: Early multi-modal recommendation models typically incorporate multi-modal information of items based on collaborative filtering models [17]. For example, the VBPR [5] model leverages item image multi-modal information to enhance the recommendation results. The ACF [2] utilizes attention mechanisms to assist multi-modal models in learning user preferences. However, collaborative filtering-based multi-modal recommendation often focuses solely on the simple user-item interactions and fails to adequately explore potential relationships between multi-modal items. In recent years, the introduction of graph structures has been prominent in recommender systems [16,23]. By constructing user-item interaction graphs and auxiliary graphs, the graph information contained in the interactions can be effectively learned, significantly enhancing the representation of items and users through information aggregation. For instance, the MMGCN[9] model constructs interaction graphs under different modalities to capture users' multi-modal preferences. GRCN[10] incorporates an adaptive improved graph structure to filter out noise items and enhance recommendation accuracy.

However, the current models overlook potential item relationships in each modality and fail to account for potential damage to sequence information during the learning of graph structure information.

3 The Proposed Method

In this section, we first formulate the problem, following with the three main components, i.e., (1) multi-modal-aware fusion layer, (2) sequence-aware learning layer, (3) prediction and optimization. The overall architecture of MMGCF is shown in Fig. 2.

3.1 Problem Statement

Let U, I represent a set of users and items, and $\mathbf{R} \in \mathbb{R}^{|U|\times|I|}$ represents the history of user-item interactions, and for each $\mathbf{R}_{ij} \in \mathbf{R}$, $\mathbf{R}_{ij} = 1$ if user i has clicked on item j. We define the user-item interaction graph $G = (U, I, E)$, where E are obtained from the user-item interactions $\mathbf{R}$. $\mathbf{u}_i, \mathbf{e}_j \in \mathbb{R}^d$ is the ID embedding of user i and item j, where d is the embedding dimension. $\mathbf{e}_j^m \in \mathbb{R}^{d_m}$ is the embedding of user i and item j under modality m, where d_m is the dimension of multi-modal feature under modality m, $m \in \mathcal{M}$. In this paper $\mathcal{M} = \{\mathrm{v}, \mathrm{t}\}$ is the multi-modal set. Given U, I, G and multi-modal features of items $\{\{e_i^m|m \in \mathcal{M}\}\,|i \in I\}$, MGCF aims to predict user's preferences.

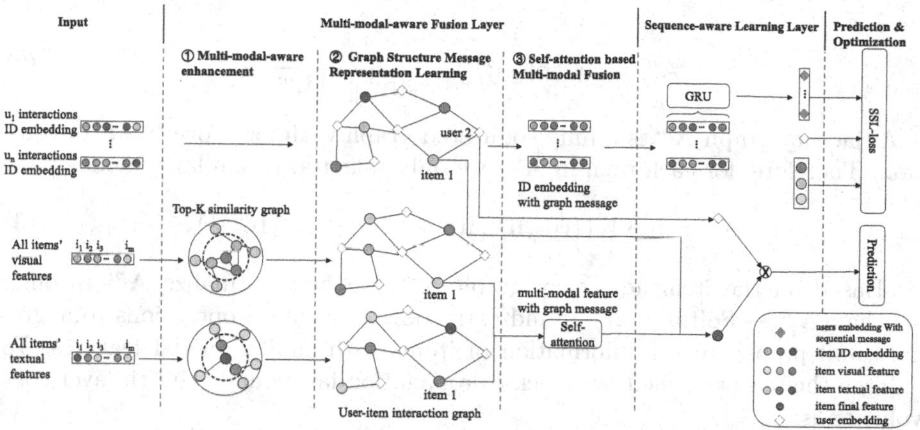

Fig. 2. Schematic illustration of MMGCF. Firstly, MMGCF constructs the item similarity adjacency graph in each modality to capture potential relationships of items and strengthens the multi-modal item representation through graph convolution. Furthermore, based on the user-item graph, the enhanced multi-modal representation and ID embedding learn the interaction information. Finally, MMGCF uses GRU and SSL to preserve the sequence information in the interactions.

3.2 Multi-modal-Aware Fusion Layer

In this layer, we establish unique dynamic adjacency graphs and perform convolution to capture the items' potential relationships within each modality in **Multi-modal-aware Enhancement** module. Furthermore, by aggregating and propagating the multi-modal information and ID information in the user-item interaction graph respectively, MMGCF learns graph message in **Graph Structure Message Representation Learning** module. Additionally, through **self-attention-based multi-modal fusion** module, we incorporate self attention network to integrate multi-modal feature.

Multi-modal-Aware Enhancement: In order to excavate the potential connections of items under each modality and enhance the expression of multi-modal information, we construct a Top-K similarity adjacency graph of items under each modality, and design graph convolution to propagate and aggregate the multi-modal information among similar items. Firstly, to align the dimensions of pre-trained multi-modal features, we project the item features of each modality into the same dimension to share the latent representation:

$$\mathbf{h}^m = \mathbf{e}^m \mathbf{W}^m + \mathbf{b}^m, \tag{1}$$

where $\mathbf{W}^m \in \mathbb{R}^{d_m \times d}$ and $\mathbf{b}^m \in \mathbb{R}^d$ control the mapping of multi-modal features. After obtaining the representation of items, we construct the item adjacency graph according to the cosine similarity between items in each modality. Therefore, we obtain the dynamic weighted adjacency graph $\mathbf{A}_{ij}^m \in \mathbb{R}^{|I| \times |I|}$ of items-items under modality m:

$$\mathbf{A}_{ij}^{m} = \cos\left(\mathbf{h}_i^m, \mathbf{h}_j^m\right) = \frac{\mathbf{h}_i^{m\top} \cdot \mathbf{h}_j^m}{\left\|\mathbf{h}_j^m\right\| \cdot \left\|\mathbf{h}_j^m\right\|}. \tag{2}$$

Adjacency graph $\mathbf{A}^m$ is a fully connected graph with large overhead convolution. Therefore, for each item in $\mathbf{A}^m$, we only take top-K similarity nodes:

$$\mathbf{A}_i^m = \text{top-K}\left(\{\cos(\mathbf{h}_i^m, \mathbf{h}_1^m), \cdots, \cos(\mathbf{h}_i^m, \mathbf{h}_n^m)\}\right), \tag{3}$$

Based on the item adjacency graph $\mathbf{A}^m$, we first normalize $\mathbf{A}^m$ through softmax($\mathbf{A}_i^m \leftarrow \text{Softmax}\left(\mathbf{A}_i^m\right)$) and carry out convolution operations to aggregate and propagate the information of items with similar modal features. To enlarge the receptive filed, we stack l propagation layers and the l-th layer convolution is:

$$\mathbf{e}_m^{(l)} = \mathbf{A}^m \mathbf{e}_m^{(l-1)}, \tag{4}$$

where $\mathbf{e}_m^{(l)}$ is the multi-modal representation of items at layer l. We initialize $\mathbf{e}_m^{(0)} = \mathbf{h}^m$.

Graph Structure Message Representation Learning: To acquire user preferences from various perspectives, we utilize ID embedding and multi-modal representations to capture the significant graph information in user-item interactions respectively. $\mathbf{A}^*$ is the normalized adjacency matrix of graph G through softmax($\mathbf{A}_i^* \leftarrow \text{Softmax}\left(\mathbf{A}_i^*\right)$). We adopt a potent method to propagate and aggregate graph information in user-item interactions:

$$\mathbf{E}_g^{m(l)} = \mathbf{A}^* \mathbf{E}_g^{m(l-1)}, \tag{5}$$

where $\mathbf{A}^* \in \mathbb{R}^{(|U|+|I|)\times(|U|+|I|)}$ contains user-item interaction information. Input $\mathbf{E}_g^{m(0)} \in \mathbb{R}^{(|U|+|I|)\times d_m}$ is concatenation of initialized $\mathbf{u}^m$ and multi-modal-aware enhanced $\mathbf{e}_m^{(l)}$. After the propagation on the l-th layer, we take the first $|U|$ and the last $|I|$ rows in $\mathbf{E}_g^{m(l)}$ as multi-modal user representations $\mathbf{E}_{U,g}^m$ and item representations $\mathbf{E}_{I,g}^m$ under modality m. Similarly, by taking the initialized ID embedding $\mathbf{u}$ and $\mathbf{e}$ as input and repeating Eq. (5), we obtain the ID embedding $\mathbf{E}_{U,g}$, $\mathbf{E}_{I,g}$ of users and items.

Self-attention-Based Multi-modal Fusion: To capture the effects of different modalities on user preferences and the relationships between modalities, we design the self-attention module integrating the multi-modal item representations under different modalities:

$$\mathbf{E}^F = \sum_{m\in\mathcal{M}}^{|\mathcal{M}|} \sigma\left(\frac{\left(\mathbf{E}_{i,g}^m \mathbf{W}^Q\right)^\top \cdot \left(\mathbf{E}_{i,g}^m \mathbf{W}^K\right)}{\sqrt{d}}\right) \cdot \left(\mathbf{W}^V \mathbf{E}_{i,g}^m\right), \tag{6}$$

where $\mathbf{W}^K, \mathbf{W}^Q, \mathbf{W}^V \in \mathbb{R}^{d_m \times d}$ are key learnable parameters. We end up with item representation $\mathbf{E}^F \in \mathbb{R}^{|I|\times d}$ that incorporates multi-modal information.

3.3 Sequence-Aware Learning Layer

By employing Gated Recurrent Units (GRU) [3], we preserve and capture the temporal dynamic message presenting in interaction sequence. Through GRU, we obtain the user's sequential representation, encoding the relevant information embedded within the sequence structure.

The input $\mathbf{h} = [\mathbf{h}_1, \mathbf{h}_2, \ldots, \mathbf{h}_{t-1}, \mathbf{h}_t]$ to GRU is the ID embedding set of all items clicked by the user in chronological order, where $\mathbf{h}_i \in E_{I,g}$. The GRU affects the user sequential representation $\mathbf{u}_i^s$ through both the candidate state of the current moment $\mathbf{u}_i^c$ and the state of the previous moment $\mathbf{u}_{i-1}^s$:

$$\mathbf{u}_i^s = (1 - \mathbf{z}_i)\, \mathbf{u}_{i-1}^s + \mathbf{z}_i \mathbf{u}_i^c, \tag{7}$$

$$\mathbf{u}_i^c = \tanh\left(\mathbf{W}\mathbf{h}_i + \mathbf{U}\left(\mathbf{r}_i \odot \mathbf{u}_{i-1}^s\right)\right), \tag{8}$$

$$\mathbf{z}_i = \sigma\left(\mathbf{W}_z \mathbf{h}_i + \mathbf{U}_z \mathbf{u}_{i-1}^s\right), \tag{9}$$

$$\mathbf{r}_i = \sigma\left(\mathbf{W}_r \mathbf{h}_i + \mathbf{U}_r \mathbf{u}_{i-1}^s\right), \tag{10}$$

where $\mathbf{W}, \mathbf{W}_r, \mathbf{W}_z \in \mathbb{R}^{d \times d}$ and $\mathbf{U}, \mathbf{U}_r, \mathbf{U}_z \in \mathbb{R}^{d \times d}$ control the proportion between $\mathbf{u}_{i-1}^s$ and $\mathbf{h}_i$.

3.4 Prediction and Optimization

In MMGCF, item ID representation $\mathbf{E}_{I,g}$ is combined with initialized ID embedding $\mathbf{e}$ and fused multi-modal item representation $\mathbf{E}^F$ as the final item representation $\mathbf{x}$ by hyperparameter η. The learned user ID representation $\mathbf{E}_{U,g}$ is directly used as the final user representation. By dot product operation of user and item representations, the top-k similar items are predicted as the user preferences:

$$\mathbf{x} = \mathbf{E}_{I,g} + \mathbf{e} + \eta \cdot \mathbf{E}^F, \tag{11}$$

$$\hat{y} = \mathbf{x}^\top \mathbf{E}_{U,g}. \tag{12}$$

To tackle the issues of data sparsity and insufficient training in multi-modal recommendation, we incorporate a self-supervised learning (SSL) [11,19] loss into the overall loss function, which aligns representations into a shared presentation space and captures correlations between data. We employ cosine similarity to accomplish SSL. The objective is to encourage multi-modal features of an item to be closely aligned with its corresponding ID embedding. In addition, we aim to minimize the distance between the target item representation and the user ID embedding and user sequential representation:

$$\mathcal{L} = -\cos\left(\mathbf{E}^F, \mathbf{E}_{I,g}^{\text{target}}\right) - \cos\left(\mathbf{E}_{U,g}, \mathbf{E}_{I,g}^{\text{target}}\right) - \cos\left(\mathbf{u}^s, \mathbf{E}_{I,g}^{\text{target}}\right). \tag{13}$$

3.5 Computational Complexity

In Multi-modal-aware Fusion Layer, MMGCF costs $O(|I|d_m d)$ to project the multi-modal features and $O(|I|^2 d)$ to construct the similarity graphs within each modality. MMGCF designs a simple aggregation method costing $O(|E|d/B)$ on the graph convolution, where B is batch-size. In Sequence-aware Learning Layer, MMGCF costs $O(|I|d^2)$ to capture sequence information. In Optimization, the SSL only costs $O(dB)$. The total computational cost for MMGCF is $O(|I|d_m d + |I|^2 d + |E|d/B + |I|d^2)$. In summary, MMGCF has faster graph convolution and loss function operations than other models, but consumes additional time to build the internal multi-modal similarity graphs. Besides, the similarity graphs only slightly affect the speed of model training, and MMGCF is still very fast in testing and practical applications.

4 Experiments and Analysis

We conduct extensive experiments on three real-world datasets to answer the following key questions:

- **Q1**: How does our model perform compared with other state-of-the-art recommendation models?
- **Q2**: How do the key learning components in MMGCF affect recommendation performance?
- **Q3**: How do the different hyper-parameters affect recommendation performance?

4.1 Experimental Settings

In this section, we describe the datasets, baseline methods, evaluation metrics and parameter settings employed in our experiments.

Datasets: We adopt three real-world datasets, *i.e.*, TikTok, Amazon-Baby, Amazon-Sports.

- **TikTok**. TikTok dataset contains video records clicked by users on TikTok short video platform with visual, audio and textual features of videos.
- **Amazon.** Two datasets, Amazon-Sports and Amazon-Baby, include click records of products on Amazon platform, visual and textual features of items.

Evaluation Metrics: Each dataset is divided at a ratio of 8:1:1, respectively for training, verification and testing of the model. We adopt Recall@K, NDCG@K, Precision@K to evaluate models' performance. Moreover, we set the K as 20 for the evaluation following the settings in GRCN [22].

Baseline Methods: (1) **MF-BPR** [12] uses multiple feedback sources simultaneously during training. (2) **NGCF** [17] encodes collaborative information in interactions during embedding. (3) **LightGCN** [6] lightgcn abandons useless feature transformation and nonlinear activation. (4) **VBPR** [5] combines visual features extracted from product images into matrix decomposition.

(5) **LightGCN-M** incorporates multi-modal information into the graph convolution of lightgcn. (6) **MMGCN** [21] constructs a user-item bipartite graph on each modality to enrich the representation of each node. (7) **GRCN** [22] adjusts interaction graph to remove the noise in the interaction. (8) **CLCRec** [20] designs mutual information-based contrastive learning, which solves the item cold-start issue. (9) **MMGCL** [24] uses masking and dropout to build contrast learning to train the model. (10) **SLMRec** [13] generates multiple views of the user/item through data enhancement and introduces a negative-case sampling technique.

Parameter Settings: We search dimension d in the range of $\{16, 32, 64, 128\}$ and multi-modal weight η in the range of $\{0.1, 0.2, 0.3, 0.4\}$. In most cases, $d = 64$ and $\eta = 0.2$ lead to the optimal result. Our MMGCF is implemented with Pytorch. We adopt Adam [8] to optimize model with the learning rate of 0.001. Besides, we fix the maximum training epoch at 1000. Furthermore, convolution layer and the Top-k of adjacency graph are fixed at 1 and 10.

Table 1. Overall performance comparison. We mark the best result in boldface, and the best baseline is underlined.

Baseline	TikTok			Amazon-Baby			Amazon-Sports		
	R@20	P@20	N@20	R@20	P@20	N@20	R@20	P@20	N@20
MF-BPR	0.0346	0.0017	0.0130	0.0440	0.0024	0.0200	0.0430	0.0023	0.0202
NGCF	0.0604	0.0030	0.0238	0.0591	0.0032	0.0261	0.0695	0.0037	0.0318
LightGCN	0.0653	0.0033	0.0282	0.0698	0.0037	0.0319	0.0782	0.0042	0.0369
VBPR	0.038	0.0018	0.0134	0.0486	0.0026	0.0213	0.0582	0.0031	0.0265
LightGCN-M	0.0679	0.0034	0.0273	0.0726	0.0038	0.0329	0.0705	0.0035	0.0324
MMGCN	0.0730	0.0036	0.0307	0.0640	0.0032	0.0284	0.0638	0.0034	0.0279
GRCN	0.0804	0.0036	0.0350	0.0754	0.0040	<u>0.0336</u>	0.0833	0.0044	0.0377
CLCRec	0.0621	0.0032	0.0264	0.0610	0.0032	0.0284	0.0651	0.0035	0.0301
MMGCL	0.0799	0.0037	0.0326	0.0758	0.0041	0.0331	<u>0.0875</u>	<u>0.0046</u>	<u>0.0409</u>
SLMRec	<u>0.0845</u>	<u>0.0042</u>	<u>0.0353</u>	<u>0.0765</u>	<u>0.0043</u>	0.0325	0.0829	0.0043	0.0376
MMGCF	**0.0894**	**0.0045**	**0.0385**	**0.0865**	**0.0048**	**0.0377**	**0.0941**	**0.0050**	**0.0428**
Improve	5.80%	7.14%	9.07%	13.07%	11.63%	12.20%	11.48%	8.70%	4.65%

4.2 Performance Comparison (Q1)

We contrast MMGCF with baseline models in three datasets. The performances are shown in Table 1, from which we observe:

- **MMGCF consistently outperforms all baseline models on three datasets.** MMGCF improves Recall@20 of the best baseline by 5.80%, 13.07%, 11.48% on TikTok, Amazon-Baby, and Amazon-Sports, respectively above second-best baseline, indicating better recommendation performance for our model. Experimental performances empirically certify the superiority of MMGCF.

- **Multi-modal features play an irreplaceable role in the recommendation.** The inclusion of multi-modal information in conjunction with the graph structure yields a substantial improvement in performance compared to the model that solely relies on the graph structure. This finding highlights the valuable role of multi-modal information in aiding the model to effectively learn and understand the user's underlying preferences.
- **The GNN significantly enhances recommendation capability of models.** It is true that VBPR incorporating multi-modal information, may not perform as well as LightGCN, which focuses primarily on graph structure and convolution operations. This observation highlights the significance of the graph structure and convolution operations in facilitating the learning of user preferences and item representations.
- **The SSL achieves a performance improvement by a large margin.** In the approaches, the incorporation of self-supervised learning into the multi-modal model leads to improved results by facilitating a comparison of the learned features across different modalities. This comparison enhances the representation of both items and users, effectively addressing the issue of data sparsity.

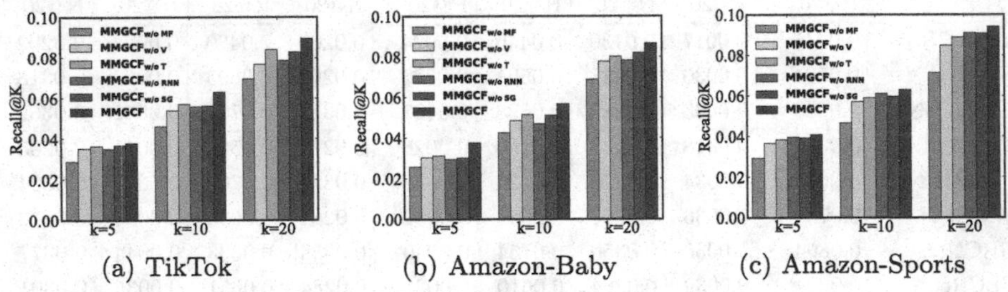

(a) TikTok (b) Amazon-Baby (c) Amazon-Sports

Fig. 3. Effect of key components on three datasets.

4.3 Ablation Study (Q2)

In order to prove the validity of key components in the model, ablation experiments are conducted on the multi-modal features, multi-modal-aware enhancement and sequence-aware learning layer in MMGCF. Specifically, we design the following three models based on original MMGCF:

- $\mathbf{MMGCF_{w/o\ V}}$: MMGCF without visual modality feature.
- $\mathbf{MMGCF_{w/o\ T}}$: MMGCF without textual modality feature.
- $\mathbf{MMGCF_{w/o\ MF}}$: MMGCF without visual and textual modality feature.
- $\mathbf{MMGCF_{w/o\ SG}}$: MMGCF without multi-modal-aware enhancement.
- $\mathbf{MMGCF_{w/o\ RNN}}$: MMGCF without GRU capturing sequential message.

Based on comparison results in Fig. 3, we have following observations: (1) **MMGCF** outperforms all models with missing parts. (2) $\mathbf{MMGCF_{w/o\ MF}}$ getting the worst result indicates that incorporating multi-modal features can significantly assist the model in learning user preferences. Moreover, by comparing single-modal feature models ($\mathbf{MMGCF_{w/o\ V}}$ and $\mathbf{MMGCF_{w/o\ T}}$), we argue that visual features have a greater impact on user preferences. (3) The performance of $\mathbf{MMGCF_{w/o\ SG}}$ suggests that through multi-modal-aware enhancement, MMGCF effectively utilizes the connections between items in each modality to gain a deeper understanding of user preferences. (4) The performance of $\mathbf{MMGCF_{w/o\ RNN}}$ demonstrates that by incorporating the sequence-aware learning layer, MMGCF captures sequential messages of interactions to achieve more accurate results.

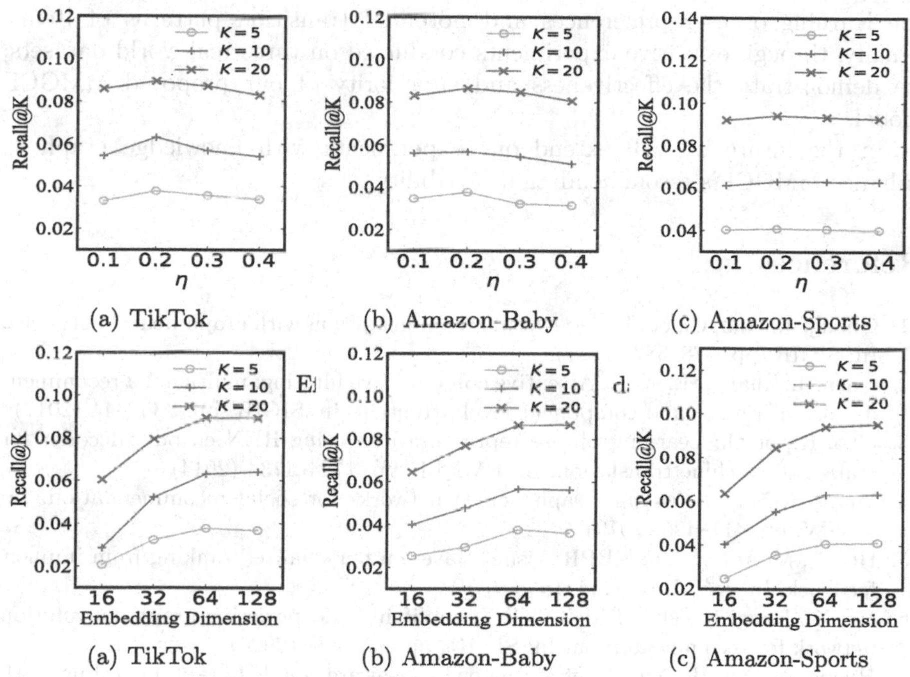

(a) TikTok (b) Amazon-Baby (c) Amazon-Sports

(a) TikTok (b) Amazon-Baby (c) Amazon-Sports

Fig. 4. Effect of different embedding dimension on three datasets.

4.4 Parameter Analyses (Q3)

We analyze key parameters including d and η in this section:

Weight Parameter η**:** We conduct experiments on three datasets with a set of representative η values 0.1, 0.2, 0.3, 0.4. The results in Fig. 4 show that when $\eta = 0.2$, the MMGCF performance reaches the best. This result shows that too much multi-modal information will overwrite the original ID embedding, resulting in user-item interaction information loss in learning.

Embedding Dimension d**:** Experiments are conducted under different embedding dimension in the range of 16 to 128. As Fig. 4 shows, MMGCF basically improves with dimension increasing. However, when dimension is 128, due to overfitting, the performance is basically the same as that of dimension 64.

5 Conclusion and Future Work

In this paper, we propose multi-modal graph and sequence fusion learning architecture for recommendation (MMGCF), an innovative approach to address the challenges in multi-modal recommendation. The proposed model aims to enhance the expression of multi-modal features by excavating fine-grained relationships between items in each modality. Furthermore, our model captures both graph information and sequence information in user-item interactions, enabling the learning of user preferences and potential transition patterns of items. Finally, through extensive experiments conducted on three real-world datasets, we demonstrate the effectiveness and superiority of our proposed MMGCF model.

In the future we will extend our experiments with knowledge graph to enhance MMGCF's recommendation capability.

References

1. Chang, J., Gao, C., et al.: Sequential recommendation with graph neural networks. In: SIGIR, pp. 378–387 (2021)
2. Chen, J., Zhang, H., et al.: Attentive collaborative filtering: multimedia recommendation with item- and component-level attention. In: SIGIR. pp. 335–344 (2017)
3. Cho, K., et al.: Learning phrase representations using RNN encoder-decoder for statistical machine translation. In: EMNLP, pp. 1724–1734 (2014)
4. Fan, W., Ma, Y., et al.: Graph neural networks for social recommendation. In: WWW, pp. 417–426 (2019)
5. He, R., McAuley, J.J.: VBPR: visual Bayesian personalized ranking from implicit feedback. In: AAAI, pp. 144–150 (2016)
6. He, X., Deng, K., et al.: LightGCN: simplifying and powering graph convolution network for recommendation. In: SIGIR, pp. 639–648 (2020)
7. Huang, C., Xu, H., Xu, Y., et al.: Knowledge-aware coupled graph neural network for social recommendation. In: AAAI, pp. 4115–4122 (2021)
8. Kingma, D.P., et al.: Adam: a method for stochastic optimization. In: ICLR (2015)
9. Liu, F., Cheng, Z., Sun, C., et al.: User diverse preference modeling by multimodal attentive metric learning. In: MM, pp. 1526–1534 (2019)
10. Liu, Q., Wu, S., Wang, L.: DeepStyle: learning user preferences for visual recommendation. In: SIGIR, pp. 841–844 (2017)
11. Liu, X., Zhang, F., et al.: Self-supervised learning: generative or contrastive. IEEE Trans. Knowl. Data Eng. **35**(1), 857–876 (2023)
12. Rendle, S., Freudenthaler, C., et al.: BPR: Bayesian personalized ranking from implicit feedback. In: UAI, pp. 452–461 (2009)
13. Tao, Z., Liu, X., et al.: Self-supervised learning for multimedia recommendation. IEEE Trans. Multimedia (2022)

14. Vaswani, A., et al.: Attention is all you need. In: NIPS, pp. 5998–6008 (2017)
15. Wang, Q., Wei, Y., et al.: DualGNN: dual graph neural network for multimedia recommendation. IEEE Trans. Multimedia **25**, 1074–1084 (2023)
16. Wang, S., Hu, L., Wang, Y., et al.: Graph learning based recommender systems: a review. In: IJCAI, pp. 4644–4652 (2021)
17. Wang, X., He, X., et al.: Neural graph collaborative filtering. In: SIGIR, pp. 165–174 (2019)
18. Wang, Z., Wei, W., et al.: Global context enhanced graph neural networks for session-based recommendation. In: SIGIR, pp. 169–178 (2020)
19. Wei, W., Huang, C., et al.: Contrastive meta learning with behavior multiplicity for recommendation. In: WSDM, pp. 1120–1128 (2022)
20. Wei, Y., Wang, X., Li, Q., et al.: Contrastive learning for cold-start recommendation. In: MM, pp. 5382–5390 (2021)
21. Wei, Y., Wang, X., et al.: MMGCN: multi-modal graph convolution network for personalized recommendation of micro-video. In: MM, pp. 1437–1445 (2019)
22. Wei, Y., Wang, X., et al.: Graph-refined convolutional network for multimedia recommendation with implicit feedback. In: MM, pp. 3541–3549 (2020)
23. Wu, S., Sun, F., et al.: Graph neural networks in recommender systems: a survey. ACM Comput. Surv. **55**(5), 97:1–97:37 (2023)
24. Yi, Z., Wang, X., et al.: Multi-modal graph contrastive learning for micro-video recommendation. In: SIGIR, pp. 1807–1811 (2022)
25. Yu, F., Zhu, Y., et al.: TAGNN: target attentive graph neural networks for session-based recommendation. In: SIGIR, pp. 1921–1924 (2020)

Co-attention Guided Local-Global Feature Fusion for Aspect-Level Multimodal Sentiment Analysis

Guoyong Cai, Shunjie Wang(✉), and Guangrui Lv

Key Laboratory of Guangxi Trusted Software, College of Computer and Information Security, Guilin University of Electronic Technology, Guangxi 541004, China
wsj12130@foxmail.com

Abstract. Aspect-level multimodal sentiment analysis is a target oriented fine-grained sentiment analysis task aimed at determining the sentiment polarity of a given aspect of a sentence in conjunction with relevant multimodal data. Multimodal alignment and fusion remains a challenge for this task, and this paper proposes to solve this issue by considering the inter-modal local interactions. Therefore, a co-attention guided local-global feature fusion (CLGFF) method is proposed. The CLGFF method mines both aspect-guided global multimodal features and local fine-grained alignment between multimodalities, and then fuses them together for better exploitation of the global-local semantic correlation. A large number of experiments are carried out on two aspect-level multimodal sentiment datasets. A series of methods are compared from the experiments, and the results show that the proposed CLGFF method can better capture the local semantic correlation within the modality and the fine-grained consistency between different modalities, thereby improving the performance of aspect-level multimodal sentiment analysis.

Keywords: aspect-level multimodal sentiment analysis · global features · local features · co-attention

1 Introduction

Nowadays, people express their sentiment in the way of multiple modalities coexistence on social media networks, no longer limited to unimodality. Usually there exists some collaboration between multimodal data in terms of sentiment and semantics [1], thus it is important to mine the synergy between multimodal data. E.g. for a post with short texts and pictures, the judgment of its sentiment may depend mainly on the pictures or the alignment relations between localized regions of the pictures and key words of the text sentence. As the research deepening in the field of sentiment analysis, the researcher's focus has shifted from coarse-grained multimodal sentiment analysis to fine-grained aspect-level multimodal sentiment analysis (AMSA), and AMSA has made progress recently [2–4].

© The Author(s), under exclusive license to Springer Nature Singapore Pte Ltd. 2024
Q. Liu et al. (Eds.): PRCV 2023, LNCS 14425, pp. 370–382, 2024.
https://doi.org/10.1007/978-981-99-8429-9_30

Unlike coarse-grained multimodal sentiment analysis, AMSA pays more attention to fine-grained inter- and intra-modal interactions, and also focuses on the guidance of aspect to sentiment information units of different modality. Therefore, the existing multimodal fusion methods (feature-level fusion [5,6], decision-level fusion [7,8] and consistent regression fusion [9,10]) are not suitable for AMSA. Recently, some studies [2–4] have explored aspect-level multimodal fusion from the perspectives of multiple interactions, global guidance, and multi-head attention, and have made some progress. However, unlike the above, the local-global perspective is used to realize the local semantic alignment between different modalities while implementing aspects for global guidance of multi-modalities in this paper. Therefore, a co-attention guided local-global feature fusion (CLGFF) method is proposed for AMSA.

The main contributions of our paper are as follows:

- An aspect-guided global co-attention (AGGC) module is proposed. Since text and images are heterogeneous, aspect-text alignment and aspect-image alignment are achieved using different attentional mechanisms to obtain aspect-guided global text representations and image representations, respectively.
- A multimodal local co-attention (MLCA) module is proposed. Since the aspect-guided global co-attention module does not realize the interactive association between multimodalities, the MLCA module realize the local fine-grained interactive alignment of text and image, so that each text word is associated with each image region correspondingly.
- A large number of experiments are performed on two public aspect-level multimodal sentiment datasets, which show that the proposed model is more superior. Furthermore, the effectiveness of each module of our proposed model is verified by a series of ablation experiments.

2 Related Work

Aspect-Based Sentiment Analysis. Aspect-level sentiment analysis has been extensively studied as a fine-grained sentiment analysis task [11]. Dong et al. [12] first introduced recurrent neural networks into this research direction. Tang et al. [13] proposed a target-dependent long and short-term memory network (TD-LSTM) to achieve the strong and weak relationship between aspect words and text. Li et al. [14] proposed a dual graph convolutional networks (Du-alGCN) model, which used two different branches of graph convolutional networks to consider the syntactic structure and semantic relevance of aspects and texts. Tang et al. [15] proposed a deep memory network (MemNet) which captured the relationship between aspects and text words based on multi-hop attention. Chen et al. [16] proposed a memory recurrent attention network (RAM) model to model distantly dependent aspect-directed text representations. However, the data processed by these studies are only text information. In our work, both text modality and image modality information are considered.

Multimodal Sentiment Analysis. Multimodal feature fusion takes into account the multimodal interactions. In recent years, multimodal feature fusion has been mainly based on deep learning methods. Zhao et al. [17] proposed the visual audio attention network (VAANet), which fused the image features containing space, channel and time and the audio features containing time by splicing. Zadeh et al. [18] proposed a method named tensor fusion network (TFN). Yu et al. [19] proposed self-supervised multi-task multimodal sentiment analysis network(Self-MM), which used a self-supervised strategy to generate text, audio and image modal labels to assist multimodal tasks to learn the consistency and difference between modalities. Zhang et al. [20] proposed the temporal sentiment localization network (TSL-Net), which combined the emotional features of visual and audio patterns through splicing to predict the temporal sentiment distribution, and uses cosine similarity to reduce over-fitting. Zhang et al. [21] proposed the cross-modal temporal erasing network to locate video and audio-related information in a weakly supervised manner. However, the above studies are coarse-grained multimodal sentiment analysis, rather than the fine-grained aspect-level multimodal sentiment analysis studied in this paper.

Aspect-Level Multimodal Sentiment Analysis. With the deepening of research, aspect-level multimodal sentiment analysis has also emerged. Yu et al. [22] proposed target-oriented multimodal BERT (TomBERT), which alignd aspect and image by stacking BERTs and captures dynamically intra-modality and inter-modality. Khan et al. [23] proposed late fusion caption transformer BERT (LF-CapTrBERT), which used non-autoregressive text generation method to generate a caption for an image, and then used a multi-layer BERT encoder to mine the association between that caption and Twitter sentence. Li et al. [24] proposed the cross-modal multitask transformer (CMMT), which learned an intra-modal representation of aspect perception by introducing visual aspect-viewpoint prediction and textual aspect-emotion extraction. Li et al. [25] proposed affective region recognition and fusion networks (ARFN), which used the Yolov5 algorithm to extract target regions in an image, and fuses the text and visual representations through a multimodal interaction layer. However, the above research does not take into account the fine-grained alignment between modalities. Unlike the above research, we propose a multimodal local co-attention mechanism to adaptively align each word in the text and each local region of the image to establish fine-grained association alignment.

3 Method

3.1 Problem Description

For a image and text dataset C, each sample $c \in C$ is composed of a sentence $T = (t_1, t_2, ..., t_n)$ and an image V and an aspect word A, where aspect words are derived from entity words in the text. The goal of aspect-level multimodal sentiment analysis is to determine the sentiment polarity corresponding to the aspect word based on the image-text pair. In addition, the sentence T is divided into left text T_l, aspect words A, and right text T_r in order to more fully exploit the aspect word's associated alignment with the text.

3.2 Proposed Method

The proposed model framework is shown in Fig. 1. Firstly, the image features and text features are extracted by image encoder and text encoder respectively. Then the extracted multimodal features are input into the aspect-guided global co-attention module, which selects the appropriate attention mechanism according to the structural features of each modality to capture aspect-image alignment and aspect-text alignment. At the same time, the extracted multimodal features are input into the multimodal local co-attention module to adaptively mine the fine-grained alignment and fusion of local features of images and texts. Finally, the local multimodal features and the aspect-guided global multimodal features are integrated, which is finally used for aspect-level sentiment judgment.

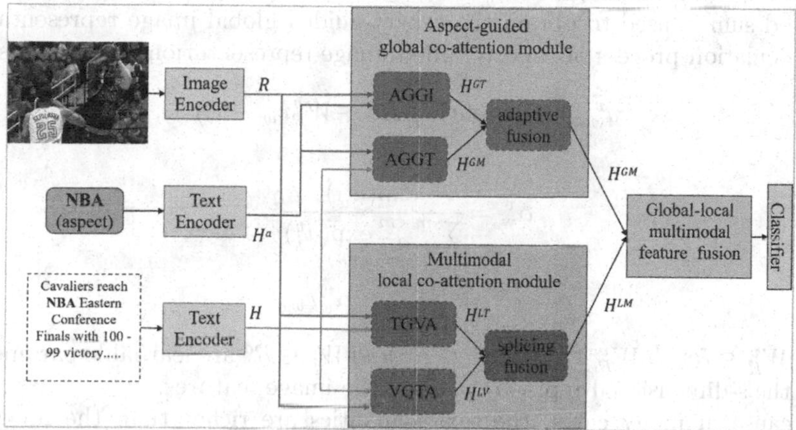

Fig. 1. The framework of co-attention guided local-global feature fusion method.

Aspect-Guided Global Co-attention Module. The aspect-guided global co-attention module is composed of aspect-guided global textual representation and aspect-guided global visual representation, which are used to explore the aspect-text global alignment and aspect-image global alignment, respectively.

(1) Aspect-Guided Global Text (AGGT) Mechanism. Since aspect and text are homogeneous, the multiplicative attention mechanism is used to achieve the association between aspect words and text words. Firstly, the hidden state representation of the aspect, left text and right text hidden state is obtained by the pre-training model BERT, and the average value of the aspect hidden state is taken as the overall representation of the aspect H^a. Then, the AGGT mechanism is used to learn the interaction between the aspect and the left text, so as to obtain the aspect-guided left text representation H^{la}. The calculation process is as follows:

$$H^{la} = W_l(\tanh(W_{l1}H^l) \circ \tanh(W_{l2}H^a)) + b_l \tag{1}$$

where H^l represent the left text hidden state representation, W_{l1}, W_{l2}, $W_l \in R^{d\times d}$ and $b_l \in R^d$ represent the trainable parameters, and $\circ$ is element-wise multiplications. Similarly, the aspect-directed right text H^{ra} can be obtained. To avoid loss of textual information, H^{la}, H^{ra} are concatenated as aspect-guided global textual representation H^{GT}.

(2) Aspect-Guided Global Image (AGGI) Mechanism. Considering that aspects and images are heterogeneous, an additive attention mechanism is used to capture the association between aspects and images. Firstly, pre-trained 152-layer ResNet [26] is used to obtain the feature representation R of image V, then R is reshaped to $R = \{r_w \in R^s, w = 1, 2, \cdots, m\}$, where m represents the number of regions in the image, and r_w represents the feature representation of each image region. The AGGI mechanism is used to calculate the attention score of each region of the image under the guidance of the aspect word, and then the weighted sum is used to obtain the aspect-guided global image representation. The calculation process of aspect-guided image representation is as follows:

$$u_{\mathrm{w}}^v = W_v \tanh(W_H^v H^a + W_R^v r_w + b^v) \tag{2}$$

$$\alpha_{\mathrm{w}}^v = \frac{\exp(u_w^v)}{\sum_{j=1}^{m\times m} \exp(u_j^v)} \tag{3}$$

$$r^{gv} = \sum_{w=1}^{m\times m} \alpha_w^v r_w \tag{4}$$

where $W_H^v \in R^{d\times d}$, $W_R^v \in R^{d\times s}$, $b^v \in R^d$ and $W_v \in R^d$ are learnable parameters. r^{gv} is the s-dimensional representation of the image features.

Because in many cases, the text semantics are richer than the image, in order to strengthen the association between r^{gv} and H^{GT}, an adaptive fusion mechanism is used to generate the aspect-guided global multimodal features.

$$c = \tanh(W_H H^{GT} + W_R r^{gv} + b^z) \tag{5}$$

$$H^{\mathrm{GV}} = c \circ r^{gv} \tag{6}$$

$$H^{GM} = W_g(\tanh(W_T H^{GT}) \circ \tanh(W_V H^{GV})) + b_g \tag{7}$$

where $W_H \in R^{d\times 2d}$, $W_R \in R^{d\times s}$ and $b^z \in R^d$ are learnable parameters, $W_T \in R^{d\times 2d}$, $W_V \in R^{d\times s}$, $W_g \in R^{d\times d}$, $b_g \in R^d$ are the learnable parameter.

Multimodal Local Co-attention Module. The previous subsection introduces the aspect-guided global co-attention module, which mines aspect-guided images and aspect-guided texts from a global perspective. However, the interaction between text and image has not been explored, but multimodal correlation alignment is a very important task. Therefore, a multimodal local co-attention module is designed to achieve fine-grained local alignment interaction between word-level text and region-level images. Firstly, in order to establish the intra-modal dependencies, multi-headed self-attention (MHSA) is used to act on the text representation H and the image feature R, respectively, so $H^T = \{h_1^T, h_2^T, ..., h_n^T\}$ and $H^I = \{h_1^I, h_2^I, ..., h_{m\times m}^I\}$ are obtained. Then, text-guided visual attention and visual-guided text attention are further designed to realize the semantic alignment and fusion of local features between images and texts.

(1) Text-Guided Visual Attention (TGVA). Due to the application of cross-modal learning in various fields, such as visual question answering [27], cross-modal retrieval [28], influenced by the above literature, the TGVA module is designed to achieve text-guided multimodal alignment and fusion. To start with the details, firstly, we focus on the j-th word in the text H^T to learn the visual features associated with that word. The calculation process is as follows:

$$x_{ij} = \tanh([W_t^{\mathsf{T}} h_j^T, W_r^{\mathsf{T}} h_i^I] + b_c) \tag{8}$$

where $[\cdot]$ denotes feature splicing, h_j^T for the j-th region in the text from H^T, h_i^I for the i-th region in the image from H^I, $W_r \in R^{d\times 1}$ and $W_t \in R^{d\times 1}$ denote the parameter matrix, T represent transpose operation. Subsequently, the fused vectors x_{ij} are linearly transformed to obtain the normalized attention score by the softmax operation:

$$\alpha_{ij} = \frac{\exp(W_m^{\mathsf{T}} x_{ij} + b_m)}{\sum_{i=1}^{m\times m} \exp(W_m^{\mathsf{T}} x_{ij} + b_m)} \tag{9}$$

where $W_m \in R^d$ and $b_m \in R^1$ represent the weight and the bias parameter, respectively. Then, the visual representation I^j of the word h_j^T is as follows:

$$I^j = \sum_{i=1}^{m\times m} \alpha_{ij} h_i^I \tag{10}$$

Since the different contribution of different modality to emotion, a gated fusion approach is proposed to adaptively fuse the representation of the word h_j^T and the visual representation I^j related to the word:

$$z_j = \text{sigmoid}(W_f[I^j, h_j^T] + b_f) \tag{11}$$

$$l_j^T = (1 - z_j) \circ h_j^T + z_j \circ I^j \tag{12}$$

where W_f and b_f denote the weight matrix and bias parameters, z_j denotes the degree of relevance of word h_j^I to the visual representation I^j of that word. After obtaining the visual contextual features $[l_1^T, l_2^T, ..., l_n^T]$ corresponding to all words in this way, the mean of these features is used as the output feature of the TGVA.

$$H^{LT} = \sum_{j=1}^{n} \frac{1}{n} l_j^T \tag{13}$$

(2) Visual-Guided Text Attention (VGTA). In order to more fully exploit the synergy between the multimodalities, the VGTA mechanism is specially designed to simultaneously explore the text words associated with each visual region. The computational process is similar to Eqs. (8)–(13) of text-guided visual attention, which is not repeated. The final available visual-guided multimodal features are H^{LV}.

Based on (1) TGVA and (2) VGTA, the text-guided local multimodal feature H^{LT} and the visual-guided local multimodal feature H^{LV} are obtained respectively, which fully exploit the alignment and fusion between multimodalities. Then H^{LT} and H^{LV} are further fused to establish the correlation between them and to facilitate their interactions. The calculation process is as follows:

$$H^{LM} = \tanh(W^c[H^{LT}, H^{LV}] + b^c) \tag{14}$$

where $W^c \in R^{2d \times d}$ and $b^c \in R^d$ are trainable parameters.

In order to prevent the loss of information, H^{GT}, H^{GV}, H^{GM} and H^{LM} are further hierarchically fused to generate global-local multimodal sentiment representations for final sentiment judgment. The calculation process is as follows:

$$H = [H^{GT}, H^{GV}, H^{GM}, H^{LM}] \tag{15}$$

Subsequently, the multimodal representation H input the softmax function for the classification of emotional polarity, as follows:

$$p(y|H) = \text{softmax}(W^{\top} H + b) \tag{16}$$

where $W \in R^{6d \times 3}$ and $b \in R^3$ are trainable parameters, y represents the classification probability.

Table 1. Statistics of the multimodal Twitter datasets

Dataset	Split	Positive	Neutral	Negative	Total	Max. lengths
Twitter-2015	Train	928	1883	368	3179	35
	Dev	303	670	149	1122	40
	Test	317	607	113	1037	36
Twitter-2017	Train	1508	1638	416	3562	39
	Dev	515	517	144	1176	31
	Test	493	573	168	1234	38

4 Experimental Analysis

Datasets. The Twitter-2015 and Twitter-2017 aspect-level multimodal English datasets are used to evaluate our CLGFF model. Each example contains a image-text pair and an aspect word and the emotional polarity corresponding to the aspect word. The details of the two datasets are shown in Table 1.

Implementation Details. In the experiment of this paper, for the images, the scale size is resized to 224×224, which is represented as a tensor of $s \times m \times m$, m is 7 and s is 2048. For text, the maximum padding length is set to 60. For CLGFF, the Adam [29] optimizer is used in the training process. In addition, the batch size is set to 10 and the value of dropout is set to 0.5. The learning rate of CLGFF is 0.0002 on the Twitter-2017 dataset, and 0.0005 on the Twitter-2015 dataset. PyTorch is used to implement the CLGFF model and ablation experiments on NVIDIA Tesla P100-PCIE GPU. The accuracy (ACC) and Macro-F1 values are used as evaluation metrics of the experiments for comparative analysis.

4.1 Experimental Result

This paper selects the following novel aspect-level sentiment analysis models as baselines to evaluate the performance of the CLGFF method.

(1) Img-Aspect, which uses visual and aspect features for classification.
(2) MemNet [15], which uses multi-hop attention mechanism is used to update the stored memory to achieve deep memory.
(3) RAM [16], which employs a multi-attention mechanism to capture long-range emotional features.
(4) MGAN [30], which uses an attention mechanism to learn the interaction between aspects and text at the lexical level, then uses attention to learn the overall interaction between aspects and text.
(5) Res-RAM-TFN and Res-MGAN-TFN are two variants of the multimodal fusion approach [18], which use a bilinear interaction operator to combine the textual representations and MGAN through a complex fusion matrix.
(6) MIMN [2], which uses multi-hop memory networks to model the interaction between aspect words, text and visual.
(7) ESAFN [3], which uses attention mechanism to align aspects with image and texts for aspect level multimodal sentiment analysis.
(8) Res-RAM, Res-MGAN, Res-ESTR are three variants of multimodal fusion methods [31] that first apply maximum pooling on visual features to obtain $g = MaxPool(R)$, and then g is used for sentiment classification.
(9) TomBERT [22], which uses the classical BERT architecture to implement aspect-guided multimodal features and multimodal fusion.
(10) LF-CapTrBERT [23], which uses non-autoregressive text generation method to generate a caption for an image, and then uses a multi-layer BERT encoder to mine the association between that caption and sentence.
(11) CMMT [24], which learns aspect-guided intra-modal representation by visual aspect-viewpoint prediction and textual aspect-emotion extraction.
(12) ARFN [25], which uses the Yolov5 algorithm to extract target regions in an image, and fuses the text and visual through a multimodal interaction layer.
(13) CLGFF: Our proposed method.

The experimental results of the CLGFF method and the comparison method are shown in Table 2, from which it is found that the CLGFF method achieves

Table 2. Performance of CLGFF and baseline models

Modality	Method	Twitter-2015		Twitter-2017	
		ACC(%)	Macro-F1(%)	ACC(%)	Macro-F1(%)
Visual	Img-Aspect	64.78	49.97	62.73	58.33
Text	MemNet	70.11	61.76	64.18	60.90
	RAM	70.68	63.05	64.42	61.01
	MGAN	71.17	64.21	64.75	61.46
Text+Visual	Res-RAM	71.55	64.68	65.40	62.23
	Res-RAM-TFN	69.91	61.49	63.45	58.92
	Res-MGAN	71.65	63.88	66.37	63.04
	Res-MGAN-TFN	70.30	64.14	64.10	59.13
	MIMN	71.84	65.69	65.88	62.99
	Res-ESTR	72.03	63.98	66.13	63.63
	ESAFN	73.38	67.37	67.83	64.22
	CMMT	-	66.5	-	68.5
	LF-CapTrBERT	76.89	72.14	68.83	66.54
	TomBERT	77.15	71.75	70.34	68.03
	ARFN	78.50	73.70	70.58	68.43
	CLGFF	**77.72**	**74.21**	**71.15**	**69.45**

better results. For our proposed CLGFF model, the F1 values on the Twitter-2015 dataset and Twitter-2017 dataset are 74.21% and 69.45%, respectively. First of all, the CLGFF method is more effective than the single-mode method, such as Img-Aspect, MGAN. In addition, the F1 value of the CLGFF method on the Twitter-2015 dataset is 2.46%, 2.07%, 7.71%, 0.51% higher than that of TomBERT, LF-CapTrBERT, CMMT and ARFN, respectively. The F1 value of the CLGFF method on the Twitter-2017 dataset is 1.42%, 2.91%, 0.95%, 1.02% higher than that of TomBERT, LF-CapTrBERT, CMMT, and ARFN, respectively. Overall, our proposed approach is more advantageous for AMSA.

4.2 Ablation Experiments

The following demonstrates the role of different modules in the CGLFF, ablation experiments are performed on the same dataset and accuracy and Macro-F1 are chosen as evaluation metrics. The comparison method of ablation experiments are as follows and the results of the experiments are shown in Table 3.

(1) **MLCA:** Removing the multimodal local co-attention module in CLGFF and keep other parameters constant.
(2) **MLCA-vgta:** In CLGFF's multimodal local co-attention module, text-guided visual attention is retained and visual-guided text attention is discarded.

(3) **MLCA-tgva:** In CLGFF's multimodal local co-attention module, visual-guided text attention is retained and text-guided visual attention is discarded.
(4) **MLCA-no-gm:** In CLGFF's multimodal local co-attention module, the fine-grained multimodal features are directly fused over the feature sequence.
(5) **AGGC-no-gm:** In CLGFF's aspect-guided global co-attention module, aspect-guided global multimodal representation are directly stitched together.
(6) **CLGFF-16:** Keeping the CLGFF network parameters unchanged, the visual features are divided into 16 visual regions.
(7) **CLGFF-36:** Keeping the CLGFF network parameters unchanged, the visual features are divided into 36 visual regions.
(8) **CLGFF-64:** Keeping the CLGFF network parameters settings unchanged, the visual features are divided into 64 visual regions.

Table 3. Ablation experiments on Twitter-2015 and Twitter-2017

Method	Twitter-2015		Twitter-2017	
	ACC(%)	Macro-F1(%)	ACC(%)	Macro-F1(%)
CLGFF	77.72	74.21	71.15	69.45
MLCA	74.83	68.88	68.88	66.98
MLCA-vgta	75.80	69.61	69.69	67.45
MLCA-tgva	77.30	73.42	70.10	68.46
MLCA-no-gm	77.34	72.77	69.77	67.65
AGGC-no-gm	77.15	72.31	69.61	67.84
CLGFF-16	77.07	69.22	69.95	64.78
CLGFF-36	76.51	68.35	70.64	66.08
CLGFF-64	76.30	68.17	70.56	65.85

From Table 3, it is found that removing TGVA mechanisms, VGTA mechanisms, or the MLCA module, the experimental effects are reduced, which suggests that fine-grained alignment of inter-modal interactions facilitates the model's understanding of aspect-directed multimodal content. In addition, the effectiveness of the gate mechanism is also found from Table 3. If the adaptive fine-grained alignment between different modalities is not adopted, the performance of MLCA-no-gm and AGGC-no-gm will decrease. Furthermore, the number of regions of the image are analyzed experimentally as well. It can be seen from Table 3 that with the increase of image regions granularity, the performance of the model increases first and then decreases. This may be due to the fact that if the image content is too rich or simple, the image regions granularity is less or more difficult to fit the alignment of text and image. In summary, the design of each module of the CLGFF model and the setting of parameters are effective.

5 Conclusion

This paper proposes co-attention guided local-global feature fusion model for aspect-level multimodal sentiment analysis task. In the CLGFF model, a multimodal local co-attention module is designed to mine the local fine-grained alignment association between text modality and image modality, and an aspect-guided global co-attention module is designed to obtain the global multimodality representation of aspect orientation. Then, the global and local multimodality features are adaptively combined to judge the sentiment of aspect polarity. The experimental results on the Twitter-2015 and Twitter-2017 datasets prove the superiority of the CLGFF model. In the future research, we will pay more attention to the heterogeneity of multimodality in terms of the gap of semantic and sentiments occurring in multimodal data, and plan to adopt approaches of contrast learning and multimodal pretraining models.

Acknowledgements. This work is supported by the National Natural Science Foundation of China (No. 62366010), The Key lab of Trusted software (k202060) and CCF-Zhipu AI Large Model Fund.

References

1. Truong, Q.T., Lauw, H.W.: VistaNet: visual aspect attention network for multimodal sentiment analysis. In: Proceedings of the AAAI Conference on Artificial Intelligence, vol. 33, no. 01, pp. 305–312 (2019)
2. Xu, N., Mao, W., Chen, G.: Multi-interactive memory network for aspect based multimodal sentiment analysis. In: Proceedings of the AAAI Conference on Artificial Intelligence, vol. 33, no. 01, pp. 371–378 (2019)
3. Yu, J., Jiang, J., Xia, R.: Entity-sensitive attention and fusion network for entity-level multimodal sentiment classification. IEEE/ACM Trans. Audio Speech Lang. Process. **28**, 429–439 (2019)
4. Gu, D., Wang, J., Cai, S.: Targeted aspect-based multimodal sentiment analysis: an attention capsule extraction and multi-head fusion network. IEEE Access **9**, 157329–157336 (2021)
5. Xu, N., Mao, W., Chen, G.: A co-memory network for multimodal sentiment analysis. In: the 41st International ACM SIGIR Conference on Research & Development in Information Retrieval, pp. 929–932 (2018)
6. Nemati, S., Rohani, R., Basiri, M.E.: A hybrid latent space data fusion method for multimodal emotion recognition. IEEE Access **7**, 172948–172964 (2019)
7. Yu, Y., Lin, H., Meng, J.: Visual and textual sentiment analysis of a microblog using deep convolutional neural networks. Algorithms **9**(2), 41 (2016)
8. Kumar, A., Srinivasan, K., Cheng, W.H.: Hybrid context enriched deep learning model for fine-grained sentiment analysis in textual and visual semiotic modality social data. Inf. Process. Manag. **57**(1), 102141 (2020)
9. Chen, F., Gao, Y., Cao, D.: Multimodal hypergraph learning for microblog sentiment prediction. In: 2015 IEEE International Conference on Multimedia and Expo (ICME), pp. 1–6 (2015)
10. Xu, J., Huang, F., Zhang, X.: Sentiment analysis of social images via hierarchical deep fusion of content and links. Appl. Soft Comput. **80**, 387–399 (2019)

11. Zhang, L., Wang, S., Liu, B.: Deep learning for sentiment analysis: a survey. Wiley Interdisc. Rev. Data Min. Knowl. Discov. **8**(4) (2018)
12. Dong, L., Wei, F., Tan, C.: Adaptive recursive neural network for target-dependent Twitter sentiment classification. In: Proceedings of the 52nd Annual Meeting of the Association for Computational Linguistics, pp. 49–54 (2014)
13. Tang, D., Qin, B., Feng, X., Liu, T.: Effective LSTMs for target-dependent sentiment classification. In: Proceedings of the 26th International Conference on Computational Linguistics: Technical Papers, pp. 3298–3307 (2016)
14. Li, R., Chen, H., Feng, F., Ma, Z., Wang, X., Hovy, E.: Dual graph convolutional networks for aspect-based sentiment analysis. In: Proceedings of the 59th Annual Meeting of the Association for Computational Linguistics and the 11th International Joint Conference on Natural Language Processing, pp. 6319–6329 (2021)
15. Tang, D., Qin, B., Liu, T.: Aspect level sentiment classification with deep memory network. In: Proceedings of the 2016 Conference on Empirical Methods in Natural Language Processing, pp. 214–224 (2016)
16. Chen, P., Sun, Z., Bing, L.: Recurrent attention network on memory for aspect sentiment analysis. In: Proceedings of the 2017 Conference on Empirical Methods in Natural Language Processing, pp. 452–461 (2017)
17. Zhao, S., et al.: An end-to-end visual-audio attention network for emotion recognition in user-generated videos. In: Proceedings of the AAAI Conference on Artificial Intelligence, pp. 303–311 (2020)
18. Zadeh, A., Chen, M., Poria, S., Cambria, E., Morency, L.P.: Tensor fusion network for multimodal sentiment analysis. In: Proceedings of the 2017 Conference on Empirical Methods in Natural Language Processing, pp. 1103–1114 (2017)
19. Yu, W., Xu, H., Yuan, Z., Wu, J.: Learning modality-specific representations with self-supervised multi-task learning for multimodal sentiment analysis. In: Proceedings of the AAAI Conference on Artificial Intelligence, pp. 10790–10797 (2021)
20. Zhang, Z., Yang, J.: Temporal sentiment localization: listen and look in untrimmed videos. In: Proceedings of the 30th ACM International Conference on Multimedia, pp. 199–208 (2022)
21. Zhang, Z., Wang, L., Yang, J.: Weakly supervised video emotion detection and prediction via cross-modal temporal erasing network. In: Proceedings of the IEEE/CVF Conference on Computer Vision and Pattern Recognition, pp. 18888–18897 (2023)
22. Yu, J., Jiang, J.: Adapting BERT for target-oriented multimodal sentiment classification. In: Proceedings of the Twenty-Eighth International Joint Conference on Artificial Intelligence, pp. 5408–5414 (2019)
23. Khan, Z., Fu, Y.: Exploiting BERT for multimodal target sentiment classification through input space translation. In: Proceedings of the 29th ACM International Conference on Multimedia, pp. 3034–3042 (2021)
24. Yang, L., Na, J.-C., Yu, J.: Cross-modal multi task transformer for end-to-end multimodal aspect based sentiment analysis. Inf. Process. Manag. **59**(5) (2022)
25. Jia, L., Ma, T., Rong, H., Al-Nabhan, N.: Affective region recognition and fusion network for target-level multimodal sentiment classification. IEEE Trans. Emerg. Top. Comput. **0**(1) (2023)
26. He, K., Zhang, X., Ren, S.: Deep residual learning for image recognition. In: Proceedings of the IEEE Conference on Computer Vision and Pattern Recognition, pp. 770–778 (2016)
27. Chen, C., Han, D., Chang, C.C.: CAAN: context-aware attention network for visual question answering. Pattern Recogn. **132** (2022)

28. Liu, Y., Liu, H., Wang, H., Meng, F., Liu, M.: BCAN: bidirectional correct attention network for cross-modal retrieval. IEEE Trans. Neural Netw. Learn. Syst. (2023)
29. Kingma, D.P., Ba, J.: Adam: a method for stochastic optimization. arXiv preprint arXiv:1412.6980 (2014)
30. Fan, F., Feng, Y., Zhao, D.: Multi-grained attention network for aspect-level sentiment classification. In: Proceedings of the 2018 Conference on Empirical Methods in Natural Language Processing, pp. 3433–3442 (2018)
31. Hazarika, D., Poria, S., Zadeh, A., Cambria, E., Morency, L.P., Zimmermann, R.: Conversational memory network for emotion recognition in dyadic dialogue videos. In: Proceedings of the Conference. Association for Computational Linguistics. North American Chapter. Meeting, p. 2122 (2018)

Discovering Multimodal Hierarchical Structures with Graph Neural Networks for Multi-modal and Multi-hop Question Answering

Qing Zhang[1,3], Haocheng Lv[1], Jie Liu[1,2(✉)], Zhiyun Chen[1], Jianyong Duan[1], Mingying Xv[1], and Hao Wang[1]

[1] School of Information Science and Technology, North China University of Technology, Beijing, People's Republic of China

[2] China Language Intelligence Research Center, Capital Normal University, Beijing, People's Republic of China

liujxxxy@126.com

[3] CNONIX National Standard Application and Promotion Lab, Beijing 100144, China

Abstract. Multimodal reasoning is a challenging task that requires understanding and integrating information from different modalities, such as text and image. Existing methods for multimodal reasoning often fail to capture the rich structural information among visual and textual semantics in different modalities, which is crucial for generating accurate answers. In this paper, we propose a novel method that leverages graph neural networks to model the structural information to enhance multimodal reasoning. Specifically, we first use a Multimodal and Multi-hop reader to attend to different chunks in the context based on the question, and then search for multi-hop candidate tokens within these chunks. Next, we construct a graph to represent the relations among the chunks. Then we apply a Sparse Matrix-Tree algorithm to learn a hierarchical informative structure. Then, we use a Hierarchy-aware Message Passing mechanism to perform multi-hop reasoning on the selected edges and update the node representations. Finally, we use a graph-selection decoder to generate the answer based on the structure-enriched chunk representation. We conduct experiments on the WebQA dataset, which is a large-scale multimodal question answering dataset [1]. The results show that our method outperforms the baseline methods in terms of reasoning and the overall answer accuracy. We also provide some qualitative analysis to illustrate how our method benefits from the structural information among different modalities.

Keywords: Multimodal · Multi-hop reasoning · Graph Neural Networks

---

This work is supported by National Key Research and Development Program of China (2020AAA0109700), National Natural Science Foundation of China (62076167).

Q. Zhang and H. Lv—These authors contributed equally to this work.

© The Author(s), under exclusive license to Springer Nature Singapore Pte Ltd. 2024
Q. Liu et al. (Eds.): PRCV 2023, LNCS 14425, pp. 383–394, 2024.
https://doi.org/10.1007/978-981-99-8429-9_31

1 Introduction

Multimodal and Multi-hop question answering (MMQA) is a task that requires understanding and integrating information from different modalities, such as text and image, to answer questions that may involve multiple steps of reasoning. MMQA is a challenging and important task, as it can enable various applications, such as visual dialog, image captioning, and visual search. However, existing methods for MMQA often fail to capture the rich structural information among visual and textual semantics in different modalities, which is crucial for generating accurate answers. For example, in Fig. 1, to answer the question "What color do the stalks of the Mycena adonis and the Stinkhorn mushroom share?", one needs to identify the relevant chunks and tokens in the text and image, such as "Stinkhorn mushroom", "Mycena adonis", and corresponding image chunk. And then infer the relation between them. Moreover, one also needs to perform multi-hop reasoning to connect the clues from different modalities. However, most existing methods for MMQA either ignore the structural information among different modalities or use a fixed and predefined structure, which may not be optimal for different questions and contexts [2,3].

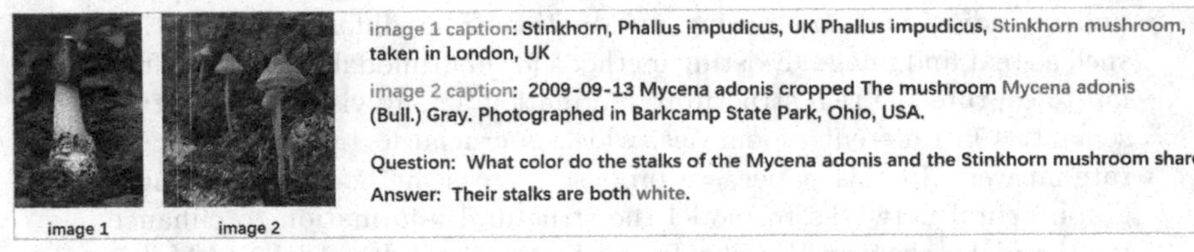

Fig. 1. Example on WebQA. A sample contains two images and corresponding image captions or multiple texts as context to answer questions, so in the webqa dataset, each question requires multiple hops of reasoning to answer correctly

In this paper, we propose a novel method that leverages graph neural networks (GNNs) to model the structural information to enhance multimodal reasoning. To reduce human annotation and error propagation related to NER, our method do not rely on the pre-extracted NERs. Instead, we extract relevant information from chunks with an attentive reader, which can avoid inevitable errors brought by off-the-shelf NER tools. Specifically, we first use a Multimodal and Multi-hop reader to select different chunks in the context attended to the question, next we use the question-aware attention to pay attention to the special tokens in each chunks to find the chunk where the intermediate entity is located, and then search for multi-hop bridge tokens within these chunks. Next, we construct a graph to represent the relations among the chunks. Next, we apply a Sparse Matrix-Tree algorithm to learn a hierarchical informative structure. Then, we use a Hierarchy-aware Message Passing mechanism to perform multi-hop reasoning on the selected edges and update the node representations [4–6]. Finally, we use a multimodal decoder to generate the answer based on the structure-enriched multi-hop tokens representation. The main contributions of our paper are as follows:

- We introduce a multi-hop and multimodal reader to find related chunk across different modal which corresponding to each multi-hop reasoning step.
- We propose a novel method that uses GNNs to discover the hierarchical structural information among different modalities for MMQA.
- We provide qualitative analysis to illustrate how our method benefits from the structural information among different modalities and how it can handle complex questions that require multi-hop reasoning.

2 Related Work

Multimodal and Multi-hop Reasoning. Multimodal And Multi-hop Reasoning is a task that requires connecting clues from different sources and modalities to answer questions that may involve multiple steps of reasoning. Multimodal And Multi-hop Reasoning is challenging and important, as it can enable various applications, such as visual dialog, image captioning, and visual search. However, there are few datasets that support multi-hop reasoning, such as WebQA and MultimodalQA [7]. WebQA is a large-scale dataset that contains questions based on web pages with both text and image content. MultimodalQA is a smaller dataset that contains questions based on Wikipedia articles with tables and images. Both datasets require retrieving relevant evidence from multiple sources and generating natural language answers.

Multi-hop Reading Comprehension. Some previous works for multi-hop RC use memory-based methods that employ attentions to iteratively refine the query and context representations based on their interactions [8–10] To explicitly model entity relations, [11] proposed to construct entity graphs and apply Graph Neural Networks for information aggregation. [12] formulated reasoning as a path-finding problem with neural encoding to score candidate paths. Path modeling was also used in [13] using pointer networks. However, these methods only focus on local information without the ability to generalize, and some of them depend on off-the-shelf NER tools. Another research direction is to split target questions into subquestions [14] which also fail to explicitly reveal the underlying logic for reasoning.

Graph Neural Networks. Graph neural networks have demonstrated their ability to capture rich dependencies in documents to be summarized. [15] use a "heterogeneous graph" with sentence nodes and concurring word nodes to capture the sentence dependencies. [16] use two separate encoders to encode the input sequence with a parsed dependency graph. [17] introduce an adaptive sparse matrix-tree construction to learn a latent hierarchical graph and a novel propagation rule. *We follow this promising direction and extend it to a multimodal reasoning scenario, where it is more challenging, since the graph is not easy to build in advance due to the inevitable errors brought by off-the-shelf NER tools and the complex clues hidden in the context.*

3 Methodology

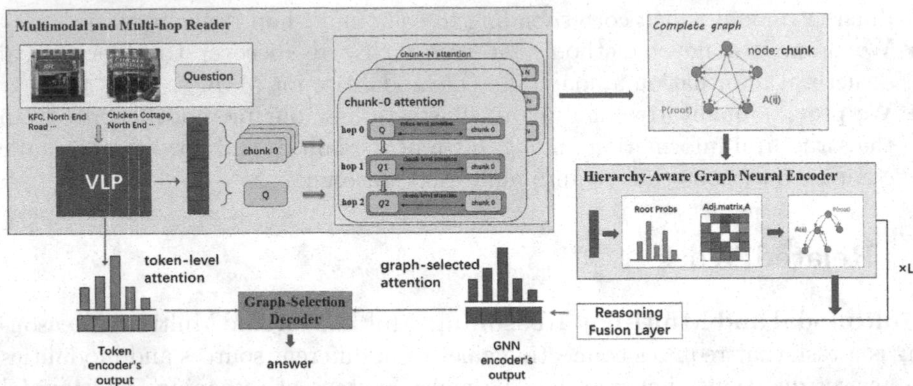

Fig. 2. The overall architecture of our method. A framework using multimodal and multi-hop reader to modeling the multi-hop representation of each chunks which concerned with the question, and then learn the Latent Hierarchical Graph via Hierarchy-Aware Graph Neural Encoder, at last generate answer by attended to original representation on multimodal and multi-hop reader and Graph representation on Hierarchy-Aware Graph Neural Encoder.

In this section, we present the details of our proposed method for multimodal reasoning. The overall architecture of our method is illustrated in Fig. 2. Our method consists of three main components: a Multimodal and Multi-hop reader inspired by [18], a hierarchy-aware graph neural encoder, and a graph-selection decoder. The multimodal and multi-hop reader obtains representations of each chunks relevant to the reasoning path by attending to the question and the context. The hierarchy-aware graph neural encoder first computes the weights of the nodes and edges of the graph by applying Sparse Matrix-Tree to the outputs of the multimodal hierarchical reader. Then, it performs reasoning by using Hierarchy-aware Message Passing. Finally, it builds the final hierarchical tree graph by using Reasoning Fusion Layer. The multimodal decoder generates the final answer by attending to the hierarchy-aware graph neural encoder.

3.1 Multimodal and Multi-hop Reader

The VLP [19] model uses different masking methods to perform bidirectional or unidirectional prediction, so we first use the VLP model to bidirectionally encode the context and question, and then use a unidirectional mask to decode the output of the VLP model in the final decoding stage. We use the setting from [18] and then propose a two-level hierarchical attention network to gradually filter out lhop (l = 0, ..., L) words that are relevant to each chunk in the question. Specifically, the goal of token-level attentions is to select tokens that are related to the question in each chunk. Then, the chunk-level attentions generate the final representations by softly attending to each chunk with different numbers of reasoning hops.

Given a question q with n_q tokens, and a multimodal context set c of length n_c, we denote by $Q \in \mathbf{R}^{n_q \times D}$ and $C \in \mathbf{R}^{n_c \times D}$ their multimodal features after VLP model, respectively. To perform multi-hop reasoning, we use different attentions to identify or reposition target words in each context. Firstly, a question-to-chunk attention is used to locate words that are similar in question and each chunk: $B^q_{ij} = \mathbf{w}_q^\top [\mathbf{Q}_i; \mathbf{C}_j; \mathbf{Q}_i \circ \mathbf{C}_j]$ where w_q is a learnable transformation vector and [;] denotes concatenations. We obtain the normalized similarity score α^q_{ij} between the i-th token in the question and the j-th token in the chunk via a softmax operation on each row of B^q. A question-aware (0-hop) context representation is produced as

$$h_q = \frac{1}{n_c}\sum_{j=1}^{n_c} \bar{\alpha}^j_q \mathbf{C}_j, \quad with \quad \bar{\alpha}^j_q = \frac{1}{n_q}\sum_{i=1}^{n_q} \alpha^{ij}_q \beta^q_i, \tag{1}$$

where β^q_i weighs the contribution of each subject token via a self-attention: β_q = softmax $(\mathbf{w}_q^\top \mathbf{Q} + b_q)$.

For (l + 1)-hop reasoning ($l \geq 0$), it is desired to relocate to intermediate (bridging) entities that are related to the l-hop entities. Hence, we adopt chunk-to-chunk attentions $B^{l+1}_{ij} = \mathbf{w}_l^\top [\mathbf{C}_i + h_l; \mathbf{C}_j; (\mathbf{C}_i + h_l) \circ \mathbf{C}_j]$ given the l-hop representation h^l where $h^0 = h^q$. We use α^{l+1}_{ij} to denote a normalized attention score between the i-th and the j-th context tokens after applying a softmax operator over each row of B^{l+1}. With $\bar{\alpha}^j_0 = \bar{\alpha}^j_q$, the (l + 1)-hop bridging context representation becomes

$$h_{l+1} = \frac{1}{n_c}\sum_{j=1}^{n_c} \bar{\alpha}^j_{l+1} \mathbf{C}_j, \quad with \quad \bar{\alpha}^j_{l+1} = \frac{1}{n_c}\sum_{i=1}^{n_c} \alpha^{ij}_{l+1} \bar{\alpha}^i_l. \tag{2}$$

When multiple contexts (images and image-captions) are available, we apply a chunk-level attention to generate the final l-hop feature representations. When l = 0, the model softly attends to each context to produce context-attended question representation as

$$\hat{h}_q = \frac{1}{K}\sum_{k=1}^{K} \bar{\gamma}^k_q h_{q,k} \tag{3}$$

where $\bar{\gamma}^k_q$ is the attention weight of context c_k obtained by normalizing over a score vector γ^q with entries $\gamma^k_q = \mathbf{v}_q^\top [q; \hat{h}_{q,k}; q \circ \hat{h}_{q,k}]$. Here $h_{q,k}$ is the question-aware context representation computed in (1) corresponding to the k-th context c_k. v_q is a trainable transformation vector. The final question representation is produced as $\bar{h}_q = \mathbf{W}_q[q; \hat{h}_q; q \circ \hat{h}_q]$ incorporating both original features and attended information. We treat $\bar{h}_q$ as 0-hop question representations.

Finally, we concat all multihop representation

$$H^{(0)} = [h^0_{q,0}, h^0_{q,1} ..., h^1_{q,0}, h^1_{q,1} ..., h^2_{q,k}]$$

as Multimodal and Multi-hop reader's final output.

3.2 Hierarchy-Aware Graph Neural Encoder

Latent Multimodal Hierarchical Graph. A complete weighted graph is used to represent the multimodal contexts, where each node corresponds to a Multimodal chunk that is relevant to the question. The marginal probability of a directional dependency between two chunks determines the edge weights. Furthermore, each node has an additional probability value, the "root probability" that signifies the hierarchical role of the chunk, such as being the lead or the most crucial reasoning clue.

Sparse Matrix-Tree Computation. We use the outputs of the Multimodal Hierarchical reader to feed the hierarchy-aware graph neural encoder. We assign the token representations of each chunk from the outputs of the Multimodal Hierarchical reader to the nodes of the graph $H^{(0)} = [s_1, ..., s_i, ..., s_M]$. This is for a context set with one or two images and their corresponding captions. then We apply two separate non-linear transformations to get a parent and a child representation for the chosen token of each chunk,

$$s(p)_i = \sigma(\mathbf{W}_p s_i + b_p) \tag{4}$$

$$s(c)_i = \sigma(\mathbf{W}_c s_i + b_c) \tag{5}$$

where W_p, W_c, b_p, b_c are parameters, σ is the ReLU activation function.

To compute a matrix $\mathbf{F} \in \mathbf{R}^{M\times M}$ with positive entries, where f_{ij} denotes the weight of the edge from node s_i to s_j; and a positive root score vector $f_{root} \in \mathbf{R}^M$. the standard matrix-tree theorem uses an exponential function. However, this leads to a dense matrix that impairs our graph reasoning module by introducing noise from redundant M sentence nodes. Inspired by sparse self-attention, we propose an adaptive approach to induce sparsity in MTC. We use ReLU$(x \in \mathbf{R}) = \max\{x, 0\}$ (or coordinate-wise for vectors) instead of the exponential function, and compute the root and edge scores by a fully connected layer and a bi-linear attention layer, respectively,

$$f^{root}_{(i)} = RELU(W_r s^{(p)}_i + b_r) + \epsilon, \tag{6}$$

$$f_{ij} = RELU({s^{(p)}_i}^\top W_{bi} s^{(c)}_j) + \epsilon, \tag{7}$$

where W_{bi}, W_r, b_r are trainable. (We use $\epsilon = 10^{-6}$ to prevent matrix singularity issues.) Unlike the exponential function, ReLU enforces F and $f^{(root)}$ to be non-negative, thus allowing zero probability and eliminating dependency edges and roots. We then insert these values into the standard MTC and compute the edge and root probabilities as the adjacency matrix A(i, j) = P$(z_{ij}$ = 1) and root probability p_i^r indicating the hierarchical role (i.e., the propensity to be a connector) of each sentence.

Reasoning by Cross-Modal Hierarchy Message Passing. We use a message-passing mechanism based on a learned hierarchical graph. This mechanism realizes cross-modal reasoning, where connectors can aggregate information from their related information nodes and propagate it to others. For the i-th sentence node, the edge marginal controls the aggregation from its K information nodes; and the root probability controls how the neighboring information is combined as the i-th node's update $u^{(l)}$ in the l-th reasoning layer,

$$\mathbf{u}_i^{(l)} = (1 - p_i^r)\mathcal{F}_r(s_i^{(l)}) + (p_i^r)\sum_{k=1}^{K} A_{ik}\mathcal{F}_n(s_k^{(l)}), \tag{8}$$

where $\mathcal{F}_r$ and $\mathcal{F}_n$ are functions with parameters. The idea is that a connector token should have high connectivity with the relevant information nodes, and collect more details. Each information node learns to either preserve the uniqueness of its information or merge the information from the connectors. To eliminate unnecessary information, we use a gated mechanism as the information filter in the node update.

$$\mathbf{g}_i^{(l)} = \sigma(\mathcal{F}_g([\mathbf{u}_i^{(l)}; \mathbf{h}_i^{(l)}])), \tag{9}$$

$$\mathbf{h}_i^{(l+1)} = LN(\mathbf{g}_i^{(l)} \odot \phi(\mathbf{u}_i^{(l)}) + (1 - \mathbf{g}_i^{(l)}) \odot \mathbf{h}_i^{(l)}), \tag{10}$$

where $\mathcal{F}_g$ is a parametric function and $\odot$ is the element-wise dot product. We use layer normalization (LN) to stabilize the output for the update function. The function σ is the sigmoid function, and ϕ can be any non-linear function.

Reasoning Fusion Layer. We build reasoning chains consisting of L hops by stacking L hierarchical tree graph blocks together. To handle cases where less than L hops are needed, we add a fusion layer to aggregate the output from each reasoning hop to produce the final output of the hierarchy-aware graph neural encoder.

$$\mathbf{h}_i^{(G)} = (W_g[\mathbf{h}_i^{(1)}, ..., \mathbf{h}_i^{(L)}] + b_g) + \mathbf{h}_i^{(0)}, \tag{11}$$

where W_g, b_g are trainable parameters. We learn the latent hierarchical graphs for each layer separately, where each message passing layer has its own graph.

3.3 Graph-Selection Decoder

Besides using token-level decoding attention, we also propose a graph-selection attention mechanism (GSA) to provide the decoder with learned hierarchical information, while achieving chunk-level content selection. In each decoding step t, our decoder first obtains a graph context vector, $\mathbf{c}_G^t$, which contains the global information of the latent hierarchical graph. We first compute the graph-level attention distribution $\mathbf{a}_G^t$ by,

$$e_{vi}^{t} = ATTN^{(G)}(\mathbf{h}^{(L)}, \mathbf{z}_t), \tag{12}$$

$$\mathbf{a}_G^{t} = SOFTMAX(\mathbf{e}^{t}), \tag{13}$$

where $ATTN^{(G)}$ is a graph attention function. The vectors $\mathbf{h}_i^{(L)} \in \mathbf{R}^d$, $\mathbf{z}_t \in \mathbf{R}^d$ are the L-th layer node embeddings for chunk i and decoding state at time t, respectively. The graph context vector $\mathbf{c}_G^t \in \mathbf{R}^d$ is finally obtained by summing all $\mathbf{h}_i^{(L)}$ weighted by $\mathbf{a}_G^t$. The value of $\mathbf{c}_G^t$ is used as an additional input for computing token-level attention,

$$e_i^{t} = ATTN^{(T)}(\mathbf{h}_{enc}(X), \mathbf{z}_t, \mathbf{c}_G^t), \tag{14}$$

$$\mathbf{a}_T^{t} = SOFTMAX(\mathbf{e}^{t}), \tag{15}$$

where $ATTN^{(T)}$ is a token-level attention function. The final context vector $\mathbf{c}_f^t$ is fused from the graph $\mathbf{c}_G^t$ and token context vectors $\mathbf{c}_T^t$ with a parametric function g_f, $\mathbf{c}_f^t = g_f(\mathbf{c}_G^t, \mathbf{c}_T^t)$.

4 Experiments

4.1 Datasets

We validate our method on the WebQA dataset. This dataset consists of multi-source and multi-form question-answer pairs, where all questions are knowledge-seeking and resemble real-world uses. 44% of image-based queries and 99% of text-based questions require two or more knowledge sources. The training, validation, and test sets contain 34K/5K/7.5K question-answer pairs. Since our method is designed based on multimodal contexts, we randomly sample 5000 queries from the original training set of WebQA as the development set for evaluation. Therefore, the final training, validation, and test sets contain 29K/5K/5K question-answer pairs, respectively. The answers in WebQA are complete and free-form natural language sentences to better assess the model's generation capability. For evaluation, WebQA uses BARTScore and keyword accuracy score. BARTScore measures the fluency between the probability of generating a candidate given a reference. Keyword accuracy measures the overlap of key entities between generation and reference. These two scores are multiplied to calculate the overall score.

4.2 Baselines

We use multimodal pre-training models, such as VLP, as the baselines for WebQA. VLP combines the input text and image and passes them to a transformer-based encoder-decoder to generate the answer. We experiment with two variants of VLP: vanilla VLP and VLP + VinVL. VinVL is the state-of-the-art method for extracting visual features. During fine-tuning, these models take the concatenation of < [CLS], S_i, [SEP], Q, [SEP] > as input and estimate the probability of S_i being selected. Then < [CLS], S, [SEP], Q, [SEP] > is fed into the model for answer generation.

4.3 Implementation Details

Our method uses VLP as the multimodal feature extractor and the final decoder. VLP is a Unified Vision-Language Pre-Training Model, during training stage, we fed < [CLS], C [SEP] Q, A [SEP] > into the VLP model and random mask tokens in A. And then during generating answer stage, we keep adding mask tags at the end of the sentence to predict the next word, then we will stop generate answer until the next predicted word belongs to special tags.

In the multimodal feature extraction stage, Text segments, including the questions, answers, and image captions, are tokenized by the Bert-base-cased tokenizer. Each image is represented by 100 regions predicted by an object detection model, which is a variant of Faster RCNN with an ResNeXt-101 FPN backbone, pretrained on Visual Genome. We take the output of fc1 layer from the object detection network an 2048-dim feature and finetune the fc2 layer.

Because our method only modifies the question answering stage and the baseline's retrieval score is relatively low, so we use the gold evidence retrieval results as our context retrieval results input. For training, we set each chunk length as 50 tokens. Then we set we use the Adam optimizer with the 10^{-5} initial learning rate and linear decay of the learning rate. These methods utilize beam search (n = 5) during decoding.

4.4 Experimental Results

We show the results of WebQA in Table 1 and Table 2. VLP (Qonly) only feeds the question into VLP. We can see that our method achieves the best performance, substantially surpassing all the baselines on QA accuracy.

Table 1. Results on the WebQA validation set on image-based queries when using model to retrieve evidence. Our method use VLP+X101fpn as evidence retrieval model

Method	Retr	FL	Accuracy	Overall
Question-Only	-	34.9	22.2	13.4
VLP+X101fpn	67.4	46.1	44.3	23.2
VLP+VinVL	**69.4**	**47.3**	47.6	26.7
Our Method	67.4	46.7	**48.6**	**27.1**

Table 2. Results on the WebQA validation set on image-based queries (Use all correct evidence from dataset).

Method	FL	Accuracy	Overall
Question-Only	34.9	22.2	13.4
VLP+X101fpn	45.6	46.9	24.2
VLP+VinVL	**46.7**	48.7	26.8
Our Method	44.1	**51.6**	**27.7**

We used two kinds of evidence retrieval results to test the model, namely the evidence retrieved by VLP+x101fpn and all the correct evidence contained in the dataset. As shown in Table 1 and Table 2, Since our model also uses VLP and x101fpn as multimodal feature encoders, the bart-score is similar to the baseline. But our method has higher accuracy than all baselines on QA, and the accuracy of QA is more significantly improved after using all the correct evidence, which proves that our model has better results on reasoning tasks especially when provided with all the correct evidence.

4.5 Ablation Study

To demonstrate the impact of Multimodal and Multi-hop reader and hierarchy-aware graph neural encoder in our approach, we perform ablation studies and report the results in Table 3.

Table 3. Ablation study of our method on the validation-set of WebQA. Reader and GNN are short for Multimodal and Multi-hop reader and Hierarchy-Aware Graph Neural Encoder.).

Method	FL	Accuracy	Overall
our method	44.1	**51.6**	**27.7**
w/o Reader	45.1	50.1	25.3
w/o GNN	43.4	49.2	26.4
w/o Reader+GNN	**45.6**	46.9	24.7

When dropping the multimodal and multi-hop reader, our method directly feeds the encoded image and text blocks into the hierarchy-aware graph neural encoder without building attention from each chunk to question, whose performance is shown on the second line of Table 3. We can see that the overall QA performance drops, demonstrating the effectiveness of using the multimodal and multi-hop reader to model multi-hop token attention. The third line of Table 3 shows the results when we drop the hierarchy-aware graph neural encoder, in which the encoded representations from the multimodal and multi-hop reader directly perform cross attention with the decoder. The overall QA performance and fluency decline, demonstrating that the hierarchy-aware graph neural encoder significantly helps the reasoning process.

4.6 Case Study

We present a examples from WebQA and explicit reasoning process in Fig. 3.

As shown in left part of Fig. 3, the model needs to locate two images through the image description, and select the same color of tablecloths from the two images, which involves multi-hop and multimodal reasoning.

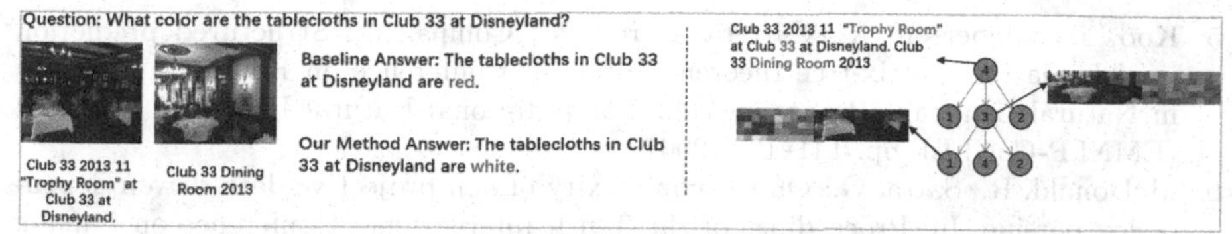

Fig. 3. Example on WebQA and our Method's reasoning process of the example.

As shown in right part of Fig. 3, We use the highest-weight node in the root node as the root node, and select three high-weight child nodes from the adjacency matrix for each layer. The orange nodes are the highest-weight nodes and the green nodes are the suboptimal nodes. The tree has three layers, matching the multi-hop reasoning steps of the reader. Our method can select the points of interest related to the question from each chunk through a multimodal reader, and find the chunk candidates for each hop by modeling the multi-hop relationship, and learn the hierarchical relationship of these multi-hop candidate chunks through GNN.

5 Conclusion

We propose a multimodal and multi-hop question answering method based on multi-hop attention and hierarchical tree graph. Our work is motivated by discovering the hierarchical structure of chunks across different modalities to perform multi-hop reasoning, to model the inter-dependent reasoning steps, and to learn a hierarchical reasoning step graph. Our experiments demonstrate the advantage of this method, as it surpasses the existing baselines on various metrics. However, modeling the cross-modal intermediate entity representations for multi-hop reasoning is still not accurate enough. It indicates that there is still room to utilize the hierarchical structure more effectively during answer generation.

References

1. Chang, Y., Narang, M., Suzuki, H., Cao, G., Gao, J., Bisk, Y.: WebQA: multihop and multimodal QA. In: Proceedings of the IEEE/CVF Conference on Computer Vision and Pattern Recognition, pp. 16495–16504 (2022)
2. Oguz, B., et al.: UniK-QA: unified representations of structured and unstructured knowledge for open-domain question answering. In: Findings of the Association for Computational Linguistics: NAACL 2022, pp. 1535–1546, July 2022
3. Qi, P., Lee, H., Sido, T., Manning, C.D.: Answering open-domain questions of varying reasoning steps from text. In: Proceedings of the 2021 Conference on Empirical Methods in Natural Language Processing, pp. 3599–3614, November 2021
4. Smith, D.A., Smith, N.A.: Probabilistic models of nonprojective dependency trees. In: Proceedings of the 2007 Joint Conference on Empirical Methods in Natural Language Processing and Computational Natural Language Learning (EMNLP-CoNLL), pp. 132–140, June 2007

5. Koo, T., Globerson, A., Carreras Pérez, X., Collins, M.: Structured prediction models via the matrix-tree theorem. In: Joint Conference on Empirical Methods in Natural Language Processing and Computational Natural Language Learning (EMNLP-CoNLL), pp. 141–150 (2007)
6. McDonald, R., Satta, G.: On the complexity of non-projective data-driven dependency parsing. In: Proceedings of the Tenth International Conference on Parsing Technologies, pp. 121–132, June 2007
7. Talmor, A., et al.: MultiModalQA: complex question answering over text, tables and images (2021). arXiv preprint arXiv:2104.06039
8. Dhingra, B., Jin, Q., Yang, Z., Cohen, W., Salakhutdinov, R.: Neural models for reasoning over multiple mentions using coreference. In: Proceedings of the 2018 Conference of the North American Chapter of the Association for Computational Linguistics: Human Language Technologies, Volume 2 (Short Papers), pp. 42–48, June 2018
9. Clark, C., Gardner, M.: Simple and effective multi-paragraph reading comprehension. In: Proceedings of the 56th Annual Meeting of the Association for Computational Linguistics (Volume 1: Long Papers), pp. 845–855, July 2018
10. Tu, M., Wang, G., Huang, J., Tang, Y., He, X., Zhou, B.: Multi-hop reading comprehension across multiple documents by reasoning over heterogeneous graphs. In: Proceedings of the 57th Annual Meeting of the Association for Computational Linguistics, pp. 2704–2713, July 2019
11. Tang, Z., Shen, Y., Ma, X., Xu, W., Yu, J., Lu, W.: Multi-hop reading comprehension across documents with path-based graph convolutional network. In: Proceedings of the Twenty-Ninth International Conference on International Joint Conferences on Artificial Intelligence, pp. 3905–3911, January 2021
12. Kundu, S., Khot, T., Sabharwal, A., Clark, P.: Exploiting explicit paths for multi-hop reading comprehension. In: Proceedings of the 57th Annual Meeting of the Association for Computational Linguistics, pp. 2737–2747, July 2019
13. Chen, J., Lin, S., Durrett, G.: Multi-hop question answering via reasoning chains. arXiv preprint arXiv:1910.02610 (2019)
14. Min, S., Zhong, V., Zettlemoyer, L., Hajishirzi, H.: Multi-hop reading comprehension through question decomposition and rescoring. In: Proceedings of the 57th Annual Meeting of the Association for Computational Linguistics, pp. 6097–6109, July 2019
15. Wang, D., Liu, P., Zheng, Y., Qiu, X., Huang, X.J.: Heterogeneous graph neural networks for extractive document summarization. In: Proceedings of the 58th Annual Meeting of the Association for Computational Linguistics, pp. 6209–6219, July 2020
16. Jin, H., Wang, T., Wan, X.: SemSUM: semantic dependency guided neural abstractive summarization. In: Proceedings of the AAAI Conference on Artificial Intelligence, vol. 34, no. 05, pp. 8026–8033, April 2020
17. Qiu, Y., Cohen, S.B.: Abstractive summarization guided by latent hierarchical document structure. In: Proceedings of the 2022 Conference on Empirical Methods in Natural Language Processing, pp. 5303–5317, December 2022
18. Wang, W., Pan, S.: Deep inductive logic reasoning for multi-hop reading comprehension. In: Proceedings of the 60th Annual Meeting of the Association for Computational Linguistics (Volume 1: Long Papers), pp. 4999–5009, May 2022
19. Zhou, L., Palangi, H., Zhang, L., Hu, H., Corso, J., Gao, J.: Unified vision-language pre-training for image captioning and VQA. In: Proceedings of the AAAI Conference on Artificial Intelligence, vol. 34, no. 07, pp. 13041–13049, April 2020

Enhancing Recommender System with Multi-modal Knowledge Graph

Chengjie Sun[1,2](✉), Weiwei Chen[2], Lei Lin[2], and Lili Shan[2]

[1] State Key Laboratory of Communication Content Cognition, People's Daily Online, Beijing 100733, China
sunchengjie@hit.edu.cn

[2] Faculty of Computing, Harbin Institute of Technology, Heilongjiang 150001, China
{wwchen,linl,shanll}@insun.hit.edu.cn

Abstract. Recommender systems have shown great potential to solve the problem of information overload and improve user experience in various online applications. To address the data sparsity and cold start problems in these systems, researchers have proposed leveraging knowledge graphs (KGs) as auxiliary information for recommendations, which contain valuable external knowledge. However, most of these works have neglected the diversity of data types, such as texts and images, in multi-modal knowledge graphs (MMKGs). In this paper, we propose the Multi-modal Knowledge Graph Attention Network to enhance recommender systems by leveraging multi-modal knowledge. Specifically, We propose a multi-modal knowledge graph attention mechanism to facilitate knowledge propagation over MMKGs. The resulting knowledge-enhanced entity embedding representations are utilized for recommendation. Extensive experiments conducted on MovieLens datasets demonstrate the rationality and effectiveness of our model.

Keywords: Recommender system · Multi-modal Knowledge Graph · Attention mechanism

1 Introduction

The explosive growth of online content and services has provided users with a vast array of choices, such as news [16], movies [19] and music [14]. Recommender systems (RS) aim to address the problem of information overload by finding a small number of items that match users' personalized interests. Among various recommendation strategies, collaborative filtering (CF) has achieved great success by considering users' historical interactions and making recommendations based on their potential common preferences [6]. However, CF-based methods

Supported by the National Key Research and Development Program of China under Grant 2020YFB1406902 and State Key Laboratory of Communication Content Cognition, People's Daily Online (No. A12003).

© The Author(s), under exclusive license to Springer Nature Singapore Pte Ltd. 2024
Q. Liu et al. (Eds.): PRCV 2023, LNCS 14425, pp. 395–407, 2024.
https://doi.org/10.1007/978-981-99-8429-9_32

are often plagued by sparsity and cold-start problems. To address these limitations, researchers have proposed incorporating auxiliary information into CF, such as social networks [5], user/item attributes [10], images [19], and context [9]. Among various auxiliary information, knowledge graph (KG) typically contains more facts and relations about items. KG is a directed heterogeneous graph, in which nodes correspond to entities and edges correspond to relations. Knowledge graph has been successfully applied in many fields, such as question answering, word embedding, and text classification.

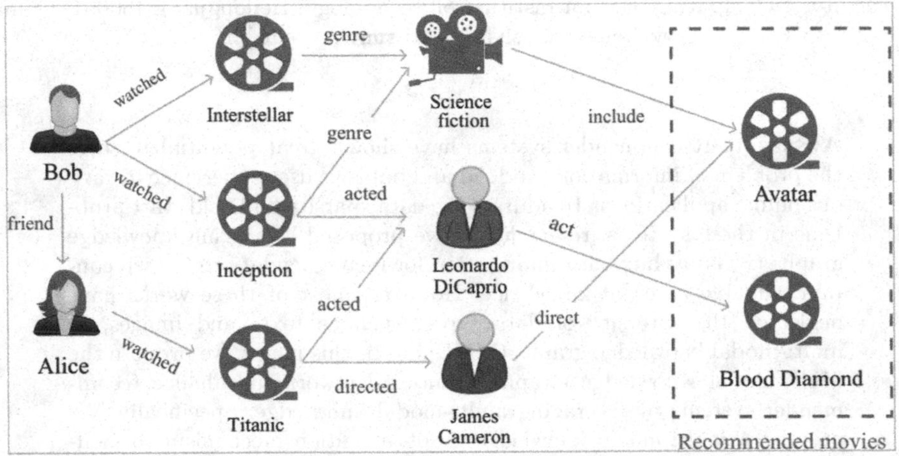

Fig. 1. Illustration of knowledge graph enhanced movie recommender systems [2]

Inspired by the successful applications of KG in various tasks, researchers also attempt to use KG to improve the performance of recommendation systems. As shown in Fig 1, KG can benefit recommendations in three ways: (1) KG introduces semantic relatedness between items, which helps to discover their potential connections and improve the accuracy of recommended items; (2) KG contains multiple types of relations, which helps to properly expand users' interests and increase the diversity of recommended items; (3) KG connects users' historical records and recommended records, thus bringing interpretability to the recommendation system.

However, existing knowledge graph-based recommendation methods often overlook multi-modal information like images and text, which can significantly influence the recommendation process. For instance, users frequently view movie trailers or read reviews before watching a movie, or assess product images and user feedback before online shopping. Thus, integrating this multi-modal data into knowledge graphs is vital. By considering images or texts as entities or entity attributes, Multi-modal Knowledge Graphs efficiently incorporate this visual or textual information.

To fully leverage multi-modal data, we propose ERMMKG, a recommendation model based on multi-modal knowledge graph. By designing specific models for diverse modalities to extract their features, and subsequently integrating them via the multi-modal knowledge graph propagation layer, we can enhance user interest preferences with this multi-modal information. Extensive experiments conducted on MovieLens datasets demonstrate the rationality and effectiveness of our model.

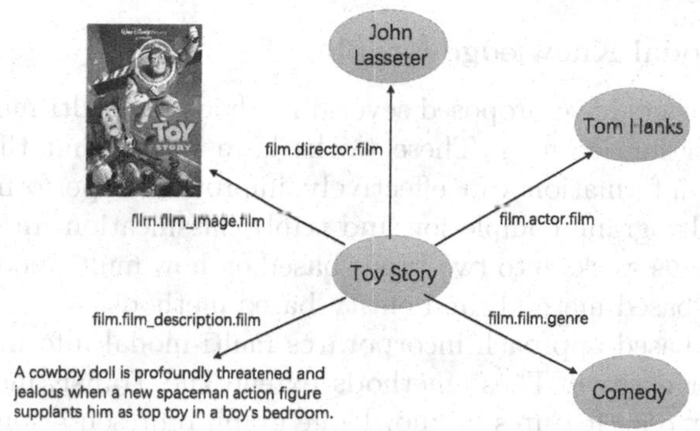

Fig. 2. Example of a multi-modal knowledge graph.

2 Related Work

2.1 KG-Based Recommendation

In general, existing KG-aware recommendation can be classified into three categories [2]:

Embedding-based methods represent the first category of knowledge graph utilization techniques in recommendation systems. These methods apply Knowledge Graph Embedding (KGE) algorithms for knowledge graph pre-process, then integrate learned entity embeddings into a recommendation framework. For instance, DKN [12] treats entity and word embeddings as separate channels, then merges them via a CNN framework for news recommendation, while CKE [19] fuses a collaborative filtering module with item's knowledge, text, and image embeddings within a unified Bayesian framework. Despite their flexibility in leveraging knowledge graphs , KGE algorithms are generally more effective for in-graph applications like link prediction.

The second category is path-based, which targeting the exploration of diverse item connection patterns within the KG. For example, PER [18] and FMG [20] regard the KG as a heterogeneous information network, employing meta-path latent features to signify the variety of relation paths/graphs linking users and items. Although path-based methods intuitively utilize the KG, they heavily depend on manually designed meta-paths, which are challenging to optimize and sometimes unfeasible to craft in specific scenarios.

The third category is unified methods. As discussed above, both approaches utilize only one aspect of information in the graph. To fully exploit the information in the KG for better recommendations, unified methods which integrate both the semantic representation of entities and relations, and the connectivity information have been proposed. These methods refine the entity representation with the guidance of the connective structure in the KG.

2.2 Multi-modal Knowledge Graphs

Recently, researchers have proposed several methods for multi-modal knowledge graph representation learning. These works have shown that the introduction of multi-modal information can effectively improve the performance of tasks such as knowledge graph completion and triple classification. In this article, we categorize previous works into two types based on how multi-modal information is used: feature-based methods and entity-based methods.

The feature-based approach incorporates multi-modal information as auxiliary features for entities. These methods extend the TransE model to include visual or other modal features in knowledge graph representation learning. One representative work in this area is IKRL [17]. This method still uses a translation-based training mode, but introduces visual information into both the head and tail entity embeddings. Specifically, two entity embeddings are created for each head and tail entity, representing structured and visual information, respectively. Similar to TransE, they define a scoring function for triplets that consists of four parts, jointly determined by the two types of entity representations.

Different from the feature-based approach, the entity-based approach treats multi-modal information as structured knowledge and directly represents it as entities in the knowledge graph. This approach introduces new relationships, such as hasImage and hasDescription . MKBE [7] employs neural encoders and decoders to enhance the initial layer of embedding-based relation models, training with DistMult and ConvE knowledge graph embedding algorithms. After learning the knowledge graph representation with ConvE and DistMult, a neural decoder employs entity embeddings to generate any missing multi-modal attributes.

3 Method

In this section, we will introduce the model ERMMKG (Enhancing Recommender System with Multi-modal Knowledge Graph), the overall framework of which is illustrated in Fig 3. The model consists of two key components: the multi-modal knowledge graph embedding module and the recommendation module.

The multi-modal knowledge graph embedding module utilizes an encoder to extract features from raw data. Following this, a knowledge graph propagation layer is implemented to generate vector representations for each user/item

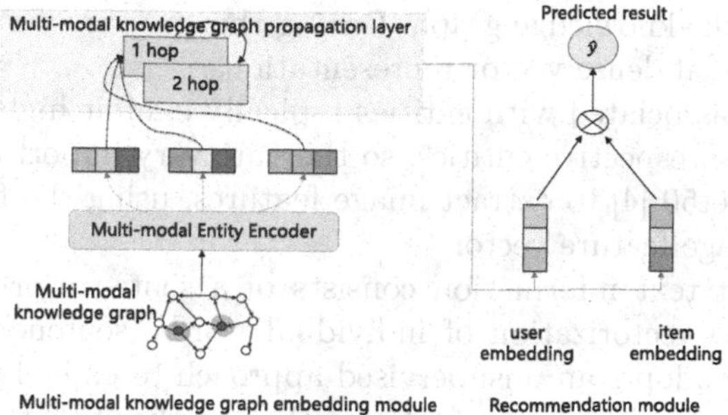

Fig. 3. Framework overview of the ERMMKG model.

that incorporate multi-modal knowledge graph information. Simultaneously, we employ knowledge graph representation learning to ensure that the entity vectors still retain the structural information of the knowledge graph.

The recommendation module draws on the augmented representations of users and items derived from the multi-modal knowledge graph module, aiming to predict user click probability accurately.

3.1 Multi-modal Entity Encoder

The multi-modal entity encoder designs distinct deep learning network architectures for entities from varying modalities, aiming to efficiently extract features. The specific structure of this design is illustrated in Fig 4.

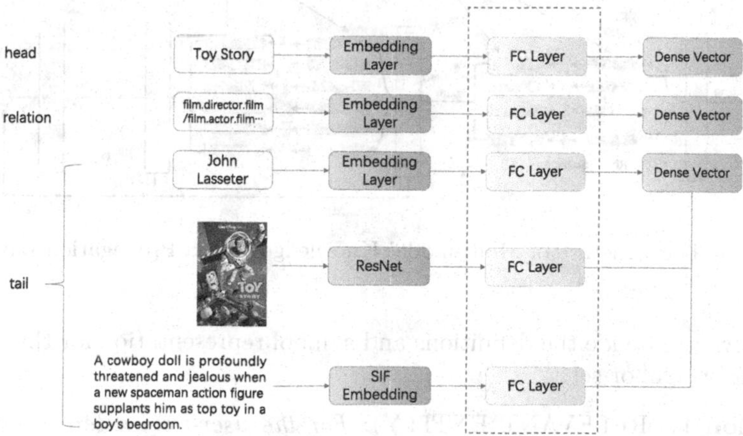

Fig. 4. Multi-modal Entity Encoder

Knowledge graph structures knowledge. In order to fully exploit the structural information , we employ KGE algorithm to vectorize the entities and relationships within the knowledge graph. During this process, we transforming IDs into low-dimensional dense vector representations.

Images. Images associated with entities typically exhibit high relevance to the information of the respective entities, so they are very important. We employ a pre-trained ResNet50 [4] to extract image features, using the final hidden layer output as the image feature vector.

Text. Given that text information consists of a sentence or a few sentences, different from the vectorization of individual words, sentence vectorization is also required. We adopts an unsupervised approach to embed text information. Specifically, we first tokenize the text and then applies SIF [1] to calculate the weighted average of the words in a sentence. Finally, we utilizes a fully connected layer to unify all modal vectors into the same dimensional space.

3.2 Multi-modal Knowledge Graph Propagation Layer

Current multi-modal knowledge graph representation learning methods mostly extract information from independent entity-relation-entity triples, which overlooks the critical process of multi-modal information fusion in multi-modal learning. To achieve the fusion of multi-modal information, we designs a multi-modal knowledge graph propagation layer, whose structure is depicted in Fig 5. The core idea of the multi-modal knowledge graph propagation layer is to model user interests by leveraging preference propagation, and fuse multi-modal information during the process of preference propagation.

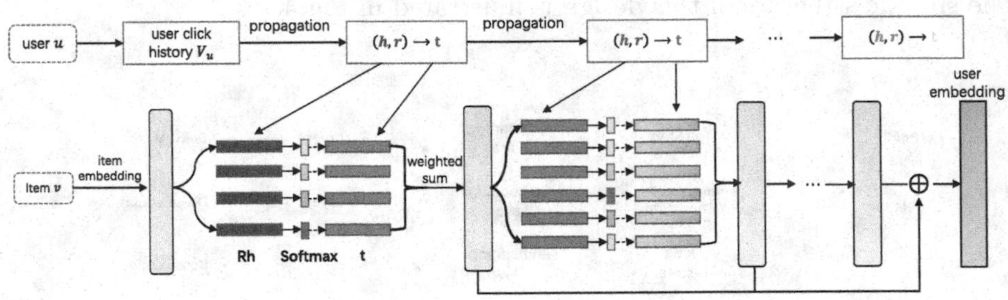

Fig. 5. The structure of Multi-modal Knowledge Graph Propagation Layer

Firstly, we provide the definitions and symbol representation for the relevant entity and neighbor set.

Definition 1. (RELEVANT ENTITY)*: For the user-item interaction matrix* **Y** *and knowledge graph $\mathcal{G}$, the k-hop relevant entity for user u is defined as in equation (1):*

$$\mathcal{E}_u^k = \{t|(h,r,t) \in \mathcal{G}, h \in \mathcal{E}_u^{k-1}\} \tag{1}$$

When $k = 0$, $\mathcal{E}_u^0 = \mathcal{V}_u = \{v|y_{uv} = 1\}$ denotes the items with which the user has historical interactions.

Definition 2. (NEIGHBOR SET)*: The k-hop neighbor set of user u is the set of triples whose head entity is in the* $(k-1)$*-hop relevant entity set* ${\mathcal{E}_u}^{k-1}$*:*

$$\mathcal{S}_u^k = \{(h,r,t) | (h,r,t) \in \mathcal{G}, h \in \mathcal{E}_u^{k-1}\} \tag{2}$$

As shown in Fig 5, each item v has a vector representation. For every user, we initially align the items from their history with entities, thereby obtaining a set of initial relevant entities for the user. From these initial relevant entities, we can define the first-hop neighbor set $\mathcal{S}_u^1$ based on the definition. For item v, each triple is assigned a relevance score α_i that signifies the relevance of item v to the head entity h_i in relation space r_i, as depicted in equation (3):

$$\alpha_i = \frac{\exp(\mathbf{v}^\mathrm{T}\mathbf{R}_i\mathbf{h}_i)}{\Sigma_{(h,r,t)\in\mathcal{S}_u^1}\exp(\mathbf{v}^\mathrm{T}\mathbf{R}\mathbf{h})} \tag{3}$$

where $\mathbf{R}_i \in \mathbb{R}^{d\times d}$ is the feature of the relation r_i, and $\mathbf{h}_i \in \mathbb{R}^d$ is the feature of the head entity h_i. The reason for incorporating the relation when calculating the relevance score α is that different relations focus on different aspects of the entity. In one relation, two entities that are similar may be entirely irrelevant in another relation.

After calculating all relevance scores in the first-hop neighbor set, we perform a weighted sum of the tail entities t_i in the triples to obtain the vector $\mathbf{o}_u^1$. This vector represents the user's interest vector after one preference propagation. The calculation formula for $\mathbf{o}_u^1$ is as follows:

$$\mathbf{o}_u^1 = \sum_{(h,r,t)\in\mathcal{S}_u^1} \alpha_i \mathbf{t}_i \tag{4}$$

where $\mathbf{t}_i \in \mathbb{R}^d$ is the feature of the tail entity t_i.Through the operations, the user's interest expands from historical interaction items to the entity set $\mathcal{E}_u^1$ via the knowledge graph, completing a full preference propagation process. In such a propagation process, we can not only obtain knowledge-enhanced user and item representations but also effectively integrate the multi-modal information of entities.

The user vector $\mathbf{o}_u^1$ can replace the item vector $\mathbf{v}$ and repeat the above preference propagation process to obtain the user's second-order vector $\mathbf{o}_u^2$. Similarly, after H rounds of preference propagation, we can obtain H user interest vectors, $\mathbf{o}_u^1, \mathbf{o}_u^2, ..., \mathbf{o}_u^H$. The final user representation is as shown in equation (5):

$$\mathbf{u} = \sum_{i=1}^{H} \mathbf{o}_u^i \tag{5}$$

The final user feature is input into the recommendation module and calculates the model-predicted click probability with the item feature as shown in Eq. 6, where $\sigma(\cdot)$ is the sigmoid function.

$$\hat{y}_{uv} = \sigma(\mathbf{u}^T\mathbf{v}) = \frac{1}{1+\exp(-\mathbf{u}^\mathrm{T}\mathbf{v})} \tag{6}$$

4 Experiments

4.1 Datasets and Settings

We selected the MovieLens-1M dataset [3] for our experiments. This dataset contains nearly a million rating records, each including a user's rating for a movie along with a timestamp.

The knowledge graph for MovieLens datasets comes from [13] which uses Microsoft Satori to construct the knowledge graph for this dataset. Base on it, we add image and text modality information to the knowledge graph. Afterwards, we add two relationships, $film.film\_image.film$ and $film.film\_description.film$, to incorporate the images and texts as entities into the initial knowledge graph, thereby forming a multi-modal knowledge graph. The statistics of MovieLens and the corresponding multi-modal knowledge graph are presented in Table 1.

Table 1. statistics of MovieLens-1M and the multi-modal knowledge graph

Category	Counts
users	6036
items	2445
interactions	753772
relations	14
entities	186901
1-hop triples	25672
2-hop triples	178049
3-hop triples	318266
4-hop triples	923718

The settings of parameters used in the proposed ERMMKG model are given in Table 2.

Table 2. The settings of parameters used in ERMMKG model

Parameter	Value
Epoch	20
batch size	1024
image vector dimension	2048
text vector dimension	300
entities' (in KG) vector dimension	16
learning rate	0.01

4.2 Baselines

We compare ERMMKG with the following state-of-the-art baselines:

- **LibFM** [8] is a commonly used feature-based factorization machine model. It can handle high-dimensional sparse data and capture interactions between features.
- **KGCN** [15] proposes a knowledge-aware Graph Convolutional Network. When calculating the representation of an entity, it samples a part of its neighbors from the KG as its receptive field and combines the neighbor information with its own representation.
- **PER** [18] designs multiple meta-paths for the knowledge graph, and represents users and items by extracting features based on different meta-paths. Moreover, it also designs a scoring function to estimate the probability of interaction between users and items.
- **DKN** [12] models user interests through user click history. It employs a knowledge-aware convolutional neural network to fuse knowledge graph entities and news headline information, and it designs an attention network to capture dynamic user interests.
- **RippleNet** [11] models user interests by utilizing the propagation of information within the knowledge graph.

4.3 Results

Performance of All the Methods. As can be seen from Table 1, each entity has numerous neighbors, with multi-modal entities being just two. Therefore, random neighbor sampling could overlook image and text entities, impeding the inclusion of multi-modal information in the recommendation system and overall performance enhancement. We propose: 1) prioritizing image or text entities during neighbor sampling if available; 2) initializing the movie entity representation as the sum of text representation and image representation. Although both methods help, the latter is less effective as shown in Table 3. Multi-modal information serves as auxiliary data, usually improving model performance, but the straightforward summation doesn't account for individual attention variations towards different information, limiting its effectiveness.

Table 3. Effect of sampling methods

	AUC	ACC
Random sampling	0.9160	0.8402
Initializing entities with image and text	0.9185	0.8429
Prioritize selection of text and image	**0.9213**	**0.8452**

Following the sampling method improvement, the performance of all models is presented in Table 4. ERMMKG outperforms the baseline methods with an

AUC of 0.9213 and ACC of 0.8452 on the movie dataset, demonstrating its effective utilization of multi-modal knowledge graph to represent user interests.

Table 4. The results comparison of different models in MovieLens-1M dataset

Model	AUC	ACC
LibFM	0.8923	0.8125
KGCN	0.8981	0.8132
PER	0.7126	0.6673
DKN	0.6552	0.5893
RippleNet	0.9146	0.8384
ERMMKG	**0.9213**	**0.8452**

Effects of Modalities. We conducted ablation experiments to assess how image and text information impact the model's performance under different knowledge graph modalities. The results are shown in Table 5, where 'base' represents original entities and relations, and 'text' and 'img' indicate text and image entities and relations.

Table 5. Effect of modalities on model

model	AUC	ACC
base	0.9142	0.8379
base+text	0.9187	0.8432
base+img	0.9201	0.8440
base+img+text	**0.9213**	**0.8452**

Findings from the results yield the following conclusions: (1)The model's performance improves dramatically when multi-modal data is incorporated. By augmenting the original movie entity representation, text and image features enhance movie modeling. Also, user modeling is based on historical movie interactions, incorporating multi-modal information improves user profiling. Hence, multi-modal data assists in modeling both users and items, leading to superior recommendations. (2) Visual information plays a more significant role than textual data in enhancing model performance. Visual elements, such as actor images and scenes on movie posters, often capture user attention and influence choices. ResNet used for image feature extraction, effectively extracts image features.

Effects of Hyper-parameter. Additionally, we investigated the effects of two hyperparameters: the neighbor set size d and the maximum number of hops H. We fixed H at 2 when evaluating d, and d at 32 when investigating H. The results are shown in Table 6.

Table 6. The effects of the neighbor set size and hop numbers.

d	8	16	32	64
AUC	0.9171	0.9193	**0.9213**	0.9179
ACC	0.8418	0.8439	**0.8452**	0.8421
H	1	2	3	4
AUC	0.9181	**0.9213**	0.9118	0.9029
ACC	0.8420	**0.8452**	0.8361	0.8242

The results of the Table 6 allow us to make the following conclusions: (1) Model performance is initially improved by larger neighbor sets due to increased knowledge capture, but declines when the set becomes overly large, likely due to noise infiltration. (2) Optimal performance is achieved when the maximum hop count is 2. Insufficient or excessive hops can limit useful information or increase noise respectively, impacting the model's ability to discern entity relationships and user interests, thus affecting performance.

5 Conclusion

In this paper, we propose an Enhanced Recommender Model based on Multi-modal Knowledge Graph (ERMMKG), which introduces a multi-modal knowledge graph into the recommendation system. The ERMMKG model uses an attention mechanism to utilize and integrate multi-modal information during the propagation of information in the knowledge graph, obtaining a knowledge-enhanced user interest representation. Experiments on the MovieLens dataset validate the rationality and effectiveness of the proposed model.

References

1. Arora, S., Liang, Y., Ma, T.: A simple but tough-to-beat baseline for sentence embeddings. In: International Conference on Learning Representations (2017)
2. Guo, Q., et al.: a survey on knowledge graph-based recommender systems. IEEE Trans. Knowl. Data Eng. **34**(8), 3549–3568 (2020)
3. Harper, F.M., Konstan, J.A.: The movielens datasets: history and context. ACM Trans. Interact. Intell. Syst. **5**(4), 1–19 (2015)

4. He, K., Zhang, X., Ren, S., Sun, J.: Deep residual learning for image recognition. In: Proceedings of the IEEE Conference on Computer Vision and Pattern Recognition, pp. 770–778 (2016)
5. Jamali, M., Ester, M.: A matrix factorization technique with trust propagation for recommendation in social networks. In: Proceedings of the Fourth ACM Conference on Recommender Systems, pp. 135–142 (2010)
6. Koren, Y., Bell, R., Volinsky, C.: Matrix factorization techniques for recommender systems. Computer **42**(8), 30–37 (2009)
7. Pezeshkpour, P., Chen, L., Singh, S.: Embedding multimodal relational data for knowledge base completion. In: Proceedings of the 2018 Conference on Empirical Methods in Natural Language Processing, pp. 3208–3218 (2018)
8. Rendle, S.: Factorization machines with libFM. ACM Trans. Intell. Syst. Technol. (TIST) **3**(3), 1–22 (2012)
9. Trouillon, T., Welbl, J., Riedel, S., Gaussier, É., Bouchard, G.: Complex embeddings for simple link prediction. In: International Conference on Machine Learning, pp. 2071–2080. PMLR (2016)
10. Wang, H., Zhang, F., Hou, M., Xie, X., Guo, M., Liu, Q.: Shine: Signed heterogeneous information network embedding for sentiment link prediction. In: Proceedings of the Eleventh ACM International Conference on Web Search and Data Mining, pp. 592–600 (2018)
11. Wang, H., Zhang, F., Wang, J., Zhao, M., Li, W., Xie, X., Guo, M.: Ripplenet: Propagating user preferences on the knowledge graph for recommender systems. In: Proceedings of the 27th ACM International Conference on Information and Knowledge Management, pp. 417–426 (2018)
12. Wang, H., Zhang, F., Xie, X., Guo, M.: Dkn: Deep knowledge-aware network for news recommendation. In: Proceedings of the 2018 World Wide Web Conference, pp. 1835–1844 (2018)
13. Wang, H., Zhang, F., Zhang, M., Leskovec, J., Zhao, M., Li, W., Wang, Z.: Knowledge-aware graph neural networks with label smoothness regularization for recommender systems. In: Proceedings of the 25th ACM SIGKDD International Conference on Knowledge Discovery & Data Mining, pp. 968–977 (2019)
14. Wang, H., Zhang, F., Zhao, M., Li, W., Xie, X., Guo, M.: Multi-task feature learning for knowledge graph enhanced recommendation. In: The World Wide Web Conference, pp. 2000–2010 (2019)
15. Wang, H., Zhao, M., Xie, X., Li, W., Guo, M.: Knowledge graph convolutional networks for recommender systems. In: The World Wide Web Conference, pp. 3307–3313 (2019)
16. Wu, C., Wu, F., Qi, T., Zhang, C., Huang, Y., Xu, T.: Mm-rec: Visiolinguistic model empowered multimodal news recommendation. In: Proceedings of the 45th International ACM SIGIR Conference on Research and Development in Information Retrieval, pp. 2560–2564 (2022)
17. Xie, R., Liu, Z., Luan, H., Sun, M.: Image-embodied knowledge representation learning. In: Proceedings of the 26th International Joint Conference on Artificial Intelligence, pp. 3140–3146 (2017)
18. Yu, X., Ren, X., Sun, Y., Gu, Q., Sturt, B., Khandelwal, U., Norick, B., Han, J.: Personalized entity recommendation: A heterogeneous information network approach. In: Proceedings of the 7th ACM International Conference on Web Search and Data Mining, pp. 283–292 (2014)

19. Zhang, F., Yuan, N.J., Lian, D., Xie, X., Ma, W.Y.: Collaborative knowledge base embedding for recommender systems. In: Proceedings of the 22nd ACM SIGKDD International Conference on knowledge Discovery and Data Mining, pp. 353–362 (2016)
20. Zhao, H., Yao, Q., Li, J., Song, Y., Lee, D.L.: Meta-graph based recommendation fusion over heterogeneous information networks. In: Proceedings of the 23rd ACM SIGKDD International Conference on Knowledge Discovery and Data Mining, pp. 635–644 (2017)

Location Attention Knowledge Embedding Model for Image-Text Matching

Guoqing Xu[1(✉)], Min Hu[1], Xiaohua Wang[1], Jiaoyun Yang[1], Nan Li[2], and Qingyu Zhang[1]

[1] Anhui Province Key Laboratory of Affective Computing and Advanced Intelligent Machine, National Smart Eldercare International Science and Technology Cooperation Base, Hefei University of Technology, Hefei 230602, China
2021171134@mail.hfut.edu.cn

[2] School of Mental Health and Psychological Sciences, Anhui Medical University, 81 Meishan Road, Shushan District, Hefei 230032, Anhui, China

Abstract. Image-text matching is the core algorithm of cross-modal retrieval, which plays a central role in connecting vision and text. Due to the well-known semantic gap between visual and textual modalities, yet image-text matching is a vital challenging task. In order to reduce the huge semantic difference between images and texts, existing methods use the consensus knowledge for image-text matching tasks. However, the consensus knowledge is only extracted based on the co-occurrence frequency of words in sentences in the corpus, and does not consider the semantic information contained in the image, resulting in a decline in semantic matching performance. To solve this issue, we propose a Location Attention Knowledge Embedding (LAKE) model to improve the consensus knowledge utilization by inferring the location of objects in an image. Specifically, our model consists of three parts: Firstly, we design a location feature extraction (LFE) module, which divides the image into blocks, uses the location attention to generate valuable location features, and then splices the location features with the extracted regional image features to obtain the image features containing location information. At the same time, text features are extracted using the BERT model. Secondly, we use a knowledge representation module to extract the consensus knowledge features. Finally, the similarity between the image and the text is calculated based on the knowledge fusion feature to complete the matching process. Quantitative and qualitative results on public datasets Flickr30k and MSCOCO demonstrate the effectiveness of the method.

Keywords: Image-text matching · Consensus knowledge · Location attention

1 Introduction

Recently, there are more and more graph-driven machine learning methods [1,2]. Inspired by these methods, many methods [3–5] introduce external knowledge to

© The Author(s), under exclusive license to Springer Nature Singapore Pte Ltd. 2024
Q. Liu et al. (Eds.): PRCV 2023, LNCS 14425, pp. 408–421, 2024.
https://doi.org/10.1007/978-981-99-8429-9_33

the image-text matching problem based on the graph structure, these methods can add an additional consensus knowledge feature space in addition to the image feature space and text feature space. By performing auxiliary matching in a unified knowledge space, the semantic differences between images and texts can be overcome to a certain extent, and the matching effect can be improved.

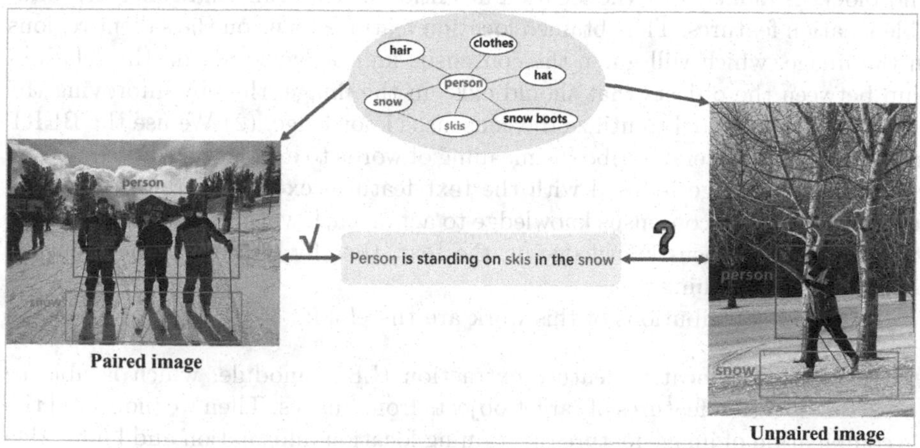

Fig. 1. In the image on the left, "person" and "snow" can be successfully identified, but "skis" is occluded. The consensus knowledge graph can provide common sense information indicating that the "person" in the "snow" should be wearing "skis", thus making the match successful. In the image on the right, according to the indication of the consensus knowledge graph, it is believed that the presence of skis in the image will lead to successful matching, but it is obvious that the image on the right does not match the text, resulting in a decline in the accuracy of image-text matching.

In the process of introducing consensus knowledge, the two most important parts are extracting knowledge features and using knowledge features. Previous methods [3–5] can do a good job in extracting knowledge features, but there are still many problems in using knowledge features. As shown in Fig. 1, in the paired image, "person", "snow", and "skis" are all key concepts for successful text matching. In the paired image, "person" and "snow" can be successfully identified, but "skis" cannot be identified because it is occluded by "snow". According to the consensus knowledge graph, the consensus knowledge indicating that the "person" in the "snow" should wear "skis" can be provided, and the hidden information in the image can be mined through the knowledge graph, so that the image and the text can be successfully matched.

But for the unpaired image, both "person" and "snow" can be successfully identified. If we simply think that there will be "skis" in the image through the consensus knowledge graph, the image that originally does not match the text will also successfully match the text. As a result, the accuracy of image-text matching decreases. Since the model simply fuses knowledge features and image features, the utilization efficiency of knowledge features will decrease, resulting in poor final matching effect.

To solve the above issue, we propose a Location Attention Knowledge Embedding (LAKE) model for image-text matching. LAKE mainly includes two innovations: (1) We propose a location feature extraction module (LFE). Firstly, the image is divided into blocks, and then the relationship between image regions and blocks is modeled by the location attention mechanism, and generates valuable location features. The obtained location features focus on the salient regions in the image, which will guide the consensus knowledge to act on the relationship between the objects that should exist in the image, thereby improving the efficiency of the model to utilize the consensus knowledge. (2) We use the BERT model that can better describe the meaning of words to extract text features. The consensus knowledge is fused with the text features extracted from the BERT model to guide the consensus knowledge to act on each word, which improves the effect of the consensus knowledge on the text, thereby improving the accuracy of image-text matching.

The main contributions of this work are threefold:

(1) We propose a location feature extraction (LFE) module, which is able to extract location features of target objects from images. Then we pioneered the combination of image features containing location information and knowledge features, to improve the effect of consensus knowledge on images.
(2) We propose to use the BERT model to extract text features, which can generate context-aware word-level feature representations. The extracted text features can contain the information of the sentence context, better guide the consensus knowledge to act between each word in the sentence, and improve the effect of the consensus knowledge on the text.
(3) Combining image feature representation with location information and text feature representation with context information, Location Attention Knowledge Embedding (LAKE) model is constructed for image-text matching tasks. A large number of experimental results on benchmark datasets prove the proposed model validity.

2 Related Work

At present, image-text matching methods have been widely and deeply researched. Image-text matching methods can be roughly divided into two categories: (1) Image-text matching based on original features and (2) Image-text matching based on knowledge features.

Image-text matching based on original features: The general solution in this field is to optimize the method of extracting global or local features of images and texts for matching. In global matching, Faghri et al. [6] improve the performance of triplet loss by attending to the hardest misaligned pairs. However, global matching ignores the fine-grained region correspondence between images and text, and cannot achieve satisfactory results. Learning semantic alignment between image regions and text words has become popular in image-text matching. Lee et al. [7] proposed a stacked cross-attention model (SCAN) that uses both image regions and words in sentences as context to discover full latent alignments and infer image-text similarity.

Image-text matching based on knowledge features: Traditional image-text matching only considers the features of images and texts, so it is difficult to overcome the natural semantic differences between images and texts. Inspired by the above research, scholars try to use the consensus knowledge to solve this problem. Shi et al. [3] proposed a scene concept map by aggregating image scene graphs and extracting frequently co-occurring concept pairs as scene commonsense knowledge. Wang et al. [4] borrowed the idea of using the concept of co-occurrence as consensus knowledge, and proposed a consensus perception model (CVSE) that directly integrates knowledge features into graphic and text features, with the help of knowledge features, it achieves better performance in the field of image matching.

3 Method

Our proposed LAKE model is shown in the Fig. 2. Firstly, we introduce the image representation module and text representation module in Sect. 3.1 and Sect. 3.2 respectively, then introduce the knowledge representation module in Sect. 3.3, then introduce the feature fusion module in Sect. 3.4, and finally introduce the image-text matching module in Sect. 3.5.

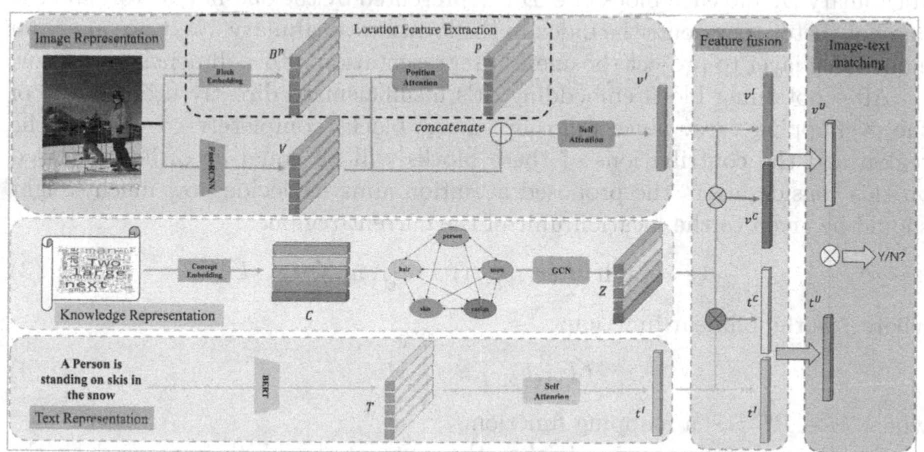

Fig. 2. Our model has five components: image representation module (blue), text representation module (green), knowledge representation module (yellow), feature fusion module (orange), and image-text matching module (gray). (Color figure online)

3.1 Image Representation

To obtain target object features from images, we adopt the pre-trained Fast R-CNN [8] model to extract image features. For each image, a set of feature representations $V = \{v_1, v_2, \ldots, v_q\}, v_i \in R^d$ can be obtained for each image, by utilizing a bottom-up attention module [9]. In an image, the position of the target object in the image is a very important clue, which can better guide the role of consensus knowledge. Inspired by this, as shown in Fig. 2, we design a location feature extraction (LFE) module to extract location features of objects in images, and fuse location information into image feature representations. Given an image, in order to reveal the relative position of the object in the whole image, the whole image is first divided into $K \times K$ blocks equally, and the position of each block is represented by the index $k\epsilon \in [1, K^2]$. Then, the number of intersecting pixels in each block and the region is calculated according to Eq. 1:

$$a_{ij} = \left|v_i \cap b_{p_{ij}}\right|, j = 1, 2, \cdots, K^2 \tag{1}$$

where v_i is the number of pixels of the i-th region, and p_{ij} is the block index of the j-th maximum overlapping with the region v_i. a_{ij} represents the number of intersection pixels between the i-th region and the j-th block.

For the region, arrange in descending order according to the number of intersecting pixels, and select the highest-ranked L blocks, as shown in Eq. 2:

$$a_{ij} = \left|v_i \cap b_{p_{ij}}\right|, j = 1, 2, \cdots, L \tag{2}$$

To obtain a more accurate description of locations, we embed block indices into the dense representation. The divided K^2 blocks are regarded as the location vocabulary B, and each block $b_i \in B$ is represented by the one-hot vector, and the one-hot vector represents the index in the location vocabulary. Next, we apply an embedding layer to project the one-hot representation into τ-dimensional vector.

After obtaining block embedding, it's insufficient to directly use the rate of the overlapping area. Since there are many blocks completely covered by the region and the contributions of these blocks will be equal accordingly. Based on this consideration, the proposed attention aims to decide how much weight should be given to the location unit of the current region:

$$\beta_{ij} = \tanh\left(f\left(v_i, b_{p_{ij}}\right)\right), i \in [1, n], j \in [1, L] \tag{3}$$

where f is the bilinear function:

$$f\left(v_i, b_{p_{ij}}\right) = v_i^T M b_{p_{ij}} \tag{4}$$

where $M \in R^{d \times \tau}$ is a mapping function.

Besides, another intuition is that the more of the block is covered by the region, the more important it should be. According to the above considerations, the calculation equation of the location feature extraction of the target is improved as shown in Eq. 5:

$$p_i = \sum_{j=1}^{L} b_{p_{ij}} \times \gamma_{ij} \tag{5}$$

and

$$\gamma_{ij} = \frac{\gamma'_{ij}}{\sum_{j=1}^{L} \gamma'_{ij}}, \text{ where } \gamma'_{ij} = \frac{\exp(\beta_{ij})}{\sum_{j=1}^{L} \exp(\beta_{ij})} \times a_{ij} \tag{6}$$

The image feature representation V and the target position feature representation P are spliced together to obtain the image features with position information $V^P = \{(v_1, p_1), (v_2, p_2), \cdots, (v_q, p_q)\}, (v_i, p_i) \in R^{d+\tau}$. Using the linear layer to adjust the dimension of V^P from $R^{q\times(d+\tau)}$ to $R^{q\times d}$, which is convenient for subsequent calculations. The image features V^P is used as the key and value items, and the global image features $\overline{V^P} = \frac{1}{q}\sum_{i=1}^{q}(v_i, p_i)$ is adopted as the query item for the attention strategy, a self-attention mechanism [10] is adopted to combine the image features V^p into the original image feature $v^I \in R^d$.

3.2 Text Representation

The consensus knowledge is obtained by the co-occurrence frequency of words in sentences, so we think that using the BERT [11] model to extract the features of each word in a sentence will improve the effect of the consensus knowledge on the text. Moreover, the obtained word features will contain the information of the sentence context, which improves the utilization efficiency of consensus knowledge on the text. Subsequent ablation experiments prove that the effect of using the BERT model to improve the network is quite obvious.

We feed the sentence into the BERT model, which can derive each word vector in the sentence. For a sentence of length s, we can represent it as follows: Obtain a set of word-level features $\{t_1, t_2, \cdots, t_s\}, t_i \in R^d$. The word-level features T is merged into the original text feature $t^I \in R^d$ through the self-attention mechanism [10].

3.3 Knowledge Representation

In this section, we provide high-level semantics for more accurate image-text matching by capturing the intrinsic connections between concepts as commonsense knowledge in human reasoning. To achieve this, we construct a knowledge representation module. It includes three key steps: knowledge screening and classification; construction of concept association graph; representation learning of concept association graph. The details of these steps are shown below.

Knowledge Screening and Classification. In this step, knowledge concepts are extracted from the image caption corpus of natural sentences in the dataset, which is the basis of knowledge representation and extraction. First, word frequencies are computed using the probability-based approach in [12]. Then the top n most frequent concepts are selected as nodes in the knowledge graph. Finally, these concepts are vectorized using the Global Vector (GloVe) technique [13], denoted as $C = \{c_1, c_2, \cdots, c_n\}, c_i \in R^d$.

Construction of Concept Association Graph. In this step, the concept association graph is constructed through the co-occurrence relationship in [5]. Specifically, two concepts c_i and c_j appearing in the same sentence are considered a co-occurring concept pair. Their occurrences are placed in the co-occurrence matrix O, where o_{ij} is the number of occurrences. To select more accurate concept pairs, the confidence of the concept pair considering the long-tailed distribution is calculated using the scaling function:

$$G = f\left(o_{ij}\right) = \begin{cases} 0, & \text{if } o_{ij} \leq \theta \\ 1, & \text{if } o_{ij} > \theta \end{cases} \tag{7}$$

where θ is the threshold used to remove some concept pairs that rarely co-occur. G is a binary matrix, where 1 represents the co-occurrence of c_i and c_j and 0 represents the lack thereof.

Representation Learning of Concept Association Graph. In this step, Graph Convolutional Network (GCN) [14] is a multilayer neural network that operates on a graph and updates a node's embedding representation by propagating information based on its neighborhood. In this section, we use two GCN layers to learn concept representations, which introduce high-order neighborhood information between concepts to model their interdependencies. Given an instantiated concept representation C and a concept association graph G, the embedded features of layer 1 are computed by Eq. 8 as:

$$H^{(1)} = \text{Sigmoid}\left(\tilde{A}H^{(0)}W^{(0)}\right) \tag{8}$$

where $H^{(0)} = C, \tilde{A} = D^{-\frac{1}{2}}GD^{-\frac{1}{2}}$ denotes a normalized symmetric matrix and $W^{(0)} \in R^{d \times d}$ denotes a learnable parameter matrix. The embedded features of layer 2 are calculated by Eq. 9 as:

$$H^{(2)} = \text{Sigmoid}\left(\tilde{A}H^{(1)}W^{(1)}\right) \tag{9}$$

where $\tilde{A} = D^{-\frac{1}{2}}GD^{-\frac{1}{2}}$, $W^{(1)} \in R^{d \times d}$ denotes a learnable parameter matrix. The output of the second layer $H^{(2)}$is the final knowledge representation $Z = \{z_1, z_2, \cdots z_n\}, z_i \in R^d, i = 1, \cdots, n$.

3.4 Feature Fusion

Calculation of Knowledge Features Based on Image and Text. In this section, we need to use the knowledge features obtained for the original image and text features, and extract knowledge from images and texts through the following methods. Image knowledge features can v^C be calculated by Eq. 10:

$$\begin{aligned} a_i^v &= \frac{\exp\left(\lambda v^I W^v z_i^T\right)}{\sum_{i=1}^n \exp\left(\lambda v^I W^v z_i^T\right)} \\ v^C &= \sum\nolimits_{i=1}^n a_i^v \cdot z_i \end{aligned} \tag{10}$$

where $W^v \in R^{d\times d}$ is the learnable parameter matrix, z_i^T which represents z_i the transpose of the knowledge feature, λ controls the smoothness of the Softmax function.

Text knowledge features are obtained by fusing text features and knowledge features t^C through Eq. 11:

$$\begin{gathered} a_j^t = \frac{\exp\left(\lambda t^I W^t z_j^T\right)}{\sum_{j=1}^n \exp\left(\lambda t^I W^t z_j^T\right)} \\ t^C = \sum\nolimits_{j=1}^n a_j^t \cdot z_j \end{gathered} \tag{11}$$

where $W^t \in R^{d\times d}$is the learnable parameter matrix, which z_j^T represents the transpose of the knowledge feature z_j, λ controls the smoothness of the Softmax function.

Fusion of Original Features and Knowledge Features. The original image features v^I and original text features t^I are integrated with the image knowledge features v^C and text knowledge features t^C to comprehensively characterize the image and text information. The fused image and text features are calculated by Eq. 12 and Eq. 13:

$$v^U = \alpha v^C + (1-\alpha)\, v^I \tag{12}$$

$$t^U = \alpha t^C + (1-\alpha)\, t^I \tag{13}$$

where α is a variable parameter that determines the relative proportion between the two types of features.

3.5 Image and Text Matching

Training. we use a bidirectional triplet ranking loss [6] for model training. Bidirectional triple ranking loss is widely used in metric learning

$$\mathrm{Loss}(v,t) = \sum_{(v,t)} \{\max\left[0, \delta - s(v,t) + s\left(v,t^-\right)\right] \quad + \max\left[0, \delta - s(v,t) + s\left(v^-,t\right)\right]\} \tag{14}$$

where δ is the predefined margin parameter and $s(i,j)$ is the cosine distance function. Given a matched image-text pair (v,t) as input, its unmatched image-text pairs are correspondingly represented by (v,t^-) and (v^-,t).

Overall, the final loss L is defined as Eq. 15:

$$L = \lambda_1 Loss\left(v^U, t^U\right) + \lambda_2 Loss\left(v^C, t^C\right) + \lambda_3 Loss\left(v^I, t^I\right) \tag{15}$$

where λ_1, λ_2, λ_3 aim to balance the weight of different loss functions.

Matching. During the matching process, we only use Eq. 12 and Eq. 13 knowledge fused image features v^U and text features t^U for computation, and use cosine distance to measure their cross-modal similarity.

4 Experiments

4.1 Datasets

To verify the effectiveness of the proposed model LAKE, we carry out extensive experiments in terms of image to text retrieval and text to image retrieval on two publicly available benchmark datasets: Flickr30k dataset and MSCOCO dataset.

Flickr30k [15] is an image caption dataset containing 31,783 images, with each image annotated with five sentences. We split the dataset into 29,783 training images, 1000 validation images and 1000 testing images. We report the performance evaluation of image-text retrieval on 1000 testing set.

MSCOCO [16] is another image captioning dataset, including a total of 123,287 images, with each image roughly annotated with five sentence-level descriptions. We split the dataset into 113,287 training images, 1000 validation images and 5000 testing images. We report experimental results on 5000 test images.

4.2 Evaluation Metrics

For both image-to-text retrieval and text-to-image retrieval tasks, we adopt the commonly used evaluation metrics Recall at K (R@K) to measure the performance of our proposed model. Specifically, we use R@1, R@5, and R@10 for all experiments. Additionally, we calculate the average of all evaluation metrics as mR to better show the effectiveness of the model.

4.3 Implementation Details

All our experiments are implemented on a GeForce GTX 3090 GPU. For image feature extraction, the number of detected regions in each image is $q = 36$, and the dimensionality of the region representation vector is $f = 2048$. Each image is split into 16×16 blocks ($K = 16$), and we set L to 15. The block index is first embedded into a 200-dimensional space, and then the original 2048-dimensional visual vector is mapped into a 1024-dimensional space through a linear projection layer together with 200-dimensional position features. For text feature extraction, we use the pre-trained weights of the BERT model, the BERT output dimension is fixed to 768, and the dimension of the word embedding space is set to 300. The dimension of the joint space is set to 1024. In knowledge representation, we use the 300-dimensional GloVe trained on the Wikipedia dataset to initialize semantic concepts. The size of the semantic concept vocabulary is $n = 300$. In Sect. 3.3, λ is set to 0.75 in Eq. 10 and Eq. 11, and α is set to 10 in Eq. 12 and Eq. 13. In Sect. 3.4, λ_1, λ_2, λ_3 is set to 3, 5, 1, respectively in Eq. 15. Our model is trained using the Adam optimizer with a minimum batch size of 128. The learning rate in the first 15 epochs was set to 0.0002, and the learning rate in the last 15 epochs was set to 0.00002. The dropout is also employed with a dropout rate of 0.4.

Table 1. Comparison of cross-modal retrieval with competing methods on the Flickr30k dataset.

Methods	Image-to-Text			Text-to-image			mR
	R@1	R@5	R@10	R@1	R@5	R@10	
SCAN [7]	67.4	90.3	95.8	48.6	77.7	85.3	77.5
VSRN [17]	71.3	90.6	96	54.7	81.8	88.2	80.4
PFAN++ [18]	70.1	91.8	96.1	52.7	79.9	87	79.6
IMRAM [19]	74.1	93	96.6	53.9	79.4	87.2	80.7
MMCA [20]	74.2	92.8	96.4	54.8	81.4	87.8	81.2
SMFEA [21]	73.7	92.5	96.1	54.7	82.1	88.4	81.3
SHAN [22]	74.6	93.5	96.9	55.3	81.3	88.4	81.7
DRSA [23]	76.5	93.9	97.2	**57.7**	82.2	88.6	82.7
GLFN [24]	75.1	93.8	97.2	54.5	82.8	89.9	82.2
CVSE [4]	73.6	90.4	94.4	56.1	83.2	90	81.3
HCGE [5]	74	91.2	94.2	56.6	**83.9**	90.1	81.7
LAKE(Ours)	**76.5**	**94.4**	**97.6**	56.3	83.5	**90.3**	**83.1**

4.4 Comparison Methods

Experimental results from the Flickr30k dataset are presented in Table 1. These results show that our model outperforms its competitors in most cases. Compared to our baseline model CVSE, for text retrieval, it improves R@1 by 2.9%, R@5 by 4.0%, R@10 by 3.2%, and for image retrieval, it improves R@1 by 0.2%,

Table 2. Comparison of cross-modal retrieval with competing methods on the MSCOCO dataset.

Methods	Image-to-Text			Text-to-image			mR
	R@1	R@5	R@10	R@1	R@5	R@10	
SCAN [7]	72.7	94.8	98.4	58.8	88.4	94.8	84.7
VSRN [17]	76.2	94.8	98.2	62.8	89.7	95.1	86.1
PFAN++ [18]	77.1	96.5	98.3	62.5	89.9	95.4	86.6
IMRAM [19]	76.7	95.6	98.5	61.7	89.1	95	86.1
MMCA [20]	74.8	95.6	97.7	61.6	89.8	95.2	86.8
SMFEA [21]	75.1	95.4	98.3	62.5	90.1	96.2	86.3
SHAN [22]	76.8	96.3	98.7	62.6	89.6	95.8	86.6
DRSA [23]	78.4	96.2	98.9	**63.5**	90.1	96	87.2
GLFN [24]	78.4	96	98.5	62.6	89.6	95.4	86.8
CVSE [4]	69.2	93.3	97.5	55.7	86.9	93.8	82.7
LAKE(Ours)	**78.7**	**96.7**	**99.1**	62.4	**90.5**	**96.3**	**87.3**

R@5 by 0.3%, and R@10 by 1.1%. In terms of the most important metric mR (average of the sum of R@1, R@5 and R@10), our model achieves a value of 83.1%, which is the best among many state-of-the-art methods.

Experimental results from the MSCOCO dataset are presented in Table 2. We can see that our LAKE outperforms the state-of-the-art on most evaluation metrics. Compared with our baseline model CVSE, for text retrieval, it improves R@1 by 9.5%, R@5 by 3.4%, R@10 by 1.6%, and for image retrieval, it improves R@1 by 6.7%, R@5 by 3.6%, and R@10 by 2.5%. On the mR indicator, our network achieved the best result, which is 4.6% higher than the CVSE model.

4.5 Ablation Experiments

To fully verify the effectiveness of our proposed model, we analyze the impact of all newly proposed modules on the model results on the Flickr30k dataset. In Table 3, origin model represents the experimental results of the baseline model used in this paper. (1) Neglecting the LFE module on our proposed model LAKE, the performance drops significantly, which verifies the effectiveness of the LFE module in improving the accuracy of image-text matching. (2) When the model uses the Bi-GRU model to extract text features instead of the BERT model, the image retrieval effect becomes worse, which verifies that BERT has a good effect in extracting text features, thus proving the importance of this module in the model. (3) When the model uses the LFE module and the BERT model at the same time, the experimental results achieve the best results.

Table 3. Compares the experimental results of the full model with those of the model with different parts removed.

Methods	Image-to-Text			Text-to-image			mR
	R@1	R@5	R@10	R@1	R@5	R@10	
Origin model	73	90.2	93.9	55.7	82.9	89.7	80.9
w/o LFE	74.2	91.8	95.2	56	83.3	89.9	81.7
w/o BERT	75.8	93.8	97.1	55.8	83	90.1	82.6
LAKE FULL	**76.5**	**94.4**	**97.6**	**56.3**	**83.5**	**90.3**	**83.1**

In Table 4, we discuss the effect of the number of blocks $K \times K$ that the image is divided into blocks on the results. We choose $K = 4$, 8, 16, and 32 as preselected parameter values to conduct experiments without changing other parameters. The experimental results on the Flickr30k test set are shown in Table 4. Compared with other cases, when we set K to 16, the effect is the best.

Table 4. The influence of the number of image segmentation blocks on the experimental results on the Flickr30k test set.

Methods	Image-to-Text			Text-to-image			mR
	R@1	R@5	R@10	R@1	R@5	R@10	
4	74.5	91.5	95.5	56.2	83.3	90.1	81.9
8	75.8	93.5	96.8	56.3	83.4	**90.4**	82.7
16	**76.5**	**94.4**	**97.6**	**56.3**	**83.5**	90.3	**83.1**
32	76	93.8	97.2	56	83.2	90.1	82.7

5 Conclusion

This paper proposes a LAKE model to solve the problem of poor utilization of consensus knowledge in image-text matching. On the one hand, we propose a location feature extraction module to extract the location features of objects in an image and fuse the positional information into the image feature representation, improving the utilization efficiency of consensus knowledge in the image. On the other hand, we use the BERT model to extract text features and generate a deep bidirectional text representation that can fuse left and right contextual information, improving the effect of consensus knowledge on text. Finally, experimental results on the Flickr30k and MSCOCO datasets demonstrate the effectiveness of our proposed method. In future work, we will continue to explore how to build a more effective knowledge graph and establish a deeper relationship between knowledge concepts to improve the accuracy of image-text matching.

Ackonwlegement. This work was supported in part by the National Natural Science Foundation of China under Grant62176084, and Grant62176083, and in part by the Fundamental Research Funds for the Central Universities of China under Grant PA2022GDSK0068 and PA2022GDSK0066.

References

1. Liu, C., Mao, Z., Zhang, T., et al.: Graph structured network for image-text matching. In: Proceedings of the IEEE/CVF Conference on Computer Vision and Pattern Recognition, pp. 10921–10930 (2020)
2. Gu, J., Zhao, H., Lin, Z., et al.: Scene graph generation with external knowledge and image reconstruction. In: Proceedings of the IEEE/CVF Conference on Computer Vision and Pattern Recognition, pp. 1969–1978 (2019)
3. Shi, B., Ji, L., Lu, P., et al.: Knowledge aware semantic concept expansion for image-text matching. In: Proceedings of the International Joint Conference on Artificial Intelligence (2019)
4. Wang, H., Zhang, Y., Ji, Z., Pang, Y., Ma, L.: Consensus-aware visual-semantic embedding for image-text matching. In: Vedaldi, A., Bischof, H., Brox, T., Frahm, J.-M. (eds.) ECCV 2020. LNCS, vol. 12369, pp. 18–34. Springer, Cham (2020). https://doi.org/10.1007/978-3-030-58586-0_2

5. Zhang, L., Li, M., Yan, K., et al.: Hierarchical knowledge-based graph embedding model for image-text matching in IoTs. IEEE Internet of Things J. **9**(12), 9399–9409 (2021)
6. Faghri, F., Fleet, D.J., Kiros, J.R., Fidler, S.: VSE++: improving visual-semantic embeddings with hard negatives. arXiv preprint arXiv:1707.05612 (2017)
7. Lee, K.-H., Chen, X., Hua, G., Hu, H., He, X.: Stacked cross attention for image-text matching. In: Ferrari, V., Hebert, M., Sminchisescu, C., Weiss, Y. (eds.) ECCV 2018. LNCS, vol. 11208, pp. 212–228. Springer, Cham (2018). https://doi.org/10.1007/978-3-030-01225-0_13
8. Ren, S., He, K., Girshick, R., et al.: Faster R-CNN: towards real-time object detection with region proposal networks. In: Advances in Neural Information Processing Systems, pp. 91–99 (2015)
9. Anderson, P., He, X., Buehler, C., et al.: Bottom-up and top-down attention for image captioning and visual question answering. In: Proceedings of the IEEE Conference on Computer Vision and Pattern Recognition, pp. 6077–6086 (2018)
10. Vaswani, A., Shazeer, N., Parmar, N., et al.: Attention is all you need. In: Advances in Neural Information Processing Systems, p. 30 (2017)
11. Devlin, J., Chang, K., Lee, K., et al.: BERT: pre-training of deep bidirectional transformers for language understanding. arXiv preprint arXiv:1810.04805 (2018)
12. Huang, Y., Wu, Q., Song, C., et al.: Learning semantic concepts and order for image and sentence matching. In: Proceedings of the IEEE Conference on Computer Vision and Pattern Recognition, pp. 6163–6171 (2018)
13. Pennington, J., Socher, R., Manning, C.D.: GloVe: global vectors for word representation. In: Proceedings of the 2014 Conference on Empirical Methods in Natural Language Processing (EMNLP), pp. 1532–1543 (2014)
14. Kipf, T.N., Welling, M.: Semi-supervised classification with graph convolutional networks. arXiv preprint arXiv:1609.02907 (2016)
15. Plummer, B., Wang, L., Cervantes, C., et al.: Flickr30k entities: Collecting region-to-phrase correspondences for richer image-to-sentence models. In: Proceedings of the IEEE International Conference on Computer Vision, pp. 2641–2649 (2015)
16. Lin, T.-Y., et al.: Microsoft COCO: common objects in context. In: Fleet, D., Pajdla, T., Schiele, B., Tuytelaars, T. (eds.) ECCV 2014. LNCS, vol. 8693, pp. 740–755. Springer, Cham (2014). https://doi.org/10.1007/978-3-319-10602-1_48
17. Li, K., Zhang, Y., Li, K., et al.: Visual semantic reasoning for image-text matching. In: Proceedings of the IEEE/CVF International conference on computer vision, pp. 4654–4662 (2019)
18. Wang, Y., Yang, H., Bai, X., et al.: PFAN++: bi-directional image-text retrieval with position focused attention network. IEEE Trans. Multimedia **23**, 3362–3376 (2020)
19. Chen, H., Ding, G., Liu, X., et al.: IMRAM: iterative matching with recurrent attention memory for cross-modal image-text retrieval. In: Proceedings of the IEEE/CVF Conference on Computer Vision and Pattern Recognition, pp. 12655–12663 (2020)
20. Wei, X., Zhang, T., Li, Y., et al.: Multi-modality cross attention network for image and sentence matching. In: Proceedings of the IEEE/CVF Conference on Computer Vision and Pattern Recognition, pp. 10941–10950 (2020)
21. Ge, X., Chen, F., Jose, J.M., et al.: Structured multi-modal feature embedding and alignment for image-sentence retrieval. In: Proceedings of the 29th ACM International Conference on Multimedia, pp. 5185–5193 (2021)
22. Ji, Z., Chen, K., Wang, H.: Step-wise hierarchical alignment network for image-text matching. arXiv preprint arXiv:2106.06509 (2021)

23. Qi, S., Yang, L., Li, C., et al.: Dual relation-aware synergistic attention network for image-text matching. In: 2022 11th International Conference on Communications, Circuits and Systems (ICCCAS), pp. 251–256 (2022)
24. Zhao, G., Zhang, C., Shang, H., et al.: Generative label fused network for image-text matching. Knowl.-Based Syst. **263**, 110280 (2023)

Pedestrian Attribute Recognition Based on Multimodal Transformer

Dan Liu[1], Wei Song[1,2,3(✉)], and Xiaobing Zhao[1,3]

[1] School of Information Engineering, Minzu University of China, Beijing 100081, China

[2] Key Laboratory of Ethnic Language Intelligent Analysis and Security Governance of MOE, Minzu University of China, Beijing 100081, China

[3] National Lauguage Resource Monitoring and Research Center of Minority Languages, Minzu University of China, Beijing 100081, China

songwei@muc.edu.cn

Abstract. Pedestrian attribute recognition (PAR) is susceptible to variable shooting angles, lighting, and occlusions. Improving recognition accuracy to suit its application in various complex scenarios is one of the most important tasks. In this paper, based on the Image-Text Multimodal Transformer, the intra-modal and inter-modal correlations are learned from pedestrian images and attribute labels. The applicability of six different multimodal fusion frameworks for attribute recognition is explored. The impact of different frameworks' fused feature division methods on recognition accuracy is compared and analyzed. The comparative experiments verify the robustness and efficiency of the Early Concatenate framework, which has achieved multiple best metric scores on the two major public PAR datasets, PA100k and RAP. This paper not only proposes a new Transformer-based high-accuracy multimodal network, but also provides feasible ideas and directions for further research on PAR. The comparative discussion based on various multimodal frameworks also provides a perspective that can be learned for other multimodal tasks.

Keywords: Pedestrian Attribute Recognition · Multimodal Learning · Transformer

1 Introduction

PAR aims to automatically and accurately identify and classify various attributes of pedestrians through computer algorithms and techniques. As an important part of intelligent video surveillance technology, it has a wide range of applications in many life scenarios and can also be used for pedestrian tracking, intelligent transportation systems, human-computer interaction, virtual reality and

---

Supported by the Graduate Research and Practice Projects of Minzu University of China (SJCX2022038).

© The Author(s), under exclusive license to Springer Nature Singapore Pte Ltd. 2024
Q. Liu et al. (Eds.): PRCV 2023, LNCS 14425, pp. 422–433, 2024.
https://doi.org/10.1007/978-981-99-8429-9_34

Fig. 1. Diagram of pedestrian images in the public PAR dataset.

gaming, etc. Significant progress has been made in PAR with the development of deep learning algorithms and the establishment of large-scale labeled datasets. However, factors such as diverse pedestrian appearance, posture changes, occlusion, and illumination changes will have a negative impact on the accuracy of recognition. Improving the accuracy and exploring more robust and effective models to deal with these challenges remains a pressing issue to be addressed.

The diagram of pedestrian images in the public dataset for PAR is shown in Fig. 1. In addition to a large amount of attribute information present in the pedestrian images, the attribute labels also contain valuable information that can be used for classification tasks, such as gender can be roughly inferred from clothing style and hair length. Multimodal Transformer refers to the application of Transformer [16] to the processing and fusion of multimodal data. It can effectively interact information from different modalities to extract a richer and more accurate cross-modal representation. Therefore, in order to improve the accuracy of PAR, this paper explores the applicability of six networks with different modal fusion architectures based on the image-text multimodal Transformer. The effect of the fused feature division method on the recognition accuracy is also discussed. The best-performing network is obtained through comparison experiments. The main contributions of the paper can be summarized as follows:

1. Novel image-text multimodal Transformer networks are proposed for PAR, and two major mainstream PAR datasets verify the efficiency and robustness of the networks.
2. This paper compares and analyzes the performance of networks based on six different modal fusion architectures, and contrasts the impact of different fused feature division methods on network accuracy, which provides new ideas and directions for further research on PAR.
3. The paper provides a reference perspective and exploration scheme for feature fusion of other multimodal tasks.

2 Related Works

DeepMAR [5] and FT-CNN [1] directly process global images, which have the advantages of simplicity, intuitiveness and efficiency. But they lack the consider-

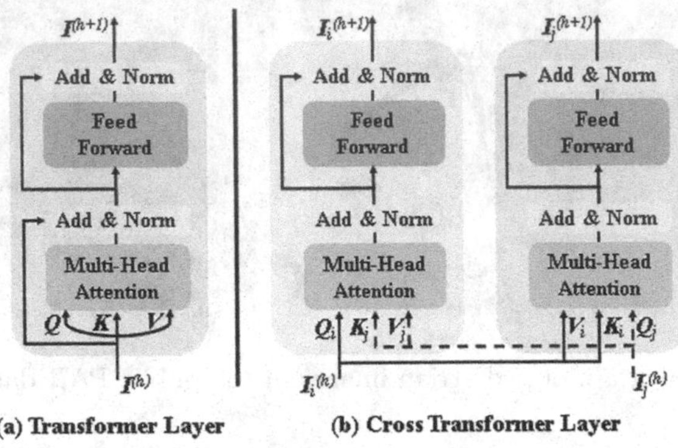

Fig. 2. Comparison of the structure of Transformer Layer and Cross Transformer Layer.

ation for fine-grained recognition. PANDA [20] uses global and fine-grained local features jointly for recognition by localizing body parts. But the final recognition results rely heavily on the accuracy of local localization and some algorithms require manual labeling of local location labels. CNN-RNN [17] proposes a framework for jointly learning image and label embeddings to describe semantic label dependencies and image-label correlations, which provides new ideas for subsequent research. LSTM-based sequence prediction models [18] have been proposed, but the time efficiency still needs to be improved due to continuous estimation. Graph model-based algorithms [8] use a graph inference framework to jointly model the spatial and semantic relationships. It's more complex and the computational efficiency is difficult to meet the required speed in practical application. Models based on visual attention [11] have been proposed, and the attention mechanism can make network focus on the parts that are more conducive to recognition. DRFormer [15] uses ViT [4] to extract semantic information from images and locate attribute-related regions to facilitate semantic and spatial modeling. VTB [2] is the first to model the recognition task as a multimodal problem. Label2label [9] proposed an image-conditioned masked language model from the perspective of language modeling.

3 Proposed Models

3.1 Cross Transformer

Transformer models sequence data by stacking multiple encoder and decoder units. The encoder consists of multiple structurally identical multi-head self-attention layers and feedforward neural networks (FFN). And the decoder consists of multi-head self-attention layers, masked multi-head self-attention layers, FFN and an output layer. In the self-attention mechanism, the linear transformation parameters of the three vectors query (Q), key (K), and value (V)

are all learnable and their dimensions are not necessarily the same. Performing different attention calculations in different dimensions allows it to process information from different modalities simultaneously and fuse information from different modalities together smoothly, which provides the basis for Transformer to excel in many multimodal learning (MML) tasks. In order to enable the fusion of information between modalities, the self-attention layer in Cross Transformer is replaced by cross-attention layer. As shown in Fig. 2, Cross Transformer receives Q from modality M_i and K, V from modality M_j for attention computation to obtain contextual global information between different modalities. It builds bridges between data from different domains, and facilitates the interaction and migration of information leading to improved model performance and generalization.

3.2 Multi-modal Transformer

In a specific MML task, for input tokens I_i and I_j from two arbitrarily different modalities, there are six mainstream Transformer-based modal fusion architectures [19] as shown in Fig. 3. Early Summation adds up the information from different modalities token by token before feeding the multimodal fusion feature into Transformer. The formula is shown in Eq. 1, where I_{ij} is the input of Transformer, O_{ij} is the output of Transformer, α and β are the weights of the two modalities, and $Trans_{ij}$ denotes the used single-stream Transformer. Q_{ij}, K_{ij} and V_{ij} correspond to the Q, K and V of $Trans_{ij}$. This fusion method is simple and intuitive without adding additional computational complexity but requires multiple tuning to obtain the most appropriate weights.

$$\begin{aligned} I_{ij} &= \alpha I_i + \beta I_j \\ O_{ij} &= Trans_{ij}(Q_{ij}, K_{ij}, V_{ij}) \end{aligned} \tag{1}$$

Early Concatenation concatenates features from two modalities directly and sends them to Transformer for feature fusion. This allows the network to encode multimodal information contexts well and learn global intra-modal and inter-modal information directly. Its computational formula is shown in Eq. 2. This method makes the input sequence I_{ij} longer, which leads to an increment in the computational complexity of the network and makes the training and inference time increase.

$$\begin{aligned} I_{ij} &= Concat(I_i, I_j) \\ O_{ij} &= Trans_{ij}(Q_{ij}, K_{ij}, V_{ij}) \end{aligned} \tag{2}$$

Multi-stream to One-stream first uses the dual-stream Transformer to extract the features of two modalities and then concatenates the features into Transformer for cross-modal interaction. The formula is shown in Eq. 3, where $DualTrans_{i,j}$ denotes the two independent dual-stream Transformers. In contrast, One-stream to Multi-stream concatenates the tokens of two modalities first and sends them to Transformer for extracting inter-modal features. The fused outputs are then split to calculate the intra-modal information separately, as shown in Eq. 4. These two frameworks maintain the independence of single-modal representation while

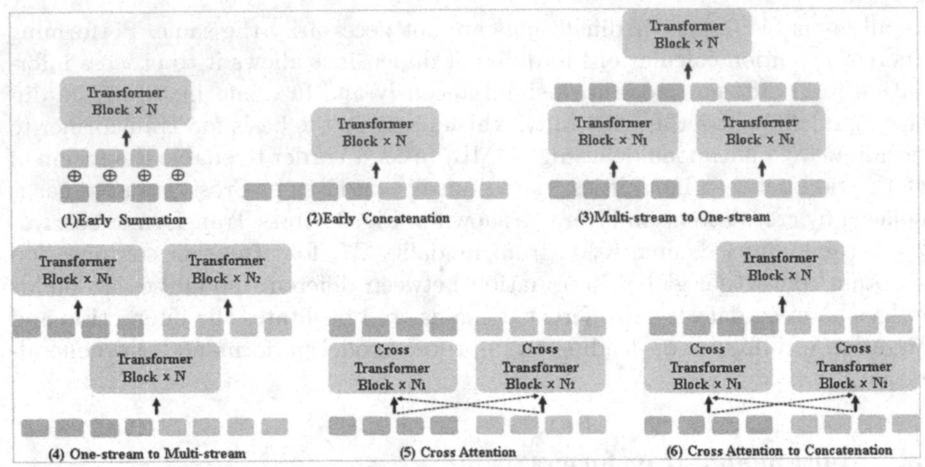

Fig. 3. Six mainstream Transformer-based multimodal fusion architectures. The different colored squares represent the information of different modalities, and N, N_1 and N_2 indicate the number of stacked blocks respectively.

achieving cross-modal interaction, but the stacking of multiple Transformer layers not only increases the number of parameters significantly but also makes the training and tuning more difficult.

$$\begin{aligned}\hat{I}_i &= DualTrans_i(Q_i, K_i, V_i)\\ \hat{I}_j &= DualTrans_j(Q_j, K_j, V_j)\\ I_{ij} &= Concat(\hat{I}_i, \hat{I}_j)\\ O_{ij} &= Trans_{ij}(Q_{ij}, K_{ij}, V_{ij})\end{aligned} \tag{3}$$

$$\begin{aligned}I_{ij} &= Concat(I_i, I_j)\\ \hat{I}_i, \hat{I}_j &= Trans_{ij}(Q_{ij}, K_{ij}, V_{ij})\\ O_i &= DualTrans_i(\hat{Q}_i, \hat{K}_i, \hat{V}_i)\\ O_j &= DualTrans_j(\hat{Q}_j, \hat{K}_j, \hat{V}_j)\end{aligned} \tag{4}$$

Cross Attention implements cross-modal perception using only a dual-stream Transformer, as shown in the formula Eq. 5, where $CrossTrans_{i,j}$ denotes the Cross Transformer. The architecture are based on Cross Transformer. It does not cause higher computational complexity, but lacks the correlation feature operations within each independent modality. Cross Attention to Concatenation concatenates the output of dual-stream Transformer and feeds it into the Transformer Layer again for feature fusion. It compensates the shortcomings of the Cross Attention to some extent, and enables the interaction of contextual information within each modality, which is shown in the formula Eq. 6.

$$\begin{aligned}O_i &= CrossTrans_i(\hat{Q}_i, \hat{K}_j, \hat{V}_j)\\ O_j &= CrossTrans_j(\hat{Q}_j, \hat{K}_i, \hat{V}_i)\end{aligned} \tag{5}$$

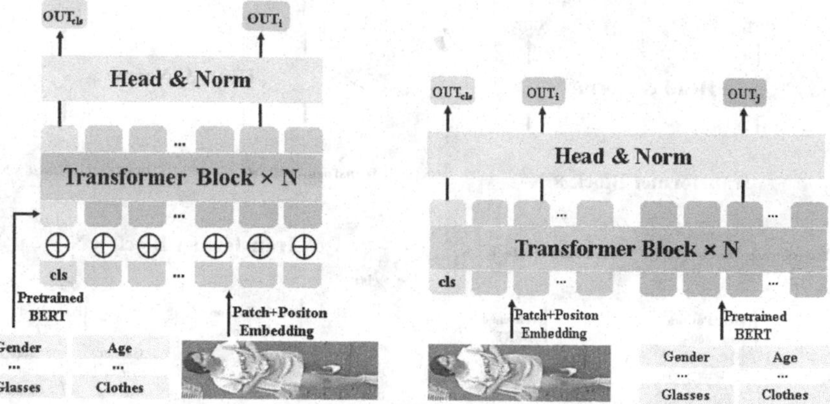

Fig. 4. ES and EC's network diagrams, N is 12 in the experiments and the yellow squares indicate cls tokens. (Color figure online)

$$\begin{aligned}\hat{I}_i &= CrossTrans_i(\hat{Q}_i, \hat{K}_j, \hat{V}_j)\\ \hat{I}_j &= CrossTrans_j(\hat{Q}_j, \hat{K}_i, \hat{V}_i)\\ I_{ij} &= Concat(\hat{I}_i, \hat{I}_j)\\ O_{ij} &= Trans_{ij}(Q_{ij}, K_{ij}, V_{ij})\end{aligned} \tag{6}$$

3.3 Multi-modal Transformer Models

In order to investigate the applicability of the above modal fusion frameworks to PAR and obtain more accurate networks, this paper builds networks based on these fusion methods for comparison experiments and comparative analysis. To keep the variables consistent, the image tokens in all networks are acquired in the same way as ViT. VilBERT [12] mentions that text should have more processing before interacting with visual features. Therefore, attribute label embeddings of all networks are encoded by pre-trained BERT [3].

The network diagrams of ES and EC that based on Early Summation and Early Concatenation framework are shown in Fig. 4. Since the image token and text embedding dimensions are not consistent, the text embedding should be zero-padding or image token must be compressed. Compressing the image token can significantly reduce the scale of the input and speed up the training speed, but many image features will be lost. The zero-padding operation does not reduce the amount of computation but retains all effective information. The networks imitates ViT for introducing cls token to image token. The fused features and cls token are classified at the classification head to obtain OUT_i and OUT_{cls}. The final output is the mean value of the two vectors. EC's image patches and label embeddings are directly concatenated and sent into the Transformer encoder. The fused output features are split into three parts: cls token, fused image features and fused text features, which are sent to the classification head to obtain OUT_{cls}, OUT_i and OUT_j, respectively. The output is their mean value.

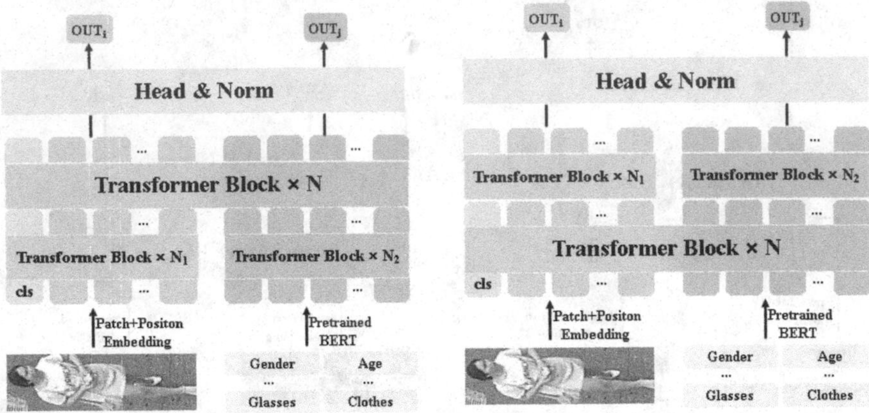

Fig. 5. MtO and OtM's network diagrams, N, N_1 and N_2 are 12 in the experiments and the yellow squares indicate cls tokens. (Color figure online)

In order to obtain higher recognition accuracy, comparative experiments of other division methods are listed in the next section.

The structure diagrams of the networks MtO and OtM based on Multi-stream to One-stream and One-stream to Multi-stream architectures are shown in Fig. 5. Their network compositions are basically the same, but the execution order of dual-stream Transformer and single-stream Transformer is different. Similarly, the fused features output of these two networks consists of two parts, which is the fused image output OUT_i and the fused text output OUT_j.

The structure diagrams of the networks CA and $CAtC$ based on Cross-Attention and Cross-Attention to Concatenation architectures are shown in Fig. 6. These two networks introduce Cross Transformer to realize the fusion of cross-modal information. The output of CA consists of three parts, while $CAtC$ does not split the fused features and directly sends the whole features to the classification head for obtaining the predicted labels.

4 Experiments

4.1 Datasets

This paper uses two mainstream general-purpose PAR datasets Richly Annotated Pedestrian (RAP) [7] and Pedestrian Attribute Dataset (PA100k) [11] to verify the effect of the network. Labels are represented in binary form, and each attribute has a label value of 1 or 0, indicating the presence or absence of that attribute.

4.2 Evaluation Metrics and Loss Function

The evaluation metrics of PAR algorithms generally contain label-based metrics and instance-based metrics. The label-based metrics includes mean accuracy

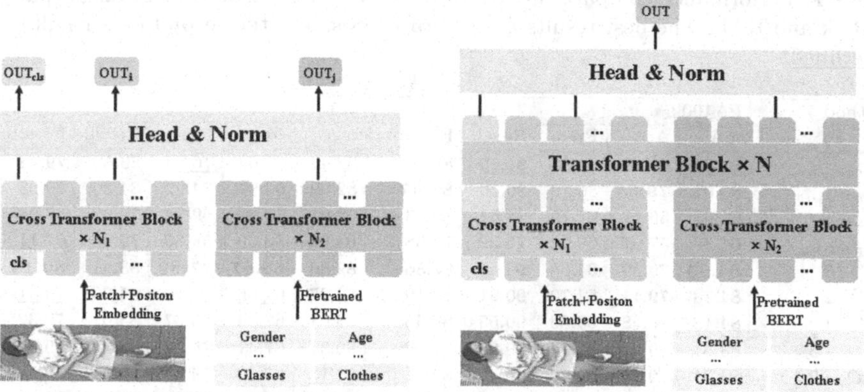

Fig. 6. CA and CAtC's network diagrams, N, N_1 and N_2 are 12 in the experiments and the yellow squares indicate cls tokens. (Color figure online)

(mA), which is the result of summing the classification accuracy of positive and negative samples and then averaging over all attributes. mA is calculated for each attribute separately, resulting in ignoring the correlation between different attributes. Instance-based metrics can better represent the consistency among attributes, which used in this paper contain Accuracy (Acc), Precision (Prec), Recall and F1 score.

The paper uses the weighted binary cross-entropy loss function to alleviate the problem of uneven distribution of pedestrian attributes in the dataset. For a given dataset $D = \{(X_i, Y_i)\}, i = 1, 2, ..., N$ with M attributes, where Y_i is the groundtruth of the pedestrian image X_i. N denotes the number of training images and the predicted labels are $\hat{Y}$. The loss function is shown in Eq. 7, where ω_j is the weight to alleviate the distribution imbalance, which is calculated by r_j based on Eq. 8, and r_j represents the positive sample ratio of jth attribute in D.

$$Loss = -\frac{1}{N}\sum_{i=1}^{N}\sum_{j=1}^{M}\omega_j(Y_{ij}log(\hat{Y_{ij}}) + (1 - Y_{ij})log(1 - \hat{Y_{ij}})) \quad (7)$$

$$\omega_j = \begin{cases} e^{1-r_j}, & Y_{ij} = 1 \\ e^{r_j}, & Y_{ij} = 0 \end{cases} \quad (8)$$

4.3 Implementation Details

The experimental device processor used in this paper is Intel® CoreTM i7-7800X CPU @3.50 GHz with a memory of 471G, and the GPU is GeForce RTX 2080Ti. All the training images are resized to 256*128. All networks use SGD optimizer with moment of 0.9 and weight decay of 0.0005. The initial learning rate is 0.001, and the networks are trained for a total of 50 epochs, where the pre-training weights are all from ViT.

Table 1. Performance comparison with different division method of fused output on PA100k and RAP. The best results are in **bold** fonts, and the second best results are <u>underlined</u>.

Method	PA100k					RAP				
	mA	Acc	Prec	Recall	F1 score	mA	Acc	Prec	Recall	F1 score
$ES_{padding}-2$	<u>84.39</u>	79.01	84.33	<u>91.15</u>	87.13	82.42	67.63	<u>76.64</u>	83.59	79.48
$ES_{padding}-1$	83.02	**79.68**	<u>86.12</u>	89.76	**87.47**	82.99	66.97	74.93	84.56	79.03
$ES_{pooling}-2$	64.63	56.77	67.79	74.64	70.33	69.01	52.00	59.14	79.23	67.08
$ES_{pooling}-1$	64.39	57.07	68.02	75.22	70.68	68.91	52.00	58.89	79.68	67.11
$EC-3$	**84.65**	79.39	84.68	**91.20**	87.36	82.30	**68.67**	**77.89**	83.41	**80.18**
$EC-2$	84.13	79.11	84.70	90.61	87.10	82.37	<u>67.66</u>	76.20	84.00	<u>79.51</u>
$EC-1$	84.12	79.58	85.36	90.55	87.43	83.01	67.34	75.37	84.61	79.32
$MtO-3$	83.31	79.01	85.14	89.93	87.03	82.37	67.41	76.22	83.54	79.31
$MtO-2$	84.31	79.14	84.89	90.41	87.11	<u>83.13</u>	67.42	74.83	85.61	79.43
$MtO-1$	83.17	<u>79.67</u>	**86.20**	89.58	<u>87.44</u>	**83.23**	67.10	73.81	86.58	79.26
$OtM-3$	83.79	78.37	84.30	90.00	86.59	83.00	66.04	72.07	**87.38**	78.52
$OtM-2$	82.59	79.15	86.09	88.85	87.05	82.90	66.88	73.45	<u>86.83</u>	79.13
$CA-3$	73.06	64.04	71.97	83.01	76.41	76.22	59.49	65.08	86.29	73.68
$CA-2$	70.53	63.36	73.17	80.33	75.86	75.93	58.90	64.75	85.59	73.18
$CAtC-3$	74.16	65.44	72.50	84.40	77.35	64.62	49.08	55.81	78.44	64.57
$CAtC-2$	72.26	66.07	75.74	81.05	77.69	78.28	61.23	67.35	85.63	74.87
$CAtC-1$	72.78	66.15	74.78	82.44	77.83	78.83	61.45	67.20	86.48	75.09

4.4 Classification Results

To explore the effect of the division method on recognition accuracy when the above six networks' fused features are fed into the classification head, multiple sets of comparison experiments were performed, and the results are shown in Table 1. The number after each network name represents the number of parts into which the fused features are divided. To ensure that the feature dimensions of the two modalities are consistent for patch-by-patch summation, $ES_{padding}$ and $ES_{pooling}$ denote the padding operation for text features and the pooling operation for image features, respectively. Since $ES_{pooling}$ results in the loss of a large amount of effective visual information, both division methods of $ES_{pooling}$ perform much less well than $ES_{padding}$ on the two datasets. $ES_{padding}-2$ achieves the second best score for two metrics on PA100k, and one for RAP. On PA100k, $ES_{padding}-1$ achieves the second best score for Prec and the best values for Acc and F1 score. The best values for both mA and Recall on PA100k were obtained by $EC-3$, which also achieved the highest values for Acc, Prec, and F1 score on RAP and had the strongest overall performance among all networks. The $EC-2$ achieved the second best values for Acc and F1 score on RAP, but was slightly inferior compared to $EC-3$. $EC-1$ did not achieve the best or second-best scores but its scores on the key metrics mA, Acc and F1 score were comparable or even better than those of $EC-2$ and $EC-3$. $MtO-2$ achieved a second best score. $MtO-1$ achieved the highest score for Prec and the second best score for Acc and F1 score on PA100k. It also obtains the best value for mA on RAP. Its performance is slightly better than $MtO-3$ and $MtO-2$, but is slightly inferior to that of the EC. The Recall of $OtM-3$ and $OtM-2$ had the best and second best scores on RAP. $OtM-2$ performs slightly better than $OtM-3$, but neither is as good as the MtO. There is a large gap between CA and the former four types of networks. The performance of $CAtC-3$ on PA100k is similar to that of $CAtC-2$ and $CAtC-1$, but its performance on RAP is far from both of them. It is possible that the network structure is complex and more output divisions lead to slower network convergence.

Experiments have verified the effectiveness of ES. Its outstanding performance benefits from the contextual interaction of the two modalities' information from an early stage and the global inter-modal and intra-modal correlations are learned. The performance of the $ES_{padding}$ and MtO is slightly inferior to that of the EC but also efficient for PAR. The simplicity and performance of the $ES_{padding}$ make it more advantageous in practical applications and under the condition of the limited experimental equipment. MtO models independent and fused modalities information successively, which can maintain the independence of modalities, but the feature extraction of independent modalities may conversely be detrimental to the contextual modeling of subsequent modal fusion. Similarly, the features fused first in the OtM may have an impact on the subsequent independent modal operations. The overall performance of CA and $CAtC$ is less satisfactory compared to other networks, probably because the pre-training weights used do not match well with Cross Transformer. Transformer is difficult to train, and training them directly for downstream tasks can easily trap the network in local optima, resulting in unsatisfactory final results. Overall, the division method does have a slight impact on the accuracy. Moreover, the division with the highest accuracy varies for different networks. In the specific implementation, it is still necessary to conduct comparative experiments to determine the division scheme.

Table 2 shows the comparison of the proposed networks with other PAR networks. On PA100k, $ES_{padding}$ achieved the second best scores on mA and Recall. EC had the best overall performance. Its mA and Recall obtained the highest values with 84.65 and 91.20 on PA100k. Its Acc and F1-score also achieved the highest scores on RAP. mA, Acc and F1-score are the more important three of the five metrics, which are sufficient to prove the robustness of EC. MtO achieved two second best metrics on PA100k. Both it and OtM achieved one best and one second-best score on RAP. Both networks also performed well but the overall metrics were slightly inferior to EC. The two networks perform almost equally but MtO has a slight dominance. CA had lower scores than $CAtC$ for all five metrics on PA100k. In contrast, CA had higher metrics than $CAtC$ on RAP. The superiority of the $ES_{padding}$, EC, MtO and OtM networks can be clearly seen from the table. In particular, EC achieves better results in both datasets.

Table 2. Performance comparison with state-of-the-art method on PA100k and RAP. * represents the reimplementation performance using the same setting. The best results are in **bold** fonts, and the second best results are underlined.

Method	PA100k					RAP				
	mA	Acc	Prec	Recall	F1 score	mA	Acc	Prec	Recall	F1 score
DeepMAR [5]	72.70	70.39	82.24	80.42	81.32	73.79	62.02	74.92	76.21	75.56
HPNet [11]	74.21	72.19	82.97	82.09	82.53	76.12	65.39	77.33	78.79	78.05
VeSPA [13]	76.32	73.00	84.99	81.49	83.20	77.70	67.35	79.51	79.67	79.59
LGNet [10]	76.96	75.55	86.99	83.17	85.04	78.68	68.00	**80.36**	79.82	80.09
PGDM [6]	74.95	73.08	84.36	82.24	85.04	74.31	64.57	78.86	75.90	77.35
ALM [14]	80.68	77.08	84.24	88.84	84.46	81.87	68.17	74.71	86.48	80.16
VTB* [2]	83.11	**80.19**	80.07	89.23	**87.75**	82.36	67.81	75.88	84.78	79.69
Lable2Lable* [9]	79.10	78.32	**87.60**	85.93	86.39	79.12	67.87	80.05	79.45	79.43
$ES_{padding}$	84.39	79.01	84.33	91.15	87.13	82.42	67.63	76.64	83.59	79.48
EC	**84.65**	79.39	84.68	**91.20**	87.36	82.30	**68.67**	77.89	83.41	**80.18**
MtO	83.17	79.67	86.20	89.58	87.44	**83.23**	67.10	73.81	86.58	79.26
OtM	83.79	78.37	84.30	90.00	86.59	83.00	66.04	72.07	**87.38**	78.52
CA	73.06	64.04	71.97	83.01	76.41	76.22	59.49	65.08	86.29	73.68
$CAtC$	72.78	66.15	74.78	82.44	77.83	78.83	61.45	67.20	86.48	75.09

5 Conclusion

Complex scenes and variations in shooting angles and lighting bring challenges to PAR. To improve the accuracy, this paper not only extracts valid information from images but also further explores the correlation between attribute labels. Multimodal Transformer enables the model to better understand the relationship between different modalities in order to extract cross-modal features. To investigate the impact of Transformer-based multimodal information fusion architecture on the accuracy of PAR, six sets of networks are designed for training, testing and comparative analysis in this paper. The impact of the fused feature's division method on the experimental results is also explored in comparison, and the networks' effects are compared with the existing algorithms in the field. The experiments verify the accuracy and robustness of Early Concatenate architecture, obtain more accurate attribute recognition results, and provide new ideas and directions for the research of PAR.

References

1. Abdulnabi, A.H., Wang, G., Lu, J., Jia, K.: Multi-task CNN model for attribute prediction. IEEE Trans. Multim. **17**(11), 1949–1959 (2015)
2. Cheng, X., Jia, M., Wang, Q., Zhang, J.: A simple visual-textual baseline for pedestrian attribute recognition. IEEE Trans. Circuits Syst. Video Technol. **32**(10), 6994–7004 (2022)
3. Devlin, J., Chang, M.W., Lee, K., Toutanova, K.: Bert: pre-training of deep bidirectional transformers for language understanding. arXiv preprint arXiv:1810.04805 (2018)
4. Dosovitskiy, A., et al.: An image is worth 16x16 words: transformers for image recognition at scale. arXiv preprint arXiv:2010.11929 (2020)

5. Li, D., Chen, X., Huang, K.: Multi-attribute learning for pedestrian attribute recognition in surveillance scenarios. In: 2015 3rd IAPR Asian Conference on Pattern Recognition (ACPR), pp. 111–115. IEEE (2015)
6. Li, D., Chen, X., Zhang, Z., Huang, K.: Pose guided deep model for pedestrian attribute recognition in surveillance scenarios. In: 2018 IEEE International Conference on Multimedia and Expo (ICME), pp. 1–6. IEEE (2018)
7. Li, D., Zhang, Z., Chen, X., Ling, H., Huang, K.: A richly annotated dataset for pedestrian attribute recognition. arXiv preprint arXiv:1603.07054 (2016)
8. Li, Q., Zhao, X., He, R., Huang, K.: Visual-semantic graph reasoning for pedestrian attribute recognition. In: Proceedings of the AAAI Conference on Artificial Intelligence, vol. 33, pp. 8634–8641 (2019)
9. Li, W., Cao, Z., Feng, J., Zhou, J., Lu, J.: Label2Label: a language modeling framework for multi-attribute learning. In: Avidan, S., Brostow, G., Cissé, M., Farinella, G.M., Hassner, T. (eds.) ECCV 2022. LNCS, vol. 13672, pp. 562–579. Springer, Cham (2022). https://doi.org/10.1007/978-3-031-19775-8_33
10. Liu, P., Liu, X., Yan, J., Shao, J.: Localization guided learning for pedestrian attribute recognition. arXiv preprint arXiv:1808.09102 (2018)
11. Liu, X., et al.: Hydraplus-net: attentive deep features for pedestrian analysis. In: Proceedings of the IEEE International Conference on Computer Vision, pp. 350–359 (2017)
12. Lu, J., Batra, D., Parikh, D., Lee, S.: Vilbert: pretraining task-agnostic visiolinguistic representations for vision-and-language tasks. Adv. Neural Inf. Process. Syst. **32** (2019)
13. Sarfraz, M.S., Schumann, A., Wang, Y., Stiefelhagen, R.: Deep view-sensitive pedestrian attribute inference in an end-to-end model. arXiv preprint arXiv: 1707.06089 (2017)
14. Tang, C., Sheng, L., Zhang, Z., Hu, X.: Improving pedestrian attribute recognition with weakly-supervised multi-scale attribute-specific localization. In: Proceedings of the IEEE/CVF International Conference on Computer Vision, pp. 4997–5006 (2019)
15. Tang, Z., Huang, J.: Drformer: learning dual relations using transformer for pedestrian attribute recognition. Neurocomputing **497**, 159–169 (2022)
16. Vaswani, A., et al.: Attention is all you need. Adv. Neural Inf. Process. Syst. **30** (2017)
17. Wang, J., Yang, Y., Mao, J., Huang, Z., Huang, C., Xu, W.: Cnn-rnn: a unified framework for multi-label image classification. In: Proceedings of the IEEE Conference on Computer Vision and Pattern Recognition, pp. 2285–2294 (2016)
18. Wang, J., Zhu, X., Gong, S., Li, W.: Attribute recognition by joint recurrent learning of context and correlation. In: Proceedings of the IEEE International Conference on Computer Vision, pp. 531–540 (2017)
19. Xu, P., Zhu, X., Clifton, D.A.: Multimodal learning with transformers: a survey. IEEE Trans. Pattern Anal. Mach. Intell. (2023)
20. Zhang, N., Paluri, M., Ranzato, M., Darrell, T., Bourdev, L.: Panda: pose aligned networks for deep attribute modeling. In: Proceedings of the IEEE Conference on Computer Vision and Pattern Recognition, pp. 1637–1644 (2014)

RGB-D Road Segmentation Based on Geometric Prior Information

Xinyi Wu, Xia Yuan(✉), YanChao Cui, and ChunXia Zhao

School of Computer Science and Engineering, Nanjing University of Science and Technology, 210094 Nanjing, China
yuanxia@njust.edu.cn

Abstract. Deep data can provide rich spatial structure information, which can effectively exclude the interference of lighting and road texture in road segmentation. This paper proposes a road segmentation model based on two kinds of a priori knowledge: disparity information, and surface normal vector information. Then, a two-branch neural network is used to process the color image and the processed depth image separately, and an effective fusion module is designed to make full use of the complementary features of the two modalities. The experimental results on the KITTI road detection dataset and Cityscape dataset show that the method in this paper has good road segmentation performance.

Keywords: Disparity · surface normal vectors · fusion · road segmentation

1 Introduction

With the development of depth sensors, RGB-D data began to be used in various vision computing tasks, such as image understanding [15], semantic segmentation [10], and action recognition [10]. Deep image can provide geometric information such as depth-of-field, shape, and boundary of a scene, which is a good complement to RGB images, thus improving the scene understanding capability of visual computing models.

Among them, road segmentation is a common perceptual approach in the field of road scene understanding. The problem of road segmentation can usually be viewed as a dichotomous problem, i.e., the distinction between road areas and non-road areas is accomplished using existing road environment data.

Traditional road area segmentation algorithms are generally based on various manually extracted features and use these features to train various classifiers to achieve road area segmentation. These feature extraction steps are very subjective and usually require a lot of experiments and labor costs to determine the selection and combination of features, and for different scenarios, the performance of various types of features also has a very large difference. Thus, stereo vision-based road segmentation was proposed. The "v-disparity" was proposed by [7] et al. to achieve the extraction of road regions from dense disparity maps

© The Author(s), under exclusive license to Springer Nature Singapore Pte Ltd. 2024
Q. Liu et al. (Eds.): PRCV 2023, LNCS 14425, pp. 434–445, 2024.
https://doi.org/10.1007/978-981-99-8429-9_35

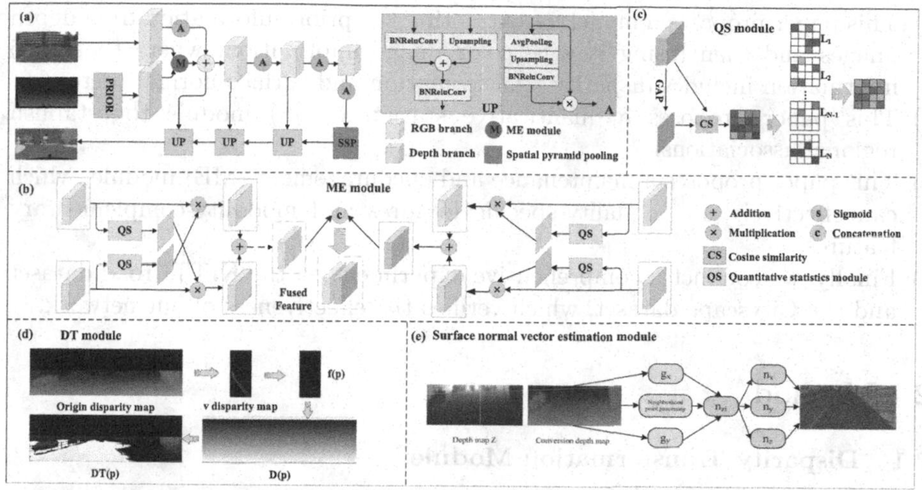

Fig. 1. (a) Overall network structure. (b) Maintenance and Encouragement Module. (c) Quantitative statistics module. (d)Disparity Transformation. (e)Surface normal vector estimation module.

[14]. Nanri et al. [8] segmented road areas by calculating the height difference of the road surface on the scan line in the dense disparity map based on the nature of the road surface height variation in the disparity map. In recent years, road segmentation based on deep learning is also basically based on the above two features. For example, SEN-RoadSeg [4] achieves high-accuracy segmentation by solving the surface normal vector.

The above network has achieved good experimental results, but there are still some problems. On the one hand, the semantic network lacks a design for improving the accuracy of road region segmentation. On the other hand, the network is not computationally efficient after highlighting the extracted road area depth data features.

Therefore, this paper explores the depth image features in terms of disparity transformation, and surface normal vectors, respectively. Then, we propose a network to strengthen features from two aspects: modality-specific features maintenance and modality-complement-ary features encouragement. Firstly, this paper performs quantitative statistics on the information of different modalities to achieve a unified representation of features. Subsequently, this paper designs the ME module to effectively integrate the complementary information of the two modes, while maintaining modality-specific features. In this module, the processed original features and the generated statistical features are used to strengthen the original feature, and the generated statistical features are used to guide another mode to complete feature interaction. The experimental results show that the road area detection accuracy has been significantly improved.

The main contributions of this work are described as follows:

(1) This paper proposes a model for extracting the priori information from depth images and then fusing it with color image information, where the priori information includes disparity transformation and surface normal vector.
(2) This paper proposes a quantitative statistics (QS) module to establish regional associations.
(3) This paper proposes a maintenance and encouragement (ME) module, which can effectively use modality-specific features and modality-complementary features.
(4) Finally, we conduct a comprehensive experiment on the KITTI road dataset and the Cityscape dataset, which verifies the effectiveness of our network.

2 Method

2.1 Disparity Transformation Module

In the disparity map, the disparity values of road areas show an approximately uniform decreasing trend from near to far because of the flatness of urban roads. Based on this prior information, this paper proposes a disparity transformation model to better utilize the geometric information in the disparity map to reduce the difference between road areas and increase the difference between road and non-road areas, to better highlight the characteristics of road areas. The disparity transformation flow designed in this paper is shown in Fig. 1(d).

As can be seen from the original disparity map, the upper half of the image is the sky area because the selected dataset is the city scene. The depth value of the sky region is infinite, which is close to no disparity value. This makes the upper half of the image contain almost no useful information, which greatly facilitates the efficiency of our use of disparity information.

According to the disparity principle, the rows in the v-disparity map correspond to the rows of the original disparity map. The value of a column in a row in a v-disparity map is the number of columns in that row whose parity value is the serial number of that column. Thus, the Structured road area is mapped as an inclined straight line in the disparity map v, and obstacles that are approximately perpendicular to the road are mapped as vertical straight lines. Therefore, the projection of the road area on the disparity map v can be represented by a linear model as follows.

$$f(p) = a_0 + a_1 v \tag{1}$$

where $a = [a_0, a_1]^T$ is the coefficient of linear model, $p = [u, v]^T$ is any one pixel in the disparity map, $f(p)$ indicates disparity value of p, v is the vertical coordinate corresponding to the pixel.

Then, as shown in Eq. (2), we convert the difference between the original disparity map $D(p)$ and $f(p)$ into a new disparity map $DT(p)$, which can better distinguish the road area from the non-road area and reduce the difference of the road area.

$$DT(p) = D(p) - f(p) \tag{2}$$

In this paper, two methods are used to estimate the coefficients of the linear model in the equation: LSD and OLS. The Line Segment Detector (LSD) starts from the conversion process. The linear model of the road area on the disparity map is first calculated, and then the difference is made with the original disparity map to obtain the conversion result. The Ordinary Least Square(OLS) starts from the conversion result. First, assuming that the road area in the original image is the main part of the image, then the average difference value of the converted disparity map is smaller when the model is similar to the actual road area. So the solution of $f(p)$ is converted to solve the minimum value of $DT(p)$. The calculation first crops the road scene image so that the main body of the computed region is the road, and the details are not eliminated for the region that is not the main body of the road, so it does not affect the conversion performance.

2.2 Surface Normal Vector Estimation Module

Traditional methods for estimating surface normal vectors can be achieved by inverse depth image fetching or difference image filtering, which includes gradient filtering and mean filtering in both horizontal and vertical directions. However, in practical applications, the depth data often have some voids or inaccurate depth values. For this reason, in this paper, on the basis of the original surface normal vector solution, the neighborhood points are further optimized by thresholding and fractional order differentiation, which improves the accuracy of the surface normal vector solution. The flow of the SNE model is shown in Fig. 1(e).

The neighborhood points around the point are used in estimating the surface normal vector, and the final surface normal vector of the point is obtained by averaging the normal vectors solved for the eight domains. Due to the error of the depth data itself, empty point or point with large errors is filtered by setting a threshold value for the larger ones. For pixel p's neighborhood point q, if $|p - q| < p * 0.1$, and $q \neq 0$, then the domain point is retained. And for pixels with small errors, we optimize them by fractional order differentiation.

Combining the correspondence between the image coordinate system and the spatial coordinate system as well as the property that the surface normal vector is perpendicular to the plane where the current point is located, we can solve to get the estimated expression of the surface normal vector as:

$$n_i = \left[f_x g_x, f_y g_y, \frac{f_\Delta X_i g_x + f_y \Delta Y_i g_y}{\Delta Z_i} \right]^T \tag{3}$$

Normalizing the above equation to the unit normal vector in the spherical coordinate system, it can be expressed as:

$$\varphi = arctan\left(\frac{f_x g_x}{f_y g_y}\right) \tag{4}$$

$$\theta = arctan\left(\frac{\sum_{i=1}^{k}\bar{n}_x cos\varphi + \sum_{i=1}^{k}\bar{n}_y sin\varphi}{\sum_{i=1}^{k}\bar{n}_{Zi}}\right) \tag{5}$$

According to [9], the order G-L fractional order differentiation of the signal can be defined as:

$$D_{S(t)}^{v} = \lim_{n\to 0} h^{-v}\sum_{m=0}^{\frac{t-a}{h}}(-1)^m\frac{\Gamma(-v+1)}{m!\Gamma(-v+m+1)}s(t-mh) \tag{6}$$

where, the range of $s(t)$ is $[a,b], m = (t-a)/h$,and h is the step size. $\Gamma(\cdot)$ represents the gamma function, which is defined as:

$$\Gamma(v) = \int_0^0 e^{-t}t^{v-1}dt \tag{7}$$

For images, the minimum scale is one pixel, so$h = 1$. Therefore, the difference expression for fractional differentiation can be expressed as:

$$\begin{aligned}\frac{\partial^v s(t)}{\partial t^v} \approx s(t) + (-v)s(t-1) + \frac{(-v)(-v+1)}{2!}s(t-2)\\ + \cdots + \frac{\Gamma(-v+1)}{n!\Gamma(-v+n+1)}s(t-n)\end{aligned} \tag{8}$$

where n is calculated from the function and corresponds to the nth neighborhood value. For image $f(x,y)$, the above equation is extended to two dimensions. To simplify the calculation, we use the first three terms of (10) to approximate it. The numerical realizations of the fractional partial differentiation in the x and y directions are denoted as

$$\frac{\partial^v f(x,y)}{\partial x^v} \approx f(x,y) + (-v)f(x-1,y) + \frac{(-v)(-v+1)}{2!}f(x-2,y) \tag{9}$$

$$\frac{\partial^v f(x,y)}{\partial y^v} \approx f(x,y) + (-v)f(x,y-1) + \frac{(-v)(-v+1)}{2!}f(x,y-2) \tag{10}$$

Therefore, by combining the definitions of the equation and the image and mask convolution, we can construct the masks corresponding to the fractional order differentiation of both images in the negative x and y directions.

2.3 Road Segmentation Network

The structure of the prior knowledge-based road segmentation network (Cross-Modality Feature Maintenance and Encouragement Network, or MENet) designed in this paper is shown in Fig. 1(a). The model consists of two parts, one is the data pre-processing layer, which is to transform the depth data using the prior knowledge model proposed in this paper. The second is the semantic segmentation layer, which consists of two branches: the RGB branch and the depth branch. The spatial pyramid module is used in the output layer part of the RGB branch to generate feature maps with multi-scale information. Finally, we add upsampling after each layer to recover the resolution of these feature maps by jumping over the connections from the RGB branch.

Fusion Module. To effectively fuse RGB and depth data information, we designed the fusion module in MENet, as shown in Fig. 3(b). The fusion module consists of two branches, one is multimodal feature interaction, which enhances complementary features by multiplying a weight vector with another input feature mapping, and the other is a self-modal enhancement, which strengthens each modal feature by multiplying its own weight vector with the input feature mapping. Finally, the two features are connected by Concat.

To solve the weight vector in the fusion module, we design the quantile statistics module, as shown in Fig. 1(c). Considering the different physical meanings of different pattern data, the feature vectors are redescribed using cosine similarity.

Assume that the input feature $F \in R^{C\times H\times W}$, and use a global average pooling(GAP) to get the global averaged feature $F_g \in R^{C\times 1\times 1}$. For each spatial position $F_{i,j}(i \in [1, W], j \in [1, H])$ on feature F, we calculate the cosine similarity between F_g and $F_{i,j}$ to obtain similarity matrix $S \in R^{1\times H\times W}$. Each position $S_{i,j}$ of S is denoted as:

$$S_{i,j} = \frac{F_g \cdot F_{i,j}}{\|F_g\|_2 \cdot \|F_{i,j}\|_2} \tag{11}$$

Then, we quantize the similarity matrix S into N levels according to the value of S and obtain quantitative level $L = [L_1, L_2, ..., L_N]$. The n-th level of L is calculated as:

$$L_n = \frac{\max(S) - \min(S)}{N} \cdot n + \min(S) \tag{12}$$

Finally, according to L, we reconstruct S to quantitative statistical feature E by reassigning the value of S, which is denoted as:

$$E_{i,j} = \begin{cases} 1, & \text{if } n = 1 \\ n, & \text{if } L_{n-1} < S_{i,j} \leq L_n, n > 1 \end{cases} \tag{13}$$

Attention Module. We introduce an attention mechanism in MENet so that valid features can be given higher weights. The attention module first uses global average pooling as the channel descriptor and then adds a convolutional layer with the same channel as the input features. Next, the convolution result is activated using a function that constrains the value of the weight vector to be between 0 and 1. Finally, the weight vector and the input feature mapping are externally multiplied to obtain the attentional features.

Table 1. Results of DTNet on KITTI dataset and Cityscape dataset.

Dataset	Method	PRE (%)	REC (%)	MaxF (%)	IOU (%)	FNR (%)	FPR (%)
KITTI	RGB	96.40	96.03	96.22	92.71	3.67	0.78
	MENet	96.76	96.52	96.52	93.28	3.71	0.70
	DTNet18 +LSD	**98.20**	95.68	96.92	94.03	4.32	**0.38**
	DTNet50 +LSD	**97.79**	96.29	97.03	94.24	3.71	0.47
	DTNet101 +LSD	97.01	**97.17**	**97.09**	**94.35**	**2.83**	0.65
	DTNet18+OLS	97.17	96.84	97.00	94.18	3.16	0.61
	DTNet50+OLS	97.31	96.85	97.08	94.32	3.15	0.58
	DTNet101+OLS	97.63	**97.06**	**97.34**	**94.82**	**2.95**	**0.52**
Cityscapes	RGB	99.07	98.60	98.83	97.69	1.40	0.51
	MENet	**99.21**	98.55	98.88	97.79	1.45	**0.43**
	DTNet 18+OLS	99.11	**98.81**	**98.96**	**97.94**	**1.19**	0.48

3 Experiment

3.1 Experimental Setup

Dataset. In this paper, the proposed algorithm is experimented on two datasets, KITTI and Cityscape. 289 training frames and 290 test frames are included in the KITTI dataset. The Cityscape dataset contains street scenes from 50 different cities. For the experiments, we redefined the category labels of the dataset to retain two categories, road areas, and non-road areas.

Evaluation Indicators. The road segmentation in this paper uses the Intersection over Union (IoU), maximum F-Measure (MaxF), recall(REC), precision(PRE), false positive rate (FPR), and false negative rate (FNR) as evaluation metrics.

3.2 Ablation Study

Ablation of DT Modules: To verify the validity of the prior module, we did a quantitative analysis on the KITTI dataset and the Cityscapes dataset, as shown in Table 1. On the KITTI dataset, we used resnet-18 as the baseline, and the IoU was 92.71% without the use of depth information. With the depth information (i.e. the baseline MENet presented above), the IoU is 93.28%. To this, we added the DT module for disparity transformation using LSD (Line Segment Detector) and OLS (Ordinary Least Square). As can be seen from the table, with the addition of the DT module using LSD, the IoU reached 94.03%, an improvement of 0.75%. With the DT module using OLS, the IoU reached 94.18%, with a 0.75% improvement. Also, we experimented with deeper networks (resnet50 and resnet101) (corresponding to DTNet50 and DTNet101 respectively) and there was a further increase in IoU. There was also a significant improvement in both the false alarm rate (FPR) and the missed alarm rate (FNR), with the FPR essentially shrinking to half its original size.

Fig. 2. Effect of surface normal vector.

On the Cityscapes dataset, again using resnet-18 as a baseline, the IoU was 97.69% without the use of depth information. With depth information, the IoU is 97.79%. After adding the DT module using the Ordinary Least Square (DTNet18+OLS), the IoU reached 97.85%. We can thus see that the segmentation network and the priori module proposed in this paper are fruitful. However, the accuracy improvement is not very significant on the Cityscapes dataset compared to the KITTI dataset. The analysis found that the road segmentation accuracy on the Cityscapes dataset was generally much higher than that on the KITTI dataset (above 98%). The reason for this is that road area segmentation is a binary classification and the segmentation task is relatively simple, while the Cityscapes dataset is large and the training samples are sufficient for the basic network to train a better segmentation, so the accuracy improvement is not obvious.

In summary, the road segmentation based on the DT prior module proposed in this paper can improve the segmentation accuracy in all six metrics. Among them, after the LSD disparity conversion, the pixel accuracy is greatly improved and the false positive rate (FPR) is reduced to half of the original, indicating that non-road areas are well differentiated from road areas. The OLS method is less effective than the LSD algorithm in the analysis of the conversion effect above, but its final achieved segmentation performance is still outstanding.

Ablation of SNE Module: To illustrate the effectiveness of the surface normal vector optimization method proposed in this paper, we compare the surface normal vector segmentation methods before and after optimization, as shown in Fig. 2. In Fig. 2, from top to bottom, there is the original image, the surface normal vector map solved using traditional singular value decomposition (SVD), the surface normal vector map solved by SNE, and the surface normal vector map after optimization. From the figure, we can see that there is some noise on the surface normal vector segmentation image before optimization. The road

Table 2. Results of SNENet on KITTI dataset.

Method	PRE (%)	REC (%)	MaxF (%)	IOU (%)	FNR (%)	FPR (%)
RGB	96.40	96.03	96.22	92.71	3.67	0.78
MENet	96.76	96.52	96.52	93.28	3.71	0.70
SNENet	97.13	**97.11**	97.12	94.40	**2.89**	0.62
SNENet+FD(2)	97.51	96.95	**97.23**	**94.61**	3.04	0.54
SNENet+FD(3)	**97.75**	96.81	97.17	94.51	3.19	**0.53**

region, cars, and buses appear to have uneven colors. With neighborhood point optimization, the noise on the surface normal vector map is greatly reduced and the road regions are converged.

We did a quantitative analysis on the KITTI dataset to verify that the surface normal vector can provide valid prior knowledge of the road area. Since the Cityscapes dataset lacks camera parameters to invert the corresponding 3D spatial points based on the depth map, only the KITTI dataset was used for this section, and the results are shown in Table 2 (FD in the table indicates the use of the domain point processing method, and the numbers in brackets indicate the use of several orders of fractional order differentiation). On the KITTI dataset, the IoU is 92.71% when no depth information is used. With the depth information (i.e. the baseline MENet presented above), the IoU is 93.28%.

Then, we used surface normal vector information (SNE) instead of depth information, and the IoU was improved somewhat. For ease of representation, we refer to the network after adding the SNE module as SNENet. On top of this, we filtered the neighborhood points and further improved the accuracy to 94.61% and the mIoU by 1.33% when using the 2nd order fractional order differential convolution kernel (FD(2)) for the domain points. A maximum accuracy of 97.75% was obtained using 3rd-order fractional order differentiation (FD(3)). As seen in Table 2, the conversion of depth information into surface normal quantities improves network accuracy, and both IoU and FNR are improved by fractional order differentiation optimization.

3.3 Comparison of Road Segmentation

We compared the results of our method with several contemporary state-of-the-art methods on the KITTI dataset. The experimental results are shown in Table 3. The model algorithm in this paper is higher than the advanced network in terms of segmentation accuracy and the maximum F-value is close to the highest value. This indicates that the introduction of the road prior knowledge can effectively improve the segmentation performance.

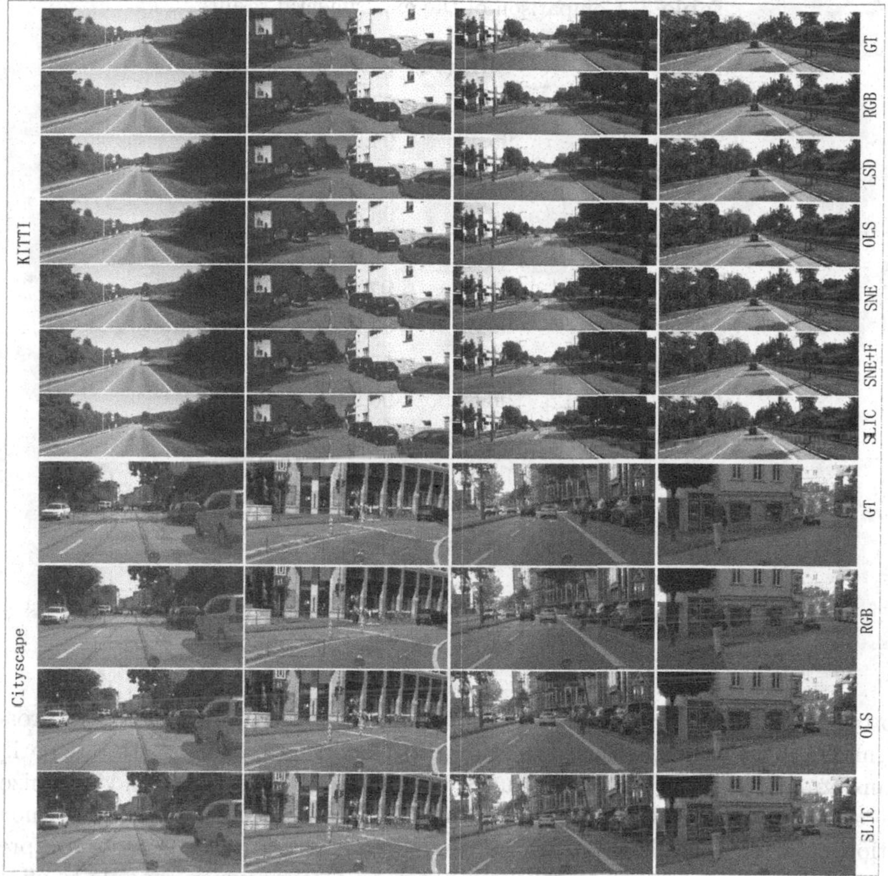

Fig. 3. Detection results of our model on the KITTI dataset and the Cityscapes dataset.

In Fig. 3, we show the visualization detection results of the model in the KITTI dataset. In the figure, green areas indicate correctly classified pixels, blue areas indicate misclassified pixels and red areas indicate pixels that misclassify non-road areas as road areas. In addition, we have marked the detection defects with red dashed boxes.

Table 3. Comparison of the experimental results.

Method	PRE (%)	REC (%)	MaxF (%)
RBNet [3]	92.81	93.60	93.21
TVFNet [6]	95.73	94.94	95.34
LC-CRF [5]	93.62	96.83	95.68
LidCamNet [1]	96.23	95.83	96.03
SNE-RoadSeg [4]	96.90	96.61	96.75
SNE-RoadSeg+ [13]	97.31	**97.49**	**97.40**
UASNet [2]	96.51	97.27	96.89
NIM-RTFNet [11]	96.43	95.62	96.02
DFM-RTFNet [12]	96.62	96.93	96.78
DTNet 18(LSD)	**98.20**	95.68	96.92
DTNet 18(OLS)	97.17	96.84	97.00
SNENet18+FD(2)	97.51	96.95	**97.23**
SNENet18+FD(3)	**97.75**	96.81	97.17

4 Conclusion

To better utilize the information of depth image, this paper proposes two road segmentation algorithms based on the priori information fusion: the disparity transformation method, and the surface normal vector method, and optimizes the segmentation results through network post-processing. In the road segmentation network, the algorithm uses a dual-stream structure to process the preprocessed RGB-D data separately and designs a feature fusion module to effectively integrate the information of the two modalities. The algorithm enhances the pavement features and obtains a high accuracy of road segmentation. However, the above models are all designed mainly for urban road scene datasets. For unstructured road scene segmentation, our model needs more exploration.

References

1. Caltagirone, L., Bellone, M., Svensson, L., Wahde, M.: Lidar-camera fusion for road detection using fully convolutional neural networks. Robot. Auton. Syst. **111**, 125–131 (2018)
2. Chang, Y., Xue, F., Sheng, F., Liang, W., Ming, A.: Fast road segmentation via uncertainty-aware symmetric network. In: 2022 International Conference on Robotics and Automation (ICRA), pp. 11124–11130 (2022)
3. Chen, Z., Chen, Z.: RBNet: a deep neural network for unified road and road boundary detection. In: Liu, D., Xie, S., Li, Y., Zhao, D., El-Alfy, E.-S.M. (eds.) Neural Information Processing. LNCS, vol. 10634, pp. 677–687. Springer, Cham (2017). https://doi.org/10.1007/978-3-319-70087-8_70

4. Fan, R., Wang, H., Cai, P., Liu, M.: SNE-RoadSeg: incorporating surface normal information into semantic segmentation for accurate freespace detection. In: Vedaldi, A., Bischof, H., Brox, T., Frahm, JM. (eds.) ECCV 2020. LNCS, vol. 12375, pp. 340–356. Springer, Cham (2020). https://doi.org/10.1007/978-3-030-58577-8_21
5. Gu, S., Zhang, Y., Tang, J., Yang, J., Kong, H.: Road detection through CRF based lidar-camera fusion. In: 2019 International Conference on Robotics and Automation (ICRA), pp. 3832–3838 (2019)
6. Gu, S., Zhang, Y., Yang, J., Alvarez, J.M., Kong, H.: Two-view fusion based convolutional neural network for urban road detection. In: 2019 IEEE/RSJ International Conference on Intelligent Robots and Systems (IROS), pp. 6144–6149. IEEE Press (2019)
7. Labayrade, R., Aubert, D., Tarel, J.-P.: Real time obstacle detection in stereovision on non flat road geometry through "v-disparity" representation. In: Intelligent Vehicle Symposium, 2002, vol. 2, pp. 646–651. IEEE (2002)
8. Nanri, T., Khiat, A., Furusho, H.: General-purpose road boundary detection with stereo camera. In: 2015 14th IAPR International Conference on Machine Vision Applications (MVA), pp. 361–364. IEEE (2015)
9. Scherer, R., Kalla, S.L., Tang, Y., Huang, J.: The grünwald-letnikov method for fractional differential equations. Comput. Math. Appl. **62**(3), 902–917 (2011). ISSN 0898–1221
10. Shimoda, W., Yanai, K.: Distinct class-specific saliency maps for weakly supervised semantic segmentation. In: Leibe, B., Matas, J., Sebe, N., Welling, M. (eds.) ECCV 2016. LNCS, vol. 9908, pp. 218–234. Springer, Cham (2016). https://doi.org/10.1007/978-3-319-46493-0_14
11. Wang, H., Fan, R., Sun, Y., Liu, M.: Applying surface normal information in drivable area and road anomaly detection for ground mobile robots. In: 2020 IEEE/RSJ International Conference on Intelligent Robots and Systems (IROS), pp. 2706–2711. IEEE (2020)
12. Wang, H., Fan, R., Sun, Y., Liu, M.: Dynamic fusion module evolves drivable area and road anomaly detection: a benchmark and algorithms. IEEE Trans. Cybernet. **52**(10), 10750–10760 (2021)
13. Wang, H., Fan, R., Cai, P., Liu, M.: SNE-roadseg+: rethinking depth-normal translation and deep supervision for freespace detection. In: 2021 IEEE/RSJ International Conference on Intelligent Robots and Systems (IROS), pp. 1140–1145. IEEE (2021)
14. Zhao, H., Shi, J., Qi, X., Wang, X., Jia, J.: Pyramid scene parsing network. In: 2017 IEEE Conference on Computer Vision and Pattern Recognition (CVPR), pp. 6230–6239 (2017)
15. Zhu, J.-Y., Wu, J., Wei, Y., Chang, E., Tu, Z.: Unsupervised object class discovery via saliency-guided multiple class learning. In: 2012 IEEE Conference on Computer Vision and Pattern Recognition, pp. 3218–3225 (2012). https://doi.org/10.1109/CVPR.2012.6248057

Contrastive Perturbation Network for Weakly Supervised Temporal Sentence Grounding

Tingting Han[1](✉), Yuanxin Lv[1], Zhou Yu[1], Jun Yu[1], Jianping Fan[1], and Liu Yuan[2]

[1] Hangzhou Dianzi University, Hangzhou, China
{ttinghan,yuz,yujun,fanjianping}@hdu.edu.cn

[2] China Academy of Electronics and Information Technology of CETC, Beijing, China
yuanliu@cetc.com.cn

Abstract. The purpose of temporal sentence grounding is to find the most relevant temporal period corresponding to the natural language query in an unmodified video. In recent years, the weak supervision paradigm, which does not require tedious annotations of starting and ending positions of the corresponding video segments, has gained significant attention due to its low annotation cost and reasonable efficiency. However, its effectiveness is seriously affected by the low-quality negative samples generated with random strategies. In this paper, we propose a Contrastive Perturbation Network (CPN), which introduces perturbation schemes into contrastive learning of weak supervised temporal sentence grounding. The perturbation involves both the proposal generation module and the reconstruction module of the CPN. In the proposal generation module, we introduce the KL divergence loss to minimize the distribution differences between the perturbed positive and real positive proposals, to force the network to be robust to the redundant information and learn fine-grained alignments between the text and video modalities. The reconstruction module leverages the perturbed features to generate a highly challenging negative proposal and strengthens the supervision to the proposal generation module by distinguishing the positive and negative proposals with the use of contrastive learning. Extensive experiments on two public benchmarks, i.e., ActivityNet Captions and Charades-STA, demonstrate that the proposed CPN could effectively improve the performance of weakly supervised temporal sentence grounding.

Keywords: Temporal Sentence grounding · Perturbation · Contrastive learning · Cross-modal analysis

1 Introduction

Videos have emerged as a crucial medium of information in the information age due to the rapid development of multi-terminal devices in the field of multimedia technology. Compared with traditional media such as text and audio, videos

© The Author(s), under exclusive license to Springer Nature Singapore Pte Ltd. 2024
Q. Liu et al. (Eds.): PRCV 2023, LNCS 14425, pp. 446–460, 2024.
https://doi.org/10.1007/978-981-99-8429-9_36

often contain more complex dynamic activities, richer semantic information, and simpler and more intuitive transmission of information, which make the analysis of videos a very challenging task. In terms of the temporal sentence grounding, which focuses on locating the most semantically relevant temporal period corresponding to the natural language query in an untrimmed video, has very important applications in video surveillance [7,12], video summarization [22] and human-computer interaction [25].

Recently, great efforts have been made in full supervised paradigm [13,23,30, 31,41,44], but they require laborious manual annotation of temporal boundaries for each query, limiting their scalability and practicality in real applications. Therefore, weakly supervised learning schemes that only require video and natural language queries during training have received more attention due to their low annotation cost. Reconstruction-based [18,42,43] approaches solve the task by ranking the reconstruction losses of proposal candidates, assuming that proposal that best matches the text should best reconstruct the entire query. Lin et al. [18] proposed a Semantic Completion Network (SCN), which used the generated proposal to match some words in the corresponding masked query to obtain a new supervision signal. While Zheng et al. [42] directly used a one-subtracted positive Gaussian mask as the negative mask and trained the CNM network in a contrastive learning way. However, the negative samples were too discriminative from the positive proposal to provide a strong supervisory signal. Zheng et al. [43] combined the above methods and proposed CPL, employing multiple learnable Gaussian functions to generate the proposals' boundaries.

In this paper, we focus on exploring highly confusing negative samples to encourage the model to learn fine-grained cross-modal alignments. We propose a Contrastive Perturbation Network (CPN) which introduces the perturbation mechanism into both the proposal generation and reconstruction modules. In the proposal generation module, we compare perturbed positive masks with real positive masks using the KL divergence loss, enabling the network to possess better robustness to redundant video content and the capability of capturing fine-grained cross-modal interactions. In the reconstruction module, we generate a highly confusing perturbed negative proposal, a reference sample and a series of mined negative samples from easy to hard. We utilize these negative proposals to enhance the network's supervision through contrastive learning. We demonstrate the effectiveness of the proposed CPN on two benchmark datasets, i.e., ActivityNet Captions [17] and Charades-STA [13], where superior performance is achieved and extensive ablations are further performed.

Our contributions can be summarized as follows:

- We design a perturbation mechanism to participate in both the processes of proposal generation and reconstruction modules, enabling the proposed Contrastive Perturbation Network (CPN) to distinguish highly confusing scenes and improve the grounding performance.
- We adopt a fine-grained cross-modal structure, constructing features of different modalities with the 2-dimensional map (2D Map), to better align the text and video information in the semantic space.

- Experiments on the Charades-STA and ActivityNet Captions datasets show that the proposed contrastive perturbation network method significantly outperforms existing weakly supervised methods.

2 Related Work

2.1 Fully Supervised Temporal Sentence Grounding

The fully supervised temporal sentence grounding task has made remarkable progress [1,3,13,35,37,38]. Starting with CTRL [13] and MCN [1], researchers explored the possibility of cross-modal fusion of text and video and introduced the traditional convolutional neural network [2,28] for processing video features. Gao et al. [13] proposed to use a classifier to determine whether candidate video segments match text query sentences, and adjust the boundaries of candidate video segments through a regressor, setting a paradigm for subsequent research. Subsequent work primarily focused on improving the alignment of text and video semantics. Zhang et al. [37] explored the possibility of convolution and used multi-scale convolution to encode the video while using each sentence query as a filter for embedding into each convolution process. Additionally, the graph convolution network was employed to enhance the expressive ability of features.

2.2 Weakly Supervised Temporal Sentence Grounding

The weakly supervised paradigm has attracted increasing attention for temporal sentence grounding tasks [5,20,21,27].The motivation behind using a weakly supervised approach is to leverage the readily available low-quality data for training the model. In the context of weakly supervised temporal sentence, weak supervision means that the sentence is only known to which video it belongs, without information about its precise temporal location. The key to weakly supervised temporal sentence grounding is to leverage effective supervision for model optimization, and various mechanisms have been proposed to address this challenge. Multiple instance learning (MIL) [14,16,20,21] and reconstruction-based methods [18,43,43] are most widely explored method in recent years.

2.3 Multiple Instance Learning

Multiple instance learning (MIL) has been widely used in video-language tasks [14,16,20,21]. In weakly supervised temporal sentence task, MIL-based methods learn the latent cross-modal semantic space by mining positive and negative samples. By aligning video and text, latent cross-modal semantic relationships are learned. As a typical MIL-based method, TGA [21] learned visual-text alignment at the video-level by maximizing the matching scores between videos and their corresponding descriptions, while minimizing the matching scores between videos and descriptions of other videos. The Cross-Relation Mining (CRM) method [16] employed cross-sentence relations to explore connection between sentences at the

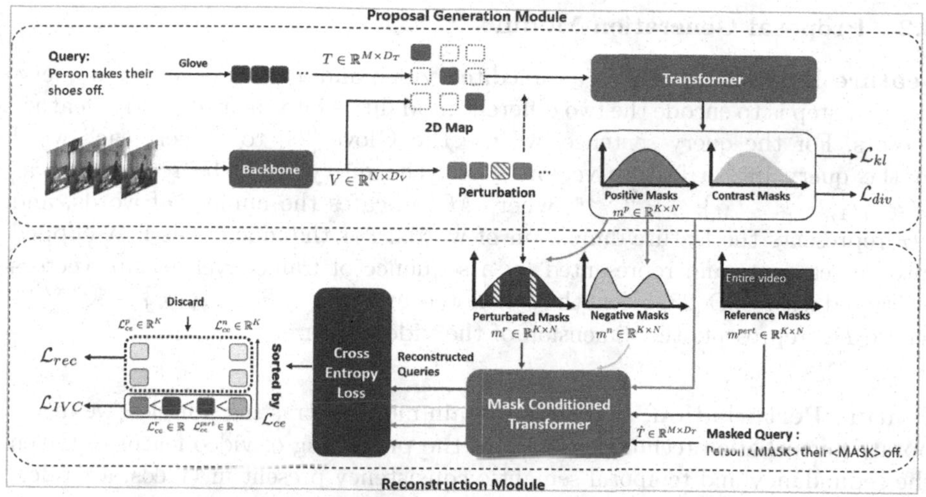

Fig. 1. The proposed Contrastive Perturbation Network (CPN). It consists of a proposal generation module and a reconstruction module. The proposal generation module generates potential positive proposals based on both the original and perturbed video features and provides fine-grained optimization guidance with the KL divergence. The reconstruction module compares the ability of positive and negative proposals to reconstruct the masked query, providing stronger supervision for the network training.

paragraph-level scope and proposed a new coarse-to-fine approach. The multiple instance learning could effectively increase the diversity of the data and improve the accuracy of sentence grounding.

3 Proposed Method

3.1 Overall Framework

The proposed Contrastive Perturbation Network (CPN) is comprised of two modules, as depicted in Fig. 1: the proposal generation module and the reconstruction module. Motivated by Luo et al. [19], we introduce the perturbation mechanism into both the two modules of CPN. The proposal generation module predicts potential positive proposals based on both the original and perturbed video features and possesses robustness to the redundant information and better temporal sentence grounding capability by minimizing the KL divergence between real positive masks and the contrast (perturbed positive) masks. The reconstruction module generates multiple negative samples, including a highly confusing perturbed negative sample, and compares their abilities to reconstruct the masked query through contrastive learning, providing stronger supervision signals for the weakly supervised network training.

3.2 Proposal Generation Module

Feature Extraction. Given an unedited video and a natural language query, the first step is to encode the two different modalities into their respective feature vectors. For the query sentence, we use the Glove [24] to convert each word in the query into a feature vector. Thus, the query could be represented as $T = \{t_1, t_2, ..., t_M\} \in \mathbb{R}^{M \times D_T}$, where M indicates the number of words, and D_T represents the feature dimension of words. For the video, it is first clipped into N segments and represented by a sequence of frame-level feature vectors extracted with C3D [28] of length N, denoted by $V = \{v_1, v_2, ..., v_N\} \in \mathbb{R}^{N \times D_V}$, where D_V represents the dimension of the video feature.

Feature Perturbation. In order to obtain more contrasting samples, we incorporate perturbation techniques [19] into the processing of video features. Given the redundancy and temporal semantic consistency present in videos, a moderate feature perturbation will not significantly alter the video's semantics, but it can effectively enhance the model's ability to extract valuable information for grounding textual queries. Therefore, we apply perturbation operations to a specific segment of v_k to obtain:

$$v_k^* = \Pi(v_k), \tag{1}$$

where v^* represents the perturbed features, $\Pi(\cdot)$ represents a specific disturbance operation, and we experiment with and compare four types of disturbance operations in total. For instance, certain dimensions of v are randomly assigned values between 0 and 1. By examining the effects of different disturbance operations in our experiments, we gain a comprehensive understanding of their outcomes. Thus, the perturbed features of the video V are represented by $V^* = \left\{v_1, v_2^*, ..., v_{N-2}, v_{N-1}^*, v_N\right\} \in \mathbb{R}^{N \times D_V}$. To prevent excessive information loss due to perturbation, we introduce a perturbation operation ratio γ to limit the number of perturbations. For practical purposes, we set γ to 0.3 as it produces relatively better results.

Cross-Modal Interactions. We enhance the expressive ability of the features by constructing a 2-dimensional (2D) map that incorporates features from two modalities (see Fig. 2). In the 2D map, each unit-length feature from one modality interacts with all features from the other modality, enabling the features to gather global semantic information more effectively. To achieve this, we align the textual features with $T_N \in \mathbb{R}^{N \times D_T}$ through padding, resulting in the 2D map representation $M \in \mathbb{R}^{D_V \times D_T}$ and the perturbed 2D map $M^* \in \mathbb{R}^{D_V \times D_T}$. These can be computed using the following formulas:

$$M = V \cdot T_N, \; M^* = V^* \cdot T_N. \tag{2}$$

Next, instead of employing the traditional Transformer [29], which only processes one-dimensional sequence input, we utilize the vision transformer (VIT)

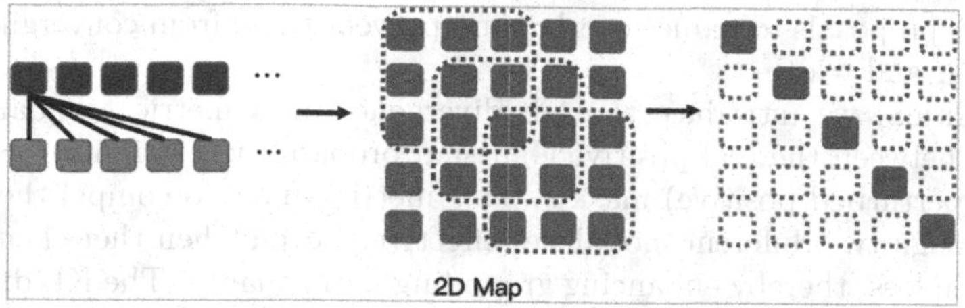

Fig. 2. We perform a dot product operation between the features of different modalities to obtain interactive information. This results in a cross-modal matrix, which we then convolve along the diagonal to obtain a mixed feature of length N.

[9] to handle the 2D map. In order to capture cross-modal interactions, we have made some modifications to the patch idea. Specifically, we apply a convolution operation with a window size of 3×3 to each element along the diagonal, resulting in a set of mixed features with a length of N, as depicted in Fig. 2. This ensures that the pre-processed features possess certain semantic interaction capabilities before being fed into the transformer.

Then, we employ the Vision Transformer (VIT) [9] to represent the multi-modal interactions between video features and textual features and extract deep latent semantic features denoted as $H = \{h_1, h_2, ..., h_N, h_{\text{cls}}\}$. Here, h_{cls} [8] represents the feature that interacts with all features and is utilized to compute the proposal location. The computation of H can be expressed as follows:

$$H = D(patch(M), E(patch(M))), \tag{3}$$

where $E(\cdot)$ and $D(\cdot)$ represent the encoder and decoder of the Transformer, respectively, and the $patch(\cdot)$ indicates the patch operation. Similarly, we utilize the identical operation to process M^* constructed using perturbed features, resulting in H^*. To obtain the probability masks required by the reconstruction module, we use the mixed semantic features, h_{cls} and h^*_{cls} to calculate the distribution center, $c^p, \in \mathbb{R}^K$, and the distribution width, $w^p, \in \mathbb{R}^K$. Subsequently, we generate proposals represented by the following distributions:

$$\begin{aligned} m^p_{ki} &= \frac{1}{\sqrt{2\pi}(w^p_k/\sigma)} exp(-\frac{(i/N - c^p_k)^2}{2(w^p_k/\sigma)^2}), \\ k &= 1, ..., K; \quad i = 1, ..., N. \end{aligned} \tag{4}$$

Loss Function. In order to generate more differentiated positive proposals, we apply the diversity loss to m^p which is obtained by the following formula:

$$\mathcal{L}_{div} = \left\| m^p m^{p\top} - \lambda I \right\|_F^2, \tag{5}$$

where $|| \cdot ||_F$ represents the Frobenius norm of the matrix, and $\lambda \in [0, 1]$ is a hyper-parameter that controls the degree of overlap between proposals. The loss encourages proposals to reduce overlap and prevent them from converging to the same center and width.

In addition, we introduce the KL divergence as a metric to quantify the similarity between the real positive Gaussian probability distribution p and the contrast (perturbed positive) mask q. This metric serves to compel the VIT [9] model to align two different modalities effectively, even when there is moderate information loss, thereby enhancing grounding performance. The KL divergence loss could be represented as:

$$\mathcal{L}_{KL} = \Sigma_{i=1}^{n} p(x) log \frac{p(x)}{q(x)}. \tag{6}$$

3.3 Reconstruction Module

Negative Proposal Generation. In this module, we employ the obtained positive Gaussian distribution to generate various samples (negative, perturbed negative) and a reference sample used for contrastively training the proposal generation module. We contend that the different samples possess varying degrees of efficacy in reconstructing the query, thus establishing a ranking. As a result, we could acquire novel and more robust supervision signals to gauge the model's output ability.

Negative Mining. Following [43], we collect negative samples from easy to hard to relieve the pressure on the optimizer to judge positive and negative samples in the early stages. When negative samples are close to positive samples, they are more likely to be confusing due to similar backgrounds and semantics.

Reference Sample. To help the model distinguish distinct proposals, we introduce the entire video as a reference sample m^r. Since the entire video contains ground-truth segments as well as a large amount of redundant information, we could infer that the reconstruction quality of the reference sample is inferior to that of the positive sample and better than that of any negative sample.

Perturbed Negative Proposal. The perturbed features, obtained through the application of Eq. 1, are utilized to generate a perturbed negative proposal according to the most confident positive mask. Since only a small ratio of random information is lost, the proposal retains the majority of semantically relevant information. Consequently, the perturbed negative proposal should better reconstruct the query compared to the reference sample. Therefore, we observe the following relation:

$$\mathbf{R}\left(m^{p}, Q\right) > \mathbf{R}\left(m^{pert}, Q\right) > \mathbf{R}\left(m^{r}, Q\right) > \mathbf{R}\left(m^{n}, Q\right), \tag{7}$$

where $R(\cdot)$ is a function that measures the semantic similarity between the query Q and the proposal specified by the mask m.

Mask Conditioned Transformer. To compare positive and negative proposals, we use a masked conditional reconstruction completion module inspired by SCN [18] and CPL [43] to measure the semantic relevance between proposals and queries. We embed masked queries using GloVe [24] and reconstruct the original query with the positive and negative proposals. The mask-conditioned transformer reconstructs the masked query utilizing various proposal masks and the reconstructed queries achieved through different proposal masks exhibit varying levels of reconstruction ability (as shown in Eq. 7). Finally, the cross-entropy loss is employed to quantify the similarity between the reconstructed and original queries. The cross-entropy losses for the positive sample, negative samples, reference, and perturbed negative samples are denoted as $\mathcal{L}_{ce}^{p}$, $\mathcal{L}_{ce}^{n1}$, $\mathcal{L}_{ce}^{n2} \in \mathbb{R}^K$ and $\mathcal{L}_{ce}^{r}, \mathcal{L}_{ce}^{pert} \in \mathbb{R}^K$. And the cross-entropy loss can be calculated using the following formula:

$$\mathcal{L}_{ce} = -\sum_{i=1}^{M} \log p\left(q_{i+1} \mid q_{1:i}\right), \tag{8}$$

where, $\mathcal{L}_{ce}$ represents the reconstruction loss based on the proposal context, and $p(q_{i+1}|q_{1:i})$ represents the probability of predicting the $i+1$ word vector from the 1st to the i-th word vector, aiming to maximize the probability of the predicted word vector and improve the prediction accuracy.

Loss Function. In order to provide effective supervision by contrastive learning, we calculate the reconstruction loss and the IVC loss to compare positive proposals and multiple negative proposals to force the proposal generation module to output more accurate proposals.

Reconstruction Loss. We choose the k'th proposal that is most relevant to the query to compute the reconstruction loss $\mathcal{L}_{ce}^{p}[k']$, where the $k' = \arg\min_k \left(\mathcal{L}_{ce}^{p}[k]\right)$. In addition, we involve the reference sample $\mathcal{L}_{ce}^{r}$ and the perturbed negative sample $\mathcal{L}_{ce}^{pert}$ to the operation since they all contain relevant information matched with the query. Therefore, the final reconstruction loss $\mathcal{L}_{rec}$ can be expressed as:

$$\mathcal{L}_{rec} = \mathcal{L}_{ce}^{p}[k] + \mathcal{L}_{ce}^{r} + \mathcal{L}_{ce}^{pert}. \tag{9}$$

IVC Loss. Given the semantic similarity relationship between positive samples, negative samples, reference and perturbed samples discussed above, we use Intra-video Contrastive loss $\mathcal{L}_{IVC}$ to contrast between the four type of samples, where β_1 and β_2 are hyper-parameters that satisfy $\beta_1 < \beta_2$:

$$\begin{aligned} \mathcal{L}_{IVC} = & \max\left(\mathcal{L}_{ce}^{p}[k'] - \mathcal{L}_{ce}^{r} + \beta_1, 0\right) \\ & + \max\left(\mathcal{L}_{ce}^{p}[k'] - \mathcal{L}_{ce}^{n_1}[k'] + \beta_2, 0\right) \\ & + \max\left(\mathcal{L}_{ce}^{p}[k'] - \mathcal{L}_{ce}^{n_2}[k'] + \beta_2, 0\right) \\ & + \max\left(\mathcal{L}_{ce}^{p}[k'] - \mathcal{L}_{ce}^{pert} + \beta_2, 0\right). \end{aligned} \tag{10}$$

3.4 Model Training

Our network includes four parts of the loss function used as a supervisory signal. $\mathcal{L}_{rec}$ helps the model reconstruct the query through the video features we generate to enhance the alignment between video and text. $\mathcal{L}_{IVC}$ contrast positive and negative samples to encourage the proposal generation module output proposals that are semantically relevant to the query with more accurate boundaries. $\mathcal{L}_{div}$ generates multiple different positive proposals with different video content to improve the recall. $\mathcal{L}_{KL}$ measures the distance between the real positive and perturbed positive masks and promote the fine-grained alignment between the text and video modalities. Finally, we compute the multi-task loss to train our network in an end-to-end manner, which can be expressed as:

$$\mathcal{L} = \mathcal{L}_{rec} + \alpha_1 \mathcal{L}_{IVC} + \alpha_2 \mathcal{L}_{div} + \eta \mathcal{L}_{KL}, \tag{11}$$

where α_1, α_2, and η are the hyper-parameters that we set.

4 Experiments

4.1 Datasets

To evaluate the effectiveness of the proposed CPN, we conducted experiments on two benchmark datasets: two benchmark datasets- Charades-STA [13] and ActivityNet Captions [15,17]. The Charades-STA dataset comprises 5338/1334 videos and 12,408/3720 video-query pairs for training and testing purposes, respectively. Our evaluation is performed on the test split. ActivityNet Captions dataset consists of 10,009/4,917/5,044 videos and 37,417/17,505/ 17,031 video-query pairs for training, validation, and testing, respectively. Our results are reported on the validation split 2 of the ActivityNet Captions dataset.

4.2 Evaluation Metric

To evaluate the performance of our proposed method, we adopt the evaluation metric of "R@n, IoU = m", as used in previous studies [16,18]. Here, m is a predefined temporal Intersection over Union (IoU) threshold, and n represents the recall rate of top-n predictions. This metric measures the percentage of predicted moments that have an IoU value greater than m among our top-n predictions. Specifically, we report "$R@n, IoU = m$" with $n \in \{1, 5\}$, $m \in \{0.1, 0.3, 0.5\}$ for the ActivityNet Captions and $m \in \{0.3, 0.5, 0.7\}$ for the Charades-STA.

4.3 Implementation Details

Data Preprocessing. To prepare the data for training and evaluation, we downsample each video every 8 frames and extract its visual features using the C3D [28] model for ActivityNet Captions and the I3D [2] model for Charades-STA. For word embeddings, we use the pre-trained GloVe [24] word2vec model to process each word token. We constrain the maximum description length to 20, and the vocabulary size to 8000.

Table 1. Evaluation results on the Charades-STA dataset($n \in \{1,5\}$ and $m \in \{0.3, 0.5, 0.7\}$). The numbers in bold are the best results and the underlines indicate the results outperform the baseline method CPL.

Method	R@1			R@5		
	IoU = 0.3	IoU = 0.5	IoU = 0.7	IoU = 0.1	IoU = 0.3	IoU = 0.5
TGA [21]	32.14	19.94	8.84	86.58	65.52	33.51
CTF [6]	39.8	27.3	12.9	–	–	–
SCN [18]	42.96	29.35	9.97	95.56	71.80	38.87
WSTAN [32]	43.39	29.35	12.28	93.04	76.13	41.53
BAR [34]	44.97	27.04	12.23	–	–	–
VLANet [20]	45.24	31.83	14.17	95.70	82.85	33.09
LoGAN [27]	48.04	31.74	13.71	89.01	72.17	37.58
MARN [26]	48.55	31.94	14.81	90.70	70.00	37.40
WSRA [11]	50.13	31.20	11.01	86.75	70.50	39.02
CCL [40]	–	33.21	15.68	–	73.50	41.87
CRM [16]	53.66	34.76	16.37	–	–	–
VCA [33]	58.58	38.13	19.57	**98.08**	78.75	37.75
LCNet [36]	59.60	39.19	18.87	94.78	80.56	45.24
RTBPN [39]	60.04	32.36	13.24	97.48	71.85	41.18
CNM [43]	60.39	35.43	15.45	–	–	–
CPL [43]	66.02	49.15	22.61	96.74	84.52	51.65
CPN (ours)	**67.41**	**50.41**	**24.05**	97.07	**85.28**	**52.25**

Table 2. Evaluation results on the ActivityNet dataset($n \in \{1,5\}$ and $m \in \{0.1, 0.3, 0.5\}$). The numbers in bold are the best results and the underlines indicate the results outperform the baseline method CPL.

Method	R@1			R@5		
	IoU = 0.1	IoU = 0.3	IoU = 0.5	IoU = 0.1	IoU = 0.3	IoU = 0.5
WS-DEC [10]	62.17	41.98	23.34	–	–	–
VCA [33]	67.96	50.45	31.00	92.14	71.79	53.83
EC-SL [4]	68.48	44.29	24.16	–	–	–
MARN [26]	–	47.01	29.95	–	72.02	57.49
SCN [18]	71.48	47.23	29.22	90.88	71.56	55.69
BAR [34]	–	49.03	30.73	–	–	–
RTBPN [39]	73.73	49.77	29.63	93.89	79.89	60.56
CTF [5]	74.2	44.3	23.6	–	–	–
WSLLN [14]	75.4	42.8	22.7	–	–	–
CCL [40]	–	50.12	31.07	–	77.36	61.29
CRM [16]	**81.61**	55.26	32.19	–	–	–
LCNet [36]	78.58	48.49	26.33	**93.95**	**82.51**	62.66
WSTAN [32]	79.78	52.45	30.01	93.15	79.38	**63.42**
CNM [43]	78.13	55.68	**33.33**	–	–	–
CPL [43]	79.86	53.67	31.24	87.24	63.05	43.13
CPN (ours)	77.99	**55.71**	29.02	92.19	79.11	58.97

Model Settings. Based on the experiments, we set the optimal patch size to 5, and the disturbance coefficient γ to 0.3. Regarding the hyper-parameters, we set σ to 9, λ to 0.135, β_1 to 0.1, β_2 to 0.15, and η to 10 for both datasets.

Table 3. The effect of different losses in the proposal generation module.

Loss		R@1			
$\mathcal{L}_{div}$	$\mathcal{L}_{kl}$	IoU = 0.3	IoU = 0.5	IoU = 0.7	mIoU
✗	✗	46.35	21.39	5.28	22.34
✓	✗	61.78	44.84	20.17	40.43
✓	✓	**67.41**	**50.41**	**24.05**	**43.87**

4.4 Comparisons to the State-of-the-Art

Table 1 and Table 2 present a comparison of CPN performance with previous works. The findings from our study highlight the following results: (1) On the Charades-STA dataset, the model's performance demonstrates a significant improvement compared to the baseline model CPL and outperforms all previous state-of-the-art (SOTA) methods across most metrics. (2) On the ActivityNet Captions dataset, we have surpassed CPL [43] in all "R@5" metrics by a significant margin and achieved SOTA in the "R@1, IoU = 0.3" metric. These experimental results demonstrate the effectiveness of the proposed CPN.

4.5 Ablation Study

Effectiveness of Different Losses. We evaluate the impact of two losses, specifically $\mathcal{L}_{kl}$ and $\mathcal{L}_{div}$, in the proposal generation module, as shown in Table 3. Our findings indicate that the design of both losses is significant for our network. Furthermore, we observe that the diversity loss can significantly improve the model's performance, while KL divergence can effectively generate better distributions. This suggests that generating multiple different positive proposals is beneficial for more accurately locating the event of interest in long videos.

Table 4. The effect of different perturbation strategies.

Perturbation Form	R@1			
	IoU = 0.3	IoU = 0.5	IoU = 0.7	mIoU
set zeros	**67.41**	50.41	**24.05**	**43.87**
normal distribution	62.35	44.78	19.16	40.53
interpolation(set $\alpha_p = 0.3$)	65.44	47.13	25.07	41.49
repeat	66.83	**50.89**	23.27	42.56

Effect of Perturbation Strategies. To verify the effectiveness of perturbation features in the model, we conduct ablation studies using different perturbation forms. We experiment with four types of perturbation features v^* using

Formula1, as follows: (1) set zeros: $v_k^* = [0, 0, 0, ...]$; (2) normal distribution: $v_k^* = random()$; (3) interpolation: $v_k^* = \alpha_p v_{k-1} + (1 - \alpha_p) v_{k+1}$; and (4) repeat: $v_k^* = v_{k-1}^*$. Our experiments reveal that the best performance is achieved when perturbation features are set to 0, as shown in Table 4.

5 Conclusion

In this study, we introduce a novel Contrastive Perturbation Network, applying perturbation techniques to enhance the proposal generation capability. Minimizing the KL divergence between the real positive Gaussian masks and the perturbed positive masks endows the proposal generation module with stronger robustness to the information redundancy. Moreover, in the reconstruction module, the more confusing perturbed negative proposal is involved in contrastive learning, which provides stronger supervision signals and improves the grounding performance. Our experiments yield promising results on the Charades-STA and ActivityNet Captions datasets. Additionally, ablation studies substantiate the effectiveness of our proposed additive approach.

Acknowledgments. This work is partly supported by the Zhejiang Province Natural Science Foundation of China (No. LQ21F020014 and No. LZ23F020007) and the National Natural Science Foundation of China (No. 62002091).

References

1. Anne Hendricks, L., Wang, O., Shechtman, E., Sivic, J., Darrell, T., Russell, B.: Localizing moments in video with natural language. In: Proceedings of the IEEE International Conference on Computer Vision, pp. 5803–5812 (2017)
2. Carreira, J., Zisserman, A.: Quo vadis, action recognition? a new model and the kinetics dataset. In: Proceedings of the IEEE Conference on Computer Vision and Pattern Recognition, pp. 6299–6308 (2017)
3. Chen, J., Chen, X., Ma, L., Jie, Z., Chua, T.S.: Temporally grounding natural sentence in video. In: Proceedings of the 2018 Conference on Empirical Methods in Natural Language Processing, pp. 162–171 (2018)
4. Chen, S., Jiang, Y.G.: Towards bridging event captioner and sentence localizer for weakly supervised dense event captioning. In: Proceedings of the IEEE/CVF Conference on Computer Vision and Pattern Recognition, pp. 8425–8435 (2021)
5. Chen, Z., Ma, L., Luo, W., Tang, P., Wong, K.Y.K.: Look closer to ground better: weakly-supervised temporal grounding of sentence in video. arXiv preprint arXiv:2001.09308 (2020)
6. Chen, Z., Ma, L., Luo, W., Wong, K.Y.K.: Weakly-supervised spatio-temporally grounding natural sentence in video. arXiv preprint arXiv:1906.02549 (2019)
7. Collins, R.T., et al.: A system for video surveillance and monitoring. VSAM Final Report **2000**(1–68), 1 (2000)
8. Devlin, J., Chang, M.W., Lee, K., Toutanova, K.: Bert: pre-training of deep bidirectional transformers for language understanding. arXiv preprint arXiv:1810.04805 (2018)

9. Dosovitskiy, A., et al.: An image is worth 16x16 words: transformers for image recognition at scale. arXiv preprint arXiv:2010.11929 (2020)
10. Duan, X., Huang, W., Gan, C., Wang, J., Zhu, W., Huang, J.: Weakly supervised dense event captioning in videos. Adv. Neural Inf. Process. Syst. **31** (2018)
11. Fang, Z., Kong, S., Wang, Z., Fowlkes, C., Yang, Y.: Weak supervision and referring attention for temporal-textual association learning. arXiv preprint arXiv:2006.11747 (2020)
12. Foroughi, H., Aski, B.S., Pourreza, H.: Intelligent video surveillance for monitoring fall detection of elderly in home environments. In: 2008 11th International Conference on Computer and Information Technology, pp. 219–224. IEEE (2008)
13. Gao, J., Sun, C., Yang, Z., Nevatia, R.: Tall: temporal activity localization via language query. In: Proceedings of the IEEE International Conference on Computer Vision, pp. 5267–5275 (2017)
14. Gao, M., Davis, L.S., Socher, R., Xiong, C.: Wslln: weakly supervised natural language localization networks. arXiv preprint arXiv:1909.00239 (2019)
15. Heilbron, F.C., Escorcia, V., Ghanem, B., Niebles, J.C.: Activitynet: a large-scale video benchmark for human activity understanding. In: 2015 IEEE Conference on Computer Vision and Pattern Recognition (CVPR), pp. 961–970. IEEE (2015)
16. Huang, J., Liu, Y., Gong, S., Jin, H.: Cross-sentence temporal and semantic relations in video activity localisation. In: Proceedings of the IEEE/CVF International Conference on Computer Vision, pp. 7199–7208 (2021)
17. Krishna, R., Hata, K., Ren, F., Fei-Fei, L., Carlos Niebles, J.: Dense-captioning events in videos. In: Proceedings of the IEEE International Conference on Computer Vision, pp. 706–715 (2017)
18. Lin, Z., Zhao, Z., Zhang, Z., Wang, Q., Liu, H.: Weakly-supervised video moment retrieval via semantic completion network. In: Proceedings of the AAAI Conference on Artificial Intelligence, vol. 34, pp. 11539–11546 (2020)
19. Luo, F., Chen, S., Chen, J., Wu, Z., Jiang, Y.G.: Self-supervised learning for semi-supervised temporal language grounding. IEEE Trans. Multim. (2022)
20. Ma, M., Yoon, S., Kim, J., Lee, Y., Kang, S., Yoo, C.D.: VLANet: video-language alignment network for weakly-supervised video moment retrieval. In: Vedaldi, A., Bischof, H., Brox, T., Frahm, J.-M. (eds.) ECCV 2020, LNCS, vol. 12373, pp. 156–171. Springer, Cham (2020). https://doi.org/10.1007/978-3-030-58604-1_10
21. Mithun, N.C., Paul, S., Roy-Chowdhury, A.K.: Weakly supervised video moment retrieval from text queries. In: Proceedings of the IEEE/CVF Conference on Computer Vision and Pattern Recognition, pp. 11592–11601 (2019)
22. Narasimhan, M., Rohrbach, A., Darrell, T.: Clip-it! language-guided video summarization. In: Advances in Neural Information Processing Systems, pp. 13988–14000 (2021)
23. Otani, M., Nakashima, Y., Rahtu, E., Heikkilä, J.: Uncovering hidden challenges in query-based video moment retrieval. arXiv preprint arXiv:2009.00325 (2020)
24. Pennington, J., Socher, R., Manning, C.D.: Glove: global vectors for word representation. In: Proceedings of the 2014 Conference on Empirical Methods in Natural Language Processing (EMNLP), pp. 1532–1543 (2014)
25. Singha, J., Roy, A., Laskar, R.H.: Dynamic hand gesture recognition using vision-based approach for human-computer interaction. Neural Comput. Appl. **29**(4), 1129–1141 (2018)
26. Song, Y., Wang, J., Ma, L., Yu, Z., Yu, J.: Weakly-supervised multi-level attentional reconstruction network for grounding textual queries in videos. arXiv preprint arXiv:2003.07048 (2020)

27. Tan, R., Xu, H., Saenko, K., Plummer, B.A.: Logan: latent graph co-attention network for weakly-supervised video moment retrieval. In: Proceedings of the IEEE/CVF Winter Conference on Applications of Computer Vision, pp. 2083–2092 (2021)
28. Tran, D., Bourdev, L., Fergus, R., Torresani, L., Paluri, M.: Learning spatiotemporal features with 3d convolutional networks. In: Proceedings of the IEEE International Conference on Computer Vision, pp. 4489–4497 (2015)
29. Vaswani, A., et al.: Attention is all you need. Adv. Neural Inf. Process. Syst. **30** (2017)
30. Wang, H., Zha, Z.J., Chen, X., Xiong, Z., Luo, J.: Dual path interaction network for video moment localization. In: Proceedings of the 28th ACM International Conference on Multimedia, pp. 4116–4124 (2020)
31. Wang, H., Zha, Z.J., Li, L., Liu, D., Luo, J.: Structured multi-level interaction network for video moment localization via language query. In: Proceedings of the IEEE/CVF Conference on Computer Vision and Pattern Recognition, pp. 7026–7035 (2021)
32. Wang, Y., Deng, J., Zhou, W., Li, H.: Weakly supervised temporal adjacent network for language grounding. IEEE Trans. Multim. **24**, 3276–3286 (2021)
33. Wang, Z., Chen, J., Jiang, Y.G.: Visual co-occurrence alignment learning for weakly-supervised video moment retrieval. In: Proceedings of the 29th ACM International Conference on Multimedia, pp. 1459–1468 (2021)
34. Wu, J., Li, G., Han, X., Lin, L.: Reinforcement learning for weakly supervised temporal grounding of natural language in untrimmed videos. In: Proceedings of the 28th ACM International Conference on Multimedia, pp. 1283–1291 (2020)
35. Xu, H., He, K., Plummer, B.A., Sigal, L., Sclaroff, S., Saenko, K.: Multilevel language and vision integration for text-to-clip retrieval. In: Proceedings of the AAAI Conference on Artificial Intelligence, vol. 33, pp. 9062–9069 (2019)
36. Yang, W., Zhang, T., Zhang, Y., Wu, F.: Local correspondence network for weakly supervised temporal sentence grounding. IEEE Trans. Image Process. **30**, 3252–3262 (2021)
37. Zhang, D., Dai, X., Wang, X., Wang, Y.F., Davis, L.S.: Man: moment alignment network for natural language moment retrieval via iterative graph adjustment. In: Proceedings of the IEEE/CVF Conference on Computer Vision and Pattern Recognition, pp. 1247–1257 (2019)
38. Zhang, S., Peng, H., Fu, J., Luo, J.: Learning 2d temporal adjacent networks for moment localization with natural language. In: Proceedings of the AAAI Conference on Artificial Intelligence, vol. 34, pp. 12870–12877 (2020)
39. Zhang, Z., Lin, Z., Zhao, Z., Zhu, J., He, X.: Regularized two-branch proposal networks for weakly-supervised moment retrieval in videos. In: Proceedings of the 28th ACM International Conference on Multimedia, pp. 4098–4106 (2020)
40. Zhang, Z., Zhao, Z., Lin, Z., He, X., et al.: Counterfactual contrastive learning for weakly-supervised vision-language grounding. Adv. Neural. Inf. Process. Syst. **33**, 18123–18134 (2020)
41. Zhao, Y., Zhao, Z., Zhang, Z., Lin, Z.: Cascaded prediction network via segment tree for temporal video grounding. In: Proceedings of the IEEE/CVF Conference on Computer Vision and Pattern Recognition, pp. 4197–4206 (2021)
42. Zheng, M., Huang, Y., Chen, Q., Liu, Y.: Weakly supervised video moment localization with contrastive negative sample mining. In: Proceedings of the AAAI Conference on Artificial Intelligence, vol. 36, pp. 3517–3525 (2022)

43. Zheng, M., Huang, Y., Chen, Q., Peng, Y., Liu, Y.: Weakly supervised temporal sentence grounding with gaussian-based contrastive proposal learning. In: Proceedings of the IEEE/CVF Conference on Computer Vision and Pattern Recognition, pp. 15555–15564 (2022)
44. Zhou, H., Zhang, C., Luo, Y., Chen, Y., Hu, C.: Embracing uncertainty: decoupling and de-bias for robust temporal grounding. In: Proceedings of the IEEE/CVF Conference on Computer Vision and Pattern Recognition, pp. 8445–8454 (2021)

MLDF-Net: Metadata Based Multi-level Dynamic Fusion Network

Feng Li[1], Enguang Zuo[2,3], Chen Chen[2,3], Cheng Chen[1], Mingrui Ma[1,2], Yunling Wang[4], Xiaoyi Lv[1,2,3,5(✉)], and Min Li[2,5(✉)]

[1] College of Software, Xinjiang University, Urumqi 830046, China
xjuwawj01@163.com

[2] College of Information Science and Engineering, Xinjiang University, Urumqi 830046, China
MinLi927@163.com

[3] Xinjiang Cloud Computing Application Laboratory, Karamay 834099, China

[4] The First Affiliated Hospital of Xinjiang Medical University, Urumqi 830054, XinJiang, China

[5] Key Laboratory of Signal Detection and Processing, Xinjiang University, Urumqi 830046, China

Abstract. Computer-aided diagnosis has been widely used in the medical field, and one of the current research hotspots for aiding diagnosis is how to effectively fuse heterogeneous data such as image data and metadata. Most recent multi-modal skin cancer diagnosis models are only fused at the feature level or decision level and have not yet paid attention to the differential influence of metadata on image features under dynamic guidance, which has limited the ability of metadata to improve the predictive performance of the model. Therefore, this paper proposed a multi-level dynamic fusion network (MLDF-Net) based on metadata guidance, which attempted to dynamically fuse relevant metadata features in the image feature extraction stage to achieve the purpose of metadata-guided image features. Firstly, we designed a feature selection block (FS Block) to suppress the influence of noise in metadata and enhance the metadata feature representation associated with images. Secondly, the filtered metadata and images are fused in the feature extractor at multiple levels, and the metadata dynamically guides the network to extract more representative image features. Lastly, the experimental results showed that MLDF-Net achieved 81.3% accuracy compared with other classification studies using the same dataset, which verified the feasibility and advancement of the multi-level dynamic fusion strategy based on metadata guidance.

Keywords: Skin cancer · Metadata · Dynamic fusion · Multilevel fusion

This work was supported through Central Guidance on Local Science and Technology Development Fund Projects (grant No. ZYYD2023D02) and Autonomous Region Science and Technology Support Xinjiang Project Plan (2021E02078).

© The Author(s), under exclusive license to Springer Nature Singapore Pte Ltd. 2024
Q. Liu et al. (Eds.): PRCV 2023, LNCS 14425, pp. 461–473, 2024.
https://doi.org/10.1007/978-981-99-8429-9_37

1 Introduction

According to the World Health Organization 2020, skin cancer accounts for one-third of all cancer types [1], and early diagnosis and treatment are vital for skin cancer patients. However, diagnosis often relies on the extensive experience of specialists and expensive dermoscopic detection equipment. The high incidence and limited resources make computer-aided diagnosis systems for skin cancer an urgent need for clinical diagnosis.

It is known from research that age and genetics are often important factors in triggering skin cancer. At the same time, lesion area and cancer history are also important clues for clinical experts in the diagnosis process, and these data are called metadata. Doctors usually screen skin lesion areas based on the combination of patients' clinical information in the diagnostic approach to make more accurate diagnoses. When relying solely on dermoscopic images for diagnosis, it can easily cause a lack of clinical information about the patient. Therefore, it forces us to integrate multi-modal information further to enhance the accuracy of computer-aided diagnosis of skin cancer models.

Image-based unimodal skin cancer-assisted diagnosis systems have been extensively investigated. Codella [2] et al. combined deep learning features and hand-coded features, and the proposed method classified two types of melanomas or moles for detection with 76% accuracy. Elattar [3] et al. completed dermatological diagnosis using Resnet50. Chatterjee [4] et al. combined ABCD rules for dermoscopic image classification, and this work used physicians' knowledge to facilitate model classification. Liu [5] et al. proposed a clinically inspired network (CI Net) that incorporated physicians' learning strategies and diagnostic processes and achieved 90.3% accuracy in the melanoma classification task. Arjun [6] et al. proposed a transfer learning deep learning model for accurately classifying benign and malignant skin lesions using deep learning techniques. Zhu [7] et al. achieved good results in the intelligent diagnosis of skin diseases based on VIT. Chen [8] et al. designed a feature combination method to extract robust features for skin disease images. Yang [9] et al. used a federation learning approach to balance performance. Moldovanu [10] et al. studied 60 different architectures of feedforward-feedback propagation networks (FFBPN) based on shape asymmetry and explored the optimal structural design from different perspectives. Wang et al. [11] implemented inter-network parameter transfer in a knowledge distillation manner to reduce the number of diagnostic model parameters. All of the works mentioned above focus on a single type of image for diagnosis, ignoring the critical role of metadata in the diagnosis process.

Several studies have attempted to introduce metadata into automatic skin cancer diagnosis work. kharazmi [12] et al. used sparse autoencoder (SAE) to combine skin lesion images with metadata. Liu [13] et al. used a cascade approach to fuse metadata features with image features to support skin disease diagnosis decisions. Pacheco [14] et al. proposed combining different deep-learning models to fuse metadata and skin lesion images in a multi-modal fusion approach. However, the above work has yet to consider the potential relationship between metadata and image features. Two papers, MetaBlock [14] and MetaNet [15] addressed this issue by using metadata to enhance image features. However, there are still three problems: 1. The guiding role of metadata in image feature extraction has yet to be considered. 2. The different effects of differentiated features among metadata on diagnostic models are ignored. If fusing metadata containing noise with images is prone to adversely affect diagnostic results. 3. Ignoring the differences in image features under different feature extraction stages.

To realize the fusion between metadata and image features under different feature extraction stages, we design a multi-modal fusion module (MG Block) based on metadata guidance, which can be flexibly embedded into the feature extractor to complete the work of metadata to guide the image features. Meanwhile, a feature selection method is designed to dynamically filter the metadata features most relevant to image features in different feature extraction stages. The contributions of the article can be summarized as follows:

1. We propose a metadata-based feature extraction guidance module (MG Block), which is plug-and-play and can be flexibly embedded in the baseline model, effectively realizing multi-level modal feature fusion of metadata and images for guiding the baseline model to extract finer features.
2. We design a dynamic feature selection method (FS Block). This module can efficiently calculate the image and metadata correlation to achieve dynamic adaptation of the two, enhance the feature expression of the metadata most relevant to the image features, and suppress the invalid metadata feature expression to a certain extent.
3. We performed a detailed validation on the PAD-UFES-20 [16], and our proposed method significantly outperformed current representative multimodal fusion work under all evaluation metrics.

2 Related Works

Metadata is a tabular structure used to express the patient's clinical background. In the PAD-UFES-20 [16], metadata contains 21 patient clinical characteristics (e.g., age, gender, anatomical region, cancer history, family background, etc.). In the process of clinical diagnosis, the patient's clinical background information is an important basis for the doctor's diagnosis, along with the clinical image information. When fusing multiple data, multi-modal deep learning models make medical diagnoses more accurate, predictable, and interpretable. The rationale for using this approach is that combining heterogeneous data can provide complementary information and improve decision diagnosis accuracy [17]. Heterogeneous data fusion is classified from low to high according to the level of fusion as pixel-level fusion, feature-level fusion, and decision-level fusion [18], with the most widely used feature-level fusion, where features extracted from different source data are combined using predefined methods [19].

Kawahara [20] et al. utilized multiple loss functions to adapt different combinations of input modalities. A related study introduced GCN into the field of multi-modal fusion by using multiclass features as graph node attributes to accomplish fusion [21,23]. Yan [22] et al. extracted richer multi-level feature representations of pathology images from multiple convolutional layers to reduce information loss of modalities before data fusion. Li [24] et al. designed a decoupled feature alignment and fusion (DeAF) framework for inter-modal alignment. Zhou [25] et al. implemented electronic medical records (EMR) and chest computed tomography (CT) image fusion for severity prediction of patients with no coronary pneumonia. Pacheco [14] et al. proposed an attention mechanism-based approach to fuse images with metadata. Tang [26] et al. proposed a multi-stage fusion approach to achieve the fusion of three modalities for multi-label dermatological classification. Wang [27] et al. fused multi-source features at the mutual likelihood level, thus solving the problem of severe size bias of image and non-image information.

Although the works mentioned above optimize multi-modal fusion work from different perspectives and achieve excellent results, they ignore the differences in image features under different feature extraction stages. They cannot play the role of metadata in the image feature extraction process. Therefore, we proposed an MG block to promote the multi-level image and metadata fusion. In particular, metadata features differ in their ability to be expressed at different levels of the fusion process, so we designed an FS Block for the MG block to filter metadata features.

3 Methodology

In this section, we detail the multi-level dynamic fusion network (MLDF-Net) based on metadata guidance, which consists of two parts: the MG Block and an FS Block. MG Block is embedded in the feature extractor to guide different feature extraction stages using metadata. Meanwhile, to be able to select relevant features dynamically, we design FS Block for MG Block to optimize the metadata input. The overall structure is shown in Fig. 1:

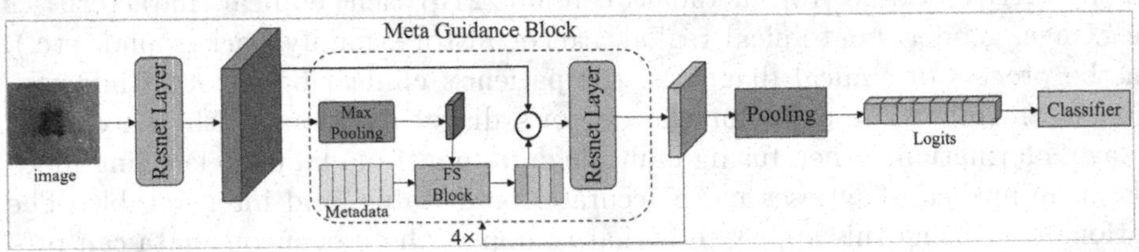

Fig. 1. The structure of the MLDF-Net. The rectangle in the Figure represents the feature map of the image, and the part in the dashed box is the MG Block. We embed the MG Block into the feature extractor (Resnet50 [27] is used as an example), and the feature map is outputted for classification after four iterations of the MG Block.

3.1 Metadata Guidance Module

Skin cancer lesions are characterized in diverse ways and are easily influenced by background and other factors. Simply extracting the color, texture, and other features of the lesion area may confuse the classifier. For example, basal cell carcinoma located in lesion areas with thin skin such as the face, ears, or nose and frequently stimulated by external forces is often accompanied by bleeding symptoms. In contrast, basal cell carcinoma in the back does not show bleeding symptoms under normal circumstances. Therefore, in this paper, MG Block is designed to fuse metadata at different stages and reduce the influence of confounding factors to optimize the image features in the network layer subsequent input.

Inspired by MetaNet [15], we propose a new metadata and image fusion approach: a metadata-based multi-level fusion method. The method uses metadata to guide the network to focus on more generalized features, and it consists of a feature selection block, convolutional blocks, and element operations. The specific structure of the MG Block is shown in Fig. 2. The metadata is mapped through the convolution block to the same dimension as the image feature map, and the mapped metadata feature vector is multiplied with the feature map of the image in the channel dimension to output the metadata-guided features. Our MG Block can be mathematically expressed as:

$$X'_{meta} = FS(X_{meta}, X_{img}) \tag{1}$$

$$X'_{img} = \varphi_{meta}(X'_{meta}) \bullet \varphi_{img}(X_{img}) \tag{2}$$

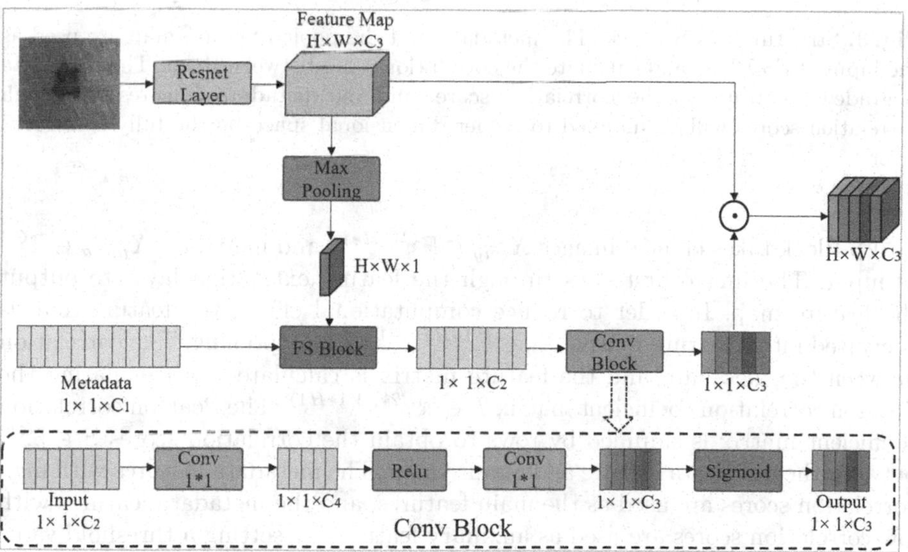

Fig. 2. Structure of MG Block. The feature map after feature extraction is pooled and input to the feature selection block together with the metadata and the filtered metadata is mapped by Conv Block to generate the channel attention vector of the image. FS Block will be introduced in detail later.

where X_{meta} represents the metadata of the patient, X'_{meta} represents the metadata feature after feature selection, X'_{img} represents the image feature map output by MG Block and φ is the feature extraction layer consisting of convolution and activation functions.

3.2 Dynamic Feature Selection Module

The model is susceptible to confusion by the hair when extracting texture features of the image. It occurs mostly in the feature extraction process of the model for images of the leg and head regions. In this case, the lesion region feature should be expressed more strongly than other metadata features. To dynamically select relevant features to guide the feature extractor, we designed the FS Block. The structure is shown in Fig. 3.

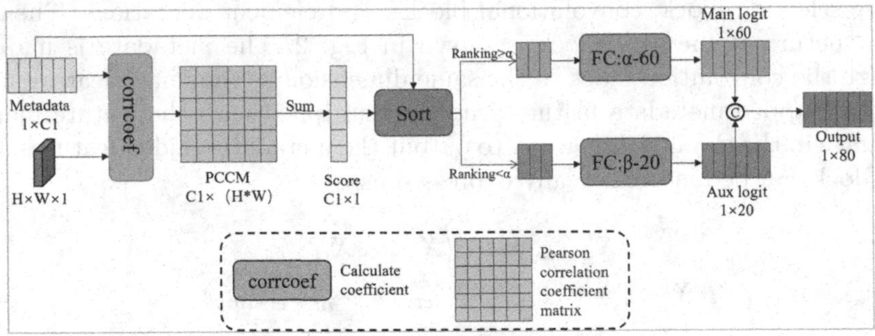

Fig. 3. Structure of FS Block. The metadata and the pooled feature map are used as the input of FS Block and calculate the correlation score between them. The metadata is graded according to the correlation score, and the metadata features with high correlation scores will be mapped to higher dimensional space by the fully connected layer.

FS Block takes clinical images $X_{img} \in \mathbb{R}^{C1*H*W}$ and metadata $X_{meta} \in \mathbb{R}^{C2}$ as input. The image first goes through the feature extraction layer to output the feature map. In order to reduce computational effort, the feature map is converted into a feature matrix $X_p \in \mathbb{R}^{1*H1*W1}$ by max pooling. The correlation between the metadata and the feature matrix is calculated by calculating the Pearson correlation coefficient matrix $T \in \mathbb{R}^{C2*(W1*H1)}$. The Pearson correlation coefficient matrix is summed by rows to obtain the correlation scores $S \in \mathbb{R}^{C2}$ between the metadata feature and feature map. The metadata features with high correlation scores are used as the main features, and the metadata features with low correlation scores are used as auxiliary features by setting a threshold value for the scores. The main features are mapped to the higher dimension, and the auxiliary features are mapped to the lower dimension. Then concatenate both features to get the metadata features after feature selection $X^{'}_{meta} \in \mathbb{R}^{C3}$. The feature selection block can be formulated as follows:

$$T_{i,j} = Corr(X^i_{meta}, \ pooling(X_{img})^j) \tag{3}$$

where T represents the correlation matrix between the metadata and the feature map. Corr() for the calculation of the Pearson correlation coefficient, the X^i_{meta} and $pooling(X_{img})^j$ represent the ith feature of the metadata and the jth pixel of the feature map after pooling, respectively.

$$S_i = \sum_{i=1}^{C2} T_i \tag{4}$$

$$R_i = Rank(X^i_{meta}, S_i) \tag{5}$$

$$X^i_{meta} = \begin{cases} x_{main} \ R_i < \omega \\ x_{aux} \ \ R_i \geq \omega \end{cases} \tag{6}$$

where S_i represents the correlation score of the ith feature of the metadata with the feature map. R_i denotes the correlation rank between the ith feature and the feature map according to S_i. The metadata features are graded according to the relevance ranking, and the features ranked above the threshold are used as main logits X_{main} and vice versa as auxiliary logits X_{aux}.

$$X'_{meta} = Cat(\phi_{main}(X_{main}), \ \phi_{aux}(X_{aux})) = FS(X_{meta}, \ X_{img}) \tag{7}$$

The main and auxiliary logits will be mapped to different dimensions by encoder ϕ . Concatenating two vectors together to output the metadata logits X'_{meta} after feature selection.

4 Experimental Details

4.1 Dataset and Evaluation Metrics

This experiment was based on PAD-UFES-20 [16] dataset: included 2298 clinical images collected from smartphone devices, 21 patient clinical characteristics (e.g., age, gender, anatomical region, cancer history, photographic skin type, family background), and six skin lesions: basal cell carcinoma (BCC), squamous cell carcinoma (SCC), actinic keratosis (ACK), seborrheic keratosis (SEK), melanoma (MEL), and nevus (NEV). We followed the metadata preprocessing procedure in MetaBlock [14]. After that, we removed the images with significant ink annotations or semantic differences. Secondly, the images which a resolution below 100 kb was super-resolved by Real-ESRGAN [29], and the effect of super-resolution processing is shown in Fig. 4. At the same time, the images with severe hair occlusion are detailed using the black hat filter, and the hair-removed process is shown in Fig. 5. Finally, we expanded the data of each disease category in the training data to 2,000 by panning, reversing, and rotating the images. The experimental evaluation metrics are accuracy, recall, precision, F1-score, and AUC (Table 1).

Table 1. Introduction to dataset partitioning.

Dataset	ACK	BCC	MEL	NEV	SCC	SEV	Total
Train set	2530	2659	2040	2191	2147	2197	13765
Test set	107	131	7	39	27	37	348
Total	2637	2790	2047	2230	2174	2234	14113

Note: ACK: actinic keratosis; BCC: basal cell carcinoma; MEL: melanoma; NEV: nevus; SCC: squamous cell carcinoma; SEK: seborrheic keratosis.

4.2 Implementation Details

The experiments are based on Pytorch, NVIDIA 3070, using Adam optimizer [30]. The training epoch is 150, the learning rate is set to 0.001, and the batch size is 32. We set the random seed to 3407.

Fig. 4. The effect of super-resolved processing.

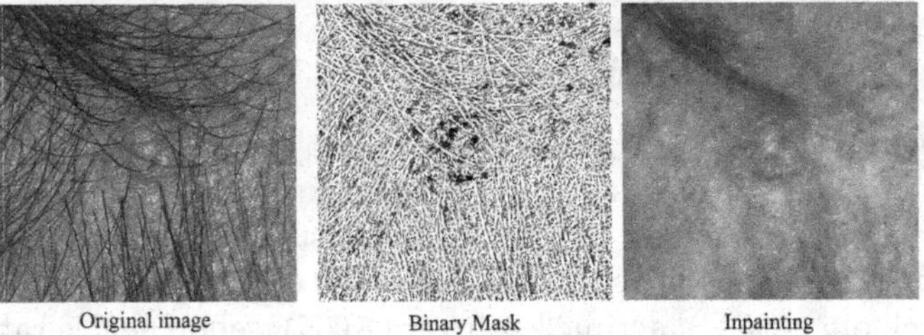

Fig. 5. Hair removal process.

Table 2. Evaluation of the effectiveness of different metadata fusion methods on classification accuracy. Note: No metadata represents using image single modal.

Fusion Method	Accuracy (%)	Precision (%)	Recall (%)	F1-score (%)	AUC
No Metadata	74.1	75.7	74.1	74.5	0.917
Cat	76.6	79.7	76.6	77.2	0.923
MetaNet [15]	75.0	77.8	75.0	75.9	0.939
MetaBlock [14]	77.5	80.0	77.5	78.2	0.933
MDF-Net [8]	79.3	80.4	79.3	79.7	0.942
DeAF [24]	78.1	79.5	78.1	78.3	0.950
MLDF-Net(No Fs)	77.9	81.4	77.9	78.9	0.951
MLDF-Net(Ours)	81.3	83.4	81.3	82.0	0.955

4.3 Experimental Results

As shown in Table 2, we compared MLDF-Net with other multi-modal fusion methods using the PAD-UFES-20 [16]. Among them, MG Block performs best in all evaluation metrics, and MLDF-Net achieves an accuracy of 81.3%. All multi-modal fusion methods outperformed metadata-free fusion in all evaluation metrics, and our proposed MLDF-Net obtained the best results. Compared with the metadata-free fusion, MLDF-Net improved the accuracy by 7.2%, the AUC by 0.038, the precision by 7.7%, the recall by 7.2%, and the F1-score by 7.5%. Meanwhile, compared with the traditional fusion method MetaBlock [14], our MLDF-Net accuracy improves by 3.8%, AUC value improves by 0.021, precision by 3.4%, recall by 3.8%, and F1-score by 3.8%. Compared with the sub-optimal fusion method MDF-Net [8], our MLDF-Net accuracy improves by 3.8%, AUC value improves by 0.016, precision by 3.4%, recall by 3.8%, and F1-score by 3.8%. Besides, MLDF-Net with the feature selection block performed better than without the feature selection block under all evaluation metrics. MG Block with the feature selection block improved AUC by 0.013, accuracy by 3.4%, precision by 3.0%, recall by 3.4%, and F1-score by 3.1% compared with MG block without the feature selection block.

Figure 6 depicts the confusion matrix of multiple multi-modal fusion methods. The results show that the proposed MLDF-Net achieves optimal performance in four of the six dermatological disease categories. From Fig. 6(a), it can be obtained that the multi-modal diagnostic approach outperforms the unimodal diagnostic approach without metadata fusion for most disease categories, most notably the multi-modal diagnostic performance on MEL identification, which improves the accuracy by 40% compared to the unimodal diagnostic. It is worth noting that all existing models are prone to confusion between SCC and BCC, with the optimal MetaBlock(d) achieving only 54% SCC recognition accuracy. In contrast, our MLDF-Net(e) achieves a breakthrough with an accuracy of 70%. Compared with the Metanet(c), which is the best performer in ACK recognition among other multi-modal fusion methods, the accuracy of MLDF-Net is improved by 5%, further enhancing the performance of multi-modal diagnosis in ACK recognition.

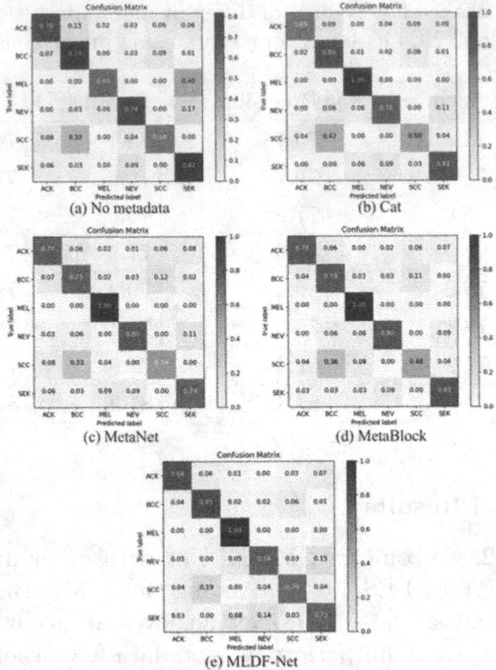

Fig. 6. Confusion matrix for different fusion methods. Note: ACK: actinic keratosis; BCC: basal cell carcinoma; MEL: melanoma; NEV: nevus; SCC: squamous cell carcinoma; SEK: seborrheic keratosis.

4.4 Ablation Experiment

As shown in Fig. 7, we plotted ROC curves for MLDF-Net with and without the feature selection block, respectively, to verify the enhancement effect of FS Block on MLDF-Net. Our MLDF-Net achieves excellent results in all kinds of disease recognition. In particular, MLDF-Net with feature selection block Fig. 7(b) improves the AUC-MacroAVG by 0.02 and AUC-MicroAVG by 0.01. Meanwhile, the dynamic feature selection block makes the model improve the AUC on SCC disease classification by 0.09, which is an impressive breakthrough.

Figure 8 presents the confusion matrix of MLDF-Net with and without the feature selection block. As can be seen in Fig. 8(b), the superior performance of MLDF-Net in SCC and BCC disease recognition is ly attributed to our FS Block, which improves the accuracy of SCC disease classification by 14% and significantly reduces the probability of misclassifying SCC as BCC. FS Block also enhanced the model's performance in ACK disease identification, with a 5% improvement in accuracy.

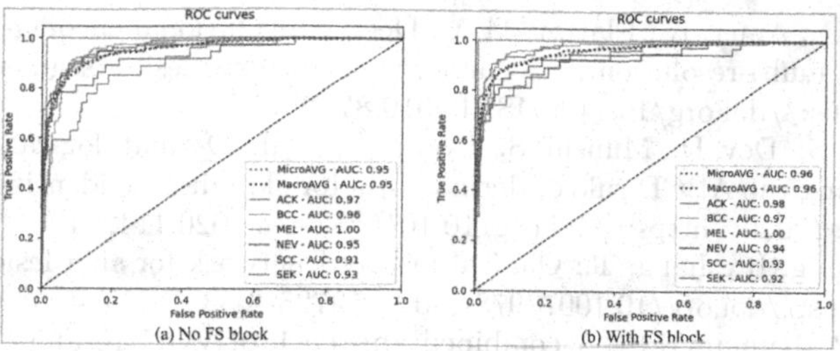

Fig. 7. ROC curves with and without FS Block. Note: ACK: actinic keratosis; BCC: basal cell carcinoma; MEL: melanoma; NEV: nevus; SCC: squamous cell carcinoma; SEK: seborrheic keratosis.

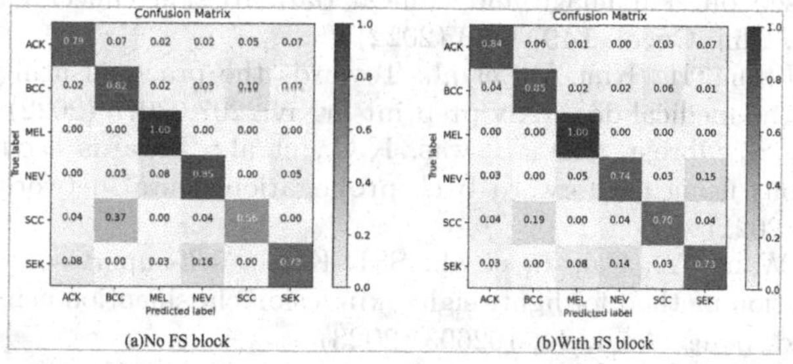

Fig. 8. Confusion matrix with and without FS Block. Note: ACK: actinic keratosis; BCC: basal cell carcinoma; MEL: melanoma; NEV: nevus; SCC: squamous cell carcinoma; SEK: seborrheic keratosis.

5 Summary

In this paper, we design a metadata-based multi-level dynamic fusion Network (MLDF-Net) to implement metadata to collaborate with images for skin cancer diagnosis. Specifically, we embed the MG Block into Resnet50 to fuse metadata and image to refine image features. We have designed an FS Block for MG Block that can dynamically adapt and select relevant metadata features, achieving dynamic adaptive matching between metadata and image features. The experimental results show that MLDF-Net outperforms the state-of-the-art multi-modal fusion methods under all evaluation metrics.

References

1. WHO: Ultraviolet (UV) radiation and skin cancer, World Health Organization (WHO) (2020). Accessed 22 Aug 2020. https://www.who.int/news-room/q-a-detail/ultraviolet-(uv)-radiation-and-skin-cancer
2. Codella, N.C.F., Nguyen, Q.B., Pankanti, S., et al.: Deep learning ensembles for melanoma recognition in dermoscopy images. IBM J. Res. Dev. **61**(4), 5:1–5:15 (2017). https://doi.org/10.1147/JRD.2017.2708299

3. Shahin, A.H., Amer, K., Elattar, M.A.: Deep convolutional encoder-decoders with aggregated multi-resolution skip connections for skin lesion segmentation. IEEE (2019). https://doi.org/10.1109/ISBI.2019.8759172
4. Chatterjee, S., Dey, D., Munshi, S., Gorai, S., et al.: Dermatological expert system implementing the ABCD rule of dermoscopy for skin disease identification. Expert Syst. Appl. (2020). https://doi.org/10.1016/j.eswa.2020.114204
5. Liu, Z., Xiong, R., Jiang, T.: Clinical-inspired network for skin lesion recognition (2020). https://doi.org/10.1007/978-3-030-59725-2_33
6. Arjun, K.P., Kumar, K.S.: A combined approach of VGG 16 and LSTM transfer learning technique for skin melanoma classification
7. Cai, G., Zhu, Y., Wu, Y., et al.: A multi-modal transformer to fuse images and metadata for skin disease classification. Vis. Comput. **39**, 1–13 (2022)
8. Chen, Q., Li, M., Chen, C., et al.: MDFNet: application of multi-modal fusion method based on skin image and clinical data to skin cancer classification. J. Cancer Res. Clin. Oncol. **149**, 1–13 (2022)
9. Yang, S., Hwang, H., Kim, D., et al.: Towards the practical utility of federated learning in the medical do. arXiv preprint arXiv:2207.03075 (2022)
10. Moldovanu, S., Obreja, C.D., Biswas, K.C., et al.: Towards accurate diagnosis of skin lesions using feedforward back propagation neural networks. Diagnostics **11**(6), 936 (2021)
11. Wang, Y., Wang, Y., Cai, J., et al.: SSD-KD: a self-supervised diverse knowledge distillation method for lightweight skin lesion classification using dermoscopic images. Med. Image Anal. **84**, 102693 (2023)
12. Kharazmi, P., Kalia, et al.: A feature fusion system for basal cell carcinoma detection through data-driven feature learning and patient profile. Skin Res. Technol. Off. J. Int. Soc. Bioeng. Skin (ISBS) **24**, 256–264 (2018). International Society for Skin Imaging (ISSI)
13. Liu, Y., Jain, A., Eng, C., et al.: A deep learning system for differential diagnosis of skin diseases (2019). https://doi.org/10.1038/s41591-020-0842-3
14. Pacheco, A.G.C., Krohling, R.: An attention-based mechanism to combine images and metadata in deep learning models applied to skin cancer classification. IEEE J. Biomed. Health Inform. **PP**(99), 1 (2021). https://doi.org/10.1109/JBHI.2021.3062002
15. Li, W., Zhuang, J., Wang, R., et al.: Fusing metadata and dermoscopy images for skin disease diagnosis. In: IEEE International Symposium on Biomedical Imaging. IEEE (2020). https://doi.org/10.1109/ISBI45749.2020.9098645
16. Pacheco, A.G.C., et al.: PAD-UFES-20: a skin lesion dataset composed of patient data and clinical images collected from smartphones. Mendeley Data **V1** (2020). https://doi.org/10.17632/zr7vgbcyr2.1
17. Xu, Z., Luo, J., Yan, J., et al.: Adversarial uni- and multi-modal stream networks for multi-modal image registration (2020). https://doi.org/10.1007/978-3-030-59716-0_22
18. Borràs, E., Ferré, J., Boqué, R., et al.: Olive oil sensory defects classification with data fusion of instrumental techniques and multivariate analysis (PLS-DA). Food Chem., 314–322 (2016). https://doi.org/10.1016/j.foodchem.2016.02.038
19. Pacheco, A.G.C., Krohling, R.A.: The impact of patient clinical information on automated skin cancer detection (2019). https://doi.org/10.1016/j.compbiomed.2019.103545
20. Kawahara, J., Daneshvar, S., Argenziano, G., et al.: Seven-point checklist and skin lesion classification using multitask multi-modal neural nets. IEEE J. Biomed. Health Inform., 538–546 (2018). https://doi.org/10.1109/JBHI.2018.2824327

21. Cao, M., Yang, M., Qin, C., et al.: Using DeepGCN to identify the autism spectrum disorder from multi-site resting-state data. Biomed. Signal Process. Control **70**, 103015 (2021). https://doi.org/10.1016/j.bspc.2021.103015
22. Yan, R., Zhang, F., Rao, X., et al.: Richer fusion network for breast cancer classification based on multi-modal data. BMC Med. Inform. Decis. Mak. **21**(1), 1–15 (2021)
23. Cui, H., et al.: Co-graph attention reasoning based imaging and clinical features integration for lymph node metastasis prediction. In: de Bruijne, M. (ed.) MICCAI 2021. LNCS, vol. 12905, pp. 657–666. Springer, Cham (2021). https://doi.org/10.1007/978-3-030-87240-3_63
24. Li, K., Chen, C., Cao, W., et al.: DeAF: a multi-modal deep learning framework for disease prediction. Comput. Biol. Med. **156**, 106715 (2023)
25. Zhou, J., Zhang, X., Zhu, Z., et al.: Cohesive multi-modality feature learning and fusion for COVID-19 patient severity prediction. IEEE Trans. Circuits Syst. Video Technol. **PP**(99), 1 (2021) https://doi.org/10.1109/TCSVT.2021.3063952
26. Tang, P., Yan, X., Nan, Y., et al.: FusionM4Net: a multi-stage multi-modal learning algorithm for multi-label skin lesion classification. Med. Image Anal. **76**, 102307 (2022). https://doi.org/10.1016/j.media.2021.102307
27. Wang, Y., Haq, N.F., Cai, J., et al.: Multi-channel content based image retrieval method for skin diseases using similarity network fusion and deep community analysis. Biomed. Signal Process. Control **78**, 103893 (2022)
28. He, K., Zhang, X., Ren, S., Sun, J.: Deep residual learning for image recognition. In: IEEE Conference on Computer Vision and Pattern Recognition, pp. 770–778 (2016)
29. Wang, X., Xie, L., Dong, C., et al.: Real-ESRGAN: training real-world blind super-resolution with pure synthetic data (2021). https://doi.org/10.48550/arXiv.2107.10833
30. Kingma, D.P., Ba, J.: Adam: a method for stochastic optimization. arXiv preprint arXiv:1412.6980 (2014)

Efficient Adversarial Training with Membership Inference Resistance

Ran Yan[1], Ruiying Du[1,2(✉)], Kun He[1], and Jing Chen[1,3]

[1] Key Laboratory of Aerospace Information Security and Trusted Computing, Ministry of Education, School of Cyber Science and Engineering, Wuhan University, Wuhan 430072, China
{yanran22,hekun,chenjing}@whu.edu.cn

[2] Collaborative Innovation Center of Geospatial Technology, Wuhan 430079, China
duraying@whu.edu.cn

[3] Rizhao Institute of Information Technology, Wuhan University, Rizhao 276800, China

Abstract. Deep cross-media computing faces adversarial example attacks, adversarial training is an effective approach to enhance the robustness of machine learning models via adding adversarial examples into the training phase. However, existing adversarial training methods increase the advantage of membership inference attacks, which aim to determine from the model whether an example is in the training dataset. In this paper, we propose an adversarial training framework that guarantees both robustness and membership privacy by introducing a tailor-made example, called reverse-symmetry example. Moreover, our framework reduces the number of required adversarial examples compared with existing adversarial training methods. We implement the framework based on three adversarial training methods on FMNIST and CIFAR10. The experimental results show that our framework outperforms the original adversarial training with respect to the overall performance of accuracy, robustness, privacy, and runtime.

Keywords: Adversarial training · adversarial examples · membership inference attacks

1 Introduction

Deep Neural Networks (DNNs) have been employed across various real-world applications in deep cross-media computing, such as intelligent image analysis [7, 17], natural language processing [1], and speech recognition [4]. Unfortunately, deep learning models trained by DNNs are found to be vulnerable to evasion attacks [3,12], in which an attacker misleads deep learning models by adding imperceptible perturbations to natural examples, called adversarial examples. In practice, those adversarial examples can be used for crime by tricking models deployed in daily applications, such as facial recognition [15].

© The Author(s), under exclusive license to Springer Nature Singapore Pte Ltd. 2024
Q. Liu et al. (Eds.): PRCV 2023, LNCS 14425, pp. 474–486, 2024.
https://doi.org/10.1007/978-981-99-8429-9_38

One of the promising approaches to defend against evasion attacks is *adversarial training*, which can be classified into two categories: *empirical* and *verifiable* [20]. The core idea of empirical adversarial training is to add adversarial examples in the training phase of a model to enhance the robustness [10]. To make the adversarial examples more representative of the adversarial domain, adversarial training usually employs multiple iterative processes [18], such as Projected Gradient Descent (PGD) [8]. Therefore, empirical adversarial training suffers from high runtime in the training phase. On the other hand, verifiable adversarial training estimates the adversarial domain around natural examples in the training phase, which usually does not significantly increase the runtime [5,11]. However, since the estimation is done for the worst case, verifiable adversarial training sacrifices the accuracy of the model.

In addition to the above-mentioned performance issues, existing adversarial training methods also increase the privacy risk of deep learning models [20]. Membership inference attack is a typical privacy threat, in which an attacker tries to determine whether a given example is used to train a model [16]. This kind of privacy leakage is dangerous in many applications. Taking medical recognition as an example, an attacker can infer the membership information of a patient's medical record from a special disease diagnosis model by the membership inference attack, which violates the patient's privacy. Experimental results in [20] showed that, compared with the normally trained models, both empirical and verifiable adversarial training lead to an increase in the advantage of membership inference attacks.

In this paper, we aim to reinforce the paradigm of current empirical adversarial training so as to reduce the privacy risk and runtime while maintaining robustness and accuracy. We focus on empirical adversarial training due to its high accuracy compared with verifiable methods. The key insight is that the adversarial examples generated in the training phase are clustered in specific areas, which means that the model may only focus on local features of the adversarial domain. Therefore, we introduce additional tailor-made examples to reduce the model's attention to the adversarial domain, which also reduces the privacy risk. Since those tailor-made examples can be efficiently generated and they retain the features of natural examples, we can replace parts of adversarial examples with those tailor-made examples to reduce the runtime without significantly reducing robustness and accuracy.

In summary, we make the following main contributions.

- We propose a framework that can integrate and enhance existing adversarial training methods. Moreover, we extend our framework to tune the trade-offs between accuracy, robustness, privacy, and runtime.
- To demonstrate the effectiveness of our framework, we implement the framework based on three empirical adversarial training methods on FMNIST and CIFAR10. The results show that our framework has better overall performance than the underlying methods.

- To explore the trade-offs in our framework, we implement 9 variants on FMNIST and CIFAR10. The results provide helpful guidance for developers to choose settings that meet their various requirements.

2 Background and Related Work

2.1 Workflow of Empirical Adversarial Training

Empirical adversarial training is an effective way to resist adversarial examples. The main idea of empirical adversarial training is to transform the problem of finding a robust model into an optimization problem that minimizes the combination of natural loss and robust loss.

At the beginning of the training phase, a model is initialized by determining its architecture and hyper-parameters, such as learning rate and step size. Then, the model is repeatedly trained for a certain number of training rounds. Finally, an optimal model is outputted at the end of the training phase. In each training round of adversarial training, the model F_θ learns the training dataset through the following three procedures: (1) *Sampling.* A set of $(\mathbf{x}_{nat}, y)$, i.e., batch data $(\mathbf{X}_{nat}, \mathbf{y})$, is sampled from the training dataset as same as the way in natural training. (2) *Generating.* For each $(\mathbf{x}_{nat}, y)$ in the batch data, an adversarial example $\mathbf{x}_{adv}$ is generated. (3) *Updating.* The weights θ is updated by minimizing $\sum_{\mathbf{x}_{nat} \in \mathbf{X}_{nat}} \ell(F_\theta, (\mathbf{x}_{nat}, \mathbf{x}_{adv}, y))$. More specifically,

$$\begin{aligned}\ell(F_\theta, (\mathbf{x}_{nat}, \mathbf{x}_{adv}, y)) = \alpha \cdot \ell_n(F_\theta, (\mathbf{x}_{nat}, y)) \\ + (1-\alpha) \cdot \ell_r(F_\theta, (\mathbf{x}_{nat}, \mathbf{x}_{adv}, y)),\end{aligned}$$

where ℓ_n is the natural loss as in natural training, i.e., cross-entropy loss, ℓ_r is the robust loss, and α is used to balance the natural loss and robust loss.

2.2 Membership Inference Attacks

Membership inference attacks aim at determining whether an example was in the training dataset of a given machine learning model [2,9]. This kind of attack may leak sensitive information of individuals once combined with background knowledge about the model. Yeom et al. [22] consider an example as a member of the training dataset if the final prediction corresponds with the ground-truth label. Salem et al. [14] set a threshold for the prediction confidence to determine membership, which does not require an attacker to decide the ground-truth label for target examples. Their experimental results show that this *confidence thresholding* method can obtain a similar inference accuracy with that of a complex attack methods. Song et al. [19] proposed to use the entropy of the confidence as the threshold to implement the membership inference attack. To improve the inference advantage, we use the confidence and its corresponding ground-truth label to compute cross entropy as the threshold, called *cross-entropy thresholding*, to implement the membership inference attack.

3 Our Method

We propose a privacy-enhancing adversarial training framework, called PINEAT, which is a recursive acronym for PINEAT Is Not Exactly Adversarial Training. After the design goals are stated, we give an overview of our framework and then describe the details.

3.1 Design Goals

Our framework aims to achieve the following goals.

- **Universal applicability.** Our framework should be compatible with existing (empirical) adversarial training methods. Specifically, we should not modify the generation of adversarial examples and loss functions.
- **Adversarial robustness.** The model trained by our framework should be able to resist adversarial examples. Moreover, the adversarial robustness should be similar to that of the underlying adversarial training method.
- **Membership privacy.** Compared with the underlying adversarial training method and even natural training method, the model trained by our framework should be resistant to the membership inference attack.
- **Runtime reduction.** The runtime of our framework should be lower than the runtime of the underlying adversarial training methods.

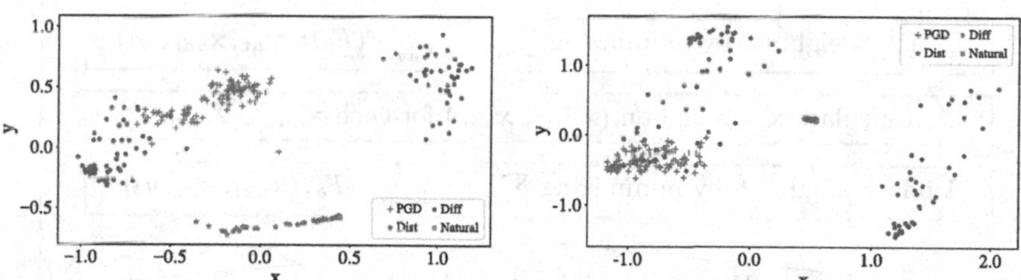

(a) The index of natural example is 300 in FMNIST. (b) The index of natural example is 304 in FMNIST.

Fig. 1. Two natural examples and all their adversarial examples via PCA dimension reduction [13].

3.2 Overview

We have analyzed the adversarial examples generated in three main adversarial training methods, including *PGD-based* [10], *Distribution-based* (Dist-based) [18], and *Difference-based* (Diff-based) [23]. The key observation is that adversarial examples for the training dataset are clustered around specific areas of natural examples, as shown in Fig. 1. Therefore, the model may overfit in those areas, i.e., adversarial domains around natural examples, resulting in an increased privacy risk of the model in adversarial domains.

To alleviate the model's excessive attention to adversarial domains, we can introduce additional examples in the training phase. These additional examples should increase the dispersion of all examples based on the same natural example used for training as much as possible. Therefore, we design a tailor-made example, called *reverse-symmetry example*, which is symmetrical with an adversarial example centered at the natural example. To keep the ground-truth label of the reverse-symmetry example consistent with that of the natural example, we require the reverse-symmetry example to satisfy the perturbation constraint as the adversarial example, i.e., $\mathbf{x}_{rs} \in \mathcal{B}_\epsilon(\mathbf{x}_{nat})$, where $\mathbf{x}_{rs}$ is the reverse-symmetry example and $\mathbf{x}_{nat}$ is the natural example.

To reduce the training runtime, we replace parts of adversarial examples whose generation is time-consuming with lightweight reverse-symmetry examples. Since reverse-symmetry examples retain the features of natural examples and enough adversarial examples are involved, this replacement will not reduce accuracy and robustness.

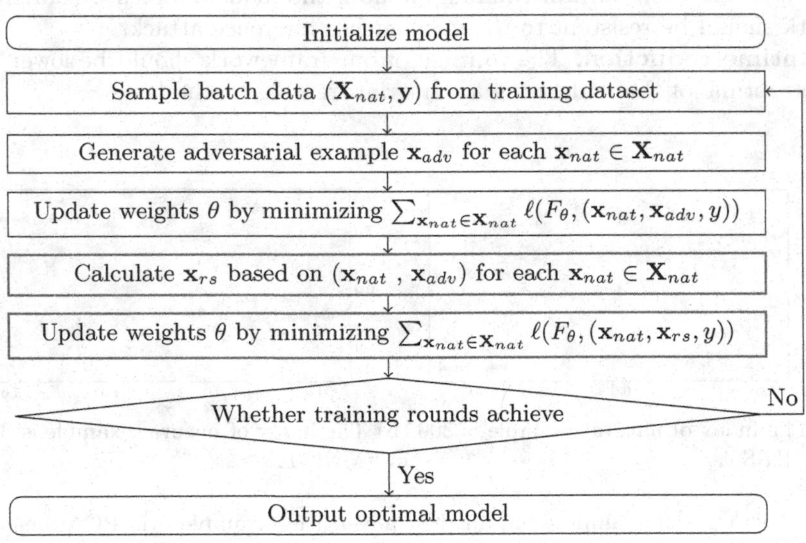

Fig. 2. Workflow of PINEAT, where the red boxes indicate the differences from empirical adversarial training. (Color figure online)

3.3 Design Details

PINEAT is a framework that extends the current empirical adversarial training paradigm, as shown in Fig. 2. Compared with existing adversarial training methods, PINEAT introduces two additional procedures to each training round.

First, an reverse-symmetry example is calculated based on the natural example and the corresponding adversarial example, satisfying $\mathbf{x}_{rs} + \mathbf{x}_{adv} = \mathbf{x}_{nat} * 2$, where $\mathbf{x}_{rs}$ is the reverse-symmetry example, $\mathbf{x}_{adv}$ is the adversarial example, and $\mathbf{x}_{nat}$ is the natural example. Since $\mathbf{x}_{adv}$ is generated under the perturbation constraint $\mathcal{B}_\epsilon(\mathbf{x}_{nat})$, $\mathbf{x}_{rs}$ also satisfies $\mathcal{B}_\epsilon(\mathbf{x}_{nat})$ and is indistinguishable from $\mathbf{x}_{nat}$ by human eye, which means that $\mathbf{x}_{rs}$ and $\mathbf{x}_{nat}$ have the same ground-truth label. Second, the model is updated again by minimizing the combination of loss on the natural examples and loss on reverse-symmetry examples. The loss function depends on the underlying adversarial training method.

Note that since the model is updated twice in a training round, our framework halves the number of training rounds required by the underlying adversarial training method. In other words, half of the adversarial examples generated in traditional adversarial training methods are replaced by reverse-symmetry examples in PINEAT. We denote this setting PINEAT-1/1, which means the ratio of the number of adversarial examples and the number of reverse-symmetry examples is 1:1.

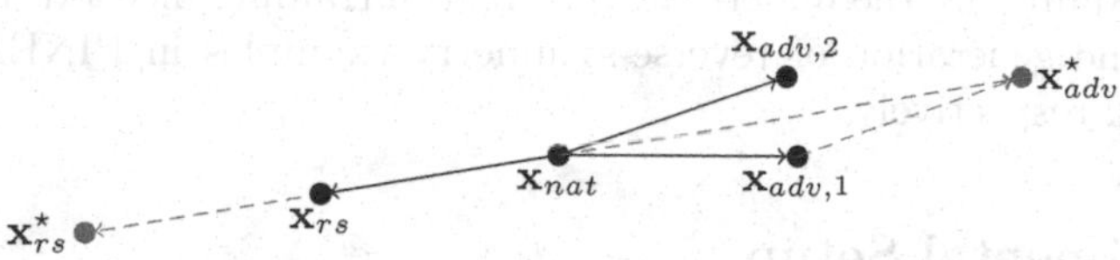

Fig. 3. Generation of reverse-symmetry examples in PINEAT-2/1. After generating two adversarial examples for the natural example, we aggregate the noises in $\mathbf{x}_{adv,1}$ and $\mathbf{x}_{adv,2}$ against the natural example and get $\mathbf{x}^\star_{adv}$. The red line represents the noise tensor $\mathbf{x}_{adv,2} - \mathbf{x}_{nat}$ that is translated from $\mathbf{x}_{nat}$ to $\mathbf{x}_{adv,1}$. Therefore, we have $\mathbf{x}^\star_{adv} - \mathbf{x}_{nat} = (\mathbf{x}_{adv,1} - \mathbf{x}_{nat}) + (\mathbf{x}_{adv,2} - \mathbf{x}_{nat})$. Then, we calculate $\mathbf{x}^\star_{rs}$ for $\mathbf{x}^\star_{adv}$ where $\mathbf{x}^\star_{rs} + \mathbf{x}^\star_{adv} = 2\mathbf{x}_{nat}$. Finally, we get $\mathbf{x}_{rs}$ by clip $\mathbf{x}^\star_{rs}$, i.e., $\mathbf{x}_{rs} = min(max(\mathbf{x}^\star_{rs}, \mathbf{x}_{nat} - \epsilon), \mathbf{x}_{nat} + \epsilon)$.

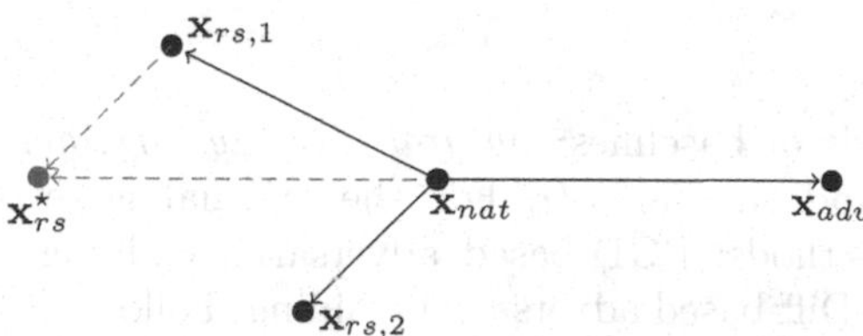

Fig. 4. Generation of reverse-symmetry examples in PINEAT-1/2. We calculate $\mathbf{x}^\star_{rs}$ which is symmetric with $\mathbf{x}_{adv}$ respect to $\mathbf{x}_{nat}$. Then, we decompose the noise in $\mathbf{x}^\star_{rs}$ into the two reverse-symmetry examples $\mathbf{x}_{rs,1}$ and $\mathbf{x}_{rs,2}$. The red line represent the noise tensor assigned to $\mathbf{x}^2_{rs}$, i.e., $\mathbf{x}^\star_{rs} - \mathbf{x}_{rs,1} = \mathbf{x}_{rs,2} - \mathbf{x}_{nat}$. In this setting, the $\mathbf{x}_{rs,1}$ and $\mathbf{x}_{rs,2}$ meet the $\mathcal{B}_\epsilon(\mathbf{x}_{nat})$ without cliping, due to $\mathbf{x}^\star_{rs} \in \mathcal{B}_\epsilon(\mathbf{x}_{nat})$.

Variants. We can adjust the adversarial/reverse-symmetry examples ratio to obtain different settings, which also means the number of model updates in a training round and the number of training rounds are changed accordingly. Specifically, in each training round of PINEAT-n/m, m reverse-symmetry examples are calculated for n adversarial examples, satisfying that $\mathbf{x}^{\star}_{adv} + \mathbf{x}^{\star}_{rs} = \mathbf{x}_{nat} * 2$, where $\mathbf{x}^{\star}_{adv}$ aggregates the noises in n adversarial examples (i.e., $\mathbf{x}^{\star}_{adv} - \mathbf{x}_{nat} = \sum_{i=1}^{n}(\mathbf{x}_{adv,i} - \mathbf{x}_{nat})$, where $\mathbf{x}_{adv,i}$ represents the i-th adversarial example) and the noise in $\mathbf{x}^{\star}_{rs}$ is decomposed into m reverse-symmetry examples. To generate adversarial examples, we use n different random seeds for one natural example. To calculate multiple reverse-symmetry examples, firstly we randomly generate $m-1$ reverse-symmetry examples under the perturbation constraint $\mathcal{B}_{\epsilon}(\mathbf{x}_{nat})$. Then, we calculate the last reverse-symmetry example by $\mathbf{x}_{rs,m} = \min(\max(\mathbf{x}^{\star}_{rs} - \sum_{j=1}^{m-1}\mathbf{x}_{rs,j}, \mathbf{x}_{nat} - \epsilon), \mathbf{x}_{nat} + \epsilon)$, where $\mathbf{x}_{rs,j}$ represents the j-th reverse-symmetry example. The function $\min(\max(\cdot, \mathbf{x}_{nat} - \epsilon), \mathbf{x}_{nat} + \epsilon)$ guarantees that the last reverse-symmetry example also satisfies the perturbation constraint $\mathcal{B}_{\epsilon}(\mathbf{x}_{nat})$. In PINEAT-$n/m$, the model is updated $n+m$ times in one training round thus our framework needs about $1/(n+m)$ of the number of training rounds required by the underlying adversarial training method. Figure 3 and 4 illustrate the generation of reverse-symmetry examples in PINEAT-2/1 and PINEAT-1/2, respectively.

4 Experimental Setup

4.1 Implementation, Dataset, and Model Architecture

The PGD-based adversarial training and Dist-based adversarial training are implemented based on TensorFlow, and the Diff-based adversarial training is implemented based on Pytorch. All implementations run on a server with Intel Xeon E5-2680, three Nvidia Tesla V100 32GB GPU, and 128G RAM. The datasets and model architectures used in our experiments are identical with those in [20]. Specifically, we adopt two standard datasets: *FMNIST* [21] and *CIFAR10* [6].

4.2 Baselines

We adopt four kinds of baselines: *natural training*, *original adversarial training*, *AdvNat-1/1*, and *AdvRan-1/1*. For the original adversarial training, we implement three methods: PGD-based adversarial training, Dist-based adversarial training, and Diff-based adversarial training. Following [20], we set the l_∞ perturbation budget ϵ to be 0.1 on FMNIST and 8/255 on CIFAR10.

For each adversarial training, we develop two additional baselines: AdvNat-1/1 and AdvRan-1/1. In AdvRan-1/1, we replace half of adversarial examples with random examples near the natural example under the perturbation constraint $\mathcal{B}_{\epsilon}(\mathbf{x}_{nat})$. In AdvNat-1/1, we use the natural example to replace the half of adversarial examples.

Table 1. Accuracy, robustness, privacy, and runtime of the model trained by different methods on FMNIST and CIFAR10, where the bold number indicates the best value between the methods on the same dataset and the underlined number indicates the best value between the methods based on the same adversarial training on the same dataset.

DS	Method		Accuracy		Robustness		Privacy			Time
			A_{train}	A_{test}	R_{train}	R_{test}	P_{Enat}	P_{Eadv}	P_{MAX}	
FMNIST	Natural		**1.0000**	0.9281	0.0534	0.0514	0.5718	**0.5011**	0.5718	1.00 h
	PGD	Original	0.9993	0.9088	**0.9692**	0.6776	0.5910	0.6493	0.6493	6.18 h
		AdvNat-1/1	**1.0000**	0.9160	0.9148	0.6737	0.5877	0.6282	0.6282	3.88 h
		AdvRan-1/1	0.9999	0.9129	0.9661	0.6730	0.6008	0.6506	0.6506	3.93 h
		PINEAT-1/1	0.9977	0.9136	0.8791	0.6824	0.5683	0.6052	0.6052	3.90 h
	Dist	Original	0.9795	0.9084	0.6768	0.5139	0.5975	0.6043	0.6043	6.78 h
		AdvNat-1/1	**1.0000**	**0.9333**	0.5073	0.4494	0.5819	0.5562	0.5819	3.85 h
		AdvRan-1/1	0.9731	0.9090	0.3409	0.3108	0.5562	0.5330	0.5562	3.90 h
		PINEAT-1/1	0.9332	0.8990	0.6175	0.5371	**0.5410**	0.5517	**0.5517**	3.82 h
	Diff	Original	0.9933	0.9082	0.9000	0.7264	0.5807	0.5936	0.5936	8.83 h
		AdvNat-1/1	0.9830	0.9093	0.8353	0.7258	0.5601	0.5614	0.5614	4.60 h
		AdvRan-1/1	0.9828	0.9095	0.8367	0.7250	0.5621	0.5625	0.5625	5.13 h
		PINEAT-1/1	0.9832	0.9102	0.8495	**0.7305**	0.5616	0.5666	0.5666	5.12 h
CIFAR10	Natural		**1.0000**	**0.9528**	0.0000	0.0000	0.5769	**0.5001**	**0.5769**	11.18 h
	PGD	Original	**1.0000**	0.8704	**0.9838**	0.4726	0.7627	0.7892	0.7892	80.75 h
		AdvNat-1/1	0.9999	0.8900	0.8993	0.4679	0.6638	0.7264	0.7264	44.24 h
		AdvRan-1/1	0.9894	0.8703	0.7763	0.4572	0.6412	0.6747	0.6747	45.25 h
		PINEAT-1/1	0.9774	0.8687	0.7365	0.4706	0.6182	0.6490	0.6490	44.07 h
	Dist	Original	**1.0000**	0.9027	0.4035	0.2686	0.6760	0.6434	0.6760	73.90 h
		AdvNat-1/1	**1.0000**	0.9259	0.3666	0.2940	0.6148	0.5957	0.6148	39.35 h
		AdvRan-1/1	0.9996	0.8950	0.3127	0.2382	0.6629	0.6273	0.6629	39.05 h
		PINEAT-1/1	0.9996	0.9039	0.3093	0.2410	0.6560	0.6242	0.6560	39.67 h
	Diff	Original	0.9975	0.8917	0.7175	**0.4749**	0.6190	0.6541	0.6541	35.18 h
		AdvNat-1/1	0.9725	0.8771	0.5502	0.4398	0.5754	0.5916	0.5916	20.67 h
		AdvRan-1/1	0.9796	0.8885	0.4555	0.3906	**0.5748**	0.5830	0.5830	22.50 h
		PINEAT-1/1	0.9941	0.9053	0.5458	0.4223	0.5899	0.6000	0.6000	22.55 h

4.3 Metrics

We evaluate the performance of our basic framework and its variants from the following four aspects.

- *Accuracy.* This metric is the proportion of examples in the dataset that are categorized to the correct class. The accuracy of the models on the natural training example is denoted as A_{train} and the accuracy of the models on the natural test example is denoted as A_{test}.
- *Robustness.* We use the ability of a model to correctly classify adversarial examples to indicate the robustness of the model. The adversarial examples are generated via PGD method for each natural example in the dataset. We denote the accuracy on the adversarial examples generated from training dataset as R_{train}, and the accuracy on the adversarial examples from test dataset as R_{test}.
- *Privacy.* We use membership inference accuracy to measure the privacy of a model. We use P_{Enat} to denote the accuracy of CrossEntropy-thresholding

method which exploits natural examples, and P_{Eadv} to denote the accuracy of the method which exploits adversarial examples generated under $\mathcal{B}_\epsilon$.
- *Runtime.* This metric refers to the time for training a model in our experimental environment.

5 Experimental Results

5.1 Performance of Our Basic Framework

We evaluate the accuracy, robustness, privacy, and runtime of our basic framework PINEAT-1/1 and the baselines. The performance comparison is shown in Table 1.

Observation 1. *In terms of accuracy, the model trained by PINEAT-1/1 in most cases performs better than the original adversarial training method and AdvRan-1/1, but performs worse than AdvNat-1/1.*

We focus on the metric A_{test} since it can better reflect the accuracy of the model in practice. Diff-based PINEAT-1/1 outperforms the corresponding original method, AdvNat-1/1, and AdvRan-1/1 on both FMNIST and CIFAR10. However, PGD-based and Dist-based AdvNat-1/1 outperform others, with an increase of at most 3.43%.

Observation 2. *In terms of robustness, the model trained by PINEAT-1/1 performs similarly to the original adversarial training method and in most cases performs better than AdvNat-1/1 or AdvRan-1/1.*

We focus on the metric R_{test} since it can better reflect the robustness of the model in practice. On FMNIST, PINEAT-1/1 outperforms the original method while the latter outperforms AdvNat-1/1 and AdvRan-1/1. On CIFAR10, the original method in most cases outperforms others while the difference between PINEAT-1/1 and the original method is at most 5.26%.

Observation 3. *In terms of privacy, the model trained by PINEAT-1/1 performs better than the original adversarial training method, similar to AdvNat-1/1, and in most cases better than AdvRan-1/1.*

We focus on the metric P_{MAX} since in practice an attacker can implement various membership inference attacks and use the best result. PINEAT-1/1 outperforms the original method on both FMNIST and CIFAR10, with an increase of at least 2% and at most 14.02%. Moreover, PINEAT-1/1 sometimes even outperforms the natural training method.

Observation 4. *In terms of runtime, PINEAT-1/1 performs better than the original adversarial training method and performs similarly to AdvNat-1/1 and AdvRan-1/1.*

The runtime depends on the number of adversarial examples generated in the training phase, which is a time-consuming operation. Therefore, the natural training method always has the best runtime while the original adversarial training method always has the worst. PINEAT-1/1, AdvNat-1/1, and AdvRan-1/1 have a similar runtime since half of the adversarial examples are replaced in these methods. AdvNat-1/1 has the least runtime, because it misses a step compared to AdvRan-1/1 (generate random examples) and PINEAT-1/1 (calculate tailor-made examples).

Observation 5. *PINEAT-1/1 achieves a good balance between accuracy, robustness, privacy, and runtime.*

This observation can be obtained directly from the previous three observations. Specifically, in terms of overall performances, the best method on FMNIST and CIFAR10 is Diff-based PINEAT-1/1.

5.2 Performance of Our Framework Variants

In PINEAT-n/m, n adversarial examples and m reverse-symmetry examples are generated in a training round and the number of training rounds is determined by $m+n$ (i.e., $1/(n+m)$ of that of the underlying adversarial training method). We focus on the accuracy, robustness, privacy, and runtime with the change of adversarial/reverse-symmetry example ratio n/m. The performance comparison is shown in the supplementary material table.

Observation 6. *Given the number of training rounds, the accuracy increases as the adversarial/reverse-symmetry example ratio decreases in most cases.*

We focus on the metric A_{test} as in Observation 1. This observation applies to most cases, except for PGD-based and Dist-based PINEAT-1/m on FMNIST. We believe that when there are few adversarial examples, reverse-symmetry examples may disturb the model boundary. We think this phenomenon is related to the loss function. The loss function of Diff-based method contains the natural loss ($\alpha = 1/2$) while PGD-based and Dist-based methods only have the robust loss ($\alpha = 0$), thus the model boundary in the latter two methods is susceptible to interference of reverse-symmetry examples' bias when the example space is small. In addition, most (58.3%) variants perform better than the underlying adversarial training method and some variants are even close to the natural training method, such as Dist-based PINEAT-2/3 on FMNIST and Diff-based PINEAT-1/3 on CIFAR10.

Observation 7. *Given the number of reverse-symmetry examples calculated in a training round, the robustness decreases as the adversarial/reverse-symmetry example ratio decreases in most cases.*

We focus on the metric R_{test} as in Observation 2. This observation applies to most cases, except for four variants. In Dist-based PINEAT-1/2 and PINEAT-1/3 on CIFAR10, the robustness of models does not fall but increases. Dist-based adversarial examples are not of high quality, showing that robustness of model trained by the original Dist-based is much lower than the other two original methods. Then, the random reverse-symmetry examples near the natural example maybe instead play the role of adversarial examples. In the PGD-based PINEAT-1/2 and PINEAT-1/3 on FMNIST, the robustness of models also increases. We believe that the PGD-based adversarial examples are too concentrated in specific area, and the reverse-symmetry examples hit possible adversarial area that the PGD-based adversarial examples did not consider before. In addition, most (64.4%) variants perform better than the underlying adversarial training method.

Observation 8. *Given the number of adversarial examples generated in a training round, the privacy becomes better as the adversarial/reverse-symmetry example ratio decreases.*

We focus on the metric P_{MAX} as in Observation 3. This observation applies to most cases, except for Dist-based PINEAT-1/4 on FMNIST. We believe that the special case is related to the characteristics of adversarial examples generated in Dist-based methods. The Dist-based methods tend to add noise to a small number of pixels in the image and add extremely negligible noise on the remaining pixels. In the Dist-based PINEAT-1/4 on FMNIST, the adversarial/reverse-symmetry example ratio is very small and the multiple reverse-symmetry examples are very similar to the natural examples. That means the training process is similar to natural training, but the privacy result is still better than that of natural model. In addition, all variants perform better than the underlying adversarial training method and most (55.9%) variants are even better than the natural training method.

Observation 9. *The runtime decreases as the adversarial/reverse-symmetry example ratio decreases.*

This observation can be inferred from Observation 6. In terms of runtime, our methods are superior to the underlying adversarial training method.

6 Conclusion

In this paper, we propose a framework based on empirical adversarial training, named PINEAT. With the design of reverse-symmetry examples, our framework can eliminate the negative impact of adversarial training on the model privacy while retaining its robustness advantage. Our experimental results show that the basic framework and its variants are better than the traditional adversarial training methods in terms of the overall performances.

Acknowledgements. This research was supported in part by the National Key R&D Program of China under grant No. 2022YFB3102100, the National Natural Science Foundation of China under grants No. 62076187, 62172303, the Key R&D Program of Hubei Province under grant No. 2022BAA039, and Key R&D Program of Shandong Province under grant No. 2022CXPT055.

References

1. Andor, D., et al.: Globally normalized transition-based neural networks. In: ACL (2016). https://doi.org/10.18653/v1/p16-1231
2. Carlini, N., Liu, C., Erlingsson, Ú., Kos, J., Song, D.: The secret sharer: evaluating and testing unintended memorization in neural networks. In: USENIX Security Symposium (2019)
3. Carlini, N., Wagner, D.A.: Towards evaluating the robustness of neural networks. In: S&P (2017). https://doi.org/10.1109/SP.2017.49
4. Deng, L., Hinton, G.E., Kingsbury, B.: New types of deep neural network learning for speech recognition and related applications: an overview. In: ICASSP (2013). https://doi.org/10.1109/ICASSP.2013.6639344
5. Gowal, S., et al.: Scalable verified training for provably robust image classification. In: ICCV (2019). https://doi.org/10.1109/ICCV.2019.00494
6. Krizhevsky, A., Hinton, G.: Learning multiple layers of features from tiny images. Master's thesis, University of Toronto (2009)
7. Krizhevsky, A., Sutskever, I., Hinton, G.E.: Imagenet classification with deep convolutional neural networks. In: NIPS (2012)
8. Kurakin, A., Goodfellow, I.J., Bengio, S.: Adversarial machine learning at scale. CoRR (2016). https://arxiv.org/abs/1611.01236
9. Leino, K., Fredrikson, M.: Stolen memories: leveraging model memorization for calibrated white-box membership inference. In: USENIX Security Symposium (2020). https://www.usenix.org/conference/usenixsecurity20/presentation/leino
10. Madry, A., Makelov, A., Schmidt, L., Tsipras, D., Vladu, A.: Towards deep learning models resistant to adversarial attacks. In: ICLR (2018)
11. Mirman, M., Gehr, T., Vechev, M.T.: Differentiable abstract interpretation for provably robust neural networks. In: ICML (2018)
12. Papernot, N., McDaniel, P.D., Jha, S., Fredrikson, M., Celik, Z.B., Swami, A.: The limitations of deep learning in adversarial settings. In: EuroS&P (2016). https://doi.org/10.1109/EuroSP.2016.36
13. Pearson, K.: LIII. on lines and planes of closest fit to systems of points in space. London Edinburgh Dublin Philos. Mag. J. Sci. (1901). https://doi.org/10.1080/14786440109462720
14. Salem, A., Zhang, Y., Humbert, M., Berrang, P., Fritz, M., Backes, M.: ML-leaks: model and data independent membership inference attacks and defenses on machine learning models. In: NDSS (2019)
15. Sharif, M., Bhagavatula, S., Bauer, L., Reiter, M.K.: Accessorize to a crime: real and stealthy attacks on state-of-the-art face recognition. In: CCS (2016). https://doi.org/10.1145/2976749.2978392
16. Shokri, R., Stronati, M., Song, C., Shmatikov, V.: Membership inference attacks against machine learning models. In: S&P (2017). https://doi.org/10.1109/sp.2017.41
17. Simonyan, K., Zisserman, A.: Very deep convolutional networks for large-scale image recognition. In: ICLR (2015). https://arxiv.org/abs/1409.1556
18. Sinha, A., Namkoong, H., Duchi, J.C.: Certifying some distributional robustness with principled adversarial training. In: ICLR (2018)
19. Song, L., Mittal, P.: Systematic evaluation of privacy risks of machine learning models. In: USENIX Security Symposium (2021)
20. Song, L., Shokri, R., Mittal, P.: Privacy risks of securing machine learning models against adversarial examples. In: CCS (2019). https://doi.org/10.1145/3319535.3354211

21. Xiao, H., Rasul, K., Vollgraf, R.: Fashion-MNIST: a novel image dataset for benchmarking machine learning algorithms. CoRR (2017). https://arxiv.org/abs/1708.07747
22. Yeom, S., Giacomelli, I., Fredrikson, M., Jha, S.: Privacy risk in machine learning: analyzing the connection to overfitting. In: IEEE CSF (2018). https://doi.org/10.1109/CSF.2018.00027
23. Zhang, H., Yu, Y., Jiao, J., Xing, E.P., Ghaoui, L.E., Jordan, M.I.: Theoretically principled trade-off between robustness and accuracy. In: ICML (2019). https://proceedings.mlr.press/v97/zhang19p.html

Enhancing Image Comprehension for Computer Science Visual Question Answering

Hongyu Wang, Pengpeng Qiang, Hongye Tan(✉), and Jingchang Hu

School of Computer and Information Technology, Shanxi University, Taiyuan, China
tanhongye@sxu.edu.cn

Abstract. Computer science visual question answering is a fundamental task in the intelligent education. However, current models have poor performance in this task. There are mainly two issues in these models. Firstly, they cannot accurately capture the fine-grained objects and relations in images. Secondly, these models lack the computer domain knowledge. To address the issues, we propose an Image Comprehension Enhancing Model. Specifically, it uses object detection technique to capture fine-grained features of images and utilizes Optical Character Recognition (OCR) to transform fine-grained features into the text information. The model adopts the text information to prompt the large language model, which generates the image caption with the computer domain knowledge. The fine-grained features and image caption can enhance the model's image comprehension and compensate the lack of knowledge. Additionally, the model utilizes the cross-modal attention mechanism to integrate the image features and fine-grained features of the image with the text features. The experimental results on the CSDQA dataset demonstrate that our proposed model outperforms the baselines, and the accuracy improves at least 4.80%.

Keywords: Visual Question Answering · Image Comprehension · Large Language Model

1 Introduction

The rapid advancement of computing technologies has facilitated the implementation of intelligent education applications [1]. Computer science visual question answering (CSVQA) is a complex multimodal task in the intersection of intelligent education and Visual Question Answering (VQA) [2].

The VQA task aims to get the correct answer through understanding images and questions. With the presentation of several VQA datasets, various models have also been proposed, such as BAN [3], ViLBERT [4], UNIMO [5] model, etc. However, because the CSVQA task has more challenges than the VQA task, these VQA models don't perform well in the CSVQA task. Firstly, images have

This work was supported by the National Natural Science Foundation of China (Grant No.62076155).

© The Author(s), under exclusive license to Springer Nature Singapore Pte Ltd. 2024
Q. Liu et al. (Eds.): PRCV 2023, LNCS 14425, pp. 487–498, 2024.
https://doi.org/10.1007/978-981-99-8429-9_39

fine-grained objects and complex relations. For example, Table 1 shows examples of the CSVQA task [6], the image of example (a) has many fine-grained objects: “front”, “back”, “28”, “19” and so on. And there are many relations between objects, such as the relationship between “front” and “28”, the relationship between “28” and “19”, the relationship between “A” and “B” in the image of example (b) and so on. Secondly, the image and question comprehension require computer domain knowledge. In the Table 1, the “Queue”, “Tree”, “length of the queue”, “generations of ancestor node” conceptions in images and questions need model to possess the computer domain knowledge.

Table 1. Examples of the CSVQA task.

Example	(a)	(b)
Image		
Question	What is the length of the queue ?	How many generations of ancestor nodes does node E have ?
Choice	A.3 B.4 C.5 D.6	A.3 B.5 C.2 D.4
Answer	C.5	C.2

Recently, the large language models have attracted the researchers in the VQA task. Because the large language models have strong language comprehension and knowledge base, researchers resort to it as implicit knowledge engines. Yang et al. [7] first proposed to use the image caption model to translate the image into textual caption, which helps the large language model to answer questions about the images. Thanks to the powerful knowledge comprehension of the large language models, Yang et al. proposed method achieves significant performance in the knowledge-based VQA task. But because the image caption may not accurately represent the information in the image, it will cause the large language model to predict the wrong answer.

In this paper, we propose an Image Comprehension Enhancing Model (ICEM). Specifically, we first finetune an object detection model to capture fine-grained features of images. Then, we utilize OCR to transform fine-grained features into the text information, which are used as prompt of the large language model to generate more accurate image caption with computer domain knowledge. In additional, we adopt the cross-modal attention mechanism to integrate image features with the text features. Concretely, Our main contributions can be summarized as : (1) We propose an Image Comprehension Enhancing Model. It can capture fine-grained information of images and make the large language model to accurately generate image caption with computer domain knowledge, which enhance the image comprehension and compensate the lack of knowledge. (2) We propose a new integration strategy, which uses the cross-modal attention to integrate the image features and fine-grained features of the image with the text features respectively. This strategy can help understand the semantic correlations between images and texts. (3) Experiments and ablation studies are conducted to verify the effectiveness of our model. On the Diagrams Question Answering in Computer Science (CSDQA) dataset, the accuracy improves at least 4.80% compared with baselines.

2 Related Work

2.1 Visual Question Answering

Since Agrawal A et al. [2] first proposed Visual Question Answering (VQA) task in 2015, researchers have studied this task and made some progress. In the beginning, researchers proposed joint embedding based on image and text features methods [8–10]. Malinowski et al. [8] proposed to use CNN to extract image features and utilize LSTM to represent questions. Then they used element-wise multiplication to fuse image and text features. To avoid noisy information brought by the joint embedding methods, the attentional mechanisms method was proposed. It focusses on key regions in images and key words in questions [11,12]. Yu et al. [11] designed self-attention units and guided attention units, then constructed deep modular cooperative attention networks to realize the interaction between text features and image features. Recently, pre-based multimodal representation learning method are becoming popular [4,5,13]. Researchers pre-trained on large-scale VQA datasets based on the Transformer model framework. Wei Li et al [5] proposed a UNIMO multimodal pre-training model, which could handle both unimodal and multimodal downstream tasks.

Compared with the VQA task, computer science visual question answering task focuses on images with knowledge instead of the natural images. Although current methods can perform well on conventional VQA datasets, they cannot comprehend images with knowledge effectively. The results in our baselines have proved this conclusion.

2.2 VQA in Intelligent Education

In intelligent education, researchers have also proposed a series of visual question answering tasks and methods. Four representative visual question answering tasks in intelligent education are shown in Table 2:

Table 2. VQA in intelligent education.

Task	Domain	Grades	Questions	Images
TQA [14]	Natural Science	6–8	26,260	3,455
Geome-try3K [15]	Geometry	6–12	3,002	2,342
IconQA [16]	Math	PreK-3	107,439	96,817
CSDQA [6]	Computer Science	University	3,494	1,294

Textbook Question Answering (TQA): Li et al. [17] build Contradiction Entity-Relation Graph, which utilizes semantic rules to comprehend large contexts and spatial analysis rules. Geometry Question Answering (Geometry3K): Lu et al. [15] proposed the Inter-GPS model, which uses target detection and rule matching. It uses integrated theorem knowledge as conditional rules and carried out symbolic reasoning step by step. Icon Question Answering (IconQA): Qiu et al. [16] proposed the Patch-TRM model, which generates image patches by the hierarchical pyramid layout and the text was encoded by Transformer. Finally, model use the cross-modal joint learning images-problem features. Diagrams Question Answering in Computer Science (CSDQA): Wang et al. [18] proposed a graph pairing network (DPN), which focusing on the topology of the graph, enhance the model's understanding for image.

In summary, the mainstream methods can't comprehend the knowledge in images and questions very well and the image caption model not accurately represent the information in the image. Therefore, we propose an Image Comprehension Enhancing Model, which uses two approaches to enhance image comprehension. Firstly, it uses object detection technique to capture fine-grained features of images and utilizes Optical Character Recognition(OCR) to transform fine-grained features into the text information. Secondly, the model adopts the text information to prompt the large language model, which generates image caption with computer domain knowledge.

3 Approach

The computer science vision question answering task mainly includes images, questions, and the candidate answers, and it aims to reason answer from the candidate answers. Formally, the task is defined as: given image V, question Q, the candidate answers C, the goal is to predict the correct answer $\hat{c} \in C$ according to formula (1).

$$\hat{c} = argmaxP\left(c|V,Q\right) \quad c \in C \tag{1}$$

We propose an Image Comprehension Enhancing Model (ICEM). The model framework is shown in Fig. 1, which consists of five parts: (1) Input layer;(2) Image understanding layer; (3) Encoding layer; (4) Cross-modal attention layer; (5) Output layer.

3.1 Image Understanding Layer

Image understanding layer mainly consists of two parts. Firstly, we use target detection model to capture the fine-grained objects of images and utilize OCR to transform fine-grained features into the text information. Secondly, we adopt text information to prompt the large language model, which generate image caption with computer domain knowledge. This layer enhances the understanding of the image and compensates the lack of computer domain knowledge.

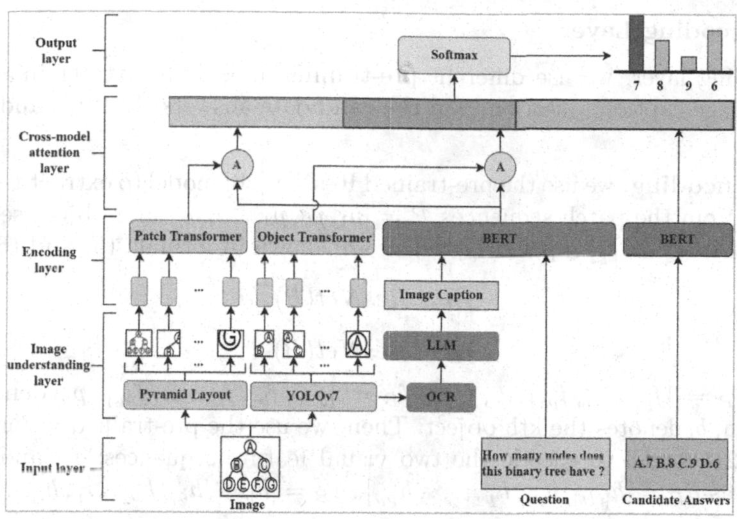

Fig. 1. Framework of Image Comprehension Enhancing Model (ICEM). The A means the cross-modal attention network.

Images of computer science have a variety of expression modes and complex logical structures, such as global features, fine-grained objects, and their relations. For the global features, we fill the image into a square in the beginning. According to the pyramid layout, we let $P = [p_1, p_2, p_3, \ldots, p_n]$ denotes the patch sequences with different scales, where p_n denotes the nth patch. For the fine-grained objects and relations features, we use the pre-trained YOLOv7 [19] model to capture them, which is shown in formula (2):

$$B = YOLOv7(V) \tag{2}$$

where, $B = [b_1, b_2, b_3, \ldots, b_k]$ denotes the detected object sequences, b_k denotes the kth object.

For image caption, Optical Character Recognition (OCR) technology is used to extract the text information from the object sequences $B = [b_1, b_2, b_3, \ldots, b_k]$, which is shown in formula (3):

$$O = OCR(B) \tag{3}$$

where, $O = [o_1, o_2, o_3, \ldots, o_g]$ denotes the detected text sequences. Then, we use the Large Language Model (LLM) to acquire the relevant knowledge of the text sequences, and generate the image caption, which is shown in formula (4):

$$D = LLM(O) \tag{4}$$

where, $D = [d_1, d_2, d_3, \ldots, d_t]$, d_t denotes the tth word in the image caption. Finally, the image caption $D = [d_1, d_2, d_3, \ldots, d_t]$ and the problem $Q = [q_1, q_2, q_3, \ldots, q_m]$ are connect: $W = [w_{[cls]]}, w_1, w_2, \ldots, w_t, \ldots, w_{t+m}]$.

3.2 Encoding Layer

In encoding layer, we use different pre-training models to extract images and texts (image caption, question, and the candidate answers) features and encode them.

Image encoding: we use the pre-trained ResNet [20] model to extract the visual features from the patch sequences $P = [p_1, p_2, p_3, \ldots, p_n]$ and object sequences $B = [b_1, b_2, b_3, \ldots, b_k]$ separately, which are shown in formula (5) and (6):

$$f_P = ResNet(P) \tag{5}$$

$$f_B = ResNet(B) \tag{6}$$

where, $f_P = [f_{p_1}, f_{p_2}, f_{p_3}, \ldots, f_{p_n}]$, $f_B = [f_{b_1}, f_{b_2}, f_{b_3}, \ldots, f_{b_k}]$, p_n denotes the nth patch, $b_k$denotes the kth object. Then, we use the pre-trained vision Transformer [21] model to encode the two visual feature sequences: f_P and f_B are denoted as $h_P = [h_{[cls]}, h_{p_1}, h_{p_2}, \ldots, h_{p_n}]$, $h_B = [h_{[cls]}, h_{b_1}, h_{b_2}, \ldots, h_{b_k}]$.

Text encoding: The image caption $D = [d_1, d_2, d_3, \ldots, d_t]$ and the problem $Q = [q_1, q_2, q_3, \ldots, q_m]$ are connected: $W = [w_{[cls]}, w_1, \ldots, w_t, \ldots, w_{m+t}]$, then the text sequences W is encoded $h_W = [h_{w_{[cls]}}, h_{w_1}, \ldots, h_{w_t}, \ldots, h_{w_{m+t}}]$ by Bert [22] model. The candidate answers are also encoded $h_C = [h_{c_{[cls]}}, h_{c_1}, \ldots, h_{c_j}]$ through the Bert model, where $C_j \in C$ denotes the jth answer in the question candidate answers.

3.3 Cross-Modal Attention Layer

In the cross-modal attention layer, we utilize the cross-modal attention mechanism to integrate the global and fine-grained features of images with the text features separately. This strategy enhances the interaction and helps understand the semantic correlations between images and texts within the model.

The input of this layer is the output of the encoding layer, which includes the patch sequences representation: h_P, the target object sequences representation: h_B and the text sequences representation: h_W. Firstly, h_p, h_b *and* h_w are mapped as patch matrix $h_P \in R^{n \times d}$, target object matrix $h_B \in R^{k \times d}$ and text matrix $h_W \in R^d$ by the fully connected layer, where n and k denote the number of sequences and d denotes the dimension of the matrix. Then we utilize the cross-modal attention mechanism to learn their joint representations, which are shown in formula (7), (8), (9) and (10):

$$a_P = softmax(W_P h_P \circ W_W h_W) \tag{7}$$

$$a_B = softmax(W_B h_B \circ W_W h_W) \tag{8}$$

$$h_{v_P} = \sum_i^n a_P(i) \times h_{P_i} \tag{9}$$

$$h_{v_P} = \sum_i^k a_P(i) \times h_{B_i} \tag{10}$$

3.4 Output Layer

In output layer, we map h_{v_P} and h_{v_B} into the candidate answers space and feed them into the classifier to predict the correct answer.

Firstly, the candidate answers representation h_C is mapped to the answer's matrix: $h_C \in R^{m \times d}$ by the fully connected layer, where m denotes the number of the candidate answers and k denotes the dimensionality of the matrix. Next, the h_{v_P} and h_{v_B} are connected with the answers matrix h_c, and finally the connect results are fed into the classifier for predicting the correct answer, which is shown in formula (11):

$$P_{ans} = softmax(W_a\left([h_{v_P}, h_{v_B}, h_C]\right) + b_c) \tag{11}$$

4 Experiments

4.1 CSDQA Dataset

We evaluate our model ICEM on the Computer Science Diagram Question Answering (CSDQA) dataset, which are publicly available at the Chinese Open Knowledge Graph (OpenKG.CN) . This dataset comes from educational scenarios such as text-books, encyclopedias, and blogs, which contains five undergraduate computer courses (Data Structure, Principles of Computer Network, Computer Architecture, Digital logic circuits, and Computer Operating System). It consists of 12 categories , comprising 1,294 diagrams and 3,494 question pairs, which includes 1,747 multiple-choice questions and 1,747 True/False questions. For our experiments, we use 70% of the data as the training set, 15% as the validation set, and the remaining 15% as the test set.[1]

4.2 Implementation Details

Pre-trained Target Detection Model: We finetune the YOLOv7 model as our target detection model. To make the model better detect abstract element in images, we labeled 12 categories of diagrams, which included 700 images and more than 2500 target objects and relations. We also modified the target box regression formula to improve the detection accuracy of small objects.

Large Language Model: For large language model, we use the ChatGPT model. The quality of the prompt plays a crucial role in the image caption task. Firstly, we prompt ChatGPT to assume the role of a computer science teacher and provide it with information about image and question. Then, through conversational interaction, we guide ChatGPT to learn about the information and retrieve relevant knowledge. Finally, we provide ChatGPT with examples of image captions and instruct it to generate captions based on the given information about image and question.

[1] OpenKG.CN: http://openkg.cn/dataset/csdqa.

Relevant Parameters: We use the AdaMax [23] optimizer with an optimal learning rate of 8×10^{-4} as the optimization algorithm. We train the multiclass classifier by binary cross-entropy loss with a batch size of 64 and a maximum number of rounds of 100. When the validation accuracy stops improving for 20 consecutive rounds, training is stopped. Finally, the model parameters that perform optimally on the validation set are saved.

4.3 Baselines

We compare our-model with several well-known models designed for conventional VQA and VQA in intelligent education task:

MUTAN [24]**:** It uses a tensor-based multimodal tucker decomposition to efficiently parameterize the bilinear interactions between visual and textual representations. The model can easily control the representation of the model and decompose the interaction tensor into interpretable elements.

MFB [25]**:** It uses a multimodal decomposition bilinear pooling approach to fuse visual features in images with text features in questions. Based on the MFB module, a collaborative attention learning architecture with image attention and problem attention co-learning was designed.

BAN [3]**:** The model considers the bilinear interaction between two sets of input channels, found that the bilinear attention distribution can seamlessly exploit the given visual-verbal information.

MCAN [11]**:** It introduces the idea of modularity into visual quizzing, designs self-attention units and guided attention units and constructs deep modular cooperative attention networks to realize the interaction between text features and image features.

Patch-TRM [16]**:** The model uses a hierarchical pyramid layout of images to generate image blocks, text is encoded by Transformer, and finally cross-modal joint learning of graph-question features to obtain the final answer.

DPN-QA [18]**:** It uses a graph pairing network (DPN) enhancement model focusing on analyzing the topology and textual information of graphs for the graph comprehension of computer textbooks.

4.4 Experimental Results and Analysis

Table 3 shows the experimental results of ICEM on the test set of the CSDQA dataset. we can see that ICEM consistently performs better than all baselines in judgment and multiple-choice questions. It indicates that our model is effective in understanding the images and answering the questions. Compared to the best performing baseline model Patch-TRM, our model improves 4.80% in all accuracy. We also find that the accuracy of all models is lower than multiple-choice questions in the CSDQA dataset. The main reason: multiple-choice questions require model with better images comprehension and more robustness than judgment questions. Furthermore, the incorporation of knowledge from both images and questions is essential, image caption can compensate the lack of computer domain knowledge to some extent.

Table 3. Experimental results (% accuracy) of different types of questions on the test set in CSDQA task. T/F denotes True/False questions and MC denotes multiple-choice questions.

Model	T/F	MC	ALL
Random	50.00	25.00	37.50
MUTAN [24]	54.43	27.86	41.15
MFB [25]	56.51	30.21	43.36
BAN [3]	57.26	27.34	43.32
MCAN [25]	59.64	29.17	44.41
Patch-TRM [16]	53.53	37.30	45.68
DPN-QA [18]	58.85	31.77	45.31
ICEM	**60.97**	**39.29**	**50.48**

4.5 Ablation Studies

In order to analyze the contributions of each module of our-model framework, we ablate our model ICEM as follows:

w/o Image Caption: The method eliminates the image caption to verify that it can provide useful information to help comprehend the image.

w/o Object Sequence: The method eliminates the fine-grained object and relational features of the image to verify that it can provide valid graph information to help comprehend the image.

w/o Cross-modal Attention: The method replaces the cross-modal attention mechanism with a simple stitching approach to verify whether global and fine-grained features of the image are effective for image-text interaction by fusing them with text features, respectively.

w/o ResNet Pre-train: The method replaces the pre-trained RestNet model that has not been pre-trained with images from the CSDQA dataset to verify that it can accurately extract feature information from the images.
The Table 4 shows the performance of four variations on the test set. Our model w/o Image Captioning achieves 47.62% accuracy, which is 2.86% lower than the whole model. The results show that the image caption with computer domain knowledge has a large impact on the model performance. The main reason: the large language model can automatically generate code based on the dialogue,

Table 4. Ablation experiment results (% accuracy).

Model	ALL	Δ
ICEM	**50.48**	–
w/o Image Caption	47.62	–2.86
w/o Object Sequence	48.46	–1.72
w/o Cross-modal Attention	49.31	-1.17
w/o ResNet Pre-train	49.87	–0.61

	Example (a)	Example (b)	Example (c)	Example (d)	Example (e)
Image					
Question	What is the length of the queue?	How many generations of ancestor nodes does node E have?	What are the elements with blue dot inside allocated to?	Which address will the Top element point to after a push operation?	What is the four node of the binary tree traversed in in order?
Caption	This is a queue, elements have 28, 19, 45, 13, 7. The front element is 28 and the back element is 7.	This is a binary tree. Its' nodes A, B, C, and E. Node A is the parent node of node B, and node B is the parent node of nodes E and C.	This is a directed graph, which there are nodes P1, P2, P3 and two empty nodes. Nodes P1, P2, and P3 are connected to the empty node.	This is an array, which have element 1. Top = 0 points to 1.	This is a binary tree, which have nodes 6, 5, 9, 1, 3. Node 6 is the parent node of nodes 5 and 9, and node 5 is the parent node of nodes 1 and 3.
Choices	A.3 B.4 C.5 D.6	A.3 B.5 C.2 D.4	A. p1 B. p2 C. p3 D. p1,p2	A. a[0] B. a[1] C. a[2] D. a[3]	A.6 B.5 C.3 D.9
Answer	C.5	C.2	D.p1,p2	A. a[0]	D.9

Fig. 2. Explanation examples got by our model. Incorrect predicted answers are marked in red. (Color figure online)

thereby it has a certain knowledge of computer domain. we prompt the large language model to understand the information in the image and generate image caption. This method not only enhances the model's image comprehension and make up for the lack of computer knowledge to some extent.

Our model w/o Object Sequence decreases significantly by 1.72%, indicating that fine-grained target detection plays a crucial role in image comprehension. And it makes the large language model to accurately generate image caption. Without Cross-modal Attention drops from 50.48% to 49.31%, indicating that the cross-modal attention mechanism is effective for interaction between images and texts. In addition, the pre-trained ResNet model is also helpful for extracting the image features.

4.6 Case Study

We performed a human analysis on the experimental samples. As shown in Fig. 2, it can be observed that there are correct predictions in some samples, such as "length of queue" in sample (a) and "how many generations of ancestor nodes" in sample (b) do not directly get the correct answer from the image and the question (the baseline model predicts incorrectly), but by capturing the fine-grained features of the images and generating captions, the model can predict the correct answer. However, we find the wrong captions in sample (c), which our model incorrectly identifies "deadlocked graph" as "directed graph". In the Sample (d) and Sample (e), model cannot understand of the "a push operation" and "in order" in the questions so that those errors cause the model to predict the wrong answer.

5 Conclusions

In this paper, we propose an Image Comprehension Enhancing Model (ICEM). It can capture fine-grained information of images and make the large language model to accurately generate image caption with computer domain knowledge, which enhance the image comprehension and compensate the lack of knowledge. In addition, we propose a new integration strategy, it can help model to understand the semantic correlations between images and texts. Finally, the experimental results demonstrate that our model performs better than the baselines, and the accuracy improves at least 4.80%.

In the future, we will consider trying to build knowledge prompt enhancement method. To compensate for the lack of knowledge well, we will use the large language models to build dynamic textual knowledge and build graph structures to enhance Integration of knowledge with images and questions.

References

1. Hwang, G.J., Xie, H., Wah, B.W., Gašević, D.: Vision, challenges, roles and research is-sucs of artificial intelligent in education. Comput. Educ. Artif. Intell. 1, 100001 (2020)
2. Antol, S., et al.: VQA visual question answering. In: Proceedings of the IEEE International Conference on Computer Vision, pp. 2425–2433 (2015)
3. Kim, J. H., Jun, J., Zhang, B.T.: Bilinear attention networks. In: Proceedings of the 32nd International Conference on Neural Information Processing Systems, pp. 1571–1581 (2018)
4. Lu, J., Batra, D., Parikh, D., Lee, S.: ViLBERT: pretraining task-agnostic visio linguistic representations for vision-and-language tasks. In: Advances in Neural Information Processing Systems, vol. 32 (2019)
5. Li, W., et al.: UNIMO: Towards UNI-FIED-modal understanding and generation via cross-modal contrastive learning. arXiv pre-print arXiv:2012.15409 (2020)
6. Wang, S., et al.: CSDQA: diagram question answering in computer science. In: Qin, B., Jin, Z., Wang, H., Pan, J., Liu, Y., An, B. (eds.) CCKS 2021. CCIS, vol. 1466, pp. 274–280. Springer, Singapore (2021). https://doi.org/10.1007/978-981-16-6471-7_21
7. Yang, Z., et al.: An empirical study of GPT-3 for few-shot knowledge-based VGA. In: AAAI, pp 3081–3089 (2022)

8. Malinowski, M., Rohrbach, M., Fritz, M.: Ask your neurons: a neural-based approach to answering questions about images. In: Proceedings of the IEEE International Conference on Computer Vision, pp. 1–9 (2015)
9. Gao, P., You, H., Zhang, Z., Wang, X., Li, H.: Multi-modality latent interaction network for visual question answering. In: Proceedings of the IEEE/CVF International Conference on Computer vision, pp. 5825–5835 (2019)
10. Kim, J.H., et al.: Multi-modal residual learning for visual GA. In: Advances in Neural Information Processing Systems, vol. 29 (2016)
11. Yu, Z., Yu, J., Cui, Y., Tao, D., Tian, Q.: Deep modular co-attention networks for visual question answering. In: Proceedings of the IEEE/CVF Conference on Computer Vision and Pattern Recognition, pp. 6281–6290 (2019)
12. Anderson, P., et al.: Bottom-up and top-down attention for image captioning and visual question answering. In: Proceedings of the IEEE Conference on Computer Vision and Pattern Recognition, pp. 6077–6086 (2018)
13. Wang, P., et al.: OFA: Unifying architectures, tasks, and modalities through a simple sequence-to-sequence learning framework. In: International Conference on Machine Learning, pp. 23318–23340. PMLR (2022)
14. Kembhavi, A., Seo, M., Schwenk, D., Choi, J., Farhadi, A., Hajishirzi, H.: Are you smarter than a sixth grader? textbook question answering for multimodal machine comprehension. In: Proceedings of the IEEE Conference on Computer Vision and Pattern recognition, pp. 4999–5007 (2017)
15. Lu, P., et al.: Inter-GPS: Interpretable geometry problem solving with formal language and symbolic reasoning. arXiv preprint arXiv:2105.04165 (2021)
16. Lu, P., et al.: IconQa: a new benchmark for abstract diagram understanding and visual language reasoning. arXiv preprint arXiv:2110.13214 (2021)
17. Li, J., Su, H., Zhu, J., Wang, S., Zhang, B.: Textbook question answering under instructor guidance with memory networks. In: Proceedings of the IEEE Conference on Computer Vision and Pattern Recognition, pp. 3655–3663 (2018)
18. Wang, S., et al.: Computer science diagram understanding with topology parsing. ACM Trans. Knowl. Disc. Data (TKDD) **16**(6), 1–20 (2022)
19. Wang, C.Y., Bochkovskiy, A., Liao, H.Y.M.: YOLOv7: trainable bag-of-freebies sets new state-of-the-art for real-time object detectors. In: Proceedings of the IEEE/CVF Conference on Computer Vision and Pattern Recognition, pp. 7464–7475 (2023)
20. He, K., Zhang, X., Ren, S., Sun, J.: Deep residual learning for image recognition. In: Proceedings of the IEEE Conference on Computer Vision and Pattern Recognition, pp. 770–778 (2016)
21. Vaswani, A., et al.: Attention is all you need. In: Proceedings of the 31st International Conference on Neural Information Processing Systems, pp. 6000–6010 (2017)
22. Kenton, J.D.M.W.C., Toutanova, L.K.: BERT: pre-training of deep bidirectional transformers for language understanding. In: Proceedings of NAACL-HLT, pp. 4171–4186 (2019)
23. Kingma, D. P., Ba, J.: Adam: a method for stochastic optimization. arXiv preprint arXiv:1412.6980 (2014)
24. Ben-Younes, H., Cadene, R., Cord, M., Thome, N.: MUTAN: multimodal tucker fusion for visual question answering. In: Proceedings of the IEEE International Conference on Computer Vision, pp. 2612–2620 (2017)
25. Yu, Z., Yu, J., Fan, J., Tao, D.: Multi-modal factorized bilinear pooling with co-attention learning for visual question answering. In: Proceedings of the IEEE International Conference on Computer Vision, pp. 1821–1830 (2017)

Cross-Modal Attentive Recalibration and Dynamic Fusion for Multispectral Pedestrian Detection

Wei Bao[1,2], Jingjing Hu[1(✉)], Meiyu Huang[2], and Xueshuang Xiang[2]

[1] Beijing Institute of Technology, Beijing, China
{baowei,hujingjing}@bit.edu.cn
[2] Qian Xuesen Laboratory of Space Technology, China Academy of Space Technology, Beijing, China

Abstract. Multispectral pedestrian detection can provide accurate and reliable results from color-thermal modalities and has drawn much attention. However, how to effectively capture and leverage complementary information from multiple modalities for superior performance is still a core issue. This paper presents a Cross-Modal Attentive Recalibration and Dynamic Fusion Network (CMRF-Net) to adaptively recalibrate and dynamically fuse multi-modal features from multiple perspectives. CMRF-Net consists of a Cross-modal Attentive Feature Recalibration (CAFR) module and a Multi-Modal Dynamic Feature Fusion (MDFF) module in each feature extraction stage. The CAFR module recalibrates features by fully leveraging local and global complementary information in spatial- and channel-wise dimensions, leading to better cross-modal feature alignment and extraction. The MDFF module adopts dynamically learned convolutions to further exploit complementary information in kernel space, enabling more efficient multi-modal feature aggregation. Extensive experiments are conducted on three multispectral datasets to show the effectiveness and generalization of the proposed method and the state-of-the-art detection performance. Specifically, CMRF-Net can achieve 2.3% mAP gains over the baseline on FLIR dataset.

Keywords: Multispectral pedestrian detection · Cross-modal attentive feature recalibration · Multi-modal dynamic feature fusion

1 Introduction

Pedestrian detection is an important computer vision task and serves many human-centric applications, such as autonomous driving, security surveillance, etc. However, pedestrian detection in a single modality still suffers from inadequate information. The visible modality is the most intuitive modality for humans but is susceptible to illumination conditions, while the thermal modality is robust to low illumination but lacks texture information.

Supported by the National Key R&D Program of China under Grant 2020YFB1709503.

© The Author(s), under exclusive license to Springer Nature Singapore Pte Ltd. 2024
Q. Liu et al. (Eds.): PRCV 2023, LNCS 14425, pp. 499–510, 2024.
https://doi.org/10.1007/978-981-99-8429-9_40

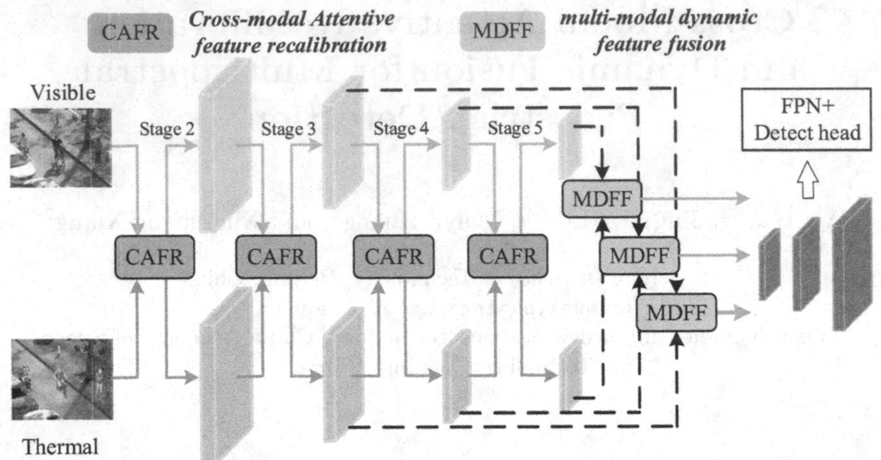

Fig. 1. The structure of CMRF-Net. The CAFR module adaptively recalibrate features for better cross-modal feature alignment and extraction, which the MDFF module dynamicly fuse features by fully leveraging complementary information from two modalities.

Multispectral pedestrian detection can provide accurate and reliable results from visible modality (RGB) and thermal modality and has drawn much attention. How to effectively utilize and integrate complementary information from two modalities is an open issue in this field. On the one hand, there are diverse noises and uncertainties in multiple modalities. The visible images are generally noisy in the poor illumination environment, and the thermal noise manifests as the objects lack texture details and present a similar appearance to their surroundings. It is expected that the noisy information can be filtered and recalibrated by using features coming from another modality. On the other hand, visible and thermal images belong to different modalities and show essential differences in terms of pedestrian morphology, texture, and properties. The visible image reflects color information while the thermal image reflects temperature information. Therefore, how to identify the difference between modalities and unify them into an effective representation is another problem. Recently, many excellent works [1–17] have been proposed to solve the above problems for better complementary information exploitation. However, these methods recalibrate and fuse multi-modal features from a single perspective, and lack diversity and flexibility.

Based on the previous analysis, we design a Cross-Modal Attentive Recalibration and Dynamic Fusion Network (CMRF-Net) to adaptively recalibrate and dynamically fuse multi-modal features from multiple perspectives as depicted in Fig. 1. CMRF-Net consists of a Cross-modal Attentive Feature Recalibration (CAFR) module and a Multi-Modal Dynamic Feature Fusion (MDFF) module in each feature extraction stage. (1) The CAFR module aims to adaptively recalibrate multi-modal features by fully leveraging local and global complementary information in spatial- and channel-wise dimensions. Such a feature rectification eliminates the uncertainty and noise from different modalities and alleviates modality heterogeneity, leading to better multi-modal feature alignment and extraction. (2) The MDFF module adopts dynamically learned convolutions to further exploit complementary information across multiple modalities in kernel space and produce enhanced fused features. Such a feature fusion learns the diverse type of complementary information in kernel space and combines them in a dynamic manner, which is beneficial to efficient multi-modal feature aggregation. Thereby, our introduced comprehensive interactions lie in multiple levels, including channel- and spatial-wise rectification from the global and local perspective, as well as aggregation in kernel space from the dynamic perspective. Finally, CMRF-Net can obtain more accurate and reliable detection results based on refined and fused features. The main contributions of our work can be summarized as follows:

1. The CMRF-Net is proposed to adaptively recalibrate and dynamically fuse multi-modal features from multiple perspectives in each feature extraction stage for multispectral pedestrian detection;
2. The CAFR module is designed to recalibrate features by fully leveraging local and global complementary information in spatial- and channel-wise dimensions, leading to better cross-modal feature alignment and extraction;
3. The MDFF module adopts dynamically learned convolutions to further exploit complementary information in kernel space, enabling more efficient multi-modal feature aggregation;
4. Various experiments are conducted on three datasets [1,6,18] to demonstrate the effectiveness and generalization of the proposed method and the state-of-the-art detection performance.

The rest of this paper is organized as follows. Section 2 introduces some related works, and Sect. 3 introduces our methods in detail. Section 4 provides the experimental settings and results analysis. Finally, some conclusions and future works are drawn in Sect. 5.

2 Related Work

2.1 Multispectral Pedestrian Detection

Traditional multispectral pedestrian detection simply aggregates multi-modal features via element-wise addition/concatenation operation. Recently, many works aim to exploit multi-modal information by leveraging the attention mechanism. MBnet [7] incorporates cross-modality complementarity with local channel-wise differential weighting. BAANet [15] proposes to distill the informative features and recalibrate the representations asymptotically in spatial-wise and channel-wise dimensions from a local view. CFT [9] integrates multi-modal features guided by the global context dependencies modeling from Transformer. Differently, our methods focus on the channel, spatial and kernel-wise complementarity from the global, local, and dynamic perspectives, which is more comprehensive.

2.2 Attentive Feature Refinement

Designing attentive feature refinement modules to improve the performance of a CNN has been widely studied in recent years. SENet [19] uses a seminal channel attention module termed Squeeze-and-Excitation to exploit the interdependencies between the channels of convolutional features. CBAM [20] combines the channel attention module with the spatial attention module. SKNets [21] add an attention design over two branches with different-sized convolutions to fuse multi-scale feature outputs. AFF [22] proposes a multiscale channel attention module, which addresses issues that arise when fusing features given at different scales from local and global views.

2.3 Dynamic Weight Networks

Making the weights of a neural network to be sample-adaptive via dynamic mechanisms has shown great potential for boosting model capacity and generalization. CondConv [23] and DyConv [24] both propose to use the linear combination of parallel static kernels, where the linear scale is dynamically aggregated via a function of individual inputs, to improve the representation capability. ODConv [25] leverages a novel multi-dimensional attention mechanism with a parallel strategy to learn complementary attentions for convolutional kernels along all four dimensions of the kernel space.

3 Methodology

3.1 Overview

The overview of CMRF-Net for multispectral detection is shown in Fig. 1. As usual, the CMRF-Net adopts two-stream backbones to extract visible and thermal features from inputs. The CAFR module adaptively recalibrates features in four feature extraction stages by fully exploiting complementary information in channel-wise and spatial-wise dimensions. The MDFF exploits complementary information in kernel space via dynamically learned convolutions to fuse multimodal features in the last three stages. Finally, the CMRF-Net can produce more accurate detection based on the more discriminative output features.

3.2 Cross-Modal Attentive Feature Recalibration

The information from different modalities is usually complementary but contains noise and uncertainties. For tackling modality-specific noises, we propose the CAFR module which performs feature rectification by using the other modality at each stage during feature extraction. CAFR processes the input features in two dimensions, including channel-wise and spatial-wise feature recalibration, which together offer a holistic calibration, enabling better cross-modal feature alignment and extraction.

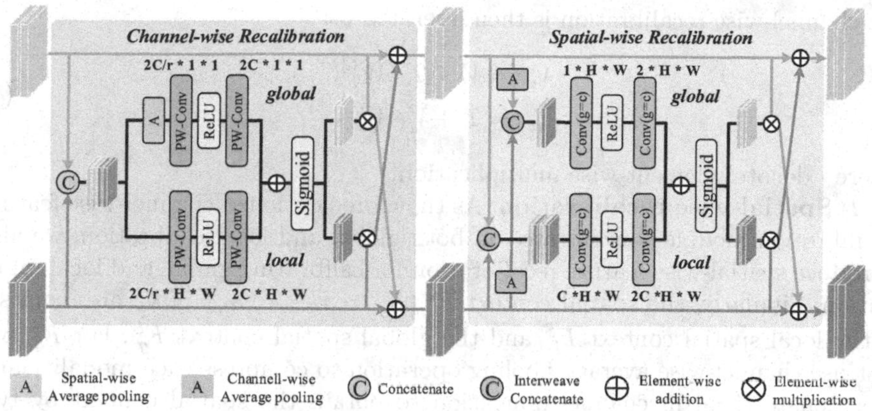

Fig. 2. The structure of CAFR module in CMRF-Net.

1) **Channel-wise recalibration**. We embed visible feature $V \in \mathbb{R}^{C\times H\times W}$ and thermal feature $T \in \mathbb{R}^{C\times H\times W}$ along the spatial axis into two attention vectors to capture channel dependency. Differently, we add the local context to the global context inside the attention module to retain more information. We choose point-wise convolution (PWConv) as the local channel context aggregator to highlight point-wise channel interactions for each spatial position. Concretely, the formula to obtain the channel attention vectors $W_V^C \in \mathbb{R}^{C\times H\times W}$ and $W_T^C \in \mathbb{R}^{C\times H\times W}$ of multi-modal features can be defined as follows:

$$W_V^C, W_T^C = F_{split}(\sigma(F^C(X))) = F_{split}(\sigma(F^C(V||T))), \tag{1}$$

where $||$ represents the concatenation operation, σ is the sigmoid-activated function to normalize the attention output in the range of (0, 1), and F_{split} is the split operation which is the inverse process of concatenation. $F^C(X) = F_l^C(X) + F_g^C(X)$ consists of the local channel context F_l^C and the global channel context F_g^C. For F_l^C, we use the standard SE attention [19] which can be defined as:

$$F_l^C(X) = PWConv_{l2}(\text{ReLU}(PWConv_{l1}(F_{swap}(X)))), \tag{2}$$

where the F_{swap} is spatial-wise average pooling and ReLU is the relu excited function. $PWConv$ is point-wise convolution and the kernel sizes of $PWConv_{l1}$ and $PWConv_{l2}$ are $2C/r \times 2C \times 1 \times 1$ and $2C \times 2C/r \times 1 \times 1$, respectively. r is the ratio of channel compression and we set it as 4 by default. As for F_g^C, we obtain it as follows:

$$F_g^C(X) = PWConv_{g2}(\text{ReLU}(PWConv_{g1}(X))). \tag{3}$$

It is noteworthy that $F_g^C(X)$ has the same shape as the input feature, which can preserve and highlight the subtle details in the low-level features. Finally, the channel-wise recalibration is then operated as:

$$\begin{aligned} V_{cr} &= V + T \cdot W_T^C, \\ T_{cr} &= T + V \cdot W_V^C. \end{aligned} \tag{4}$$

where $\cdot$ denotes element-wise multiplication.

2) **Spatial-wise recalibration**. As the aforementioned channel-wise feature recalibration module concentrates on both global and local calibration, we also introduce spatial-wise feature recalibration for calibrating global and local information. Similarly, the spatial context $F^S(X) = F_l^S(X) + F_g^S(X)$ also consists of the local spatial context F_l^S and the global spatial context F_g^S. For F_g^S, we first use channel-wise average pooling operation to compress each modality and concatenate them in channel dimension to obtain the spatial context by two convolution layers. The process of F_g^S can be defined as:

$$F_g^S(V, T) = Conv_{g2}(\text{ReLU}(Conv_{g1}(F_{cwap}(V) || F_{cwap}(T)))), \tag{5}$$

where the F_{cwap} is channel-wise average pooling. The kernel sizes of $Conv_{g1}$ and $Conv_{g2}$ are $1 \times 2 \times 3 \times 3$ and $2 \times 1 \times 3 \times 3$, respectively. As for F_l^S, we leverage the group convolution and interweave concatenation operation to capture local spatial context to maintain it as lightweight as possible. The process of F_l^S can be defined as:

$$F_l^S(V, T) = GConv_{l2}(\text{ReLU}(GConv_{l1}(V \hat{||} T))), \tag{6}$$

where the $\hat{||}$ is an interweaving concatenation operation to concatenate the visible and thermal feature channel by channel as depicted in Fig. 2. $GConv$ is the group convolution with the group size C to capture the subtle local information. The kernel sizes of $GConv_{l1}$ and $GConv_{l2}$ are $C \times 2 \times 3 \times 3$ and $2C \times 1 \times 3 \times 3$, respectively. Hence, the spatial attention vectors $W_V^S \in \mathbb{R}^{C \times H \times W}$ and $W_T^S \in \mathbb{R}^{C \times H \times W}$ can be defined by:

$$W_V^S, W_T^S = F_{split}(\sigma(F^S(V, T))), \tag{7}$$

Finally, the spatial-wise recalibration can be operated as:

$$\begin{aligned} V_{sr} &= V + T \cdot W_T^S, \\ T_{sr} &= T + V \cdot W_V^S. \end{aligned} \tag{8}$$

Moreover, we use the sequential arrangement of the spatial and channel recalibration modules for better performance.

3.3 Multi-modality Dynamic Feature Fusion

After obtaining the feature maps at each layer, we build an MDFF module to fuse features from two modalities into a single feature map as depicted in Fig. 3. The traditional method to merge fuse features via the convolutional

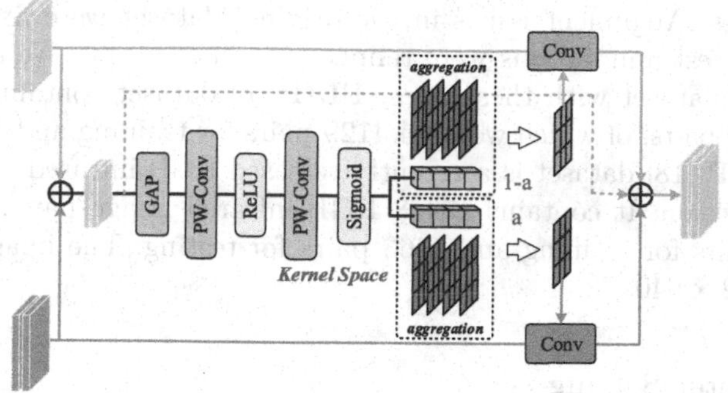

Fig. 3. The structure of MDFF module in CMRF-Net.

kernel of standard convolution is independent of the image content and unable to explore the diversely complementary information across two modalities. Inspired by recent data-driven convolution operations, we adopt the idea of learning a linear combination of multiple static convolutional kernels weighted with their input-dependent attentions to dynamically fuse multi-modal features for better performance.

Specifically, given the input visible feature V and thermal feature T, the standard convolution fuse them as follows:

$$Y = w^v V + w^t T \tag{9}$$

where w^v and w^t are static kernels sliding in V and T (We ignore the bias of convolution). To dynamically and complementarily fuse two features, we use attention mechanism to learn a group of kernel weights a as follows:

$$Y = W^v V + W^t T = \sum_i^K (a_i * w_i^v) V + \sum_i^K ((1 - a_i) * w_i^t) T, \tag{10}$$

where w^v and w^t are static kernels sliding in V and T (We ignore the bias of convolution). To dynamically and complementarily fuse two features, we use the attention mechanism to learn a group of kernel weights a as follows:

$$a = \sigma(PWConv_{d2}(\text{ReLU}(PWConv_{d1}(F_{swap}(V + T))))), \tag{11}$$

where the kernel sizes of $PWConv_{d1}$ and $PWConv_{d2}$ are $2C/r \times 2C \times 1 \times 1$ and $K \times 2C/r \times 1 \times 1$, respectively. To save parameters, we use group convolution to replace standard convolution in experiments. Moreover, we also use the residual connection to avoid degradation as depicted in Fig. 3.

4 Experiments

4.1 Datasets

KAIST [1] is the commonly used dataset for multispectral pedestrian detection. It contains 95,328 color-thermal image pairs. We evaluated a test set that consists of 2,252 images. Alignment issues in the original dataset were fixed by Zhang et al. [5] and test annotations were refined by Liu et al. [3]. We use the latest version of the dataset with these fixes. **FLIR** [6] dataset contains 5142 RGB-infrared image pairs, of which we used 4129 pairs for training and 1013 pairs for testing. **LLVIP** [18] dataset is a recently released RGB-infrared paired dataset for low-light vision. It contains 15488 RGB-infrared image pairs, of which we used 12025 pairs for training and 3463 pairs for testing. The image size for all datasets is 512×640.

4.2 Parameter Setting

A popular object detection toolbox mmyolo based on PyTorch is used to implement our method. For all experiments, our method is trained on one NVIDIA GeForce RTX 3090 GPU. We use YOLOv5 [26] as the basic detection framework and use the pre-trained CSPDarknet-53 on the MS-COCO [27] to initialize the backbone network for fast convergence. More specifically, CMRF-Net is trained with stochastic gradient descent (SGD) for 20 epochs with a total of 16 images per minibatch. The initial learning rate is set as 0.005. The weight decay is 0.0005 and the SGD momentum is 0.9. The Intersection over Union (IoU) threshold is set as 0.5 when training and testing for rigorous filtering of the bounding boxes with low precision.

4.3 Evaluation Metrics

For the KAIST dataset, all models are evaluated with the usual pedestrian detection metric for fair comparison: log- average Miss Rate over the range of $[10^{-2}, 10^{0}]$ false positives per image (FPPI) under a "reasonable" setting [28], i.e., only pedestrians taller than 50 pixels under no or partial occlusions are considered. While for the other two datasets, models are evaluated with the usual object detection metric introduced with MS-COCO [27]: the mean Average Precision (mAP) averaged over ten different IoU thresholds.

4.4 Results Analysis

Overall Performance. Table 1 reports the overall performance of CMRF-Net and the comparison with other state-of-the-art multispectral detectors on three datasets. Table 1(a) and Table 1(b) show the comparison of the KAIST dataset under the Miss Rate metric, and the FLIR and LLVIP dataset under the mAP metric, respectively. It can be observed that CMRF-Net can improve the detection performance on all datasets and finally outperforms all other detectors by a large margin except the performance under the mAP_{50} metric for the LLVIP dataset. By contrast, our method outperforms the CFT [9] in terms of $mAP_{0.75}$ and mAP metrics, which indicates that our approach can achieve better localization accuracy. For the KAIST dataset, CMRF-Net gets 7.55%, 5.86%, and 7.26% miss rates in Day, Night, and All subsets, respectively. More

Table 1. Comparison between state-of-the-art multispectral pedestrian detection methods and ours on KAIST [1], FLIR [6] and LLVIP [18] datasets.

(a) Overall performance on **KAIST** dataset under **Miss Rate** metric.

KAIST under **Miss Rate** ↓			
Methods	Day	Night	All
MSDS-RCNN [4]	8.09	5.92	7.49
AR-CNN [5]	9.94	8.38	9.34
MBnet [7]	8.28	7.86	8.13
Kim [11]	8.18	6.95	7.89
MLPD [10]	7.95	6.95	7.58
CMPD [12]	8.77	7.31	8.16
RISNet [13]	7.61	7.08	7.89
MuFEm [14]	8.16	7.51	8.07
BAANet [15]	8.37	6.98	7.92
CMRF-Net	**7.55**	**5.86**	**7.26**

(b) Overall performance on **FLIR** and **LLVIP** dataset under **mAP** metric.

FLIR under **mAP** ↑			
Methods	$mAP_{0.5}$	$mAP_{0.75}$	mAP
GAFF [8]	72.9	32.9	37.5
CFT [9]	77.7	34.8	40.0
CMDet [16]	78.6	–	–
AMSF [17]	78.9	35.5	40.9
CMRF-Net	**80.4**	**35.7**	**41.3**
LLVIP under **mAP** ↑			
CFT [9]	**97.5**	72.9	63.6
CMDet [16]	96.3	–	–
AMSF [17]	97.0	74.0	64.5
CMRF-Net	97.3	**74.7**	**65.0**

specifically, CMRF-net achieves up to 1.5% $mAP_{0.5}$ gains over the second-best (78.9% for AMSF [17]) on FLIR dataset. The quantitative detection performance increase demonstrates that the proposed CMRF-net can adaptively recalibrate and dynamically fuse multi-modal features for better feature alignment and extraction, aggregating in more robust output features.

Ablation Study. The proposed CMRF-Net consists of the CAFR and MDFF modules. To analyze the influence of each designed component, a series of ablation experiments are implemented on the FLIR dataset as depicted in Table 2. As for the ablation study for the CAFR module, we use CW-CAFR and SW-CAFR to represent channel-wise and spatial-wise recalibration, respectively. It can be seen that both CW-CAFR and SW-CAFR can help to improve the detection performance, and adopting two feature recalibrations together can further enhance the performance. Finally, the CAFR gets 1.9%, 2.0%, and 1.6% performance gains in terms of $mAP_{0.5}$, $mAP_{0.75}$ and mAP metrics, respectively. Meanwhile, the parameters and FLOPs increase by 1.5M and 0.9G, respectively, leading to the 5.4 fps decrease. As for the MDFF module, it can also improve the detection performance with the parameters and FLOPs of the model increase. Finally, our CMRF-Net achieve 2.3%, 2.4% and 1.9% improvement under $mAP_{0.5}$, $mAP_{0.75}$ and mAP metrics, respectively. All these phenomena illustrate the ablity of feature recalibration and integration for CAFR and MDFF module, respectively.

Hyper-parameters. We conduct a series of experiments to verify the influence of the number of kernels K in the MDFF module. As depicted in Table 3, as the K becomes larger, the detection performance in terms of APs metrics gradually increases. However, the parameters and FLOPs also slightly increase and the

Table 2. Ablation study for CMRF-Net on FLIR [6] dataset. CW-CAFR and SW-CAFR represent channel-wise and spatial-wise recalibration, respectively.

CW-CAFR	SW-CAFR	MDFF	$mAP_{0.5}$	$mAP_{0.75}$	mAP	params(M)	FLOPs(G)	FPS
✗	✗	✗	78.1	33.4	39.4	**72.7**	**71.0**	54.2
✓	✗	✗	79.5	34.6	40.1	74.1	71.7	49.9
✗	✓	✗	78.9	34.1	39.8	72.8	71.2	53.0
✓	✓	✗	80.0	35.4	41.0	74.2	71.9	48.8
✗	✗	✓	79.8	34.7	40.4	89.9	76.3	49.5
✓	✓	✓	**80.4**	**35.7**	**41.1**	91.4	77.2	45.3

Table 3. Hyper-parameters for the MDFF module on FLIR [6] dataset.

K	$mAP_{0.5}$	$mAP_{0.75}$	mAP	params(M)	FLOPs(G)	FPS
1(static)	78.1	33.4	39.4	**72.7**	**71.0**	54.2
2	79.1	34.2	40.1	81.7	76.3	51.5
4	79.8	**34.7**	**40.4**	89.9	76.3	49.5
6	**79.9**	34.6	40.3	98.2	76.3	47.7

Fig. 4. Visualization results for two scenes on LLVIP [18] dataset. The red circles represent the missing pedestrians.

inference speed also becomes slower. We adopt $K = 4$ in all experiments to achieve the trade-off between the speed and accuracy of the proposed method.

Visualization. In addition to quantitative comparisons, we also visualize some detection results for two scenes in Fig. 4 to show an intuitive understanding of our proposed methods. We can see that the baseline misses one pedestrian for two scenes. In contrast, the proposed CMRF-Net completely detects all ground truths without any error, which illustrates its effectiveness. The superiority can be further demonstrated by all these improvements.

5 Conclusion

This paper presents a CMRF-Net to adaptively recalibrate and dynamically fuse multi-modal features from multiple perspectives. The CAFR module in CMRF-Net recalibrates features by fully leveraging local and global complementary information in spatial- and channel-wise dimensions. The MDFF module CMRF-Net adopts dynamically learned convolutions to further exploit complementary information in kernel space. Extensive experiments arc conducted to show the effectiveness of the proposed method. In the future, we will consider how to better recalibrate features and align object features for more consistent modalities. In the future, we will consider integrating these perspectives more effectively and unifying a more lightweight plugin for better complementarity exploitation.

References

1. Hwang, S., Park, J., Kim, N., Choi, Y. and So Kweon, I.: Multispectral pedestrian detection: benchmark dataset and baseline. In: Proceedings of the IEEE Conference on Computer Vision and Pattern recognition, pp. 1037–1045 (2015)
2. Wagner, J., Fischer, V., Herman, M., Behnke, S., et al.: Multispectral pedestrian detection using deep fusion convolutional neural networks. ESANN **587**, 509–514 (2016)
3. Liu, J., Zhang, S., Wang, S., Metaxas, D.N.: Multispectral deep neural networks for pedestrian detection, arXiv preprint arXiv:1611.02644 (2016)
4. Li, C., Song, D., Tong, R. and Tang, M.: Multispectral pedestrian detection via simultaneous detection and segmentation, arXiv preprint arXiv:1808.04818 (2018)
5. Zhang, L., Zhu, X., Chen, X., Yang, X., Lei, Z., Liu, Z.: Weakly aligned cross-modal learning for multispectral pedestrian detection. In: Proceedings of the IEEE/CVF International Conference on Computer Vision, pp. 5127–5137 (2019)
6. Zhang, H., Fromont, E., Lefevre, S., Avignon, B.: Multispectral fusion for object detection with cyclic fuse-and-refine blocks. In: 2020 IEEE International Conference on Image Processing (ICIP), pp. 276–280. IEEE (2020)
7. Zhou, K., Chen, L., Cao, X.: Improving multispectral pedestrian detection by addressing modality imbalance problems. In: Vedaldi, A., Bischof, H., Brox, T., Frahm, J.-M. (eds.) ECCV 2020. LNCS, vol. 12363, pp. 787–803. Springer, Cham (2020). https://doi.org/10.1007/978-3-030-58523-5_46
8. Zhang, H., Fromont, E., Lefévre, S., Avignon, B.: Guided attentive feature fusion for multispectral pedestrian detection. In: Proceedings of the IEEE/CVF Winter Conference on Applications of Computer Vision, pp. 72–80 (2021)
9. Qingyun, F., Dapeng, H., Zhaokui, W.: Cross-modality fusion transformer for multispectral object detection, arXiv preprint arXiv:2111.00273 (2021)
10. Kim, J., Kim, H., Kim, T., Kim, N., Choi, Y.: MLPD: multi-label pedestrian detector in multispectral domain. IEEE Robot. Autom. Lett. **6**(4), 7846–7853 (2021)
11. Kim, J.U., Park, S., Ro, Y.M.: Uncertainty-guided cross-modal learning for robust multispectral pedestrian detection. IEEE Trans. Circuits Syst. Video Technol. **32**(3), 1510–1523 (2021)
12. Li, Q., Zhang, C., Hu, Q., Fu, H., Zhu, P.: Confidence-aware fusion using dempster-shafer theory for multispectral pedestrian detection. IEEE Trans. Multimedia **25**, 3420–3431 (2022)

13. Wang, Q., Chi, Y., Shen, T., Song, J., Zhang, Z., Zhu, Y.: Improving RGB-infrared object detection by reducing cross-modality redundancy. Remote Sens. **14**(9), 2020 (2022)
14. Dasgupta, K., Das, A., Das, S., Bhattacharya, U., Yogamani, S.: Spatio-contextual deep network-based multimodal pedestrian detection for autonomous driving. IEEE Trans. Intell. Transp. Syst. **23**(9), 15940–15950 (2022)
15. Yang, X., Qian, Y., Zhu, H., Wang, C., Yang, M.: Baanet: learning bi-directional adaptive attention gates for multispectral pedestrian detection. In: 2022 International Conference on Robotics and Automation (ICRA), pp. 2920–2926. IEEE (2022)
16. Sun, Y., Cao, B., Zhu, P., Hu, Q.: Drone-based RGB-infrared cross-modality vehicle detection via uncertainty-aware learning. IEEE Trans. Circ. Syst. Video Technol. **32**(10), 6700–6713 (2022)
17. Bao, W., Huang, M., Hu, J., Xiang, X.: Attention-guided multi-modal and multi-scale fusion for multispectral pedestrian detection. In: Yu, S., et al. Pattern Recognition and Computer Vision. PRCV 2022, LNCS, vol. 13534, pp. 382–393 Springer, Cham (2022). https://doi.org/10.1007/978-3-031-18907-4_30
18. Jia, X., Zhu, C., Li, M., Tang, W. and Zhou, W.: LLVIP: a visible-infrared paired dataset for low-light vision. In: Proceedings of the IEEE/CVF International Conference on Computer Vision, pp. 3496–3504 (2021)
19. Hu, J., Shen, L., Sun, G.: Squeeze-and-excitation networks. In: Proceedings of the IEEE Conference on Computer Vision and Pattern Recognition, pp. 7132–7141 (2018)
20. Woo, S., Park, J., Lee, J.Y., Kweon, I.S.: CBAM: convolutional block attention module. In: Proceedings of the European Conference on Computer Vision (ECCV), pp. 3–19 (2018)
21. Li, X., Wang, W., Hu, X., Yang, J.: Selective kernel networks. In: Proceedings of the IEEE/CVF Conference on Computer Vision and Pattern Recognition, pp. 510–519 (2019)
22. Dai, Y., Gieseke, F., Oehmcke, S., Wu, Y., Barnard, K.: Attentional feature fusion. In: Proceedings of the IEEE/CVF Winter Conference on Applications of Computer Vision, pp. 3560–3569 (2021)
23. Yang, B., Bender, G., Ngiam, J.: Condconv: conditionally parameterized convolutions for efficient inference. In: Advances in Neural Information Processing Systems, vol. 32 (2019)
24. Chen, Y., Dai, X., Liu, M., Chen, D., Yuan, L., Liu, Z.: Dynamic convolution: attention over convolution kernels. In: Proceedings of the IEEE/CVF Conference on Computer Vision and Pattern Recognition, pp. 11030–11039 (2020)
25. Li, C., Zhou, A., Yao, A.: Omni-dimensional dynamic convolution, arXiv preprint arXiv:2209.07947 (2022)
26. Jocher, G., et al.: ultralytics/yolov5: V7. 0-yolov5 sota realtime instance segmentation, Zenodo (2022)
27. Lin, T.-Y., et al.: Microsoft COCO: common objects in context. In: Fleet, D., Pajdla, T., Schiele, B., Tuytelaars, T. (eds.) ECCV 2014. LNCS, vol. 8693, pp. 740–755. Springer, Cham (2014). https://doi.org/10.1007/978-3-319-10602-1_48
28. Dollar, P., Wojek, C., Schiele, B., Perona, P.: Pedestrian detection: an evaluation of the state of the art. IEEE Trans. Pattern Anal. Mach. Intell. **34**(4), 743–761 (2011)

Author Index

© The Editor(s) (if applicable) and The Author(s), under exclusive license to Springer Nature Singapore Pte Ltd. 2024
Q. Liu et al. (Eds.): PRCV 2023, LNCS 14425, pp. 511–513, 2024.
https://doi.org/10.1007/978-981-99-8429-9

Printed in the United States
by Baker & Taylor Publisher Services

Printed in the United States
by Baker & Taylor Publisher Services